THEORETICAL AND APPLIED MECHANICS

THEORETICAL AND APPLIED MECHANICS

Volume 42

Proceedings of the 42nd Japan National Congress for Applied Mechanics, 1993

Edited by Motoo HORI and Shoichi KOBAYASHI
for Japan National Committee for Theoretical and Applied Mechanics, Science Council of Japan

UNIVERSITY OF TOKYO PRESS

ISBN 4-13-066112-4
ISBN 0-86008-508-2
Printed in Japan

PREFACE

The 42nd Congress on Applied Mechanics (NCTAM-42) was held at the Science Council of Japan in Tokyo on January 26, 27, and 28, 1993, under the joint sponsorship of the National Committee for Theoretical and Applied Mechanics of the Science Council of Japan (NCTAM), and the following thirteen related societies: the Japan Society for Aeronautical and Space Sciences, the Japan Society of Applied Physics, the Architectural Institute of Japan, the Atomic Energy Society of Japan, the Japan Society of Civil Engineers, the Japan Society of Fluid Mechanics, the Japanese Society of Irrigation, Drainage and Reclamation Engineering, the Mathematical Society of Japan, the Japan Society of Mechanical Engineers, the Mining and Materials Processing Institute of Japan, the Society of Naval Architects of Japan, the Physical Society of Japan, and the Society of Rheology, Japan. The Japan Society of Civil Engineers, the Japan Society for Aeronautical and Space Sciences, the Japan Society of Fluid Mechanics, and the Society of Rheololgy, Japan were responsible for the Congress this year, Professor S. Kobayashi of Kyoto University, acting as chairman of the Organizing Committee.

The Congress was composed of invited lecture, several organized and technical sessions, and three panel sessions.

This 42nd volume of NCTAM Proceedings contains thirty-six contributed papers selected by the Editorial Committee.

October 1993

Yoshiyuki Yamamoto
Chairman, Publication Committee
NCTAM-42 Proceedings

Motoo Hori
Chairman, Editorial Committee
NCTAM-42 Proceedings

CONTENTS

IV. Dynamical Problems

V. Fluid Dynamics

VI. Biomechanics and Biomaterials

I
THEORETICAL AND NUMERICAL ANALYSIS IN SOLID MECHANICS

Application of Kalman Filtering to Inverse Problems

Akira MURAKAMI and Takashi HASEGAWA

Department of Agricultural Engineering, Kyoto University, Kyoto

This paper overviews the application of Kalman filtering technique to several inverse problems. Kalman filtering has proven to be applicable to various inverse problems and its applications can be clarified into three different categories. The paper refers to the structure of Kalman filter and also describes the formulations incorporated with FEM in order to identify material parameters.

I. INTRODUCTION

An increasing interest has been given to the application of the Kalman filtering technique [1] to various engineering problems. This technique has proven to be useful in the analysis of inverse problems considering noisy observations (see, for example, Hoshiya *et al.* [2], Gioda *et al.* [3] and Murakami [4]).

Thus it is meaningful to describe the applicability of Kalman filtering technique to inverse problem analyses. In the following sections, we will firstly refer to the structure of Kalman filtering and then denote the formulations for several inverse problems and clarify the problem categories. Finally numerical examples are briefly shown.

In what follows, boldface lower-case and boldface capital letters denote vectors and matrices, respectively, and *italic* lower-case and capital letters denote scalar; superscripts T and -1 mean transpose and inverse, and subscripts t and k describe the time $t = t$ in continuous time and the time $t = k\Delta t$ in discrete time, respectively.

II. KALMAN FILTERING

2.1 PROBLEM DESCRIPTION

Let us consider

How can we provide an estimate of x based on the observation y?

$$y = Hx + v \tag{1}$$

where v is a white noise with zero mean and variance of R^2.

Prior distribution on x is also given as the following probabilistic density function (p.d.f.) of x

$$p_X(x) = \frac{1}{\sqrt{2\pi\sigma_0^2}} \exp\left[-\frac{(x-\bar{x})^2}{2\sigma_0^2}\right] \tag{2}$$

which obeys to the normal distribution with the known average of x, $\bar{x}$, and variance of x, σ_0. They are given a priori.

Problem is given as:

To find x which gives the maximal value of the conditional p.d.f. of x under the observation y, $p_{X|Y}(x,y)$.

$p_{Y|X}(x,y)$ is given as follows by use of equation (1)

$$p_{Y|X}(x,y) = \frac{1}{\sqrt{2\pi R^2}} \exp\left[-\frac{(y-Hx)^2}{2R^2}\right] \tag{3}$$

assuming that v has a Gaussian distribution.

On the other hand, the Bayes' theorem can provide the following p.d.f. of x under the observation y

$$p_{X|Y}(x,y) = \frac{p_X(x)p_{Y|X}(x,y)}{p_Y(y)} \tag{4}$$

$$\propto p_X(x)p_{Y|X}(x,y) \tag{5}$$

$$\propto \exp\left[-\frac{(x-\bar{x})^2}{2\sigma_0^2} - \frac{(y-Hx)^2}{2R^2}\right] \tag{6}$$

because $p_Y(y)$ is a function of y:

$$\int_{-\infty}^{\infty} p_X(x)p_{Y|X}(x,y)dx = f(y).$$

Item in the parenthesis of equation (6) can be rewritten as

$$(*) = -\frac{R^2(x-\bar{x})^2 + \sigma_0^2(y-Hx)^2}{2\sigma_0^2 R^2} \tag{7}$$

$$= -\frac{1}{2\sigma_0^2 R^2}\{(R^2+\sigma_0^2 H^2)x^2 - 2(R^2\bar{x} + Hy\sigma_0^2)x + \cdots\} \tag{8}$$

$$= -\frac{H^2\sigma_0^2 + R^2}{2\sigma_0^2 R^2}\{x^2 - \frac{2(R^2\bar{x} + Hy\sigma_0^2)}{H^2\sigma_0^2 + R^2}x + \cdots\} \tag{9}$$

$$= -\frac{H^2\sigma_0^2 + R^2}{2\sigma_0^2 R^2}\left\{\left[x - \frac{R^2\bar{x} + Hy\sigma_0^2}{H^2\sigma_0^2 + R^2}\right]^2 + \cdots\right\} \tag{10}$$

Thus we have the estimate of x

$$\hat{x} = \frac{R^2\bar{x} + Hy\sigma_0^2}{H^2\sigma_0^2 + R^2} = \bar{x} + \frac{\sigma_0^2 H}{H^2\sigma_0^2 + R^2}(y - H\bar{x}). \tag{11}$$

Here, we can summarize the result:

$$\begin{cases} \hat{x} = \bar{x} + \sigma_e^2 H R^{-2}(y - H\bar{x}), \\ \sigma_e^2 = (\sigma_0^{-2} + H^2 R^{-2})^{-1} \end{cases} \tag{12}$$

where σ_e is the variance of estimate error. In what follows, $\hat{x}$ is denoted as an estimation of x under the known $\bar{x}$ which is an average of x.

2.2 EXPRESSION OF KALMAN FILTERING

Here we ask for the evolution of the state variable x_{k+1}. A numerical solution of a partial differential equation may provide an evolution:

$$x_{k+1} = \phi x_k + G_k w_k \tag{13}$$

where ϕ is a known coefficient, G_k is the driving matrix and w_k is a white noise whose variance is Q_k^2. Driving matrix has components of 0 and 1 and denotes the relation between state variables and system noises in a vector form. In a scalar case, G_k is taken as 1.

Noisy observations y_k can be taken at time $t = k\Delta t$, and is linear in x_k

$$y_k = Hx_k + v_k \tag{14}$$

where H is a measurement coefficient which may be variable in time and v_k has zero mean and variance of R_k^2 as in the preceding section.

We call equation (13) as the 'state equation' and equation (14) as the 'observation equation' and the noises related to both equations are called as the system noise and the observational noise, respectively.

Here we start with time $t = 0$. As appeared in the preceding section, the observation of y at time $t = 0$, y_0, can provide the estimate of x_0

$$\hat{x}_0 = \bar{x}_0 + \sigma_{e_0}^2 H R^{-2}(y_0 - H\bar{x}_0) \tag{15}$$

based on prior information on $\bar{x}_0$ and R.

Next, the following p.d.f. is considered

$$p_{X|Y}(x_1, y_0) = \frac{1}{\sqrt{2\pi\sigma_1^2}}\exp\left[-\frac{(x_1-\bar{x}_1)^2}{2\sigma_1^2}\right] \tag{16}$$

which shows the probability density of x_1 based on the observation y_0. In order to evaluate a priori known value of $\bar{x}_1$ and σ_1 based on known information, following manipulations should be made

$$\bar{x}_1 = E[x_1|y_0] = E[\phi x_0 + G_0 w_0|y_0] = \phi E[x_0|y_0] = \phi \hat{x}_0, \tag{17}$$

$$\tilde{x}_1 = x_1 - \bar{x}_1 = \phi x_0 + G_0 w_0 - \phi \hat{x}_0 = \phi(x_0 - \hat{x}_0) + G_0 w_0, \tag{18}$$

$$\begin{aligned}\sigma_1^2 &= E[\tilde{x}_1^2] = E[\phi^2(x_0 - \hat{x}_0)^2 + 2\phi G_0 w_0 (x_0 - \hat{x}_0) + G_0^2 w_0^2] \\ &= \phi^2 E[(x_0 - \hat{x}_0)^2] + G_0^2 E[w_0^2] = \phi^2 \sigma_{e_0}^2 + G_0^2 Q_0^2 \\ &= \phi^2 (1 - K_0 H)\sigma_0^2 + G_0^2 Q_0^2.\end{aligned} \tag{19}$$

We can define the p.d.f. of x_1 based on the observation y_1 as follows

$$p_{X|Y}(x_1, y_1) = \frac{p_{X|Y}(x_1, y_0) p_{Y|X}(x_1, y_1)}{\int p_{X|Y}(x_1, y_0) p_{Y|X}(x_1, y_1) dx_1} \tag{20}$$

$$\propto p_{X|Y}(x_1, y_0) p_{Y|X}(x_1, y_1) \tag{21}$$

$$\propto \exp\left[-\frac{(x_1 - \phi \hat{x}_0)^2}{2\sigma_1^2} - \frac{(y_1 - H x_1)^2}{2R_1^2}\right] \tag{22}$$

Similar calculations in the preceding section should be followed. Thus we have

$$\hat{x}_1 = \bar{x}_1 + K_1(y_1 - H\bar{x}_1), \tag{23}$$

$$\bar{x}_1 = \phi \hat{x}_0, \tag{24}$$

$$K_1 = \frac{\sigma_1^2 H}{H^2 \sigma_1^2 + R_1^2}, \tag{25}$$

$$\sigma_1^2 = \phi^2(1 - K_0 H)\sigma_0^2 + G_0^2 Q_0^2. \tag{26}$$

As the time evolves, we can deductively get the results in a similar way

$$\hat{x}_k = \bar{x}_k + K_k(y_k - H\bar{x}_k), \tag{27}$$

$$\bar{x}_k = \phi \hat{x}_{k-1}, \tag{28}$$

$$K_k = \sigma_k^2 H (H^2 \sigma_k^2 + R_k^2)^{-1}, \tag{29}$$

$$\sigma_k^2 = \phi^2 (1 - K_{k-1} H)\sigma_{k-1}^2 + G_{k-1}^2 Q_{k-1}^2. \tag{30}$$

We can also summarize the results for multiple variables in a vector form. In the case of multiple variables, coefficient ϕ in equation (13) and the covariance σ_1 should be expanded as the matrices $\boldsymbol{F}$ and $\boldsymbol{P}$, respectively.

- Filter equation:

$$\hat{\boldsymbol{x}}_{k/k} = \bar{\boldsymbol{x}}_{k/k-1} + \boldsymbol{K}_k\{\boldsymbol{y}_k - \boldsymbol{H}\bar{\boldsymbol{x}}_{k/k-1}\}, \tag{31}$$

$$\bar{\boldsymbol{x}}_{k+1/k} = \boldsymbol{f}(\hat{\boldsymbol{x}}_{k/k}) \tag{32}$$

- Kalman gain:

$$\boldsymbol{K}_k = \boldsymbol{P}_{k/k-1}\boldsymbol{H}_k^T[\boldsymbol{H}_k \boldsymbol{P}_{k/k-1}\boldsymbol{H}_k^T + \boldsymbol{R}_k]^{-1} \tag{33}$$

- Estimate error covariance matrix:

$$\boldsymbol{P}_{k+1/k} = \boldsymbol{F}_k \boldsymbol{P}_{k/k} \boldsymbol{F}_k^T + \boldsymbol{G}_k \boldsymbol{Q}_k \boldsymbol{G}_k^T, \tag{34}$$

$$\boldsymbol{P}_{k/k} = \boldsymbol{P}_{k/k-1} - \boldsymbol{K}_k \boldsymbol{H}_k \boldsymbol{P}_{k/k-1} \tag{35}$$

- Initial condition:

$$\bar{\boldsymbol{x}}_{0/-1} = \bar{\boldsymbol{x}}_0, \qquad \boldsymbol{P}_{0/-1} = \boldsymbol{\Sigma}_0. \tag{36}$$

III. EXTENDED KALMAN FILTERING

We deal with the following nonlinear system equations:

$$\boldsymbol{x}_{k+1} = \boldsymbol{f}(\boldsymbol{x}_k) + \boldsymbol{G}_k \boldsymbol{w}_k, \tag{37}$$

$$\boldsymbol{y}_k = \boldsymbol{h}(\boldsymbol{x}_k) + \boldsymbol{v}_k \tag{38}$$

where $\boldsymbol{y}_k$ is the observation $(p \times 1)$, $\boldsymbol{x}_k$ is the state vector $(m \times 1)$, $\boldsymbol{f}$ and $\boldsymbol{h}$ are nonlinear functions of $\boldsymbol{x}_k$, $\boldsymbol{G}_k$ is the driving matrix $(m \times r)$, $\boldsymbol{w}_k$ is the system noise $(r \times 1)$ and $\boldsymbol{v}_k$ is the observational noise $(p \times 1)$. These noises have the following characteristics:

$$E\{\boldsymbol{w}_k\} = E\{\boldsymbol{v}_k\} = \boldsymbol{0}, \tag{39}$$

$$E\left\{\begin{pmatrix} \boldsymbol{w}_k \\ \boldsymbol{v}_k \end{pmatrix} (\boldsymbol{w}_s^T \boldsymbol{v}_s^T)\right\} = \begin{pmatrix} \boldsymbol{Q}_k & 0 \\ 0 & \boldsymbol{R}_k \end{pmatrix} \delta_{ks}, \quad \boldsymbol{R}_k > 0, \tag{40}$$

$$E\{\boldsymbol{w}_k \boldsymbol{x}_s^T\} = E\{\boldsymbol{v}_k \boldsymbol{x}_s^T\} = 0 \quad \text{for} \quad k > s \tag{41}$$

where δ_{ks} is Kronecker delta and $\boldsymbol{Q}_k = E\{\boldsymbol{w}_k \boldsymbol{w}_k^T\}$ is the covariance matrix of an observational error.

When $\boldsymbol{f}$ and $\boldsymbol{h}$ are smooth in time, we can linearize the nonlinear system equations using the first order of the Taylor series expansion:

- State equation is expanded around $\hat{\boldsymbol{x}}_{k/k}$

$$\boldsymbol{f}(\boldsymbol{x}_k) = \boldsymbol{f}(\hat{\boldsymbol{x}}_{k/k}) + \boldsymbol{F}_k(\boldsymbol{x}_k - \hat{\boldsymbol{x}}_{k/k}) + \cdots$$

- Observational equation is expanded around $\bar{\boldsymbol{x}}_{k/k-1}$

$$\boldsymbol{h}(\boldsymbol{x}_k) = \boldsymbol{h}(\bar{\boldsymbol{x}}_{k/k-1}) + \boldsymbol{H}_k(\boldsymbol{x}_k - \bar{\boldsymbol{x}}_{k/k-1}) + \cdots$$

where $\boldsymbol{F}_k$ (the state transition matrix) and $\boldsymbol{H}_k$ (the observation matrix) are:

$$\boldsymbol{F}_k = \left(\frac{\partial \boldsymbol{f}}{\partial \boldsymbol{x}}\right)_{\boldsymbol{x}=\hat{\boldsymbol{x}}_{k/k}} \quad (m \times m), \qquad \boldsymbol{H}_k = \left(\frac{\partial \boldsymbol{h}}{\partial \boldsymbol{x}}\right)_{\boldsymbol{x}=\bar{\boldsymbol{x}}_{k/k-1}} \quad (p \times m). \tag{42}$$

Hence, we have

$$\boldsymbol{x}_{k+1} = \boldsymbol{F}_k \boldsymbol{x}_k + \boldsymbol{G}_k \boldsymbol{w}_k + \boldsymbol{f}(\hat{\boldsymbol{x}}_{k/k}) - \boldsymbol{F}_k \hat{\boldsymbol{x}}_{k/k}, \tag{43}$$

$$\boldsymbol{y}_k = \boldsymbol{H}_k \boldsymbol{x}_k + \boldsymbol{v}_k + \boldsymbol{h}(\bar{\boldsymbol{x}}_{k/k-1}) - \boldsymbol{H}_k \bar{\boldsymbol{x}}_{k/k-1}. \tag{44}$$

- **Input** $\quad \boldsymbol{u}_k \equiv \boldsymbol{f}(\hat{\boldsymbol{x}}_{k/k}) - \boldsymbol{F}_k \hat{\boldsymbol{x}}_{k/k}$
- **Observation** $\quad \boldsymbol{\eta}_k \equiv \boldsymbol{y}_k - \boldsymbol{h}(\bar{\boldsymbol{x}}_{k/k-1}) + \boldsymbol{H}_k \bar{\boldsymbol{x}}_{k/k-1}$

Thus we can denote the following linearized system equations:

$$\boldsymbol{x}_{k+1} = \boldsymbol{F}_k \boldsymbol{x}_k + \boldsymbol{u}_k + \boldsymbol{G}_k \boldsymbol{w}_k, \tag{45}$$

$$\boldsymbol{\eta}_k = \boldsymbol{H}_k \boldsymbol{x}_k + \boldsymbol{v}_k. \tag{46}$$

Such a set of the system equations can be applied to an ordinary Kalman filtering algorithm:

- Filter equation:

$$\hat{\boldsymbol{x}}_{k/k} = \bar{\boldsymbol{x}}_{k/k-1} + \boldsymbol{K}_k\{\boldsymbol{y}_k - \boldsymbol{h}(\bar{\boldsymbol{x}}_{k/k-1})\}, \tag{47}$$

$$\bar{\boldsymbol{x}}_{k+1/k} = \boldsymbol{f}(\hat{\boldsymbol{x}}_{k/k}) \tag{48}$$

- Kalman gain:

$$\boldsymbol{K}_k = \boldsymbol{P}_{k/k-1} \boldsymbol{H}_k^T [\boldsymbol{H}_k \boldsymbol{P}_{k/k-1} \boldsymbol{H}_k^T + \boldsymbol{R}_k]^{-1} \tag{49}$$

- Estimate error covariance matrix:

$$\boldsymbol{P}_{k+1/k} = \boldsymbol{F}_k \boldsymbol{P}_{k/k} \boldsymbol{F}_k^T + \boldsymbol{G}_k \boldsymbol{Q}_k \boldsymbol{G}_k^T, \tag{50}$$

$$\boldsymbol{P}_{k/k} = \boldsymbol{P}_{k/k-1} - \boldsymbol{K}_k \boldsymbol{H}_k \boldsymbol{P}_{k/k-1} \tag{51}$$

- Initial condition:

$$\bar{\boldsymbol{x}}_{0/-1} = \bar{\boldsymbol{x}}_0, \qquad \boldsymbol{P}_{0/-1} = \boldsymbol{\Sigma}_0. \tag{52}$$

IV. Application to Some Inverse Problems

In the application of the afore-mentioned filter to inverse problems, the following three problems can be solved by Kalman filtering, i.e., the identification, estimation, and hybrid identification-estimation problems (see Table 1). Before presenting detailed formulations, the present effort will overview the analytical strategy for each problem appearing in the field of inverse problems in order. A key ingredient in the formulation of each problem is an appropriate selection of equations which correspond to system equations.

Table 1. Different positions of the Kalman filtering application.

Problem	Observational Eq.	State Eq.
Identification	Physical model	Stationary condition for parameters
Estimation	Observed condition	Physical model
Identification-Estimation	Observed condition	Physical model & stationary condition for parameters

4.1 Identification Problem

The first problem, i.e. an identification problem, requires a stationary condition for parameters as the state equation. The reason for such a correspondence is that the parameters to be identified should be kept constant in time. Thus, the state transition matrix $\boldsymbol{F}_t$ is a unit matrix $\boldsymbol{I}$. On the other hand, the physical model generally describes the observation equation and often leads to nonlinearity. In this observation equation, observation matrix $\boldsymbol{H}_t$ is often time-independent.

We consider here a simple example for parameter identification. In such a case, the following stationary condition for parameters $\boldsymbol{x}$ is adopted in continuous time:

$$\dot{\boldsymbol{x}} = 0. \tag{53}$$

In discrete time, it becomes

$$\boldsymbol{x}_{t+1} = \boldsymbol{I}\boldsymbol{x}_t. \tag{54}$$

Now, consider the inverse problem of characterizing the material properties, Lamé's constants λ_i and μ_i, of the elastic medium within the framework of the finite element method.

$$\boldsymbol{x}_t = [\lambda_1, \mu_1, \ldots, \lambda_n, \mu_n]^T \tag{55}$$

The condition in equation (54) also holds for unknown Lamé's constants, because they are constant[4]. Parameters should be kept constant, such that the above condition denotes the state equation and does not include system noise.

On the other hand, a finite element equation which is written for the equilibrium solution under the plane strain condition is derived as

$$\boldsymbol{K}\boldsymbol{u} = \boldsymbol{f}. \tag{56}$$

Through this equation, the observed displacements, $\boldsymbol{u}^*$, can be rewritten as the nonlinear functions of the unknowns, i.e., Lamé's constants,

$$\boldsymbol{u}^* = \boldsymbol{h}(\boldsymbol{x}_t) + \boldsymbol{v}_t. \tag{57}$$

When substituting the above equation into the observation equation, we specify observation matrix $\boldsymbol{H}$ by a certain linearization technique. Resulting matrix $\boldsymbol{H}$ involves unknown parameters, and thus, the evaluation of parameters should be finally achieved by the iteration scheme for constant measurements.

A set of equations (54) and (57) can be applied to the extended Kalman filtering procedure. A direct formulation using equation (56) also allows for the search of parameters in another way[4]. Recently, other formulations of the extended Kalman filtering procedure for an inverse analysis have also been furnished and examined in conjunction with CSM (the Charge Simulation Method)[5] and BEM[6)-9)].

Such a technique is called as 'the extended Kalman filter-Finite Element Method'[4]. Flow of the computation is given in Figure 1 and Table 2.

Table 2. Implementation of the numerical procedure.

1.	Set the initial values of the state variable $\bar{\boldsymbol{x}}_{1/0}$ and the estimate error $\boldsymbol{P}_{1/0}$.
2.	Set $i = 1$.
3.	Set $\boldsymbol{Q}_i$ and $\boldsymbol{R}_i$. (These matrices are often assumed to be constant in time.)
4.	Observe $\boldsymbol{y}_i$.
5.	Compute displacements $\boldsymbol{u}^*$ based on identified parameters and FEM.
6.	Complete $\boldsymbol{H}_i$ matrix based on Equation (42) and FEM.
7.	Evaluate $\boldsymbol{K}_i = \boldsymbol{P}_{i/i-1}\boldsymbol{H}_i^T[\boldsymbol{H}_i\boldsymbol{P}_{i/i-1}\boldsymbol{H}_i^T + \boldsymbol{R}_i]^{-1}$
8.	Evaluate $\boldsymbol{P}_{i/i} = \boldsymbol{P}_{i/i-1} - \boldsymbol{K}_i\boldsymbol{H}_i\boldsymbol{P}_{i/i-1}$.
9.	Evaluate $\hat{\boldsymbol{x}}_{i/i} = \bar{\boldsymbol{x}}_{i/i-1} + \boldsymbol{K}_i[\boldsymbol{y}_i - \boldsymbol{H}_i\bar{\boldsymbol{x}}_{i/i-1}]$.
10.	Evaluate $\boldsymbol{P}_{i+1/i} = \boldsymbol{F}_i\boldsymbol{P}_{i/i}\boldsymbol{F}_i^T + \boldsymbol{G}_i\boldsymbol{Q}_i\boldsymbol{G}_i^T$.
11.	Evaluate $\bar{\boldsymbol{x}}_{i+1/i} = \boldsymbol{f}(\hat{\boldsymbol{x}}_{i/i})$.
13.	If the convergency of state variables can be found, stop the computation.
12.	Otherwise, set $i = i + 1$ and go to Step 3.

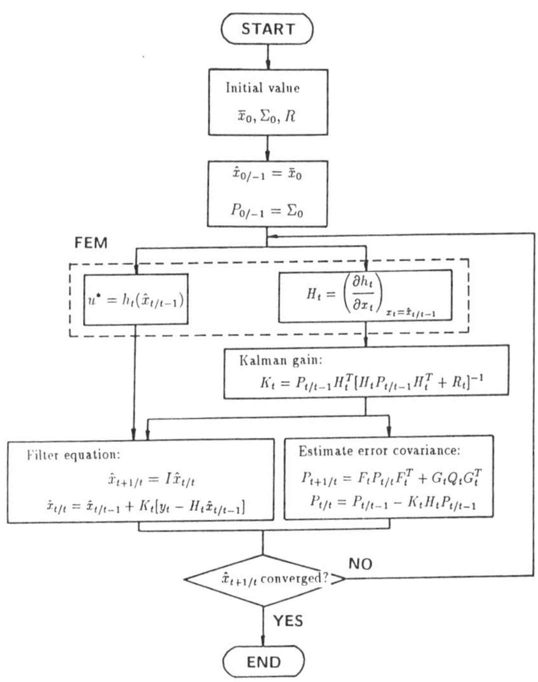

Figure 1. Flow of computation for the extended Kalman filter-Finite Element Method.

4.2 Estimation Problem

The second problem, i.e. an estimation problem, requires a physical model as the state equation. On the other hand, the observation equation denotes the observed condition. In this type of problem, we refine the prediction of the state variable the next time by using the observation at the current time, and such a refinement should be carried out each time. When we directly observe a part of the state variables, observation matrix $\boldsymbol{H}_t$ is a sparse matrix including the components of 1 (observed) and 0 (not observed) throughout the entire process, and it is time-independent.

We consider the following ordinary differential equation as an estimation problem example.

$$\dot{\boldsymbol{x}} = \boldsymbol{A}\boldsymbol{x} + g. \tag{58}$$

In discrete time, it becomes

$$\boldsymbol{x}_{t+1} = \boldsymbol{B}\boldsymbol{x}_t + \boldsymbol{C}\boldsymbol{u}_t \tag{59}$$

where $\boldsymbol{B}$ is a positive-definite matrix. This equation directly describes the state equation by adding some system noise if necessary. In this type of problem, the observation equation denotes the observed condition. For example, when we observe the i-th and j-th components of the state vector, observation matrix $\boldsymbol{H}$ is written as

$$\boldsymbol{H} = \begin{bmatrix} 0 \cdots 0 \overset{i}{1} 0 \cdots\cdots 0 \\ 0 \cdots\cdots 0 \overset{j}{1} 0 \cdots 0 \end{bmatrix} \tag{60}$$

4.3 Identification-Estimation Problem

The third problem, i.e. an identification-estimation problem, combines the above two problems. In this case, both the physical model and the stationary condition for parameters construct the state equation and the observed condition describes the observation equation. Identification and estimation reciprocally come into work through the sensitivity matrix. For the state equation, we consider equation (58) in continuous time and equation (59) in discrete time. The state vector is augmented by appending the unknown parameters for this type of problem. From the second moment analysis, a combined state transition matrix is given as follows

$$\boldsymbol{F}_t = \begin{bmatrix} \boldsymbol{B} & \boldsymbol{D} \\ 0 & \boldsymbol{I} \end{bmatrix} \tag{61}$$

where $\boldsymbol{D}$ is the 'sensitivity matrix', which dominates the interaction between identification and estimation:

$$\boldsymbol{D} = \frac{\partial}{\partial \boldsymbol{s}}(\boldsymbol{B}\boldsymbol{x}_t + \boldsymbol{C}\boldsymbol{u}_t) \tag{62}$$

where $\boldsymbol{s}$ is an unknown parameter vector.

4.4 Overview of the Applications of Kalman Filter

Here we show the results for use in various applications for the problem categories in the preceding section. Table 3 provides an overview of the applications of the Kalman filtering or extended Kalman filtering technique to several engineering problems. In what follows, we will refer to those works.

For the identification problem, Hoshiya & Saito[2),10),11)] pursued structural parameters of the linear system from seismic data by use of their own 'weighted global iteration technique (EK-WGI)'. Recently they proposed the refined numerical technique within local iteration (EK-WLI) and applied it to parameter identification[12),13)]. Hoshiya *et al.*[14),15)] also introduced the AR model for equivalent linearization of liquefaction and pursued the model parameters by using earthquake response data.

On the other hand, the authors proposed a new observational procedure including parameter identification of Asaoka's model[16)] by Kalman filtering[17),18)]. Hoshiya *et al.* applied such a technique to actual observations of organic soil[19)]. Saito *et al.*[20)] estimated lateral pressure and the coefficient of horizontal subgrade reaction by the structural model with extended Kalman filter, and presented the observational control system for executing excavation works using data based on minimum measurements. Sakai & Hoshiya[21)] developed a method that predicts and controls its future behavior on driving shields, which effectively applied Kalman filtering. Murase *et al.*[22),23),24)] applied such a procedure to characterize the material property of agricultural products and developed the measuring system.

The authors developed the formulation of the Kalman filter with FEM[25),26)]. Several researchers have followed and expanded these works. Suzuki & Ishii[27)] derived a finite element based estimation of a spatial variation of soil properties with extended Kalman filtering, where the number of unknowns is the same as the number of elements, but larger than that of measurements. Courage *et al.*[28)] had a parallel development on the same procedure. At almost the same time, Kadota *et al.*[29] identified mechanical properties of anisotropic rock mass and initial ground pressure using a similar approach to that adopted by Suzuki & Ishii. Murase *et al.*[30] combined the extended Kalman filtering with CSM in order to characterize elastic moduli. Tosaka *et al.*[6),7] and Tanaka *et al.*[8),9] also provided the extended Kalman filter with BEM aiming at the identification of material properties. Toki *et al.*[31] identified structural parameters as the masses, damping coefficients and spring constants of the linear multi-degree-of-freedom system from measured earthquake time history.

For the estimation problem, Toki *et al.*[32] proposed a procedure by which input ground motions of a multi-input system are estimated from measured responses only. Such a procedure solves the resulting ordinary differential equations by incorporating the Kalman filtering estimate for each measurement.

The interest in identification-estimation problems started with the work of Wilson *et al.*[33], who introduced the extended Kalman filtering to seepage analysis with FEM. They solved the 'interaction' between the identification of permeability and the estimation of a pressure head. Hirano[34] and Okuno *et al.*[39] also solved the similar problem for horizontally two-dimensional soil stratum.

Table 3. Applications of the Kalman filtering technique to engineering problems.

Identification		
Hoshiya & Saito[2),10),11)]	Model parameters	Linear multi-degree-of-freedom system
Hoshiya & Saito[14),15)]	Model parameters	AR model for liquefaction
Murakami & Hasegawa[17),18)]	Parameters of Asaoka's model	Asaoka's model
Hasegawa & Murakami[35),36)]	Model parameters	AR model for dam leakage
Sakai & Hoshiya[21)]	Model parameter	AR model for shields behaviour
Saito *et al.*[20)]	Lateral pressure/ coefficient of horizontal subgrade reaction	Structural model for excavation works
Murakami & Hasegawa[25),26)]	Elastic moduli	FEM
Murase *et al.*[22),23),24)]	Elastic moduli	FEM
Suzuki & Ishii[27)]	Elastic moduli	FEM
Kadota *et al.*[29)]	Elastic moduli	FEM
Murase *et al.*[30)]	Elastic moduli	BEM
Toki *et al.*[31)]	Structural parameters	Linear multi-degree-of-freedom system
Lin & Chao[37)]	Structural parameters	Linear multi-degree of-freedom system
Courage *et al.*	Elastic Moduli	FEM
Tosaka & Utani[6),7)]	Elastic Moduli	BEM
Sutoh & Hoshiya[12),13),19)]	Elastic Moduli	FEM
Hoshiya *et al.*[19)]	Parameters of Asaoka's model	Asaoka's model
Tanaka *et al.*[8)]	Heat Transfer Coefficient	BEM
Tanaka *et al.*[9)]	Location of Crack	BEM
Hoshiya & Sutoh[38)]	Elastic Moduli	FEM

Estimation		
Murakami & Hasegawa[17)]	Pore pressure	FDM
Toki *et al.*[32)]	Input ground motions	Linear multi-degree-of-freedom system
Identification-Estimation		
Wilson *et al.*[33)]	Permeability & pressure head	FEM
Hirano[34)]	Transmissivity & pressure head	FEM
Okuno & Suzuki[39)]	Permeability & pressure head	FEM

V. Numerical Example

A numerical evaluation of observed deformations is illustrated to examine the applicability of the above formulation as an example of 'the identification problem' described in the preceding section. Figure 2 describes a setup of hypothetical *in situ* loading tests. The present approach provides a solution to the inverse problem by seeking for non-homogeneous material properties from the limited number of observations due to loading. In this analysis, only Young's moduli of all the elements are considered as unknowns, whereas the value of Poisson's ratio is assumed to be known.

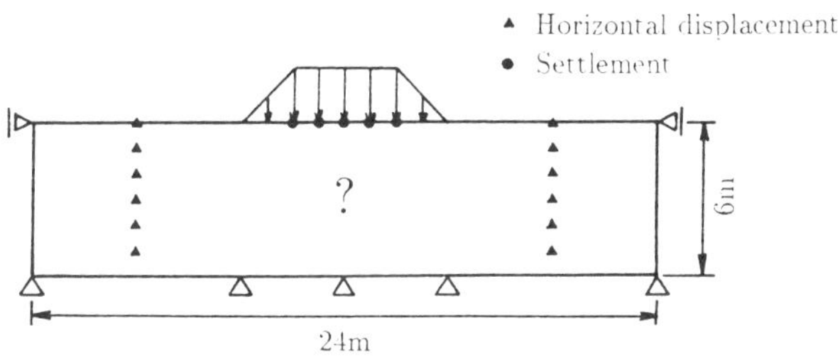

Figure 2. Hypothetical *in situ* loading test.

Figure 3 serves to illustrate a distribution of Young's moduli to be evaluated as a target. Soil deposit here includes a horizontal week layer as seen in Figure 3. Our objective is to pick up the existence of such a layer based on the results of usual loading tests.

Figure 4 shows a plot of the bird's-eye view of the identified Young's moduli based on the observed locations at the marked locations. In this analysis, the initial value of the unknowns is uniformly set to at 1,000 tf/m^2. It is seen that the existence of a week layer can be found based on the limited number of measurement devices. Figure 5 also shows the distribution of the diagonal terms of estimate error covariance. From this figure, it can be found that the estimation error is not so high.

An initial estimate is highlighted for the error covariance matrix as the key to the solution in this type of problem. The components of the matrix in this case are given according to a priori assumed probabilistic characteristics, namely,

$$\rho(\Delta x, \Delta y) = exp\left[-\left\{\left(\frac{\Delta x}{a}\right)^2 + \left(\frac{\Delta y}{b}\right)^2\right\}\right] \tag{63}$$

as denoted in Reference 5, where ρ is a correlation factor of the unknowns, Δx is the horizontal distance and Δy is the vertical distance between two arbitrary points, and a and b are the constants. The value of a is given to be larger than that of b in the horizontally layered deposit case. In the results of Figure 4, values of 5.0 and 1.0 are given to a and b, respectively. This means that the objective soil stratum is assumed to have the characteristics of a horizontally layered deposit.

Off-diagonal terms for the error covariance matrix, based on equation (63), have an important role in the identification of non-homogeneous material properties. Suitable information a priori provides successful results, as seen in Figure 4. Hence, an inappropriate estimate of this information may introduce an ill-posed identification. For example, Figure 6 reveals poor results for the same case considering only diagonal terms of the error covariance matrix in comparison with Figure 5.

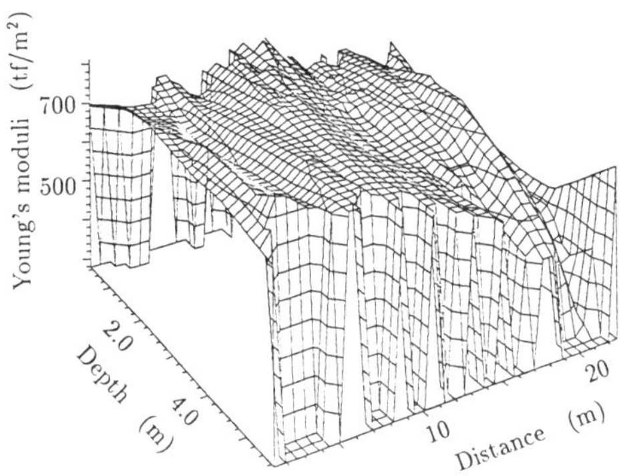

Figure 3. Distribution of Young's moduli to be estimated.

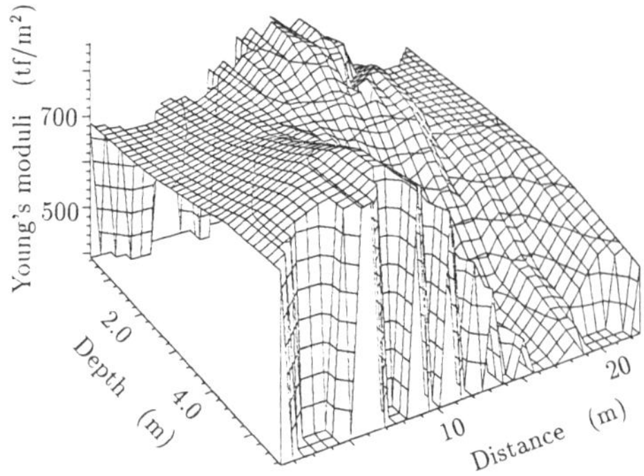

Figure 4. Distribution of identified Young's moduli considering off-diagonal terms.

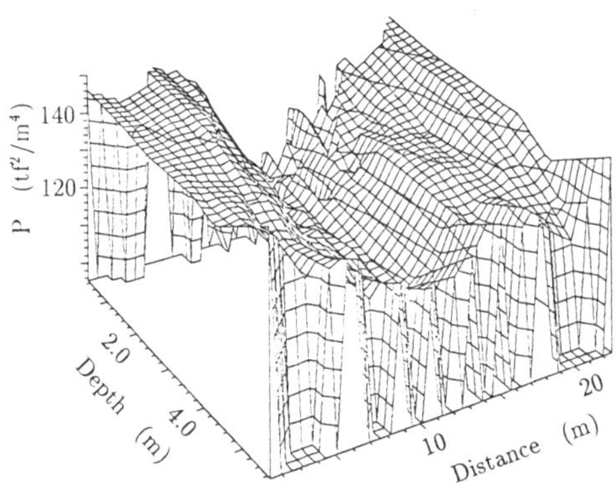

Figure 5. Distribution of estimate error covariance.

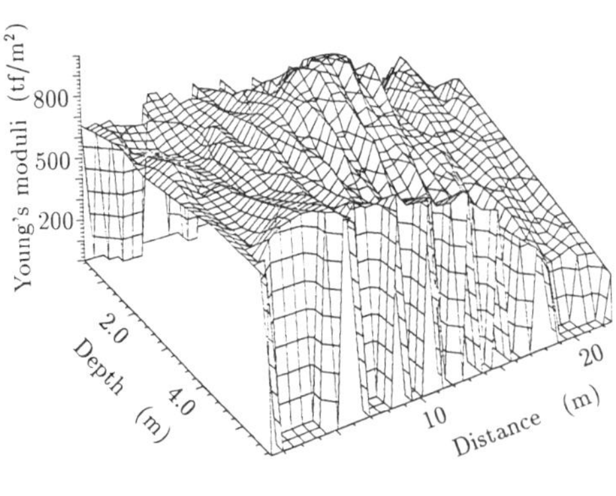

Figure 6. Distribution of identified Young's moduli without considering off-diagonal terms.

VI. Conclusion

In solving various inverse problems, the usefulness of Kalman filtering technique is demonstrated. A priori information on the initial estimate for the error covariance matrix has an important role in the identification. Several numerical performances reveal the applicability of the proposed technique.

References

1) R.E. Kalman. A new approach to linear filtering and prediction problems. *Trans. ASME, J. Basic Eng.*, 82D(1): 34–45, 1960.

2) M. Hoshiya and E. Saito. Structural identification by extended Kalman filter. *Journal of Engineering Mechanics, ASCE*, 110(12): 1757–1770, 1984.

3) G. Gioda and S. Sakurai. Back analysis procedures for the interpretation of field measurements in geomechanics. *Int. j. numer. anal. methods geomech.*, 11(6): 555–583, 1987.

4) A. Murakami. *Studies on the application of Kalman filtering to some geomechanical problems related to safety assessment.* Doctoral Dissertation, Kyoto University, 1991.

5) H. Murase, S. Koyama, and N. Honami. Kalman filter charge simulation methods for the solution of inverse problems of two-or three-dimensional elasticity. *Trans. of JSME*, 56A(531): 106–112, 1990 (in Japanese).

6) A. Utani and N. Tosaka. Identification analysis of elastic constants by extended Kalman filter–boundary element method. *Proc. 8th Japan National Symp. Boundary Element Methods*, 8: 23–28, 1991 (in Japanese).

7) N. Tosaka and A. Utani. Identification analysis of elastic constants using Kalman filter-boundary element method. *Journal of Struct. Engng., AIJ*, (446):41–50, 1993 (in Japanese).

8) M. Tanaka, M. Nakamura, and H. Ishikawa. Identification of unknowns in transient heat conduction fields by inverse analysis using extended Kalman filter and Boundary Element Method. *Proc. 9th Japan National Symp. Boundary Element Methods*, 9: 47–52, 1992 (in Japanese).

9) M. Nakamura, M. Tanaka, and R. Ochiai. Identification of unknown boundary value using elastodynamic Boundary Element Method (Consideration on information selection using sensitivities). *Proc. 70th Annual Conf. JSME*, 71–72, 1993.

10) M. Hoshiya and E. Saito. Identification problem of some seismic systems by extended Kalman filter. *Proc. JSCE*, (339): 59–67, 1982 (in Japanese).

11) M. Hoshiya and E. Saito. Estimation of dynamic properties of a multiple degree of freedom linear system. *Proc. JSCE*, (344): 289–298, 1984 (in Japanese).

12) A. Sutoh and M. Hoshiya. A basic consideration on and a new local iteration method of extended Kalman filter. *Proc. JSCE*, (437): 203–211, 1991 (in Japanese).

13) A. Sutoh and M. Hoshiya. Back analysis by finite element method using EK-WLI procedure. *Proc. JSCE*, (446): 177–185, 1992 (in Japanese).

14) M. Hoshiya and E. Saito. Equivalent linearization of liquefied sandy soil. *Proc. JSCE*, (340): 59–65, 1983 (in Japanese).

15) M. Hoshiya and E. Saito. Linearized liquefaction process by Kalman filter. *Journal of Geotechnical Engineering, ASCE*, 112(2): 155–169, 1986.

16) A. Asaoka. Observational procedure of settlement prediction. *Soils and Foundations*, 18(4): 87–101, 1978.

17) A. Murakami and T. Hasegawa. Observational prediction of settlement using Kalman filter theory. *Proc. 5th Int. Conf. on Numerical Methods in Geomechanics*, 3: 1637–1643, 1985.

18) A. Murakami and T. Hasegawa. Observational prediction of settlement using Kalman filtering. *Trans. of the Japanese Society of Irrigation, Drainage and Reclamation Engineering*, (120): 61–67, 1985 (in Japanese).

19) M. Hoshiya, K. Itabashi, T. Itoh, and A. Sutoh. Prediction of settlement on peaty soft deposits using extended Kalman filter. *Proc. JSCE*, (455): 105–114, 1992 (in Japanese).

20) E. Saito, S. Koga, and M. Kamata. Execution of excavation works based on data supported by extended Kalman filter. *Proc. JSCE*, (391): 115–124, 1987 (in Japanese).

21) K. Sakai and M. Hoshiya. Prediction and control of behaviors on driving shields using Kalman filter theory. *Proc. JSCE*, (385): 69–78, 1987 (in Japanese).

22) H. Murase, S. Oke, S. Shibusawa, and Y. Nakamura. Measuring system for mechanical properties of trouble-handling agricultural materials —Application of Kalman filter finite element inverse analysis—. *Bull. Ishikawa Agr. Coll.*, (17): 32–40, 1987.

23) H. Murase. Parameter estimation of mechanical properties of agricultural materials (Part 1) —Application of finite element inverse analysis—. *Journal of the Japanese Society of Agricultural Machinery*, 50(2): 75–83, 1988 (in Japanese).

24) H. Murase, S. Oke, S. Shibusawa, and Y. Nakamura. Parameter estimation of mechanical properties of agricultural materials (Part 2) —Measuring system—. *Journal of the Japanese Society of Agricultural Machinery*, 50(6): 69–76, 1988 (in Japanese).

25) A. Murakami and T. Hasegawa. Back analysis by Kalman filter-finite elements and a determination of optimal observed points location. *Proc. JSCE*, (388): 227–235, 1987 (in Japanese).

26) A. Murakami and T. Hasegawa. Back analysis by Kalman filter-finite elements and optimal location of observed points. *Proc. 6th Int. Conf. on Numerical Methods in Geomechanics*, 2: 2051–2058, 1988.

27) M. Suzuki and K. Ishii. Estimation of spatial variation of soil properties using extended Kalman filter algorithm. *Proc. JSCE*, (406): 71–78, 1989 (in Japanese).

28) W.M.G. Courage, P.J. Schreurs, and J.D. Janssen. Estimation of mechanical parameter values of composites with the use of finite element and system identification techniques. *Computers and Structures*, 34:231–237, 1990.

29) S. Kadota, E. Saito, A. Waku, and T. Goto. Identification method for anisotropic mechanical constants of rock mass by local iterated extended Kalman filter and application to excavation control of underground openings. *Proc. JSCE*, (406): 107–116, 1989 (in Japanese).

30) H. Murase, S. Koyama, and T. Kuwabara. Adapting the charge simulation problem to a microcomputer (III) —Inverse analysis using the extended Kalman filter—. *Journal of Japanese Society Irrigation, Drainage and Reclamation Engineering*, 57(8): 55–60, 1989 (in Japanese).

31) K. Toki, T. Sato, and J. Kiyono. Identification of structural parameters and input ground motion from response time histories. *Structural Eng./Earthquake Eng., JSCE*, 6(2): 413s–421s, 1989.

32) K. Toki, T. Sato, J. Kiyono, and H. Takeuchi. Identification of the input ground motions of a multi-input system. *Annuals, Disas. Prev. Res. Inst., Kyoto Univ.*, (32B-2): 23–38, 1989 (in Japanese).

33) J. Wilson, P. Kitanidis, and M. Dettinger. State and parameter estimation in groundwater models. *Application of Kalman filter to hydrology, hydraulics, and water resources*, 1: 657–679, Univ. of Pittsburgh, 1978.

34) F. Hirano. *Estimating spatial distribution of both transmissivity and groundwater head over a large area.* Doctoral Dissertation, Kyushu University, 1985 (in Japanese).

35) T. Hasegawa and A. Murakami. Application of Kalman filtering to the prediction of dam leakage —safety assessment for dams—. *Trans. of the Japanese Society of Irrigation, Drainage and Reclamation Engineering*, (126): 1–8, 1986 (in Japanese).

36) T. Hasegawa and A. Murakami. Prediction of dam leakage with Kalman filtering and a safety assessment for dams. *Mem. Coll. Agric., Kyoto Univ.*, (129): 1–13, 1986.

37) J.-S. Lin and B.-K. Chao. Estimation of shear moduli and damping factors of earth dam materials. *Earthquake eng. struct. dyn.*, 19(6): 891–910, 1990.

38) M. Hoshiya and A. Sutoh. Kalman filter-finite element method in identification. *Journal of Engineering Mechanics, ASCE*, 119(2): 197–210, 1993.

39) T. Okuno and M. Suzuki. Parameters identification in confined aquifer using extended Kalman filter algorithm. *Proc. JSCE*, (469): 93–102, 1993 (in Japanese).

Identification Analysis of Unknown Parameters in Elastic Fields by Kalman Filter-Boundary Element Method

Akihide UTANI* and Nobuyoshi TOSAKA**

* *Graduate School of Nihon University, Narashino, Chiba*
** *Department of Mathematical Engineering, College of Industrial Technology, Nihon University, Narashino, Chiba*

An approach based on the boundary element method in solving inverse problems of partial differential equations given as governing equations of a field is proposed in this paper. This approach is developed with the boundary element method in conjunction with the extended Kalman filter theory which has been effectively used in the fields of optimal control theory for dynamic systems. Applicability and effectiveness of the present method, which we would like to call the Kalman filter-boundary element method, are examined with identification problems of unknown quantities such as material properties in 2-D or 3-D linear elastic fields.

I. INTRODUCTION

An increasing interest in the techniques for solving problems of inverse analyses has been shown in a sphere of parameter identification in the fields governed by the partial differential equations[1)-3)]. In its sphere, the Kalman filter has been adopted as one of the powerful methods which allow error information in the observed data during a solution procedure[4)]. For instance, Murakami et al.[5)] have identified the system parameters in a discretized system such as Lamé's constants by using the finite element method incorporated with the Kalman filtering approach in 2-D domain of geomechanics field. For identification problem of material properties of 3-D domain by using the Kalman filter, Murase et al.[6)] proposed an identification scheme using the charge simulation method in conjunction with the Kalman filtering technique.

We proposed the Kalman filter-boundary element method to solve identification problems[7)-9)]. In order to examine applicability of our proposed scheme, we have solved identification problems of the concentrated heat source in 2-D domain of steady-state thermal conduction field[7)] or Lamé's constants in 2-D domain of geomechanics field[8),9)].

In this paper, we show further developments of the Kalman filter- boundary element method in identification problems of the linear elastic field. They are on the proposal of a relevant formulation in the problems and the applicability to 3-D field. A formulation for identification of the unknown boundary quantities as well as Lamé's constants in a plan elastic field is proposed. And, the applicability to the parameter identification problem of a 3-D elastic body such as a silicone gum subjected to the Hertz-type load, is also discussed. Finally, numerical examples are shown in order to demonstrate the usefulness of the proposed method.

II. KALMAN FILTER APPROACH

2.1 Linear Stochastic System

The Kalman filtering technique has an on-line data acquisition algorithm which offers successively an optimal presumption to state variables by using input data as follows: 1)dynamic characteristics of the system which generates signals, 2)statistical characteristics of noises, 3)a priori information on initial values of state variables, 4)observation data.

Let us consider the linear stochastic system in discrete time which is constituted of the state equation and the observation equation as follows:

State equation

$$z_{t+1} = \boldsymbol{\Phi}_t z_t + \boldsymbol{\Gamma}_t \boldsymbol{\omega}_t \tag{1}$$

Observation equation

$$y_t = M_t z_t + \nu_t \tag{2}$$

where $\boldsymbol{z}_t$ is a state vector$(n \times 1)$, $\boldsymbol{\Phi}_t$ is a state transition matrix$(n \times n)$, $\boldsymbol{\Gamma}_t$ is a driving matrix$(n \times m)$, $\boldsymbol{y}_t$ is an observation vector$(p \times 1)$, $\boldsymbol{M}_t$ is an observation matrix$(p \times n)$, $\boldsymbol{\omega}_t$ is a system noise vector$(m \times 1)$, $\boldsymbol{\nu}_t$ is an observation noise vector$(p \times 1)$ and the subscript t denotes the $t - th$ time step.

Both noise vectors are assumed to possess the following stochastic characteristics:

$$E(\boldsymbol{\omega}_t) = \mathbf{o}, \quad E\left[\boldsymbol{\omega}_t \boldsymbol{\omega}_s^T\right] = \boldsymbol{Q}_t \delta_{ts}, \quad E\left[\boldsymbol{\omega}_t \boldsymbol{z}_s^T\right] = \mathbf{o} \qquad for\ t \geq s \tag{3}$$

$$E(\boldsymbol{\nu}_t) = \mathbf{o}, \quad E\left[\boldsymbol{\nu}_t \boldsymbol{\nu}_s^T\right] = \boldsymbol{R}_t \delta_{ts}, \quad E\left[\boldsymbol{\nu}_t \boldsymbol{z}_s^T\right] = \mathbf{o}, \quad \boldsymbol{R}_t > 0 \qquad for\ t \geq s \tag{4}$$

in which T denotes transposition, s is the $s - th$ time step, E[$\cdot$] is an average operator, δ_{ts} is Kronecker's delta, and $\boldsymbol{Q}_t$ and $\boldsymbol{R}_t$ are the covariance matrix of a system noise and that of an observation noise, respectively.

2.2 Solution Procedure

The algorithm of the Kalman filtering for the above linear stochastic system (1) and (2) with (3) and (4) can be given as follows:

Filter equation

$$\hat{\boldsymbol{z}}_{t+1/t} = \boldsymbol{\Phi}_t \hat{\boldsymbol{z}}_{t/t} \tag{5}$$

$$\hat{\boldsymbol{z}}_{t/t} = \hat{\boldsymbol{z}}_{t/t-1} + \boldsymbol{K}_t\left[\boldsymbol{y}_t - \boldsymbol{M}_t \hat{\boldsymbol{z}}_{t/t-1}\right] \tag{6}$$

Kalman gain

$$\boldsymbol{K}_t = \boldsymbol{P}_{t/t-1} \boldsymbol{M}_t^T \left[\boldsymbol{M}_t \boldsymbol{P}_{t/t-1} \boldsymbol{M}_t^T + \boldsymbol{R}_t\right]^{-1} \tag{7}$$

Estimate error covariance matrix

$$\boldsymbol{P}_{t+1/t} = \boldsymbol{\Phi}_t \boldsymbol{P}_{t/t} \boldsymbol{\Phi}_t^T + \boldsymbol{\Gamma}_t \boldsymbol{Q}_t \boldsymbol{\Gamma}_t^T \tag{8}$$

$$\boldsymbol{P}_{t/t} = \boldsymbol{P}_{t/t-1} - \boldsymbol{K}_t \boldsymbol{M}_t \boldsymbol{P}_{t/t-1} \tag{9}$$

Initial condition

$$\hat{\boldsymbol{z}}_{0/-1} = \bar{\boldsymbol{z}}_0, \qquad \boldsymbol{P}_{0/-1} = \boldsymbol{\Sigma}_0 \tag{10}$$

in which $\bar{\boldsymbol{z}}_0$ denotes the average value vector of initial state vector and $\boldsymbol{\Sigma}_0$ is the initial value of an estimate error covariance matrix.

Further details of the filtering algorithm are available in references[10),11)].

2.3 Extended Kalman Filtering

Let us consider the following nonlinear observation equation:

$$\boldsymbol{y}_t = \boldsymbol{m}_t(\boldsymbol{z}_t) + \boldsymbol{\nu}_t \tag{11}$$

where $\boldsymbol{m}_t$ is the nonlinear mapping of state vector $\boldsymbol{z}_t$. In order to avoid nonlinearity, we may utilize the Taylor series expansion with an estimate of $\hat{\boldsymbol{z}}_{t/t-1}$ for the above observation equation. Disregarding the terms beyond the second order of this expansion, the refined observation equation can be obtained as follows:

$$\begin{aligned}
\boldsymbol{y}_t &= \boldsymbol{m}_t(\hat{\boldsymbol{z}}_{t/t-1}) + \left(\frac{\partial \boldsymbol{m}_t}{\partial \boldsymbol{z}_t}\right)_{\boldsymbol{z}_t = \hat{\boldsymbol{z}}_{t/t-1}} (\boldsymbol{z}_t - \hat{\boldsymbol{z}}_{t/t-1}) + \boldsymbol{\nu}_t \\
&= \boldsymbol{m}_t(\hat{\boldsymbol{z}}_{t/t-1}) + \boldsymbol{M}_t(\boldsymbol{z}_t - \hat{\boldsymbol{z}}_{t/t-1}) + \boldsymbol{\nu}_t \\
&= \boldsymbol{M}_t \boldsymbol{z}_t + \left\{\boldsymbol{m}_t(\hat{\boldsymbol{z}}_{t/t-1}) - \boldsymbol{M}_t \hat{\boldsymbol{z}}_{t/t-1}\right\} + \boldsymbol{\nu}_t
\end{aligned} \tag{12}$$

Moreover, we may rewrite equation (12) into

$$\boldsymbol{\eta}_t = \boldsymbol{M}_t \boldsymbol{z}_t + \boldsymbol{\nu}_t \tag{13}$$

where

$$\boldsymbol{\eta}_t = \boldsymbol{y}_t - \boldsymbol{m}_t(\hat{\boldsymbol{z}}_{t/t-1}) + \boldsymbol{M}_t \hat{\boldsymbol{z}}_{t/t-1} \tag{14}$$

The second and third terms in equation (14) can be calculated at the $(t - 1)th$ time step. Thus, the value of $\boldsymbol{\eta}_t$ can be 'observed' at time t with use of the observation vector $\boldsymbol{y}_t$. In turn, equation (13) constitutes a linear system to which the ordinary Kalman filtering algorithm given by the preceding section can be applied.

III. KALMAN FILTER-BOUNDARY ELEMENT METHOD

3.1 Statement of Identification Problem

The identification problem under consideration is such that unknown quantities, for instance, Lamé's constants or unknown boundary values, should be identified by using measured boundary responses. Figure 1 shows the statement of this identification problem. The whole geometry and the boundary conditions are assumed to be known. It is also assumed that linear boundary responses measured at some points are available for this identification analysis.

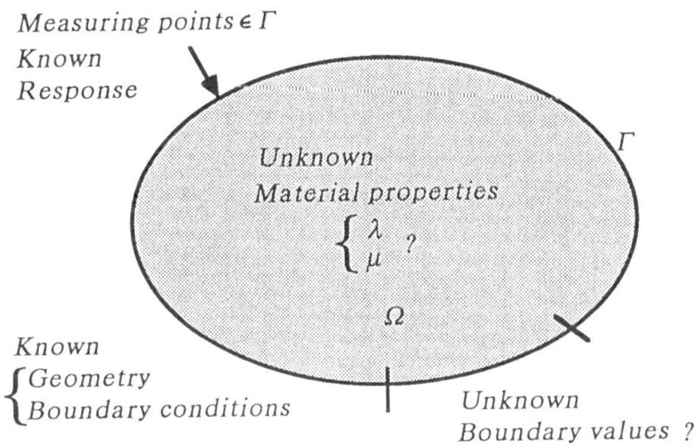

Figure 1. Statement of identification problem in elastic field.

The boundary-value problem for linear homogeneous isotropic elastic field under the no body force can be given as follows:

Governing equation(Navier equation)

$$\mu \frac{\partial^2 u_i}{\partial x_j^2} + (\lambda + \mu)\frac{\partial^2 u_j}{\partial x_i x_j} = 0 \quad in\ \Omega \tag{15}$$

Boundary conditions

$$u_i = \hat{u}_i \quad on\ \Gamma_u \quad \text{(Displacement condition)} \tag{16}$$

$$t_i \equiv t_{ij} n_j = \hat{t}_i \quad on\ \Gamma_t \quad \text{(Traction condition)} \tag{17}$$

where u_i and t_i(i=1,2 in 2-D domain and i=1 $\sim$ 3 in 3-D domain) are displacement components and traction components, respectively, λ and μ are Lamé's constants, n_j is a component of unit normal vector to the boundary, and $\hat{u}_i$ and $\hat{t}_i$ denote the prescribed values on the boundary.

3.2 Boundary Element Analysis

In this section, let us show simply the formulation of boundary element analysis for the above problem.

By taking into consideration of the linearity of Navier equation (15), we can easily transform the field equation (15) into the following boundary integral equation:

$$\frac{1}{2}\delta_{ij} u_i(\boldsymbol{y}) + \int_\Gamma t^*_{ij}(\boldsymbol{x};\boldsymbol{y}) u_i(\boldsymbol{x})\, d\Gamma(\boldsymbol{x}) = \int_\Gamma u^*_{ij}(\boldsymbol{x};\boldsymbol{y}) t_i(\boldsymbol{x})\, d\Gamma(\boldsymbol{x}) \tag{18}$$

where u^*_{ij} and t^*_{ij} are the Kelvin fundamental solution and the traction tensor, respectively, in 3-D elastostatics. They are given by

$$u^*_{ij}(\boldsymbol{x};\boldsymbol{y}) = \frac{1}{8\pi\mu(\lambda+2\mu)r}\{(\lambda+3\mu)\delta_{ij} + (\lambda+\mu)r_{,i}r_{,j}\} \tag{19}$$

and

$$\begin{aligned} t^*_{ij}(\boldsymbol{x};\boldsymbol{y}) &= \frac{-1}{4\pi(\lambda+2\mu)r^2}\left[\frac{\partial r}{\partial n}\{\mu\delta_{ij} + 3(\lambda+\mu)r_{,i}r_{,j}\}\right. \\ &+ \left.\mu(n_j r_{,i} - n_i r_{,j})\right] \qquad (r = \|\, \boldsymbol{x} - \boldsymbol{y}\,\|) \end{aligned} \tag{20}$$

Applying the boundary element discretization to the above boundary integral equation (18), we can get the following well-known discretized system:

$$\boldsymbol{H}\boldsymbol{u} = \boldsymbol{G}\boldsymbol{t} \tag{21}$$

where $\boldsymbol{H}$ and $\boldsymbol{G}$ are the influence matrices in which the unknown material properties(Lamé's constants) are included.

Here we apply the boundary conditions on Γ to this system. If the displacements are known on some boundary points, then we can determine the tractions on the corresponding points. This implies that the system of discretized equation can be reordered in such a way that all the unknowns are written on the left-hand side in terms of a vector $\boldsymbol{X}$. Then, the unknown vector can be given as

$$\boldsymbol{X} = \boldsymbol{A}^{-1}\boldsymbol{f} \tag{22}$$

If the system parameters of the field are given, the unknowns on boundary can be determined from the final form(22). This is a solution procedure of a forward analysis of the problem by using boundary element method[12].

3.3 System Equations

The formulation for identification of system parameters(λ, μ) by the Kalman filter-boundary element method is given in this section. The Kalman filtering requires two system equations which are the state equation (1) and the observation equation (2). The step-by-step measurement of $\boldsymbol{y}_t$ gives the estimation of $\boldsymbol{z}_t$ through the filtering algorithm. The key to this formulation is the constitution of the observation equation from the boundary element discretized equation.

3.3.1 State Equation

If we take into account the stationarity of physical constants such as λ and μ, we may set the following expression for the state vector as the state equation for the Kalman filtering:

$$\left\{ \begin{matrix} \lambda \\ \mu \end{matrix} \right\}_{t+1} = \boldsymbol{I} \left\{ \begin{matrix} \lambda \\ \mu \end{matrix} \right\}_t , \qquad (\boldsymbol{\Phi}_t = \boldsymbol{I}, \quad \boldsymbol{\omega}_t = \boldsymbol{o}) \tag{23}$$

The above state equation expresses such a stationary condition that the parameters to be identified should be kept constant in time. Thus, the state transition matrix $\boldsymbol{\Phi}_t$ reduces to a unit matrix $\boldsymbol{I}$, and the system noise $\boldsymbol{\omega}_t$ is not included. The suffix t in this situation does not indicate the time axis but the iterative axis[7)–9)].

3.3.2 Observation Equation

An observation equation can be derived from the boundary element discretized equation (21) in which the state vector $\boldsymbol{z}_t$ is included. Here, each component of the influence matrices $\boldsymbol{H}$ and $\boldsymbol{G}$ in equation (21) is given in the nonlinear form of λ and μ. In this case, we can apply the extended Kalman filtering approach shown in the subsection 2.3.

First of all, we rewrite equation (21) in the following form by separating the measured displacements $\boldsymbol{u}_y^*$ from $\boldsymbol{u}$:

$$\begin{bmatrix} \boldsymbol{H}_{yy} & \boldsymbol{H}_{yu} & \boldsymbol{H}_{yk} \\ \boldsymbol{H}_{uy} & \boldsymbol{H}_{uu} & \boldsymbol{H}_{uk} \\ \boldsymbol{H}_{ky} & \boldsymbol{H}_{ku} & \boldsymbol{H}_{kk} \end{bmatrix} \left\{ \begin{matrix} \boldsymbol{u}_y^* \\ \boldsymbol{u}_u \\ \bar{\boldsymbol{u}}_k \end{matrix} \right\} = \begin{bmatrix} \boldsymbol{G}_{yy} & \boldsymbol{G}_{yk} & \boldsymbol{G}_{yu} \\ \boldsymbol{G}_{ky} & \boldsymbol{G}_{kk} & \boldsymbol{G}_{ku} \\ \boldsymbol{G}_{uy} & \boldsymbol{G}_{uk} & \boldsymbol{G}_{uu} \end{bmatrix} \left\{ \begin{matrix} \bar{\boldsymbol{t}}_y \\ \bar{\boldsymbol{t}}_k \\ \boldsymbol{t}_u \end{matrix} \right\} \tag{24}$$

where $\bar{\boldsymbol{t}}_y$ is the known tractions on the boundary measuring points, $\boldsymbol{u}_u$ and $\boldsymbol{t}_u$ are the unknown displacements and tractions, respectively, on the nodal points of Γ_t except the measuring points, and $\bar{\boldsymbol{u}}_k$ and $\bar{\boldsymbol{t}}_k$ are the known displacements and tractions, respectively, on Γ_u.

We can transform the above system (24) into the following system by eliminating the unknowns $\boldsymbol{u}_u$ and $\boldsymbol{t}_u$:

$$\boldsymbol{A}_\alpha \boldsymbol{u}_y^* = \boldsymbol{f}_\alpha \tag{25}$$

where $\boldsymbol{A}_\alpha$ and $\boldsymbol{f}_\alpha$ are given by

$$\boldsymbol{A}_\alpha = \boldsymbol{H}_{yy} - [\boldsymbol{H}_{yu} - \boldsymbol{G}_{yu}] \begin{bmatrix} \boldsymbol{H}_{uu} & -\boldsymbol{G}_{ku} \\ \boldsymbol{H}_{ku} & \boldsymbol{G}_{uu} \end{bmatrix}^{-1} \begin{bmatrix} \boldsymbol{H}_{uy} \\ \boldsymbol{H}_{ky} \end{bmatrix} \tag{26}$$

$$\begin{aligned} \boldsymbol{f}_\alpha &= (\boldsymbol{G}_{yy} - [\boldsymbol{H}_{yu} - \boldsymbol{G}_{yu}] \begin{bmatrix} \boldsymbol{H}_{uu} & -\boldsymbol{G}_{ku} \\ \boldsymbol{H}_{ku} & \boldsymbol{G}_{uu} \end{bmatrix}^{-1} \begin{bmatrix} \boldsymbol{G}_{ky} \\ \boldsymbol{G}_{uy} \end{bmatrix})\bar{\boldsymbol{t}}_y + ([\boldsymbol{G}_{yk} - \boldsymbol{H}_{yk}] \\ &- [\boldsymbol{H}_{yu} - \boldsymbol{G}_{yu}] \begin{bmatrix} \boldsymbol{H}_{uu} & -\boldsymbol{G}_{ku} \\ \boldsymbol{H}_{ku} & \boldsymbol{G}_{uu} \end{bmatrix}^{-1} \begin{bmatrix} \boldsymbol{G}_{kk} & -\boldsymbol{H}_{uk} \\ \boldsymbol{G}_{uk} & -\boldsymbol{H}_{kk} \end{bmatrix}) \left\{ \begin{matrix} \bar{\boldsymbol{t}}_k \\ \bar{\boldsymbol{u}}_k \end{matrix} \right\} \end{aligned} \tag{27}$$

The above system (25) is solved for the observed vector $\boldsymbol{u}_y^*$, and the system can be expressed by the following form in which the state vector $\boldsymbol{z}_t$ is included:

$$\boldsymbol{y}_t = \boldsymbol{u}_y^* = \boldsymbol{A}_\alpha^{-1}(\boldsymbol{z}_t)\boldsymbol{f}_\alpha(\boldsymbol{z}_t) \tag{28}$$

We can consider this equation as the nonlinear observation equation (11) by adding the noise term. In this case, the sensitivity matrix corresponding to a linearized system can be defined as follows:

$$\boldsymbol{M}_t = \left[\frac{\partial \boldsymbol{u}_y^*}{\partial \lambda}, \ \frac{\partial \boldsymbol{u}_y^*}{\partial \mu}\right] \tag{29}$$

Whose components of equation (29) are determined with

$$\begin{aligned}\frac{\partial \boldsymbol{u}_y^*}{\partial \lambda} &= \frac{\partial\left\{\boldsymbol{A}_\alpha^{-1}(\boldsymbol{z}_t)\boldsymbol{f}_\alpha(\boldsymbol{z}_t)\right\}}{\partial \lambda} \\ &= -\boldsymbol{A}_\alpha^{-1}(\boldsymbol{z}_t)\frac{\partial \boldsymbol{A}_\alpha(\boldsymbol{z}_t)}{\partial \lambda}\boldsymbol{A}_\alpha^{-1}(\boldsymbol{z}_t)\boldsymbol{f}_\alpha(\boldsymbol{z}_t) + \boldsymbol{A}_\alpha^{-1}(\boldsymbol{z}_t)\frac{\partial \boldsymbol{f}_\alpha(\boldsymbol{z}_t)}{\partial \lambda}\end{aligned} \tag{30}$$

$$\begin{aligned}\frac{\partial \boldsymbol{u}_y^*}{\partial \mu} &= \frac{\partial\left\{\boldsymbol{A}_\alpha^{-1}(\boldsymbol{z}_t)\boldsymbol{f}_\alpha(\boldsymbol{z}_t)\right\}}{\partial \mu} \\ &= -\boldsymbol{A}_\alpha^{-1}(\boldsymbol{z}_t)\frac{\partial \boldsymbol{A}_\alpha(\boldsymbol{z}_t)}{\partial \mu}\boldsymbol{A}_\alpha^{-1}(\boldsymbol{z}_t)\boldsymbol{f}_\alpha(\boldsymbol{z}_t) + \boldsymbol{A}_\alpha^{-1}(\boldsymbol{z}_t)\frac{\partial \boldsymbol{f}_\alpha(\boldsymbol{z}_t)}{\partial \mu}\end{aligned} \tag{31}$$

3.3.3 Alternative Formulation

In this subsection, we propose an alternative formulation by which the unknown displacements and tractions as well as Lamé's constants can be identified simultaneously. In this case, the state vector $\boldsymbol{z}_t$ should be expressed as follows:

$$\boldsymbol{z} = (\lambda, \mu, \boldsymbol{u}_u^T, \boldsymbol{t}_u^T)^T \tag{32}$$

We set the following equation as the state equation in terms of the state vector $\boldsymbol{z}_t$:

$$\boldsymbol{z}_{t+1} = \boldsymbol{I}\boldsymbol{z}_t, \qquad (\boldsymbol{\Phi}_t = \boldsymbol{I}, \ \boldsymbol{\omega}_t = \boldsymbol{o}) \tag{33}$$

By using the separated system (24), the observation equation is given by

$$\begin{aligned}\boldsymbol{y}_t = \boldsymbol{u}_y^* &= \boldsymbol{H}_{yy}^{-1}(\boldsymbol{z}_t^{\lambda,\mu})\left[\boldsymbol{G}_{yy}(\boldsymbol{z}_t^{\lambda,\mu})\bar{\boldsymbol{t}}_y + \boldsymbol{G}_{yk}(\boldsymbol{z}_t^{\lambda,\mu})\bar{\boldsymbol{t}}_k - \boldsymbol{H}_{yk}(\boldsymbol{z}_t^{\lambda,\mu})\bar{\boldsymbol{u}}_k\right. \\ &- \left.\left[\boldsymbol{H}_{yu}(\boldsymbol{z}_t^{\lambda,\mu}) - \boldsymbol{G}_{yu}(\boldsymbol{z}_t^{\lambda,\mu})\right]\left\{\begin{array}{c}\boldsymbol{u}_u \\ \boldsymbol{t}_u\end{array}\right\}\right]\end{aligned} \tag{34}$$

in which we introduce the notation, $\boldsymbol{z}^{\lambda,\mu} = (\lambda, \mu)^T$. Equation (34) becomes to the nonlinear observation equation (11) by adding the noise term. Here we consider the following sensitivity matrix:

$$\boldsymbol{M}_t = \left[\frac{\partial \boldsymbol{u}_y^*}{\partial \lambda}, \ \frac{\partial \boldsymbol{u}_y^*}{\partial \mu}, \ \frac{\partial \boldsymbol{u}_y^*}{\partial \boldsymbol{u}_u}, \ \frac{\partial \boldsymbol{u}_y^*}{\partial \boldsymbol{t}_u}\right] \tag{35}$$

In this case, the calculations of sensitivity with respect to λ and μ can be carried out in the same way as equations (30) and (31). The sensitivity of unknown boundary values is concretely given by

$$\left[\frac{\partial \boldsymbol{u}_y^*}{\partial \boldsymbol{u}_u}, \ \frac{\partial \boldsymbol{u}_y^*}{\partial \boldsymbol{t}_u}\right] = \boldsymbol{H}_{yy}^{-1}\left[\boldsymbol{H}_{yu} - \boldsymbol{G}_{yu}\right] \tag{36}$$

3.4 Procedure for Identification Problem

We can accomplish formulations in order to apply the Kalman filtering algorithm to identification problems in a linear elastic field. Because of the steady-stateness of state equation (23) or (33) in this study, we may make use of the observation update algorithm in the Kalman filtering. The solution procedure of the identification analysis using the extended Kalman filter-boundary element method is illustrated in Figure 2. Moreover, we have practically used the U-D observed update theorem[10] based on the U-D decomposition filter to maintain stable solutions corresponding to the observation update algorithm of Kalman filtering for numerical implementation of our formulation.

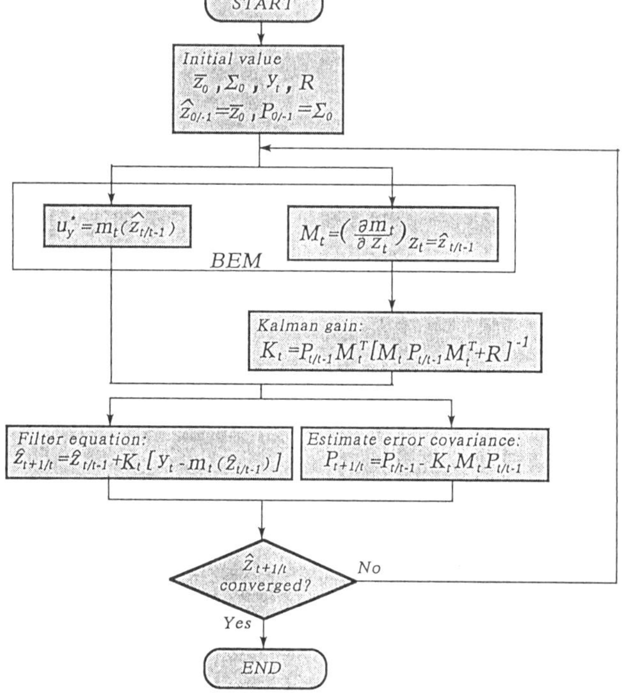

Figure 2. Computation flow chart for identification of unknown parameters.

IV. NUMERICAL RESULTS

In order to demonstrate the applicability and the effectiveness of the solution procedure proposed, numerical simulation is carried out on two models in a linear elastic field. In both identification problems, we have utilized the boundary displacements or tractions on Γ by a boundary element direct analysis as measured boundary responses. Moreover, we sssume that the initial value of the estimate error covariance is 1.0, and the covariance of observation noise is 10^{-9}.

4.1 Identification in 2-D Field

First of all, we perform an identification of elastic constants, which are Young's modulus E and Poisson's ratio ν to be determined with Lamé's constants, of 2-D elastic field. Figure 3 shows the problem statement and the boundary element mesh with twenty constant elements as well as the measuring points. All conditions and material properties in our numerical performance are also expressed in this figure.

The numerical implementation is carried out for four cases in order to examine the dependence on the location and number of observed nodal points to the identification. The difference on situation of the observation points is shown in Table 1. For all cases, we utilize both horizontal and vertical displacements on the boundary point or points to a concentrated load as the observation data. We adopt initial data for elastic constants, $E_0 = 1.0tf/m^2$, $\nu_0 = 0.1$, for cases 1 and 2, $E_0 = 150.0tf/m^2$, $\nu_0 = 0.5$, for case 3 and $E_0 = 1.0tf/m^2$, $\nu_0 = 0.1$, for case 4.

Table 1. Measured nodal points and initial values.

Case	*Measured nodal points*	*Initial values*
1	2	$E_0=1.0tf/m^2$, $\nu_0=0.1$
2	1,2,3	$E_0=1.0tf/m^2$, $\nu_0=0.1$
3	all measuring points	$E_0=150.0tf/m^2$, $\nu_0=0.5$
4	all measuring points	$E_0=1.0tf/m^2$, $\nu_0=0.1$

Figures 4 and 5 illustrate the process of identification obtained for the corresponding cases. From these figures, it is obvious that the pattern of identification process does not depend on the location or number of measuring points in this model.

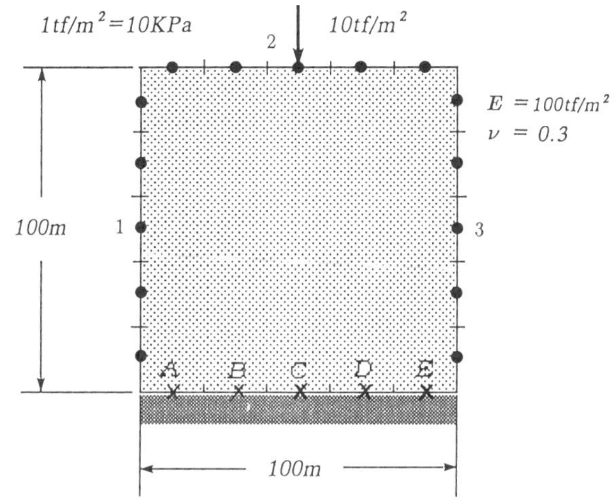

Figure 3. 2-D analytical model and boundary element mesh (• measured point).

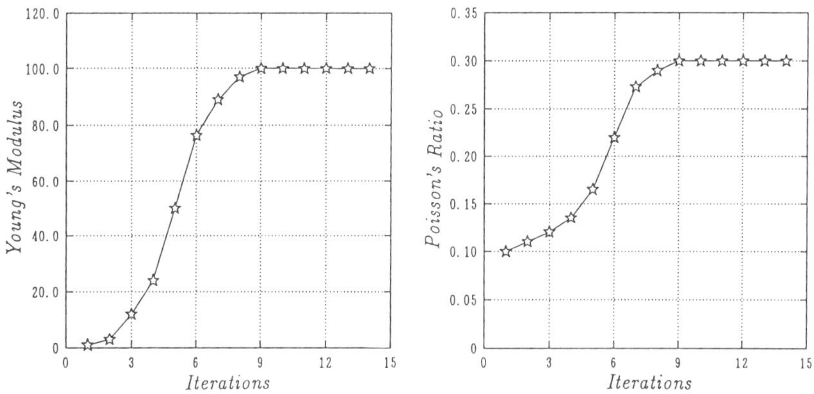

Figure 4. Process of parameter identification (Case 1).

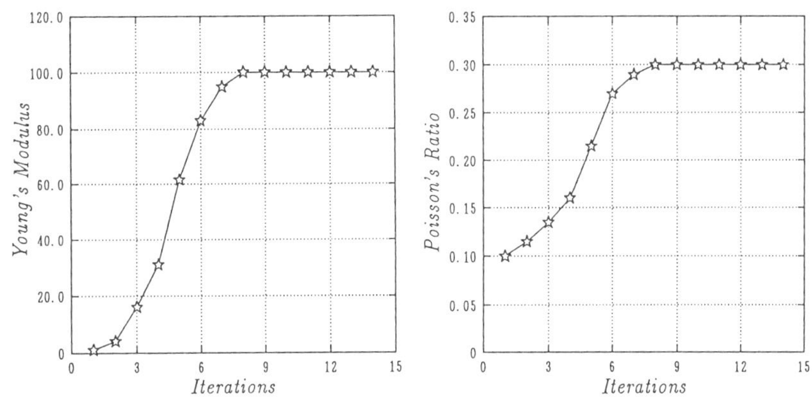

Figure 5. Process of parameter identification (Case 2).

Next, we show a numerical example in order to examine the validity of the alternative formulation developed in the subsection 3.3.3.

In this identification problem, we set only the cases where displacements are observed on boundary nodal points except fixed boundary as the location of measuring nodal points on which horizontal and vertical values are observed. However, we prepare two cases of cases 3 and 4 on the initial values of elastic constants. Initial values on tractions to be estimated are assigned the ones calculated by the boundary element direct analysis with using the above values, E_0 and ν_0, in each case.

Figures 6 and 7 show the process of identification on Young's modulus and Poisson's ratio for cases 3 and 4, respectively. Tables 2 and 3 show the estimated results of unknown tractions on five nodal points of fixed boundary for cases 3 and 4, respectively, in comparison with the true values. From these results, we consider that identification of elastic constants as well as estimation of unknown tractions for both cases 3 and 4 are in stage to be satisfied under this problem situation.

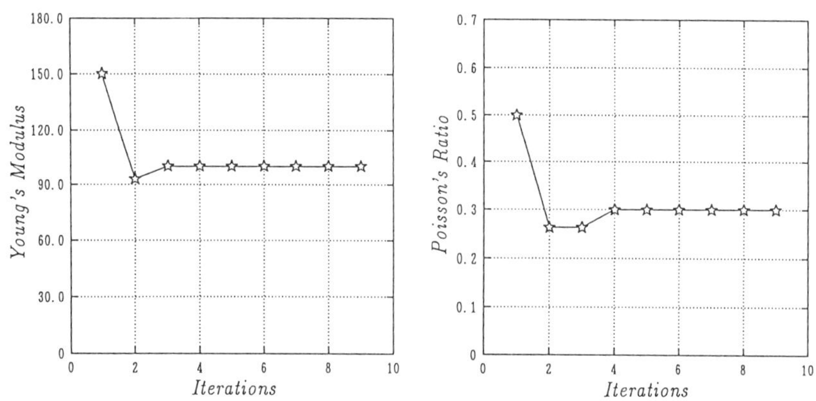

Figure 6. Process of parameter identification (Case 3).

Table 2. Results on estimation of unknown tractions (Case 3).

Nodal point		Initial	True	Calculated
A	Hor.	1.3888	0.7282	0.7282
	Ver.	2.2401	2.1852	2.1852
B	Hor.	0.4692	0.2842	0.2842
	Ver.	1.8087	1.8506	1.8506
C	Hor.	0.0	0.0	0.0
	Ver.	1.9161	1.9226	1.9226
D	Hor.	-0.4692	-0.2842	-0.2842
	Ver.	1.8087	1.8506	1.8506
E	Hor.	-1.3888	-0.7282	-0.7282
	Ver.	2.2401	2.1852	2.1852
Iterations				6

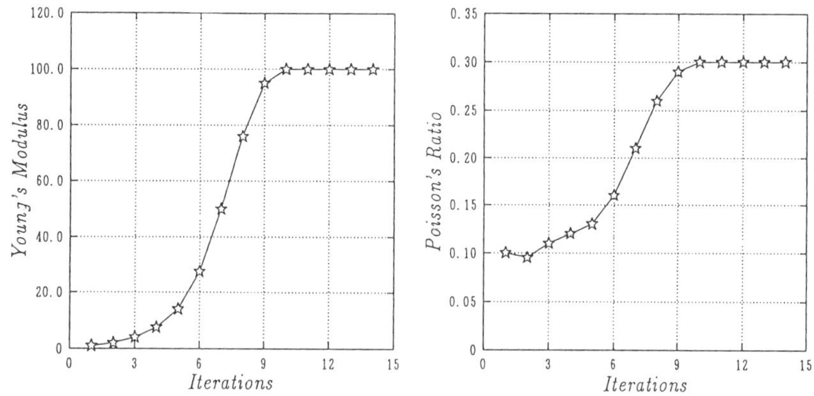

Figure 7. Process of parameter identification (Case 4).

Table 3. Results on estimation of unknown tractions (Case 4).

Nodal point		*Initial*	*True*	*Calculated*
A	*Hor.*	*0.2910*	*0.7282*	*0.7282*
	Ver.	*2.0803*	*2.1852*	*2.1852*
B	*Hor.*	*0.1591*	*0.2842*	*0.2842*
	Ver.	*1.9180*	*1.8506*	*1.8506*
C	*Hor.*	*0.0*	*0.0*	*0.0*
	Ver.	*1.9180*	*1.9226*	*1.9226*
D	*Hor.*	*-0.1591*	*-0.2842*	*-0.2842*
	Ver.	*1.9180*	*1.8506*	*1.8506*
E	*Hor.*	*-0.2910*	*-0.7282*	*-0.7282*
	Ver.	*2.0803*	*2.1852*	*2.1852*
Iterations				*11*

4.2 Identification in 3-D Field

To examine the validity of the method in 3-D field, we apply the present method to identification problem of Lamé's constants in a sample model of silicone gum.

Figure 8 shows the problem statement and Figure 9 shows the boundary element mesh in which measuring points are pointed out. All conditions and material properties in our numerical perfermance are also expressed in these figures.

Numerical calculation is carried out for two cases on the data to be measured: displacements on nodes 1 and 2 (Case 1); tractions on the base (Case 2). Measuring components on measuring points, whose displacements or tractions are the numerical input data in both cases, are listed in Table 4.

Table 4. Measuring components.

Case	*Measuring components*
1	*Direction of y,z axes (node 1) and x,z axes (node 2)*
2	*Direction of z axis*

Figures 10 and 11 illustrate the process of parameter identification starting from initial values in cases 1 and 2, and show the values transformed from Lamé's constants to Young's modulus and Poisson's ratio. It can be seen that in the case of 3-D domain, the identified results converge also to the correct values irrespective of initial values.

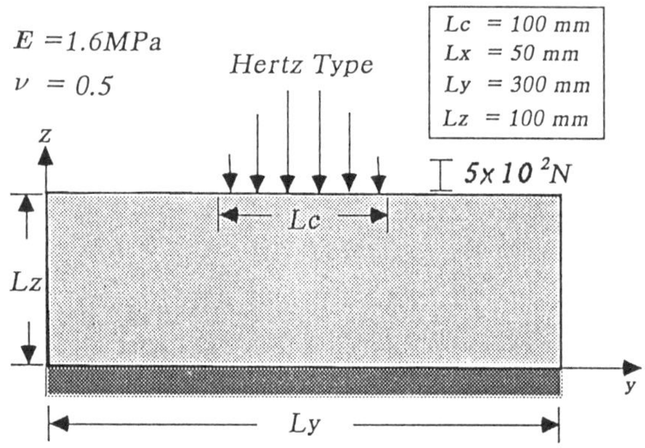

Figure 8. 3-D sample model of silicone gum.

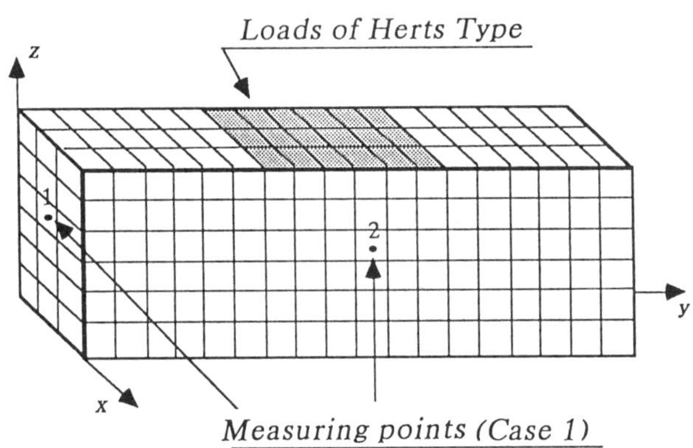

Figure 9. Boundary element mesh for sample model.

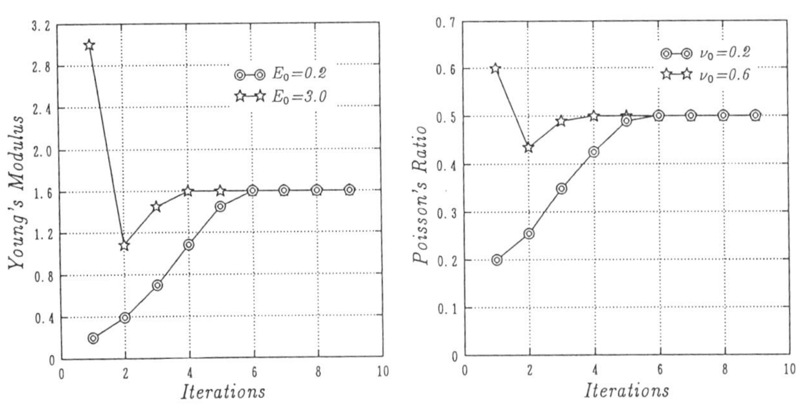

Figure 10. Process of parameter identification (Case 1).

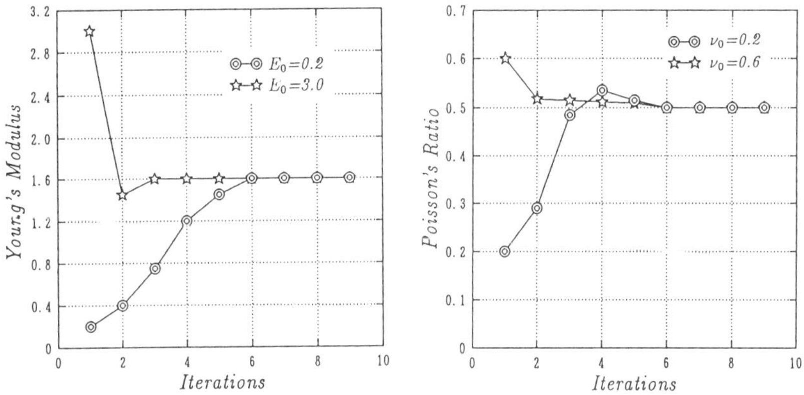

Figure 11. Process of parameter identification (Case 2).

V. CONCLUDING REMARKS

We have shown the formulation and solution procedure for identification analysis of unknown quantities by using the Kalman filter-boundary element method, and discussed the applicability and the effectiveness through numerical results of identification problems of the unknown boundary values as well as Lamé's constants in 2-D elastic field and Lamé's constants in 3-D field.

As the conclusions of this paper, the following remarks should be noted:

1) In order to identify unknown quantities based on measuring data on the boundary of a field, we constituted the state and observation equations for extended Kalman filtering by using the resulting boundary element equation from boundary element method of a field. In this place, we have shown that there exist two formulations in connection with choice of the state variables.

2) The accuracy and efficiency of the proposed procedure have been examined through numerical results of three examples. They are identification problems of elastic constants, both elastic constants and unknown boundary tractions in 2-D domain, and Lamé's constants in 3-D domain. As a result, the present method is applicable effectively to not only 2-D field but also 3-D one. A determination of optimal location of observed nodal points for inverse problem remains unsolved, although we showed one example in 4.1.

3) The one of advantages of the method shown herein is that the procedure is composed of the solution algorithm which can be taken into consideration with the observational error. To consider the observational error will be the subject of future investigation.

REFERENCES

1) Kubo, S., Ohnaka, K. and Ohji, K., Identification of heat-source and force using boundary integrals(in Japanese), Int. J. JSME, 54-501(A) (1988), pp.1329-1334.

2) Ohnaka, K. and Uosaki, K., Identification of the external input of distributed-parameter systems by the boundary element approach, Int. J. Control, 43-4 (1986), pp.1125-1133.

3) Arai, K., Ohta, H. and Yasui, T., Simple optimization techniques for evaluating deformation moduli from field observation, Soils and Foundations, 23(1) (1983), pp.107-113.

4) Murakami, A. and Hasegawa, T., Application of Kalman filtering to structural and geotechnical engineering problems(in Japanese), Journal of the Japanese Society of Irrigation, Drainage and Reclamation Engineering, 158 (1992), pp.95-104.

5) Murakami, A. and Hasegawa, T., Back analysis by Kalman filter- finite elements and optimal location of observed points, Proc. 6th Int. Corf. on Numerical Methods in Geomechanics, Innsbruck (1988), pp.2051-2058.

6) Murase, H., Koyama, S. and Honami, N., Kalman filter charge simulation methods for the solution of inverse problems of two or three dimensional elasticity(in Japanese), Proc. Symp. JSME (1989), pp.181-186.

7) Utani, A. and Tosaka, N., Identification analysis of distributed- parameter systems by using Kalman filter-boundary element method, Proc. Int. Union of Theoretical and Applied Mechanics, Inverse Problems in Engineering Mechanics (1992), pp.347-356, Springer-Verlag.

8) Tosaka, N. and Utani, A., On the Kalman filter-boundary element method in inverse problem, Boundary Elements XIV, ed. by Brebbia, C. A., Dominguez, J. and Paris, F., Computational Mechanics Publications and Elsevier Science Publishers (1992), pp.707-721.

9) Tosaka, N. and Utani, A., Identification analysis of elastic constants using Kalman filter-boundary element method (in Japanese), Journal of Structural and Construction Engineering (Transactions of AIJ), No.446 (1993), pp.41-50.

10) Katayama, T., Applied Kalman Filter(in Japanese), Asakura-shoten (1983).

11) Catlin, D.E., Estimation, Control, and the Discrete Kalman Filter, Springer-Verlag (1989).

12) Brebbia, C. A., The Boundary Element Method for Engineers, Pentech Press (1978).

Inverse Analyses of Truss-Type Structures Using Neural Networks

Kazuo YAMAMOTO

Headquarters Engineering, Department Mitsubishi Electric Corporation, Amagasaki, Hyogo

Inverse problems were formulated using the neural network in combination with the direct analysis and optimization analysis of computational mechanics focussing on the truss-type structure. The nonlinear mapping relation between the design variable and structural behavior was constructed in the feed-foward multi-layer neural network by the learning based on the computational mechanics data. The learning is carried out incrementally one by one for each learning data using the error back propagation scheme. The present method was applied to two problems. The former is identification of the position for lacked member based on the shape errors in the two dimensional pretensioned cable net structure. The latter is optimization of the structural configuration for various external load conditions in the variable geometry truss. Successful results have shown the effectiveness and efficiency of the method.

I. INTRODUCTION

Truss-type structures are defined as the structures in which the force equilibrium is attained only by the axial force of their members. They contain the pretensioned cable structure and the variable geometry truss(VGT). In this kind of structures, their configuration(shape) is determined by the equilibrium of the load, and arbitrary configuration can be realized by controlling the length of the member. So, they are often used as the structural form of the adaptive structure for space use.[1]

In order to design these structures, following items are important: (1)high precision calculation method which can give the configuration in the equilibrium state using the given dimensions, properties and external environment (Direct problem), and (2)the estimation method which determine the dimensions or properties of the member, environment and so on, in the condition that the configuration or behaviour of the interested portion of the structure is assigned (Inverse problem). In the inverse problem, much volume of tremendous iterative calculations such as the optimization analysis based on the nonlinear programming method is commonly required.

In this study, from the view point of solving the inverse problem effectively and efficiently, the method is applied which combines the computational mechanics (direct analysis results or usual optimization analysis) and the nonlinear mapping generation function of the neural networks based on the learning.

II. DEFINITION OF THE INVERSE PROBLEM

Figure 1 shows the mathematical model of the truss-type structures which is used in the study. The model consists of the node points or joints which have no volume and are distributed in the three dimensional space, and the truss or cable members which connect the node points or joints to each other and deform elastically in the axial direction. The configration and the load state of the structure can be uniquely determined by the force equilibrium condition when the dimensions and properties of the all members, the external forces or other environmental conditions are given. That is, adopting the three dimensional location of the node points as the unknown variables, we can obtain a nonlinear force equilibrium equation system which is needed to be solved using an appropriate iterative procedure.[2),3)]

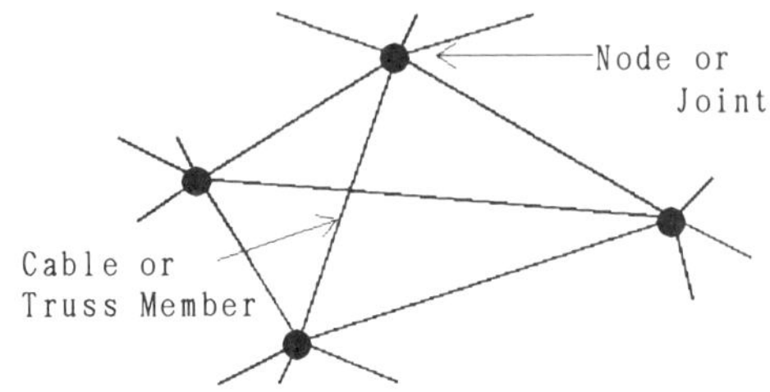

Fig. 1. Mathematical model of truss-type structures.

For this structure, inverse problems are considered as the subject such that, the unknown dimensions or properties of arbitrary members are designed so that the configuration of the interesting portion of the structure may be equal to any specified condition or the load state of the structure may be minimized(maximized). Usually, various calculation schemes based on the nonlinear planning method are used to solve this problem. In the present study the iterative optimization procedure by using the sensitivity analysis[2),3)] is refered to for reference and comparison.

III. APPLICATION OF THE NEURAL NETWORKS

The above mensioned inverse problems can be understood to be the subject obtaining a multi-to-multi nonlinear continuous or discrete mapping relation between the design variables and structural behavior. Consequently, we construct the mapping relation in the neural network by the learning procedure. We used the feed fored type three layer neural network shown in Fig.2. The network consists of an input, an intermediate and an output layers. The processing unit of network is of multi-input single-output type and its response function f is the Sigmoid function which changes its value continuously in the range [0,1]. The input/output relations are described as follows;

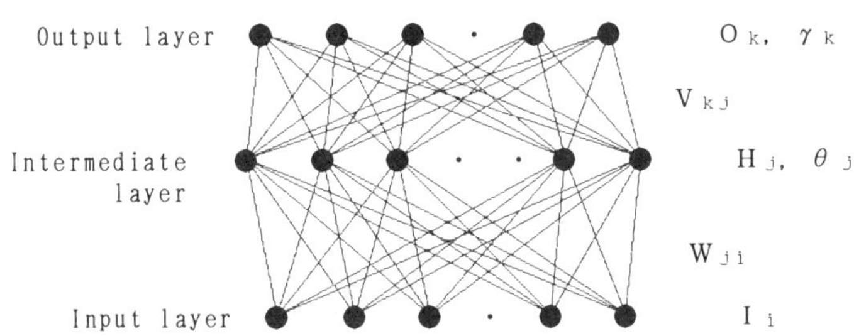

Fig. 2. Multi-layer neural network model.

$$
\begin{aligned}
H_j &= f(U_j) = 1/\{1+\exp(-2U_j/U_0)\} \\
U_j &= \Sigma_{(i)} W_{ji}\cdot I_i - \theta_j \\
O_k &= f(S_k) = 1/\{1+\exp(-2S_k/S_0)\} \\
S_k &= \Sigma_{(j)} V_{kj}\cdot H_j - \gamma_k
\end{aligned}
\tag{1}
$$

where

n_i : number of units in the input layer
n_h : number of units in the intermediate layer
n_o : number of units in the output layer
I_i : output of i-th unit in the input layer
H_j : output of j-th unit in the intermediate layer
O_k : output of k-th unit in the output layer
W_{ji} : connection coefficient from input i-th unit to intermediate j-th unit

V_{kj} : connection coefficient from intermediate j-th unit to output k-th unit
θ_j : offset of intermediate j-th unit(threshold)
γ_k : offset of output k-th unit(threshold)
U_0 : Sigmoid parameter of units in the intermediate layer
S_0 : Sigmoid parameter of units in the output layer

The error back propagation scheme was employed for the learning algorithm,[4] that is, the connection coefficients and offsets between output and intermediate layers and between intermediate and input layers are modified successively so that the square sum E of the errors between the outputs and teacher signals may be minimized using the steepest descent method. Modification of the connection coefficients and offsets is done by use of the moment method in which the modified values one step before are also used.

$$E = \Sigma_{(k)} \{T_k - O_k\}^2 / 2 \tag{2}$$

$$\begin{aligned} \triangle W_{ji}(t) &= -\alpha\,(\partial E / \partial W_{ji}) + m \cdot \triangle W_{ji}(t-1) \\ \triangle V_{kj}(t) &= -\alpha\,(\partial E / \partial V_{kj}) + m \cdot \triangle V_{kj}(t-1) \\ \triangle \theta_j(t) &= -\beta\,(\partial E / \partial \theta_j) + m \cdot \triangle \theta_j(t-1) \\ \triangle \gamma_k(t) &= -\beta\,(\partial E / \partial \gamma_k) + m \cdot \triangle \gamma_k(t-1) \end{aligned} \tag{3}$$

where

T_k : teacher signal in k-th unit in the output layer
$\triangle W_{ji}$: modification amount for connection coefficient W_{ji}
$\triangle V_{kj}$: modification amount for connection coefficient V_{kj}
$\triangle \theta_j$: modification amount for offset θ_j
$\triangle \gamma_k$: modification amount for offset γ_k
α : modification parameter of connection coefficient
β : modification parameter of offset
m : weight factor of modified value one step before
t : time or modification step number

IV. NUMERICAL SIMULATIONS AND DISCCUSSION

4.1 Position identification of lacked member in the cable net

Figure 3 shows a mathematical model of two-dimensional cable net structure, in which the problem for position identification of lacked member was investigated. The model consists of node points and elastic cable net members which are located like a go board in xz-plane. The node points in its left, right and bottom edges are fixed rigidly, and those in the top edge are free. When the length of each cable member is properly designed, the node points on the top edge are located in a parabollic curve. But, if there is the lack of any member, the distribution of node position errors from the given parabollic curve will appear in the top edge of the cable network.

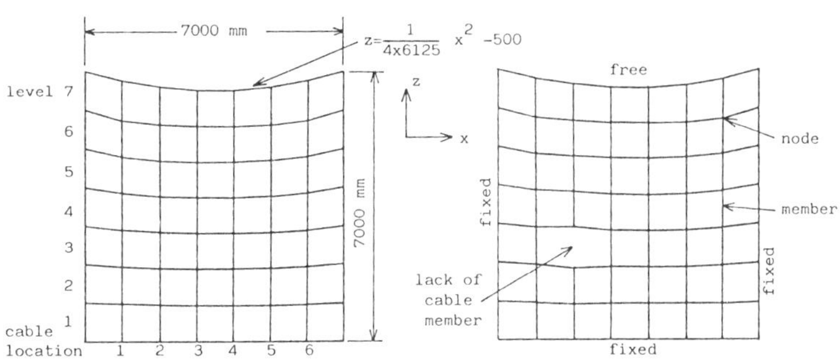

Fig. 3. Two dimensional model of cable net structure.

Suppose the problem in which the position for lacked cable member is identified from the information for error distribution of the node points on the top edge. For simplicity of the problem, we assume that the lacked position is only in the longitudinal (paralell to z axis) member and only one in number, and the node point errors are evaluated only in the z direction. The position of all longitudinal members is defined by location number and level number. For example, when the cable lack exists in the longitudinal member positioned in location 2 and level 3, estimation results by using the sensitivity analysis are shown for comparison in Fig.4 and Table 1. In order to obtain the solution uniquely the number of design variables to be estimated (stiffness of cable member) must be same as the number of variables which are included in the object function (root mean square of node position errors), and the search was done level by level of the cable net structure. When the lacked position is included in the serching range, the corresponding cable stiffness E comes near zero without limit. Thus, the position of lacked member can be identified.[2]

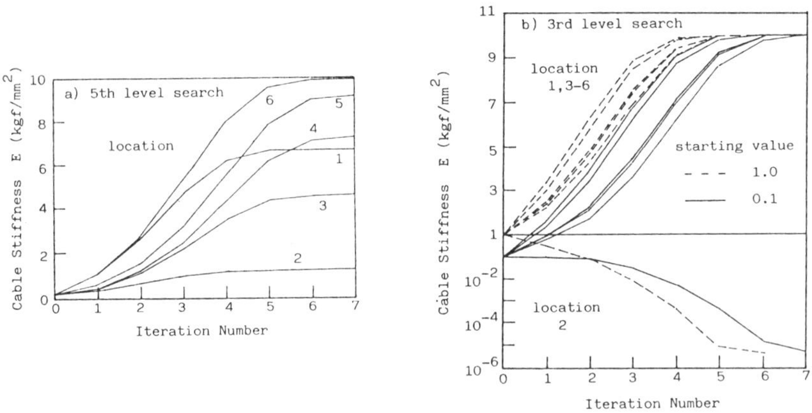

Fig. 4. Convergence of stiffness estimation by optimization.

Table 1. Converged value of stiffness estimation.

Search level	Cable location 1	2	3	4	5	6
7	3.342	1.172	2.730	4.585	6.750	8.784
6	5.650	1.330	3.546	5.922	8.226	9.644
5	6.657	1.148	4.529	7.263	9.138	9.862
4	7.793	0.680	6.162	8.701	9.723	9.964
3	9.999	4.9E-6	9.999	9.998	9.999	10.000

(Unit: kgf/mm^2)

The neural network is applied to this problem. The network model used in the study consists of 6 input, 20 intermediate and 2 output layers, and the parameters are set as follows: $\alpha=\beta=0.1$, $m=0.05$, $u_0=s_0=0.5$. These values were selected so that they may be near the optimum in consideration of the effectiveness and the precision of learning. The sets of lacked position (cable location and level) and position errors for top edge node points in z direction were employed as the learning pattern. That is, the learning was done with the input of node position errors and with the teacher signal of lacked cable position. Two combinations of learning patterns (case1 and case2) were chosen and the corresponding evaluation patterns (lacked cable position A, B and C) in each case were employed as shown in Fig.5. We show the evaluation results after learning in Table 2. In case 1, the learning was done with 16 patterns and up to 50000 times. In case 2, that was done with 22 patterns and up to 10000 and 50000 times.

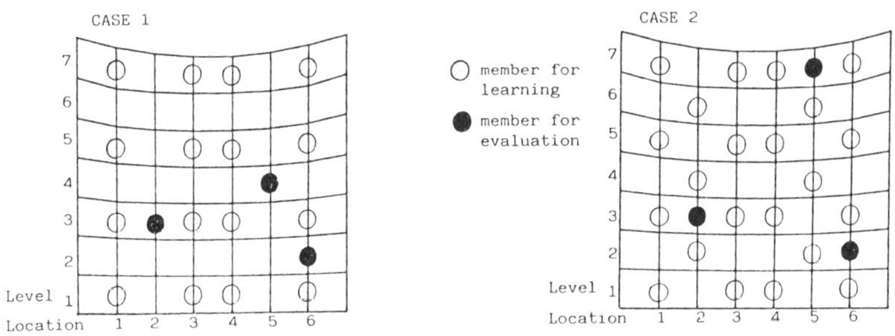

Fig. 5. Learning patterns and evaluation patterns.

Table 2. Evaluated results after learning.

	Pattern A		Pattern B		Pattern C	
	Loc.	Lev.	Loc.	Lev.	Loc.	Lev.
Case 1 [16 patterns]						
Neural-50	0.99	-0.74	5.04	-1.15	6.07	1.97
Estimation	1	?	5	?	6	2
True	2	3	5	4	6	2
Case 2 [22 patterns]						
Neural-10	1.85	3.21	5.03	6.14	5.76	1.49
Estimation	2	3	5	6	6	1
Neural-50	2.04	2.99	5.23	6.70	6.06	1.99
Estimation	2	3	5	7	6	2
True	2	3	5	7	6	2

In case 1, estimation of lacked position A and B, in which there is no learning pattern in the same x location as the evaluation pattern, was made unsuccessfully even after 50000 times learning. On the other hand, estimation of lacked position C, in which there are learning patterns in the same x location, was made successfully with good precision. This may be because the structural coupling between adjacent cable in x direction is quite little due to the characteristics of this cable net structure and estimation toward x direction is extremely difficult.

In case 2, estimation of three lacked positions was made with comparatively low precision after 10000 times learning. Especially for pattern B, in which there is a learning pattern only in next down side level to the evaluation pattern, estimation error in z direction is still large. After 50000 times learning, estimation of all lacked positions was successfully obtained with excellent precision.

4.2 Shape optimization of the variable geometry truss

We consider the shape optimization problem in which the length of specified members is determined so that the mean member load may be minimized in a static force application at the top of variable geometry truss (VGT)[5] shown in Fig.6. The VGT is a statically determinate periodical structure which has three lateral members and six diagonal members as a hexagonal unit truss. By expanding or contracting the member in appropriate position, we can realize various configurations of the entire truss structure in the three-dimensional space. The model consists of five units with the bottom nodes fixed and the top nodes applied by a static load. The natural length of all members is 1.0m in the initial state, and the young's modulus and crosssectional area of all members are 1.0×10^4kgf/mm^2, 100mm^2 respectively. When the static force with the specified angle θ_F is applied to the top nodes, the optimum

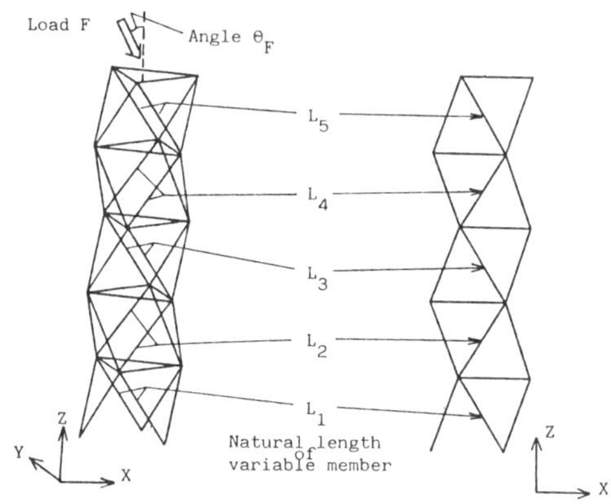

Fig. 6. Configuration of VGT.

configuration of the truss is determined so that the mean member load may be minimized over the whole truss structure. Five pairs of pararell diagonals located in the direction pararell to the xz-plane are chosen as the variable member (design variable). Because the force F pararell to xz-plane is applied and the structure is symmetrical for xz-plane, the length of two members in each pair is supposed to be identical to each other. The lengths of all variable members are limited between 0.7 m and 1.3m, which was selected in the range of geometrical condition, that is, the ratio of maximum and minimum length is smaller than 2. Table 3 and Fig.7 shows the optimum lengths of variable members and the optimum configuration of the truss for various load angles which were calculated using the optimizing method with the sensitivity analysis.[3)]

Table 3. Load angle and optimum length of variable member.

Load Angle	Length of Actuating Member (m)					Truss Incl.
θ_F	L_1	L_2	L_3	L_4	L_5	θ_T
30.0°	1.3	0.7	1.3	0.7	1.3	24.80°
22.5°	1.3	0.776	1.3	0.782	1.3	22.58°
15.0°	1.217	0.868	1.221	0.870	1.208	15.16°
7.5°	1.127	0.956	1.132	0.957	1.117	7.68°
0.0°	1.036	1.044	1.043	1.046	1.024	0.20°
-7.5°	1.944	1.133	0.953	1.135	0.932	-7.29°
-15.0°	0.853	1.221	0.864	1.223	0.840	-14.77°
-22.5°	0.759	1.3	0.768	1.3	0.742	-22.24°
-30.0°	0.7	1.3	0.7	1.3	0.7	-24.73°

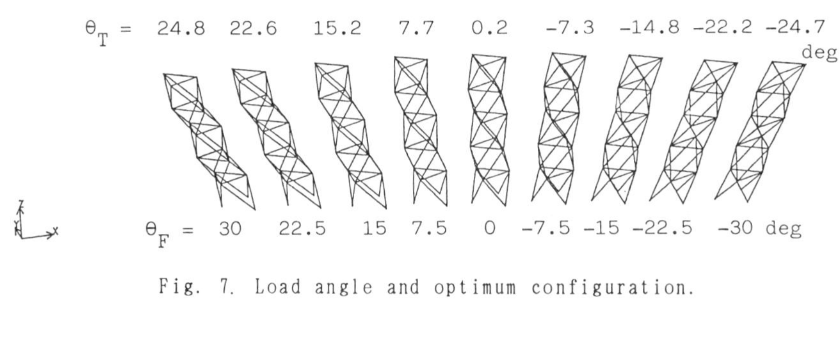

Fig. 7. Load angle and optimum configuration.

The neural network with the learning based on the back propagating errors was applied to this problem. For the unit formation the number of the input, intermediate and output layers are 1, 20 and 5 respectively. The parameters for the unit and the learning used in this problem are as follows: $\alpha=\beta=0.1$, $m=0.1$, $u_0=s_0=0.5$. These values were selected in the range near the optimum considering the effectiveness and precision of the learning. Several combinations in the set of load angle and variable member lengths shown in Table 3 were employed as the learning data, that is, the learning was done supposing the load angle as the input and the corresponding optimum variable member length as the teacher signal in the output of the network. Table 4 shows the three classes of combination of learning pattern (Neural-3, -5, -9), and the one set of evaluation pattern. The number of learning is limited to 50000 in all classes. Consequently, it can be considered that the precision of learning is best in the class Neural-3 with the least number of learning patterns, being worst in the class Neural-9 with a number of learning patterns. The evaluated results after learning is summarized in Table 5. In case of Neural-5, the configuration of VGT

Table 4. Learning patterns and evaluation patterns.

Class	Learning pattern	Evaluation pattern
Neural-3	-22.5° 0° 22.5°	-5° 10° 35°
Neural-5	-30° -15° 0° 15° 30°	
Neural-9	-30° -22.5° -15° -7.5° 0° 7.5° 15° 22.5° 30°	

Table 5. Evaluated results after learning.

Load Angle θ_F	Class	Angle Diff. $\triangle\theta$	Length of Actuating Members (m) L_1	L_2	L_3	L_4	L_5	Error (%) Max	Mean
35°	Neural-3	12.5°	1.3846	0.6884	1.4328	0.6844	1.3536	10.2	5.86
	Neural-5	5°	1.3162	0.6650	1.3236	0.6584	1.3128	5.9	3.64
	Neural-9	5°	1.3270	0.6698	1.3342	0.6668	1.3252	4.3	3.35
	Optimum	-	1.3	0.7	1.3	0.7	1.3	-	-
10°	Neural-3	10°	1.1674	0.9118	1.1590	0.9190	1.1696	1.6	1.27
	Neural-5	5°	1.1604	0.9230	1.1640	0.9266	1.1538	0.6	0.34
	Neural-9	2.5°	1.1662	0.9224	1.1700	0.9266	1.1590	1.0	0.67
	Optimum	-	1.1573	0.9265	1.1623	0.9281	1.1474	-	-
-5°	Neural-3	5°	0.9664	1.1102	0.9828	1.1114	0.9490	1.4	0.81
	Neural-5	5°	0.9768	1.1150	0.9888	1.1160	0.9616	1.0	0.69
	Neural-9	2.5°	0.9746	1.1174	0.9856	1.1188	0.9588	1.2	0.80
	Optimum	-	0.9745	1.1037	0.9827	1.1057	0.9624	-	-

in the evaluation pattern is shown comparing with the usual optimization analysis in Fig. 8. First of all, in case of the load angle θ_F= 35deg, we have the mean output error with several parcents for the lengths of variable members in all classes. The precision of estimation is insufficient. It may be because this load angle is out of the range for learning pattern (-30～30deg) and the extrapolation of variable member lengths of which the change rate for the angle has discontinuity due to the limitation of member length (0.7～1.3m) is quite difficult. Next, considering the influence of the number of learning patterns, the precision of estimation is not good both for the case of too few (Neural-3) and too many (Neural-9) learning patterns. If the number of learnings is not limited to a constant, it may be said that much more number of learning patterns brings more acculate estimation. But, in the present case, as the number of learnings is fixed to 50000, the number of learnings per pattern is not sufficient for the class Neural-9, and consequently precise estimation can not be obtained. It can be considered that optimum number of learning patterns exists in the intermediate range of those. Finally, It may be said that the angle difference between the learning pattern and the evaluation pattern also affects the precision of estimation. But the data contains also the influence of the number of learning patterns and the property of load angle. Influence of only the angle difference $\triangle\theta$ can not be clarified using the present data.

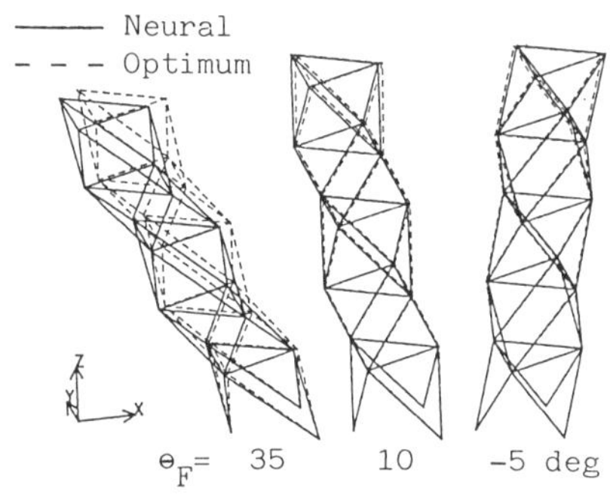

Fig. 8. Comparison of evaluated VGT configuration.

V. CONCLUDING REMARKS

Application of layered-type neural networks was investigated for inverse analyses of truss-type structures of which the configuration is determined by the equilibrium of only axial force for the member. First, the problem for identification of lacked member was solved with combination of the direct analysis and neural network learning focussing on a plane cable net structure. When the number of lacked member is only one, the proposed method brings the solutions of good precision. Second, the problem for optimum configuration under a given load condition was solved with combination of usual optimization analysis results and the neural network learning using the variable geometry truss. Estimation of optimum configration was successfully made within the range of learning patterns. In both cases, volume of computer CPU time neccessary for the learnings was consumed. But after the learning the time for estimation was quite little. It can be concluded that the proposed method is effective for these inverse problems.

REFERENCES

1) Wada, B.K., Adaptive Structures, AIAA 89-1160-CP, 30th SDM, 1989.
2) Yamamoto, K., Investigations of An Inverse Problem for Cable Truss Structures, The 4th Computational Mechanics Conference, The Japan Society of Mechanical Engineers, 1991, pp313-314. (in Japanese)
3) Yamamoto, k., Shape Optimization of Truss-Typetructures Using Sensitivity Analysis, Theoretical and Applied Mechanics, Vol.41, University of Tokyo Press, 1992, pp.77-85.
4) Rumelhart, D.E., Hinton, G.E. and Williams, R.J., Learning Representations by Back-Propagating Errors, Nature, 323, 1986, pp.533-536.
5) Miura, K. and Furuya, H., An Adaptive Structure Concept for Future Space Applications, IAF-85-211, 36th Congress of the IAF, 1985.

Design Aspect of Structure and Control System Simultaneous Optimization under Several Objective Functions

Takuzo IWATSUBO, Shozo KAWAMURA, and Kazuhiko ADACHI

Department of Mechanical Engineering, Kobe University, Kobe, Hyogo

The effect of the objective functions on the results of the structure and control system simultaneous optimization problem is discussed, comparing the optimum designs obtained by minimizing several different types of objective functions. These objective functions are defined as a linear combination of the structural and the control objective functions. One of the objective functions is a structural weight, another is a feedback gain norm and the others are transient characteristics values which are expressed as the quadratic form. A flexible cantilever beam is designed in the numerical example, and the results indicate that the usefulness of the structure and control system simultaneous optimization depends on the objective function for the example.

I. INTRODUCTION

Optimum design of the structure and its control system for the active vibration control of the structure has been investigated using two approaches; one is the traditional individual optimization in which the control system is designed after the structure has been designed, and the other is the structure and control system simultaneous optimization in which both the structure and the control system are designed simultaneously. There is interaction between the structural and the control system in their optimum design processes, because both the structural dynamic behavior and the control performance depend on the structural design variables and the control design variables. There is a possibility of attaining a superior optimum design by applying the simultaneous optimization, so that a large number of studies are currently in progress[1)~3)].

In the authors' previous works, we pointed out the following problem. The results obtained by the simultaneous optimization are superior to that obtained by the individual optimization in the case of using some objective functions[4)]. But in the case of using other objective functions, both results obtained by the simultaneous and the individual optimization are almost the same[5)]. In this paper, the effect of the objective function on the result of the structure and control system simultaneous optimization is discussed, comparing the optimum designs obtained by minimizing some different types of the objective functions.

II. BASIC EQUATION

In this paper, the system is a linear time-invariant (LTI) and single-input single-output (SISO) system with collocated sensor and actuator. The equations of motion for a structure with no external dynamic disturbance are given by following state space representation.

$$\dot{Z} = AZ + Bu \tag{1}$$

where Z is the state vector which is defined as

$$Z \equiv [\eta^T \quad \dot{\eta}^T]^T \in \mathbf{R}^{2n}, \tag{2}$$

in which $\eta \in \mathbf{R}^n$ is the modal displacement vector, and $u \in \mathbf{R}^1$ is the control input for vibration control. It is assumed that the actuator do not excite rigid-body modes of the structure, and we neglect the rigid-body modes. Thus, only n elastic modes are considered.

$$A \equiv \begin{bmatrix} \mathbf{0} & \boldsymbol{I}_n \\ -\boldsymbol{\Omega} & -\boldsymbol{\Delta} \end{bmatrix} \in \mathrm{R}^{2n\times 2n} \;, \quad \boldsymbol{B} \equiv \begin{bmatrix} \mathbf{0} \\ \boldsymbol{\Phi}^T \boldsymbol{L} \end{bmatrix} \in \mathrm{R}^{2n} \tag{3}$$

where $\boldsymbol{\Phi} \in \mathrm{R}^{n\times n}$ is the modal matrix whose columns are the eigenvectors normalized with respect to the mass matrix of the structure, $\boldsymbol{L} \in \mathrm{R}^n$ is the input vector, and

$$\boldsymbol{\Delta} \equiv \mathrm{diag}\,(2\zeta_1\omega_1, 2\zeta_2\omega_2, \cdots, 2\zeta_n\omega_n) \in \mathrm{R}^{n\times n} \;, \quad \boldsymbol{\Omega} \equiv \mathrm{diag}\left(\omega_1^2, \omega_2^2, \cdots, \omega_n^2\right) \in \mathrm{R}^{n\times n}. \tag{4}$$

$\boldsymbol{\Delta}$ and $\boldsymbol{\Omega}$ are obtained by the modal analysis, ζ_i and ω_i are the modal damping ratio and the natural circular frequency, respectively. All real parts of the eigenvalues of the matrix $\boldsymbol{A}$ are assumed to be negative.

The control system is designed for the following reduced-order model by using a linear regulator.

$$\dot{\boldsymbol{Z}}_m = \boldsymbol{A}_m \boldsymbol{Z}_m + \boldsymbol{B}_m u \tag{5}$$

Subscript m denotes the elastic mode of the reduced-order model. It is assumed that all state variables are obtained without considering the observer. The following quadratic performance index can be defined.

$$J_C = \int_0^\infty ({\boldsymbol{Z}_m}^T \boldsymbol{Q}_m \boldsymbol{Z}_m + u^2 R_m)dt \;, \quad \boldsymbol{Q}_m = {\boldsymbol{Q}_m}^T > \mathbf{0} \;, \quad R_m > 0 \tag{6}$$

where $\boldsymbol{Q}_m$ and R_m are chosen as

$$\boldsymbol{Q}_m = \begin{bmatrix} \boldsymbol{I}_m & \mathbf{0} \\ \mathbf{0} & \boldsymbol{\Omega}_m^{-1} \end{bmatrix} \in \mathrm{R}^{2m\times 2m} \;, \quad R_m = 1 \;. \tag{7}$$

The optimum state feedback control law which minimizes the quadratic performance index is given by

$$u = -R_m^{-1} {\boldsymbol{B}_m}^T \tilde{\boldsymbol{P}} \boldsymbol{Z}_m \;. \tag{8}$$

Thus the optimum feedback gain vector $\boldsymbol{G} \in \mathrm{R}^{1\times 2m}$ is

$$\boldsymbol{G} = -R_m^{-1} {\boldsymbol{B}_m}^T \tilde{\boldsymbol{P}} \tag{9}$$

where $\tilde{\boldsymbol{P}} \in \mathrm{R}^{2m\times 2m}$ is the unique symmetric positive definite solution of the following Riccati equation.

$$\tilde{\boldsymbol{P}} \boldsymbol{B}_m R_m^{-1} {\boldsymbol{B}_m}^T \tilde{\boldsymbol{P}} - {\boldsymbol{A}_m}^T \tilde{\boldsymbol{P}} - \tilde{\boldsymbol{P}} \boldsymbol{A}_m - \boldsymbol{Q}_m = \mathbf{0} \tag{10}$$

III. SIMULTANEOUS OPTIMIZATION PROBLEM FORMULATION

The structure and the control system are designed by using the optimization technique which is based on the objective function minimization. Generally, the structure and control system simultaneous optimization is formulated as follows.

The structural design variables $\boldsymbol{a}_S$ and the control design variables $\boldsymbol{a}_C$, are optimized to minimize an objective function

$$F = F(f_S(\boldsymbol{a}_S, \boldsymbol{a}_C), f_C(\boldsymbol{a}_S, \boldsymbol{a}_C)) \;, \tag{11}$$

subject to equality constraints

$$\boldsymbol{h}(\boldsymbol{a}_S, \boldsymbol{a}_C) = \mathbf{0} \tag{12}$$

such as the equations of motion, the control law etc., and inequality constraints

$$\boldsymbol{g}(\boldsymbol{a}_S, \boldsymbol{a}_C) \leq \mathbf{0} \tag{13}$$

such as the upper and the lower bounds of the design variables, the stress limitation and so on.

In Eq.(11), F is a composite objective function which is defined as a linear combination of the structural objective function f_S and the control objective function f_C. These objective functions are defined in the next section. In this formulation, the cross-sectional area, Young's modulus, material density of the structure are taken as the structural design variables $\boldsymbol{a}_S$, and the sensor/actuator location is taken as the control design variable $\boldsymbol{a}_C$.

IV. DEFINITION OF OBJECTIVE FUNCTIONS

The following two objective functions for the structure design are defined. One is the structural weight expressed as follows.

$$J_1 \equiv \sum_{i=1}^{N} \rho_i \cdot A_i \cdot l_i \tag{14}$$

where ρ_i, A_i and l_i are the material density, the cross-sectional area and the length of the i-th element, respectively. The structural weight does not include the sensor/actuator weight. The other objective function is the measure of the transient response of the open loop system. It has the form

$$J_2 \equiv \int_0^{\infty} \boldsymbol{Z}_m{}^T \boldsymbol{Q}_m \boldsymbol{Z}_m dt \ . \tag{15}$$

The minimum value of J_2 is given by

$$J_2 = \boldsymbol{Z}_0{}^T \boldsymbol{P} \boldsymbol{Z}_0 \tag{16}$$

where $\boldsymbol{Z}_0 \in \mathbf{R}^{2m}$ is the initial condition of the open loop system and $\boldsymbol{P} \in \mathbf{R}^{2m\times 2m}$ is the unique symmetric positive definite matrix which satisfies the following Lyapunov equation.

$$\boldsymbol{A}_m{}^T \boldsymbol{P} + \boldsymbol{P}\boldsymbol{A}_m = -\boldsymbol{Q}_m \tag{17}$$

The following two objective functions for the control system design are defined. One is the well known quadratic performance index of the linear regulator that is defined by Eq.(6).

$$J_3 \equiv \int_0^{\infty} (\boldsymbol{Z}_m{}^T \boldsymbol{Q}_m \boldsymbol{Z}_m + u^2 R_m) dt \tag{18}$$

The minimum value of J_3 is given by

$$J_3 = \tilde{\boldsymbol{Z}}_0^T \tilde{\boldsymbol{P}} \tilde{\boldsymbol{Z}}_0 \tag{19}$$

where $\tilde{\boldsymbol{Z}}_0 \in \mathbf{R}^{2m}$ is the initial condition of the closed loop system, and $\tilde{\boldsymbol{P}}$ is the unique symmetric positive definite solution of the Riccati equation (10). The other objective function is the Frobenious norm of the feedback gain vector[6] which is defined as

$$J_4 \equiv \mathrm{trace}\,(\boldsymbol{G}^T R_m \boldsymbol{G})\,. \tag{20}$$

Substituting (9) into the right side of the definition (20), it leads

$$J_4 = \mathrm{trace}\left(\tilde{\boldsymbol{P}} \boldsymbol{B}_m R_m^{-1} \boldsymbol{B}_m{}^T \tilde{\boldsymbol{P}}\right) \tag{21}$$

where *trace* of square matrix is the sum of all the elements along the principal diagonal.

The composite objective function for the structure and control system simultaneous optimization is defined the following form.

$$F = \sum_{k=1}^{4} \left(C_k \cdot J_k\right) , \quad \forall C_k \geq 0 \tag{22}$$

where C_k is the weighting coefficient. Each $\left(C_k \cdot J_k\right)$ is a dimensionless quantity.

V. NUMERICAL EXAMPLE

The flexible cantilever beam shown in Figure 1 is designed. The beam is modeled by twenty finite elements in which the nodal displacement and the nodal angular displacement are considered. The parameters of the beam are given in Table 1. In this example, beam thickness T, Young's modulus E and material density ρ have constant values, and only cross-sectional area A is considered as the structural design parameter. It is assumed that the distributions of the cross-sectional area $A(x)$ is expressed as the finite power series of x, which is the distance from the fixed end of the beam. This formulation is proposed in the authors' previous work[4]. For convenience, the following nondimensional distance is introduced.

$$X = x/L \tag{23}$$

where L is the overall length of the beam. The cross-sectional area $A(X)$ is given by

$$A(X) = B(X) \cdot T \,, \tag{24}$$

and the beam width $B(X)$ is expressed as following third order power series.

$$B(X) = B^* \cdot \left(\sum_{j=0}^{3} \widetilde{B}_j \cdot X^j \right) \tag{25}$$

where B^* is also given in Table 1. The nondimensional coefficients $\widetilde{B}_j$ become the structural design variables. The beam width $B(X)$ which is constrained on the shape given by Eq.(25) and the sensor/actuator location X_{SA} are optimized under the initial dynamic disturbance which is transverse 0.1N impulse applied at the free end of the beam. Tables 2 and 3 show the optimization problems under several conditions. The optimization problems are solved by using Powell's algorithm[7].

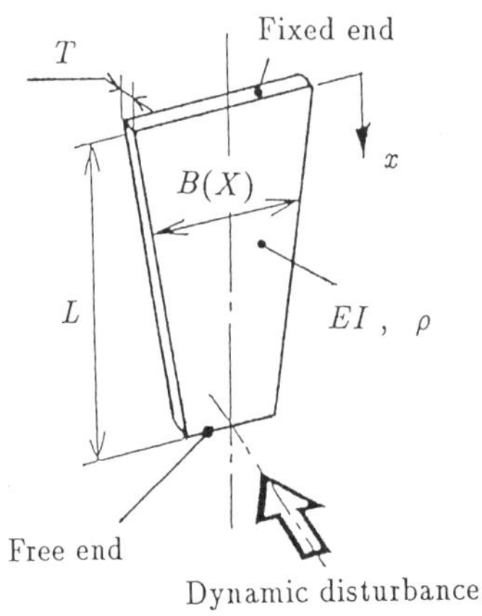

Fig.1. Flexible cantilever beam.

Table 1. Parameters of beam.

$L = 0.9$m	$E = 2.0594 \times 10^5$MPa
$T = 1.2 \times 10^{-3}$m	$\rho = 7.86 \times 10^3$kg/m^3
$B^* = 6.0 \times 10^{-2}$m	$\zeta = 0.001$
Modeling Modes :	$1st \sim 40th$ Mode $(n = 40)$
Control Modes :	$1st \sim 5th$ Mode $(m = 5)$

Table 2. Objective functions and design variables I.

Option	Objective Functions	Design Variables
Case 1	J_1	$\widetilde{B}_j$
Case 2	J_2	$\widetilde{B}_j$
Case 3	J_3	$\widetilde{B}_j$
Case 4	J_4	$\widetilde{B}_j$

In all cases listed in Table 2, the sensor/actuator location X_{SA} is fixed at the free end of the beam. The beam width is designed to minimize the structural objective function in Case 1 and 2, and the beam width is designed to minimize the control objective function in Case 3 and 4.

Figure 2 shows the optimum beam shape for each condition listed in Table 2. The optimum beam shapes can be classified into the following three different types: narrow shape obtained by Case 1, wide shape obtained by Case 2 and 3, and other shape obtained by Case 4. Table 4 indicates the values of $J_1 \sim J_4$ under the optimum condition for Case 1 to 4.

Table 3. Objective functions and design variables II.
– Kinds of simultaneous and individual optimization problems –

Design	Approach of Optimum Design		Objective Functions	Design Variables
A	Simultaneous		$C_1 \cdot J_1 + C_3 \cdot J_3$	$\tilde{B}_j$, X_{SA}
	Individual	Structure	J_1	$\tilde{B}_j$
		Control System	J_3	X_{SA}
B	Simultaneous		$C_1 \cdot J_1 + C_4 \cdot J_4$	$\tilde{B}_j$, X_{SA}
	Individual	Structure	J_1	$\tilde{B}_j$
		Control System	J_4	X_{SA}
C	Simultaneous		$C_2 \cdot J_2 + C_3 \cdot J_3$	$\tilde{B}_j$, X_{SA}
	Individual	Structure	J_2	$\tilde{B}_j$
		Control System	J_3	X_{SA}
D	Simultaneous		$C_2 \cdot J_2 + C_4 \cdot J_4$	$\tilde{B}_j$, X_{SA}
	Individual	Structure	J_2	$\tilde{B}_j$
		Control System	J_4	X_{SA}

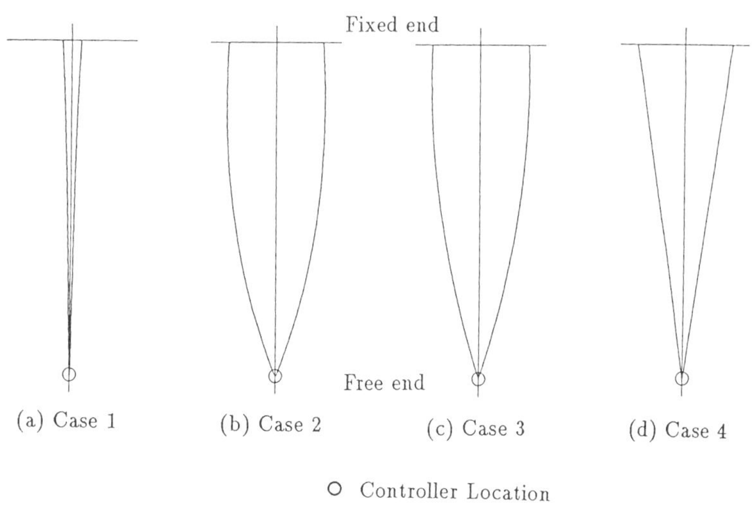

Fig.2. Optimum beam shape for Case 1 to 4.

Following observations can be made: The beam shape and structural weight depend on the structural objective function J_1. If the composite objective function includes J_1, the optimum beam shape becomes narrow and the values of $J_2 \sim J_4$ increase. The iteration histories for Case 1 to 4 are shown in Figure 3. From the iteration history of Case 1, the values of $J_2 \sim J_4$ increase corresponding to the decreasing of the value of J_1. On the contrary, the values of J_1 for Case 2 to 4 increase corresponding to the decreasing of the values of $J_2 \sim J_4$. The optimum design obtained by minimizing J_2 is the same as that obtained by minimizing J_3. The values of $J_2 \sim J_4$ obtained by minimizing J_4 are similar to that obtained by minimizing J_2 or J_3, but the value of J_1 obtained by minimizing J_4 is not.

Table 4. Values of J_1 to J_4 at optimized conditions for Case 1 to 4.

Quantity	Case 1	Case 2	Case 3	Case 4
J_1	0.123	1.139	1.063	0.756
J_2	0.5525	0.0283	0.0288	0.0397
J_3	0.002033	0.000375	0.000361	0.000386
J_4	0.012724	0.001209	0.001047	0.000890

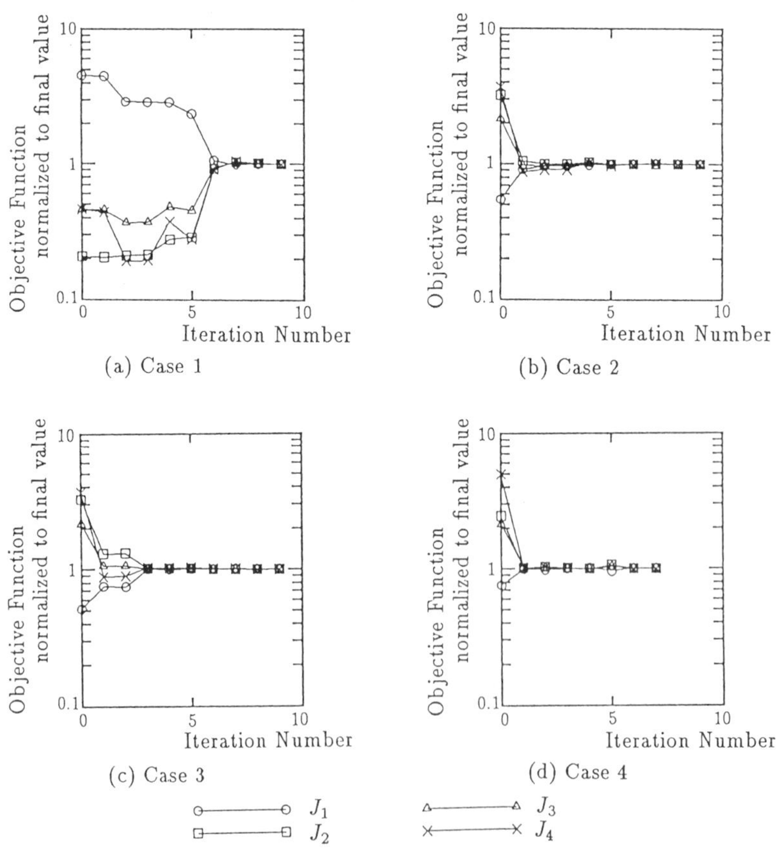

Fig.3. Iteration histories for Case 1 to 4.

Next, we compare the simultaneous optimization with the individual optimization for the several optimization problems listed in Table 3. For example, in design A, J_1 and J_3 are taken as the structural and the control objective functions, respectively. In the case of the simultaneous optimization in design A, both the structure and the control system are designed simultaneously by minimizing the composite objective function $C_1 \cdot J_1 + C_3 \cdot J_3$. In the case of the individual optimization in design A, the control system is designed by minimizing J_3 after the structure has been designed by minimizing J_1. $C_1 \sim C_4$ are weighting coefficients whose values are obtained from a certain initial beam shape and sensor/actuator location.

The optimum beam shapes for several optimization problems listed in Table 3, are shown in Figure 4. The optimum beam shapes can be classified into the two different types that are narrow shapes which are obtained by design A and B, and wide shapes which are obtained by design C and D. Table 5 indicates the values of the composite objective functions for both optimum designs obtained by the simultaneous and the individual optimization.

Following observations can be made: As we pointed out in Case 1 to 4, taking J_1 for the structural objective function causes the narrow beam shape. In the case of design A, the value of the composite objective function obtained by the simultaneous optimization is smaller than that obtained by the individual optimization. But in the other cases, both values of the composite objective functions obtained by the two optimization are almost the same. It shows that there is no advantage of using the simultaneous optimization in the cases of design B to D from the point of view of the objective function minimization.

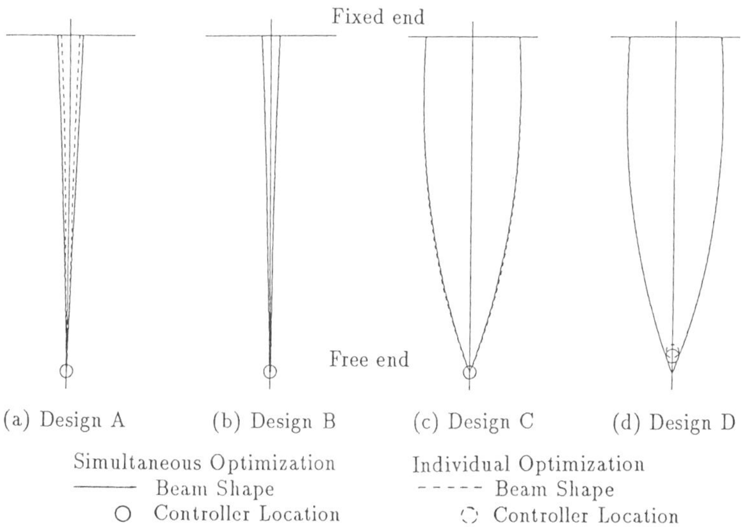

Fig.4. Optimum beam shape and controller location obtained by simultaneous and individual optimization.

Table 5. Values of composite objective functions at optimized condition for Design A to D.

Design	Simultaneous	Individual	$\frac{\text{Simultaneous}}{\text{Individual}}$ (%)
A	0.7653	0.8472	90.3
B	0.3307	0.3277	100.9
C	0.4422	0.4447	99.4
D	0.3273	0.3194	102.5

VI. CONCLUSIONS

The effect of the objective functions on the results of the structure and control system simultaneous optimization is discussed, comparing the optimization results numerically. From the results of the numerical example, we may obtain following conclusions. The minimization of the structural weight causes the increase of the values of the other objective functions. On the other hand, the minimization of the transient characteristics values causes the increase of the structural weight. The comparisons of the results obtained by the simultaneous and the individual optimization indicate that the usefulness of the simultaneous optimization depends on the objective function for the example.

REFERENCES

1) Onoda, J. and Haftka, R. T., "An Approach to Structure/Control Simultaneous Optimization for Large Flexible Spacecraft," *AIAA Journal*, 25-8(1987), pp.1133-1138.

2) Salama, M., Garba, J., Demsetz, L. and Udwadia, F., "Simultaneous Optimization of Controlled Structures," *Computational Mechanics*, 3(1988), pp.275-282.

3) Kajiwara, I. and Nagamatsu, A., "An Approach of Optimal Design for Simultaneous Optimization of Structure and Control Systems Using Sensitivity Analysis," *Transaction of the Society of Instrument and Control Engineers*, 26-10(1990), pp.1140-1147, (in Japanese).

4) Iwatsubo, T., Kawamura, S. and Adachi, K., "Simultaneous Optimum Design of Structure-Control Combined System – Compared with Individual Optimum Design –," In *First International Conference on Motion and Vibration Control*, K. Seto, K. Yoshida and K. Nonami(Editors), The Japan Society of Mechanical Engineers, 1(1992), pp.128-133.

5) Iwatsubo, T., Kawamura, S., Adachi, K. and Ikeda, M., "Simultaneous Optimum Design of Structural and Control System," In *First Joint U.S./Japan Conference on Adaptive Structures*, B. K. Wada, J. L. Fanson and K. Miura(Editors), Technomic Publishing Co., Inc., Lancaster, PA, (1990), pp.117-140.

6) Khot, N. S., Öz, H., Grandhi, R. V., Eastep, F. E. and Venkayya, V. B., "Optimal Structural Design with Control Gain Norm Constraint," *AIAA Journal*, 26-5(1988), pp.604-611.

7) Zangwill, W. I., "Minimizing a Function without Calculating Derivatives," *Computer Journal*, 10(1967), pp.293-296.

Homology Design of Frame Structure by Finite Element Method

Nobuhiro YOSHIKAWA and Shigeru NAKAGIRI

Institute of Industrial Science, University of Tokyo, Tokyo

A formulation is proposed for homology design, in which a geometrical property is maintained before, during and after deformation of a structure. The finite element sensitivity analysis is employed to approximate the relationship between the behavior change and parameter change of the structure. The governing equation of the structural behavior is partitioned with respect to the constraint for the homology, result in the governing equation of the structural parameters, whose coefficient matrix is rectangular. The structural parameters for the homology design are determined by use of the Moore-Penrose generalized inverse. The numerical examples in regard to homologous deformation and homologous vibration mode of a planar lattice frame verify the validity of the proposed method.

I. INTRODUCTION

New breed structures such as adaptive structure and intellectual structure need sophisticated design methods in order to bring their functions into full play, even when the stiffness and strength of the structures are assured duly. These structures are equipped with control system sometimes so as to meet accurately delicate demands imposed in terms of movement, location and direction, for instance. A characteristic of the sophisticated design methods seems to be the attention paid primarily to geometric quantities such as deformations in addition to the consideration of mechanical quantities such as stresses. Apart from so-called optimal design to find minimum-weight structures in the classic sense[1)], attempt has been made to design structures that satisfy requirements set for deformation. For example, tailoring design is devised as a candidate method of structural synthesis of fiber-reinforced plastic (FRP) structures by means of controlling the compliance of FRP structures[2)]. Structural design aiming at homologous deformation to maintain parabola was discussed since the performance of tiltable telescope is governed by shape accuracy of the parabolic antenna[3),4)]. When a certain geometrical relationship is maintained in a part or whole of a structure before, during and after deformation, the deformation is called homologous[3)]. In the case of the tiltable telescope, what is essential is to maintain the parabola of the antenna at any tilting angle, homologous to the undeformed state, and it is not essential to restrict the magnitude of the deformation[4)]. Such concept of homologous deformation can be extended to homologous eigenmode to be employed as a basic technique to adjust vibration behavior of structures. In this study homologous vibration mode is defined as the eigenmode of vibration for which a certain geometrical relationship is kept in a part or whole of a structure both at a standstill and in vibrant state.

Hangai proposed a method to realize a homologous deformation of truss structure modeled with finite elements[5)]. In his method, the governing equation of the structural parameters employed to afford the prescribed homologous deformation was derived from the necessary and sufficient condition of the solution of the stiffness equation to exist, whose stiffness matrix is rectangular as a result of introduction of the constraint condition prescribing the homologous deformation, and the Moore-Penrose generalized inverse is used to tackle with the rectangular matrix. It is needed to compute the sensitivities of the generalized inverse with respect to the structural parameters, giving rise to long CPU time consumed.

This paper presents a new method for homology design in problems of static deformation and undamped vibration eigenmode of linear elastic body under small displacements. The method is formulated on the basis of the finite element sensitivity analysis of the nodal displacements, not of the generalized inverse. The validity of the proposed formulation is examined through the numerical examples in regard to a planar lattice frame and is verified sufficient when the homology design itself is possible.

II. FORMULATION FOR HOMOLOGY DESIGN

2.1 Homologous Deformation

Suppose we have a baseline design of a linear and elastic structure whose stiffness equation is given as follows, after the boundary condition is dealt with,

$$[K]\{U\}=\{F\} \tag{1}$$

where $[K]$ denotes the global stiffness matrix determined by the structural parameters of the baseline design, $\{U\}$ the unknown nodal displacement vector of N degrees of freedom (DOF) and $\{F\}$ the nodal force vector. The force vector and boundary condition are assumed to be unchanged, though the structural parameters are changed for the homology design. The nodal displacement vector is partitioned into two parts of the dependent term $\{U_h\}$ of J DOF, on which a prescribed geometrical relationship for the objective homologous deformation is imposed, and the independent term $\{U_i\}$ of N-J DOF to impose the homologous deformation so as to $\{U\}^T=\{U_i \;\; U_h\}^T$. The subscript h denotes the term related to the homologous deformation, i the independent term, and the superscript T denotes matrix transpose hereafter. The stiffness equation (1) is partitioned in the form of Eq.(2) according to the nodal displacement partitioning.

$$\begin{bmatrix} K_{ii} & K_{ih} \\ K_{hi} & K_{hh} \end{bmatrix} \begin{Bmatrix} U_i \\ U_h \end{Bmatrix} = \begin{Bmatrix} F_i \\ F_h \end{Bmatrix} \tag{2}$$

In this study we suppose that the displacement constraint for the homologous deformation is expressed in the linear form of Eq.(3).

$$\{U_h\}=[C]\{U_i\} \tag{3}$$

The expression of the constraint condition is neither unique for an objective homologous deformation nor linear with respect to $\{U_i\}$, in general. The dependent term of the displacement $\{U_h\}$ should be chosen adequately for the homologous deformation. The ingredients of the constraint matrix $[C]$ are determined according to the objective homologous deformation. They are constant in the case of linear homologous deformation, but are function of the nodal displacements in the case of nonlinear homologous deformation. Substituting the expression of Eq.(3) into Eq.(2), we have the separated stiffness equations (4) and (5),

$$[K_s]\{U_i\}=\{F_i\} \tag{4}$$

$$[K_r]\{U_i\}=\{F_h\} \tag{5}$$

where

$$[K_s]=[K_{ii}]+[K_{ih}][C] \tag{6}$$

$$[K_r]=[K_{hi}]+[K_{hh}][C] \tag{7}$$

$[K_s]$ is a $(N$-$J)$x$(N$-$J)$ square and asymmetric matrix, and $[K_r]$ is a Jx$(N$-$J)$ rectangular matrix. The unknown displacement vector $\{U_i\}$ can be obtained by usual means of solving Eq.(4) because $[K_s]$ is square and non-singular. In doing so, the asymmetry of $[K_s]$ should be taken into account. When Eq.(5) holds after the substitution of the vector $\{U_i\}$ thus obtained, it means that the structural parameters employed in the baseline design afford the objective homologous deformation, and there takes place no need to change the baseline design.

When Eq.(5) does not holds for the vector $\{U_i\}$ obtained from Eq.(4), as is expected in usual, the structural parameters are to be changed so that we have such matrices $[K_s]$ and $[K_r]$ that satisfy Eqs.(4) and (5) simultaneously for a same vector $\{U_i\}$. The search for the suitable structural parameters is formulated based on the first-order approximation of the structural behavior change with respect to M parameters properly chosen as described in the following. The change of m-th parameter P_m is expressed by introducing design variable α_m as follows.

$$P_m = \overline{P}_m(1+\alpha_m) \tag{8}$$

The upper bar means the term related to a current baseline design hereafter. The first-order approximation, in the form of Eqs.(9) and (10), is used to express the change of the stiffness matrices $[K_s]$ and $[K_r]$, respectively, according to the change of the structural parameters in the vicinity of a current baseline design,

$$[K_s] = [\overline{K}_s] + \sum_{m=1}^{M} [K^I_{sm}]\alpha_m \tag{9}$$

$$[K_r] = [\overline{K}_r] + \sum_{m=1}^{M} [K^I_{rm}]\alpha_m \tag{10}$$

where the subscript m relates the design variable α_m, and the superscript I means the first-order. Corresponding to the expression of Eq.(9), we assume that the first-order approximation in the form of Eq.(11) is sufficient for the change of the displacement vector $\{U_i\}$ in the vicinity of the baseline design.

$$\{U_i\} = \{\overline{U}_i\} + \sum_{m=1}^{M} \{U^I_{im}\}\alpha_m \tag{11}$$

Then the first-order sensitivity of the displacement vector $\{U_i\}$ with respect to α_m is evaluated by Eq.(12) by the first-order perturbation technique, while we have $\{\overline{U}_i\} = [\overline{K}_s]^{-1}\{F_i\}$.

$$\{U^I_{im}\} = -[\overline{K}_s]^{-1}[K^I_{sm}]\{\overline{U}_i\} \tag{12}$$

It should be noted that the above sensitivity $\{U^I_{im}\}$ is so determined that Eq.(4) holds for the change in the vicinity of a current baseline design. Substituting Eqs.(10) and (11) after such a determination of $\{U^I_{im}\}$ into Eq.(5) and truncating the second-order term of α_m, we have Eq.(13),

$$\sum_{m=1}^{M} \left([K^I_{rm}]\{\overline{U}_i\} + [\overline{K}_r]\{U^I_{im}\}\right)\alpha_m = \{F_h\} - [\overline{K}_r]\{\overline{U}_i\} \tag{13}$$

which is rewritten as follows,

$$[A]\{\alpha\} = \{b\} \tag{14}$$

where $[A]$ is a JxM rectangular matrix, and $\{\alpha\}$ a design variable vector consisting of M unknowns. As aforementioned, all the ingredients of $[A]$ and $\{b\}$ are made known. It is obvious from Eq.(13) that the design variables are governed by the loading mode, not by the magnitude of the loading, because $\{\overline{U}_i\}$ and $\{U^I_{im}\}$ are proportional to $\{F_i\}$.

The Moore-Penrose generalized inverse $[A]^-$ is used to determine the unknown design variable vector $\{\alpha\}$ with respect to a given rectangular matrix $[A]$. When the necessary and sufficient condition for the solution to exist is satisfied, the general solution of the simultaneous linear equation, whose coefficient matrix is rectangular, is expressed as sum of the particular solution and complimentary solution including an arbitrary vector $\{h\}$ as given below[6],

$$\{\alpha\} = [A]^-\{b\} + ([I] - [A]^-[A])\{h\} \tag{15}$$

where $[I]$ is an unit matrix, and $\{h\}$ the arbitrary vector. The first term on the right hand side of Eq.(15) is the particular solution, and the second one the complimentary solution. The above expression indicates that the way to determine the change of a baseline design required to afford homologous deformation is not unique in general since indefiniteness creeps in the general solution through the arbitrary vector. We take only the particular solution, in other words, the arbitrary vector is taken equal to zero. The design variables thus determined by the particular solution gives rise to that the norm of the design variable vector is minimum, further implying that the design change from a current baseline design is made minimum, owing to the nature of the Moore-Penrose generalized inverse[6].

The determination of the design variables stated above is based on the first-order approximation of the relationship between the behavior change and parameter change. The relationship is not linear actually, however, so that it is required to renew the baseline design iteratively in order to cope with the deficiency caused by the first-order approximation until the homologous deformation is realized, that is, the right hand side of Eq.(13) is made equal to zero. The behavior analysis and sensitivity analysis should be carried out at each stage of the renewal.

2.2 Homologous Vibration Mode

The structural parameters to afford homologous vibration mode are determined by the similar formulation with that for homologous deformation. The same relationship between the structural parameters and design variables as Eq.(8) is used. Equation (16) expresses undamped linear eigenvalue problem of vibration,

$$([K] - \lambda[M])\{\Phi\} = \{0\} \tag{16}$$

where $[M]$ is the mass matrix, λ denotes the eigenvalue and $\{\Phi\}$ the eigenvector. The only mode under consideration is dealt with in this study. The dependent term of the eigenvector $\{\Phi_h\}$ of J DOF of the homologous vibration mode is related to the independent one $\{\Phi_i\}$ of N-J DOF as given below.

$$\{\Phi_h\} = [C]\{\Phi_i\} \tag{17}$$

Eq.(16) is partitioned in the form of Eq.(18) corresponding to the partitioning of the eigenvector.

$$\left(\begin{bmatrix} K_{ii} & K_{ih} \\ K_{hi} & K_{hh} \end{bmatrix} - \lambda \begin{bmatrix} M_{ii} & M_{ih} \\ M_{hi} & M_{hh} \end{bmatrix}\right)\begin{Bmatrix} \Phi_i \\ \Phi_h \end{Bmatrix} = \begin{Bmatrix} 0 \\ 0 \end{Bmatrix} \tag{18}$$

Substituting Eq.(17) into Eq.(18), we have Eqs.(19) and (20),

$$([K_s] - \lambda[M_s])\{\Phi_i\} = \{0\} \tag{19}$$

$$([K_r] - \lambda[M_r])\{\Phi_i\} = \{0\} \tag{20}$$

where

$$[M_s]=[M_{ii}]+[M_{ih}][C] \tag{21}$$

and $[M_r]$ is defined in the similar manner. The same $[K_s]$ and $[K_r]$ as given in the section 2.1 are used.

The eigenvalue problem of Eq.(19) with the asymmetric matrices of $[K_s]$ and $[M_s]$ differs from Eq.(16). The eigenvalue obtained from Eq.(19) is different from that of Eq.(16), though the same notation is used. Also $\{\Phi_i\}$ obtained from Eq.(19) is not identical with the corresponding part of $\{\Phi\}$ of Eq.(16). The identity is not necessitated. In case if Eq.(20) holds when the eigenpair λ and $\{\Phi_i\}$ obtained from Eq.(19) are substituted, the initial baseline design realizes the objective homologous vibration mode. The baseline design is to be changed when Eq.(20) is not satisfied, as is in most case.

The first-order approximation is again employed to express the change of the mass matrix and eigenpair in the vicinity of a current baseline design as follows.

$$[M_s]=[\overline{M}_s]+\sum_{m=1}^{M}[M_{sm}^{I}]\alpha_m \tag{22}$$

$$\lambda=\overline{\lambda}+\sum_{m=1}^{M}\lambda_m^{I}\alpha_m \tag{23}$$

$$\{\Phi_i\}=\{\overline{\Phi}_i\}+\sum_{m=1}^{M}\{\Phi_{im}^{I}\}\alpha_m \tag{24}$$

In the above, the sensitivities of the eigenpair λ_m^{I} and $\{\Phi_{im}^{I}\}$ are so determined that Eq.(19) holds for the change in the vicinity of a current baseline design. The Nelson's method suitable to deal with the asymmetric eigenvalue problem is employed for the computation[7)].

The change of $[M_r]$ is evaluated in the same form with Eq.(22) by the first-order approximation. Then Eq.(20) is rewritten as Eq.(25) when Eqs.(23) and (24) are substituted together with Eq.(10), and the second-order term of α_m is truncated.

$$\sum_{m=1}^{M}\left(\{[K_{rm}^{I}]-\overline{\lambda}[M_{rm}^{I}]-\lambda_{rm}^{I}[\overline{M}_r]\}\{\overline{\phi}_i\}+\{[\overline{K}_r]-\overline{\lambda}[\overline{M}_r]\}\{\phi_{im}^{I}\}\right)\alpha_m = -\left([\overline{K}_r]-\overline{\lambda}[\overline{M}_r]\right)\{\overline{\phi}_i\} \tag{25}$$

It is easily seen that Eq.(25) takes the same form as Eq.(14). This indicates the design variables for the homologous vibration mode can be determined in the similar manner with that for the homologous deformation.

III. NUMERICAL EXAMPLES

It cannot be predicted a priori whether a homology design is possible or not. The analysis of structural behavior and sensitivity are carried out at renewed baseline design in the formulation stated above. Hence, which baseline design is chosen and which structural parameters are picked up affect considerably the success of the homology design, even in case if it is possible. Then the possibility of homology design and the validity of the present formulation are examined here through the following examples in regard to a planar lattice frame shown in Fig.1. Each frame section is represented by a finite beam element with twist. The Young's modulus and modulus of rigidity of the material are taken equal to 70.0 GPa and 26.9 GPa, respectively. The mass density is taken equal to 3.0×10^3 kg/m^3.The cross-section of all sections is circular, whose diameter of the initial baseline design is 50.0 mm Twenty two design variables are assigned to the diameter of all elements. The necessary and sufficient condition for the generalized inverse solution is always satisfied in the examples. The homology design aimed at in this study is to keep straight part in a structure.

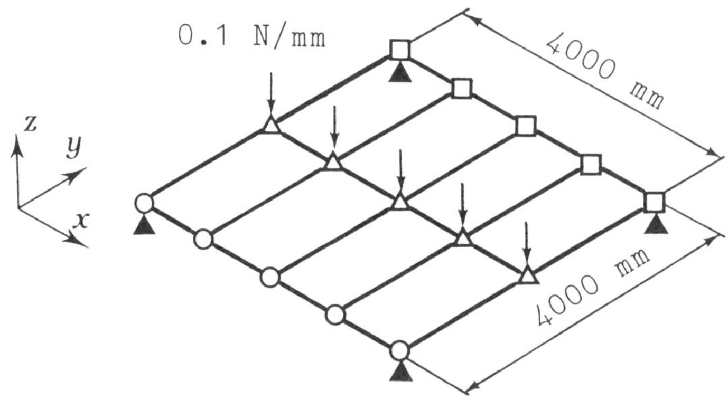

Fig.1. Planar lattice frame under distributed load.

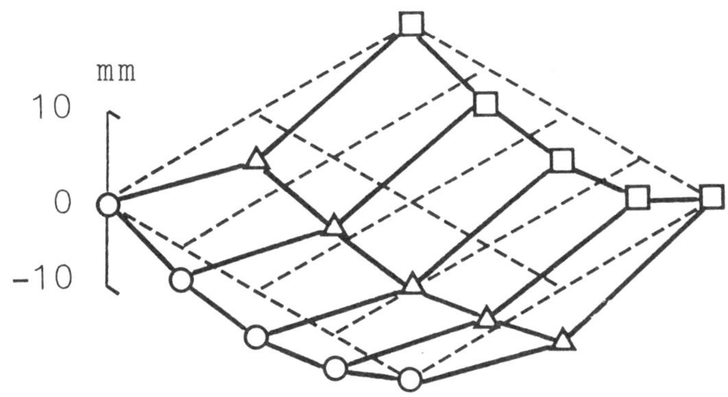

Fig.2. Deformation of initial baseline design.

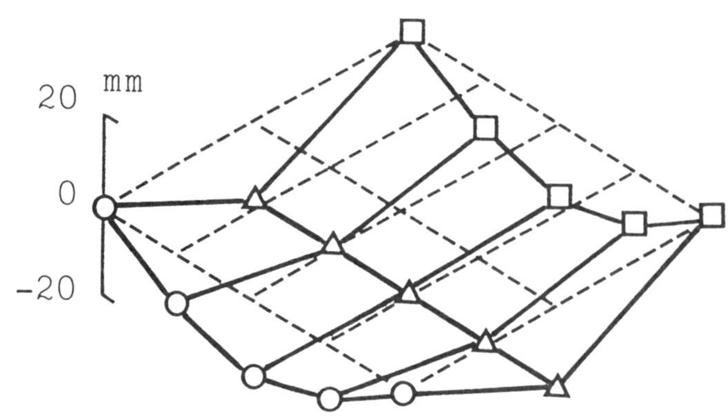

Fig.3. Homologous deformation.

3.1 How to Keep a Member Straight Against Distributed Load

Suppose that the frame in the x-y plane is supported simply at the four corner nodes and is subjected to uniformly distributed load of 0.1N/mm acting downwards vertically along the central member consisting of four sections. Then the lattice frame is deformed as shown in Fig.2, and the sagging of the central member is remarkable. Suppose that the sagging is to be avoided for such a design purpose to mount some precision instrument, for instance, while the downward and equal displacements of the central member are allowed. In this case, the homologous deformation is prescribed only for the central member to be straight and parallel to the x-y plane, being expressed by that the vertical displacements of the nodes in the central member are the same each other. The constraint matrix $[C]$ in this case is quite simple and sparse, consisting of unity and zero only.

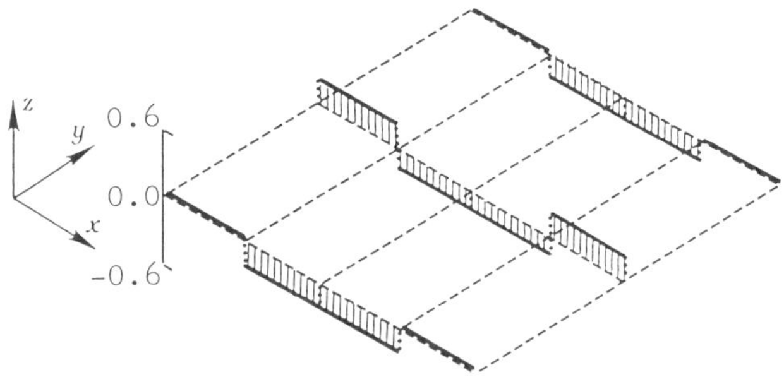

(a) x direction.

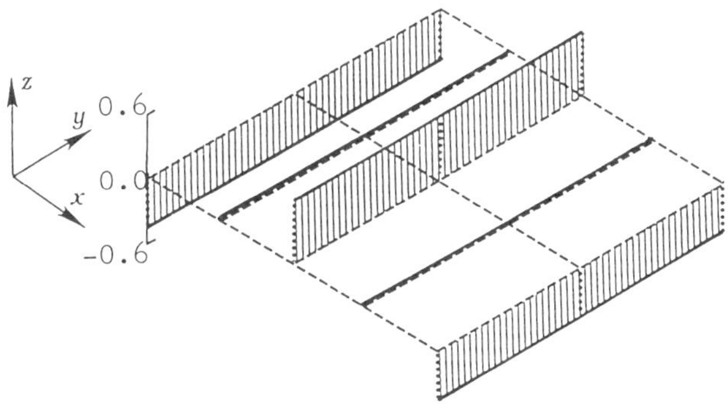

(b) y direction.

Fig.4. Amount of diameter change for homologous deformation.

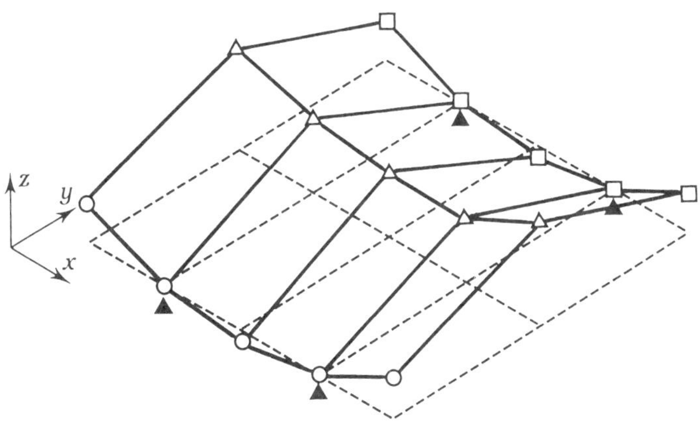

Fig.5. First mode shape of initial baseline design.

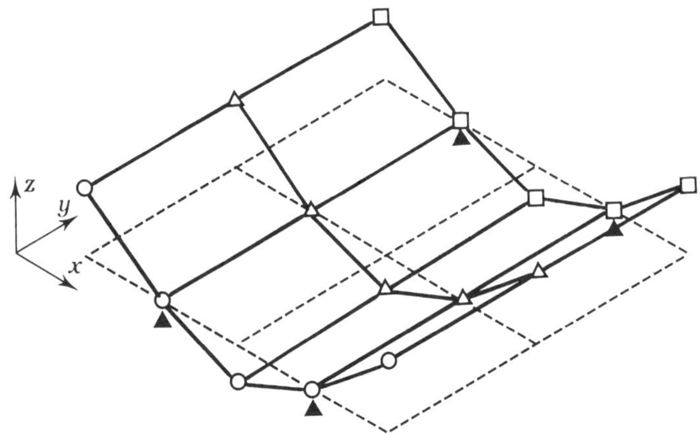

Fig.6. Homologous mode.

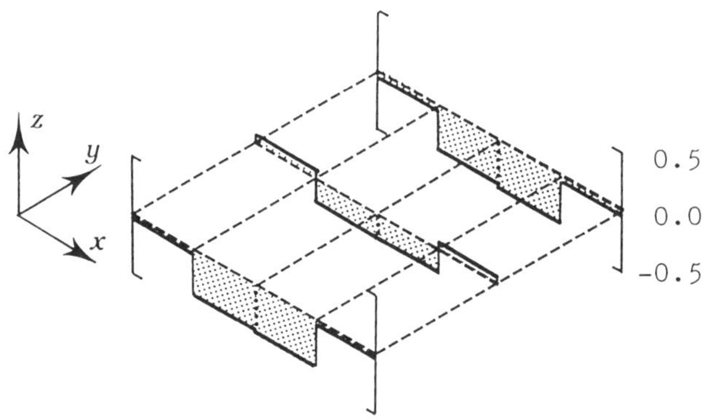

(a) *x* direction.

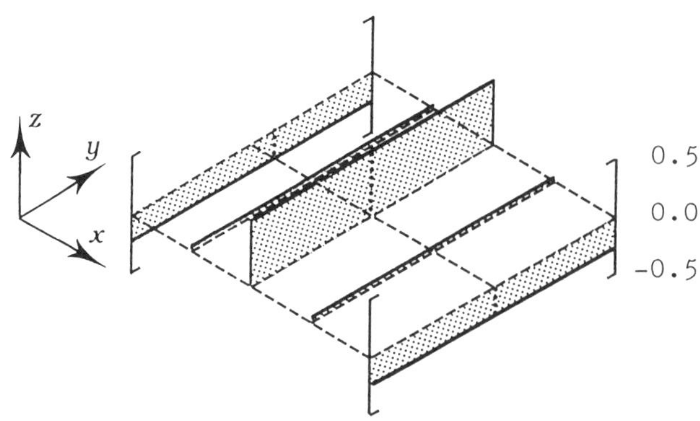

(b) *y* direction.
Fig.7. Amount of diameter change for homologous mode.

Figure 3 shows the deformed frame after the design change renewed five times for the homologous deformation. The central member is kept straight as is aimed at. It is inferred qualitatively that a simple way for keeping the central member straight against the load is to make the flexural rigidity of the four sections large extremely. This way is likely to cause more or less increase of the member weight. Figure 4 illustrates the diameter change, as the ratio to the initial diameter, obtained by the present formulation. It is seen in this figure that the homologous deformation can be achieved by means of changing subtly the diameter of all sections less than 0.6 without considerable increase of the total weight of the frame. It should be noted that the stiffness of the frame is decreased as indicated by the increase of the largest deflection from 9.6 mm to 18.4 mm for the same load though the homology design is possible in this case.

3.2 How to Keep Straight Line in Mode Shape

Figure 5 depicts the first mode shape of the out-of-plane bending vibration of the frame. The black triangles in the figure shows the simply support nodes. The corresponding natural frequency is 4.58 Hz. The mode is saddle-shape, that is, sagging in the x direction and hogging in the y direction in Fig.5. Suppose that the hogging in the y direction should be avoided to meet some design purpose. Figure 6 shows the mode shape obtained after three renewal of the baseline design, all members in the y direction being made straight, homologous to the standstill state. The natural frequency is lowered to 3.44 Hz by the design change. The required change of the diameter is shown in Fig.7 in the same manner as Fig.4. Also in this case, a simple way to eliminate the hogging mode in the y direction is to set the flexural rigidity of the sections in the same direction large extremely. In doing so, the natural frequency might be altered considerably unless adequate care is taken for the mass change. Figure 7 indicates that the homology design is made possible by rather small change of the diameter instead of making the flexural rigidity so large.

IV. CONCLUDING REMARKS

A formulation is proposed for the homology design in problems of static deformation and undamped eigenvalue problem of linear elasticity. The formulation is based on the finite element sensitivity analysis and the design change aimed at is determined by the design variables obtained by use of the Moore-Penrose generalized inverse. The numerical examples show that the present formulation is valid when the design variables are chosen properly for the homology design which is made possible owing to the initial baseline design specified suitably and the constraint imposed adequately. The numerical examples indicate that the stiffness of the structure is likely to diminish by the homology design. It is therefor desirable to incorporate some relevant conditions additionally in the homology design so as to bound displacements, stresses and natural frequency.

REFERENCES

1) Haug, E.J. and Arora, J.S., Applied optimal design, John Wiley & Sons (1979).
2) Miki, M., Onuki, M., Murotsu, Y. and Yamaguchi, T., Passive deformation control of structures with structurally-unsymmetric laminate configuration pipes, Proc. AIAA/ASME/ASCE/AHS/ASC, 34th structures, structural dynamics and material conference (1993), pp.1790-1795.
3) von Hoerner, S., Homologous deformation of tiltable telescope, Proc. ASCE, Vol.93, No.ST3 (1967), pp.461-485.
4) Morimoto, M., Kaifu, N., Takizawa, Y., Aoki, K. and Sakakibara, O., Homology design of large antenna (in Japanese), Mitsubishi electric corp., Technical report, Vol.56, No.7 (1982), pp.495-502.
5) Hangai, Y., Shape analysis of structures, Theoretical and applied mechanics, Vol.39, Proc. 39th Japan national congress for applied mechanics, Ed. JNCTAM, Science council of Japan, 1990, Univ. of Tokyo Press, pp.11-28.
6) Rao, C.R. and Mitra, S.K., Generalized inverse of matrices and its applications, John Wiley & Sons (1971).
7) Nelson, R.B., Simplified calculation of eigenvector derivatives, AIAA Journal, Vol.14, No.9 (1976), pp.1201-1205.

A Trial on the Shape Optimization of Mechanical Structures

Keishi KAWAMO* and Hiroshi HASEGAWA**

* *Department of Mechanical Control Systems, Shibaura Institute of Technology, Saitama*
** *Postgraduate Course of Mechanical Engineering, Shibaura Institute of Technology, Saitama*

A trial on the shape and topology optimization of a linear elastic structure is carried out by a method devised by the authors to testify the effectiveness of the method. The method consists of a stabilizing method which can converge the optimization process to 'the the optimal shape and a scheme which can reduce the number of iteration of the process. The validity of the method can be justified through various experiments for finite element models of two dimensional continuums.

I. INTRODUCTION

The problem of structural optimization can be classified into three different but interrelated sub-problems: the sizing optimization, shape optimization, and shape and topology optimization. In the problem of sizing optimization, design modification by the use of the sensitivity analysis method is rather simple, and a corresponding modification of a finite element model is straightforward. Therefore, the sizing optimization problems are considered to have been fully discussed today.

In the problem of shape optimization, the development of the boundary variation methodology for shape has been studied in various ways. The mathematical foundation for optimal design and design sensitivity analysis in boundary variation methods is also established. Sensitivity analysis for shape optimaization problems is thoroughly studied in various settings.[1)~5)] The state-of-the-art situation for shape design is that shape optimaization is possible on the assumption that the initial topology is fixed during iterative optimaization.

In the problem of shape and topology optimization, various methods have been studied so that the optimal topology as well as the optimal shape of a structure can be yielded. One of the methods has introduced an alternative approach by which shape optimizaiton problems are transformed to material distribution problems using composite materials. Two material constituents, substance and void, are considered, and the microscopic optimal void distribution is considered instead of shape optimization by boundary variations. The problems of the optimal topology as well as the optimal shape of a structure seem to be settled by adoption of the homogenization method which is applied to determine macroscopic constitutive equations for the material with microscopic material constituents.[6)~9)] The introduction of microstructures in the formulation of structural design problems is explained to correspond to a relaxation of the variation problem that can be formulated for the design optimization, and that can stabilize the optimal iteration.[10)] The levels of the porosity of the elements, i.e., the material properties of porous elements are extremely diverse in magnitude as shown in the examples obtained by using the method of homogenization.[9),10)] Then even by introducing the method, there still remains a problem how to determine the actual design shapes from the distribution pattern of substantial and porous elements in finite element method.

In the case of sensitivity analysis method which removes an element one by one in each step according to a criterion in the finite element model, often appear the irretrievable circumstances in which the elements that have once determined to be removed in some step of the iterative process but must be recovered in the later step in order to gain optimal shape cannot be recovered. In that case, this method does not necessarily produce the optimal shapes. A large number of iterative processes in this method are usually required until the optimal shapes are obtained for the finite element models which are divided into a large number of elements.

The authors propose two methods to solve the above-mentioned problems in the shape and topology optimization of structure. One concerns stabilization of the optimal process and the other reduction of the number of iteration in the process.

The former is similar to the homogeneous method[8),9)] in computing effective material properties (i.e., Young's moduli) by applying materials with microstructures, although the way to obtain the material properties is different. In the authors' method the material properties of the finite element are modified in each step in the way in which their strain energy is modified for each element energy to be equal to the mean value of the element energy over the domain before and after each iteration. The present method belongs to the sensitivity analysis method where each element is not removed but its material properties are modified each step.

In the latter the authors propose an approach which enables to decrease the number of the iteration. The finite elements are grouped into several classes in order of the magnitude of the stresses in the elements, and then, this method modifies Young's moduli for the materials of the elements in the classes instead of an element at the same time, using fuzzy c-means method.[11)]

The authors apply these methods to two dimensional finite element models and discuss their results.

II. GENERALIZED OPTIMAL-SHAPE-AND-TOPOLOGY PROBLEM FOR LINEAR ELASTIC CONTINUUMS

In the following a generalized formulation for optimal shape and topology problem, which is called a generalized layout problem in the present paper, is considered in linear elastic continuums.

2.1 Discrete Sensitivity Analysis Method

The present problem is to search a layout of continuum with minimum weight, i.e., minimum volume under given loads and boundary conditions, and for allowable stress σ_{al} as constraint condition.

Optimization process is carried out in the manner how to decrease volume, i.e., to remove a element to make the increment of strain energy sensitivity S_e minimum each step. S_e is obtained in discrete form instead of in continuous form in sensitivity analysis method.[8),9)]

$$S_e = \frac{1}{2}\{\delta\}_A^T[K]\{\delta\}_A. \tag{1}$$

S_e can be obtained by foundamental formulas of finite element method in the case that [D], [B] and [N] are constant in the element,

$$S_e = \frac{1}{2}\{\sigma\}_A^T([D][B][N])^{-T}[K]([D][B][N])^{-1}\{\sigma\}_A, \tag{2}$$

where [D], [B], [N], [K], -1 and -T are stress-strain matrix, stress-displacemant matrix, interpolation function matrix, stiffness matrix, inverse and inverse-transpose of matrix respectively.

Equation (2), therefore, means that the element in which strain energy is minimum, that is, the element in which the stress is minimum especially all the elements being same in the volume, should be removed each step to pursue the optimal layout. This discussion is stated in reference 4) and 5) in detail.

2.2 Relaxation Method

In the following a formulation for relaxing the variational problem in the above mentioned procedure is presented. In the case of the layout problem, elements $E^{(i)}$ takes on the form,

$$E^{(i)} = \chi^{(i)}E_0, \tag{3}$$

where E_0 is Young's modulus for the material employed for continuum, $E^{(i)}$ is Young's modulus of element i, and $\chi^{(i)}$ is an indication function for the part $\Omega^{(m)}$ of design domain Ω that is occupied by material at step m.

For the sensitivity analysis as given

$$\chi^{(i)} = \begin{cases} 1, & if \quad i \in \Omega^{(m)}, \\ 0, & if \quad i \in \Omega - \Omega^{(m)}. \end{cases} \tag{4}$$

The proposed method gives non-zero values to $\chi^{(i)}$ rather than remove the elements in the part $\Omega - \Omega^{(m)}$ as follows.

Strain energy of a design domain $\Omega^{(m)}$,

$$V_\Omega^{(m)} = \int_{\Omega^{(m)}} \frac{\sigma_i^2 t_i}{E_i} dv \tag{5}$$

at step m, where i is element number, and σ_i, E_i and t_i are the stress, Young's modulus and the thickness of element i respectively.

Strain energy of substance in the expression of finite element method is

$$V_{\Omega}^{(m)} = \sum_{i=1}^{ne} \frac{\sigma_i^2 t_i}{E_i} A_i \tag{6}$$

in domain $\Omega^{(m)}$, where ne is the number of element in $\Omega^{(m)}$.

Average of strain energy in unit volume is

$$\bar{v}^{(m)} = \frac{\sum_{i=1}^{ne} \frac{\sigma_i^2 t_i}{E_i} A_i}{\sum_{i=1}^{ne} t_i A_i} \tag{7}$$

in $\Omega^{(m)}$.

Average of strain energy of element p is

$$\bar{V}_p^{(m)} = \bar{v}^{(m)} t_p A_p, \tag{8}$$

while strain energy of element p is

$$V_p^{(m)} = \frac{\sigma_p^{(m)2} t_p}{E_p^{(m)}} A_p. \tag{9}$$

Making Eq. (8) equal to Eq. (9),

$$E_p^{(m)} = \sigma_p^{(m)2} \frac{\sum_{i=1}^{ne} t_i A_i}{\sum_{i=1}^{ne} \frac{\sigma_i^2 t_i}{E_i} A_i}. \tag{10}$$

Especially the thickness and the area of every element being equal each other, Eq. (10) is obtained as follows

$$E_p^{(m)} = \sigma_p^{(m)2} \frac{ne}{\sum_{i=1}^{ne} \frac{\sigma_i^2}{E_i}}. \tag{11}$$

As Young's modulus for the element p in Ω-$\Omega^{(m)}$ $E_p^{(m)}$ may be adopted

$$\begin{aligned} &E_p^{(m)} \text{ is given by Eq.(11)} && \text{if } E_p^{(m)} \geq E_B, \\ &E_p^{(m)} = E_B && \text{if } E_p^{(m)} < E_B, \end{aligned} \tag{12}$$

where E_B is called background Young's modulus.

2.3 Scheme to Reduce Number of Iterative

(1) Classification of stress

The elements being grouped into several classes in order of the magnitude of their stresses, the frequency disitribution of the element is obtained from the classification of the stress. It is possible roughly to understand the pattern of the stress distribution. The total number of the class N_t is determined by

$$N_t = \min(N_r, N_c), \tag{13}$$

where N_r and N_c are the number of the rows and columns by which the domain is discretized into finite elements.

The proposed method determines the classes in which Young's moduli of the elements are modified as a following equation;

$$N_{c\beta} = \beta N_{c\mu}, \tag{14}$$

where $N_{C\beta}$, $N_{C\mu}$ and β are the class number up to which the moduli are modified, class number in which the mean value of the stress exists, and a coefficient by which the number of the elements is determined for the moduli to be modified. Fig. 1 shows an example of frequency-class histogram.

(2) Choice of element whose moduli are changed

Fuzzy c-means devised by Bezdek[11] is applied to choose which elements are remained as substances (that is, the modulied not modified) or which elements are modified about their Young's moduli by using membership function as shown in Fig. 2.

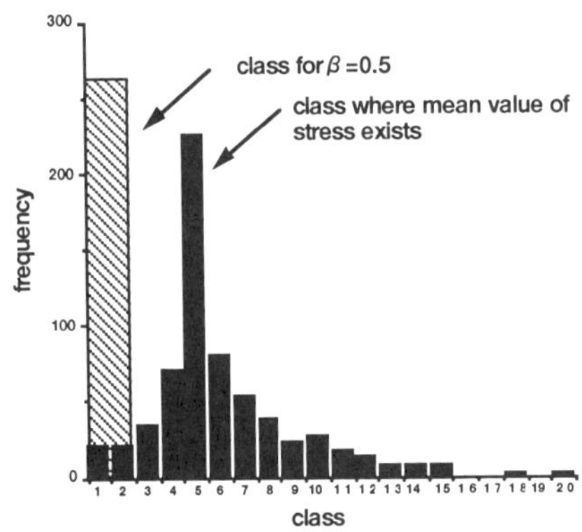

Fig. 1. Frequency-stress class histogram.

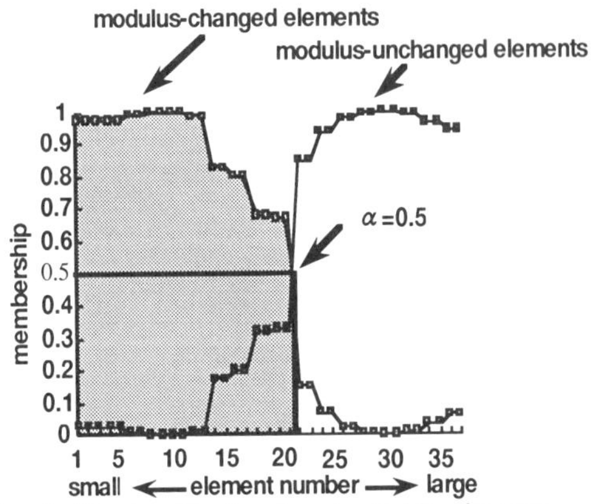

Fig. 2. Membership function to divide elements.

When the maximum stress goes up to the allowable stress with the optimization process made progress, the process is converted from fuzzy c-means method to the manner in which the element is treated one by one.

Fig. 3 shows the flow chart of the iteration process.

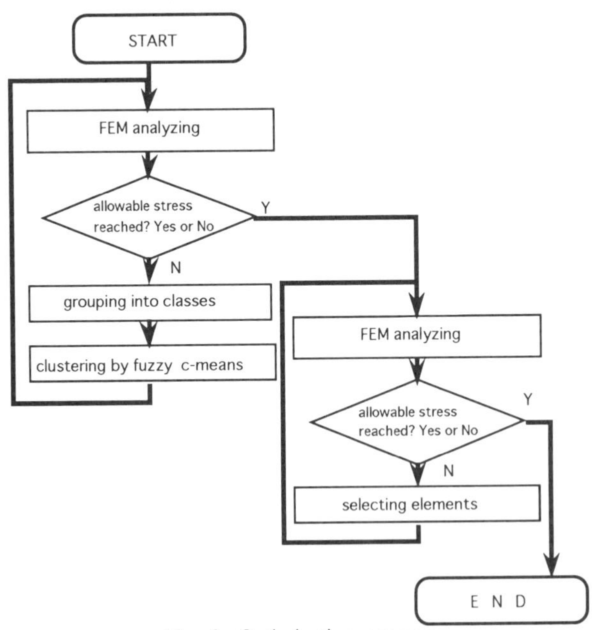

Fig. 3. Optimization process.

Ⅲ. VERIFICATION OF PRESENT METHOD

The authors examine the present method to solve the layout problem by simple problems. The problems are solved by use of 8 nodes isoparametric plane quadrilateral elements for the design domain occupied with material characterized by Young's modulus $E_0=2.1\times10^4$kgf/mm^2 and Poisson's ratio $\nu=0.3$, and by background Young's modulus $E_B=10.5$kgf/mm^2. The element stress is represented by the stress at the center of the element. All the stresses are treated as Von Mises' effective stresses.

3.1 Cantilever

(1) Relation between number of elements and coefficient β

The verificiation of the present method is to check whether the method works well, i.e., whether the configuration of the continuum obtained as the optimum layout converges to be unique as finite element meshes go to be refined, while other conditions are fixed. To understand this convergence property, the authors solve a problem in which a short cantiver is fixed at a end and subject to a vertical force 60kgf at the free end , as seen in Fig. 4 (a).

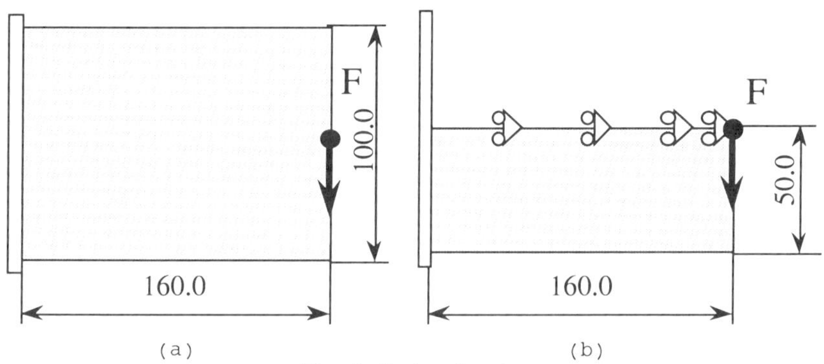

Fig. 4. Design domain.

Allowable stress is set at 12.5kgf/mm^2. The problem is solved for three types of mesh divisons 24×15, 32×20 and 50×13 discretized with equally sized finite elements covering a rectangular domain Ω (see Table 1), but only for the half of the design domain for the finest mesh divisions by use of the property of physical antisymmetry, shown in Fig. 4(b). Fuzzy c-means method is applied to the problem at the values of β, 0.3, 0.4 and 0.5.

Table 1. Finite Element Model.

model	number of elements	number of nodal point
A	24×15 =360	1159
B	32×20 =640	2025
C	50×13(26) =650(1300)	2077(4053)

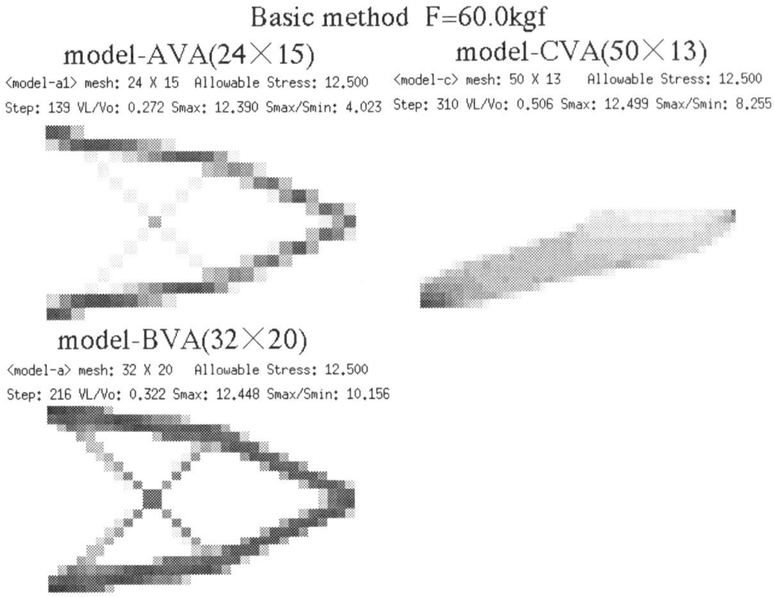

Fig. 5. Base method.

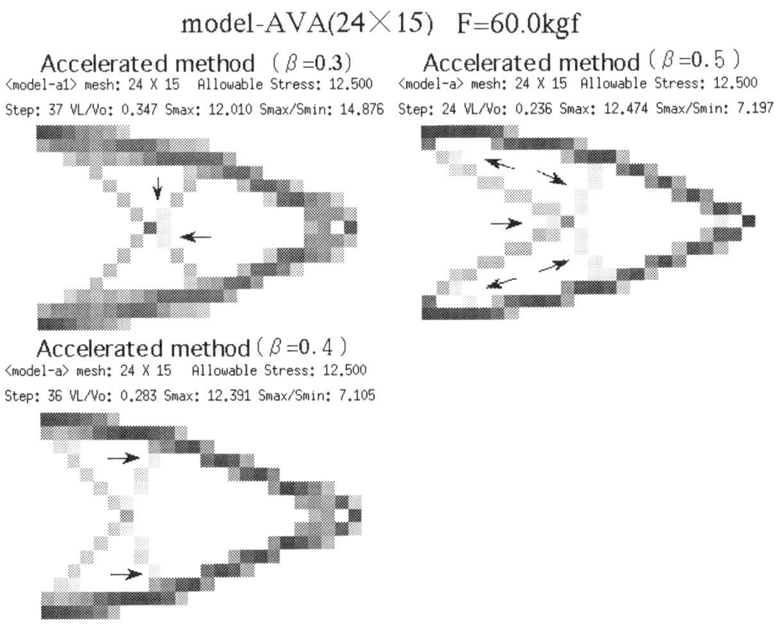

Fig. 6. Model-A.

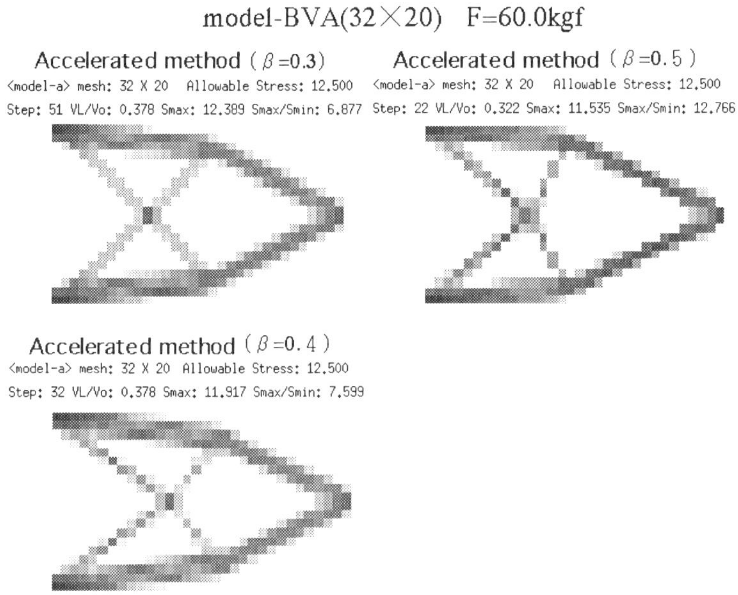

Fig. 7. Model-B.

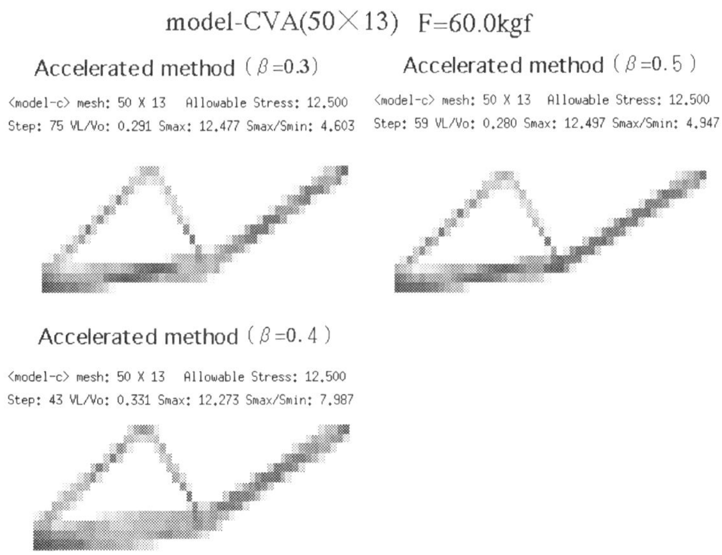

Fig. 8. Model-C.

Optimal configulations are obtained as shown in Figs. 6, 7 and 8.

Fig. 5 shows that a poor convergence only is obtained in the division 50 ×13 by the basic method which is not applied by fuzzy c-means and corresponds to one-by-one treatment by sensitivity analysis method. Figs. 6, 7 and 8 show that convergence is obtained for every β. The resultant configurations are very similar to ones obtained by homogenization method.[9] It is assumed that to modify the material properties of a large number of the element at the same time by fuzzy c-means method have a good effect on convergence. The elements pointed by arrows in Fig. 8 are that the values of their Young's moduli go down to lower levels at some step and up almost to the levels of E_0 at the final step. It is clear that the optimal configurations are convergent as the size of finite elements is reduced. Even every coarse mesh can provide a sufficient idea for the topology and shape of the optimal continumm, and a very finer meshes may be not necessary to solve.

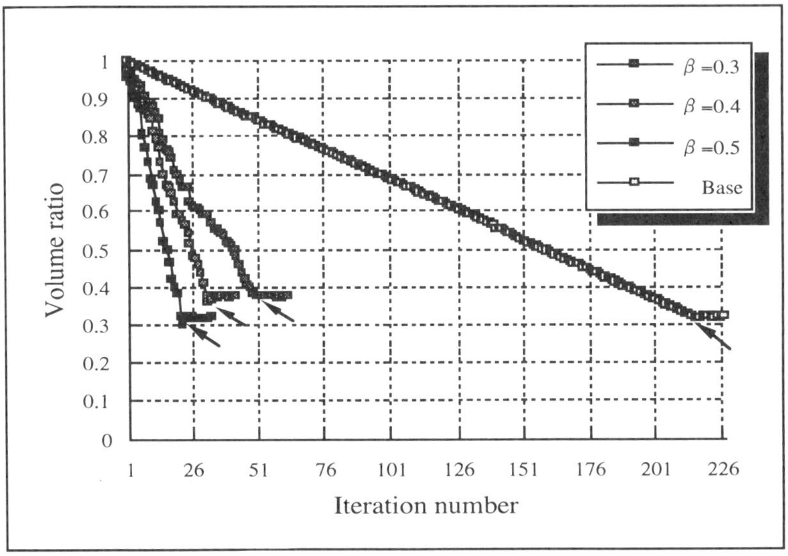

Fig. 9. Volume ratio and iteration number.

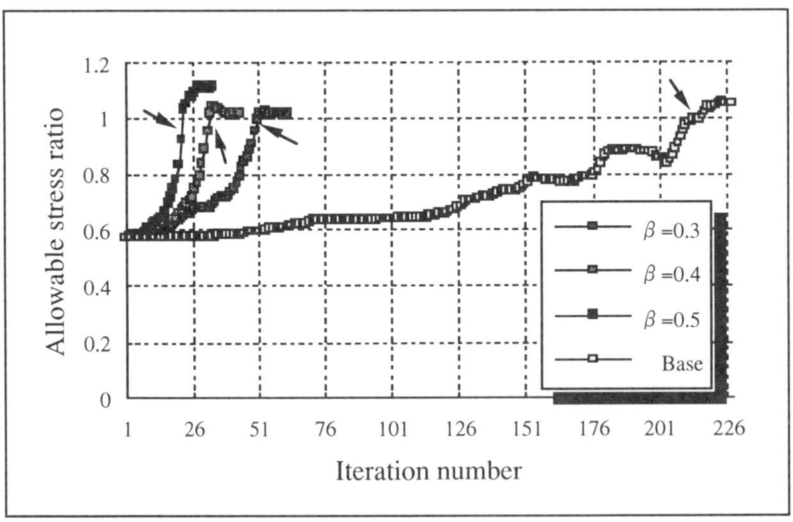

Fig. 10. Allowable stress ratio and iteration number.

Table 2, Figs. 9 and 10 shows the effectivity of fuzzy c-means application which contributes to a remarkable reduction of optimization iteration in the number. Also Table 2 shows the ratios V/V_0 of the final configuration to the initial domain.

Table 2. Choice of element coefficient β, volume ratio and stress ratio.

	iteration	allowable stress ratio (σ_{max}/σ_{al})	volume ratio (V/V_0)	miximum stress ratio ($\sigma_{max}/\sigma_{min}$)
Base	227	0.996	0.322	10.156
β=0.3	62	0.991	0.378	6.877
β=0.4	43	0.953	0.378	7.599
β=0.5	33	0.923	0.322	12.766

* : Force is applied with 60.0kgf.

** : Figues in table correspond to the values indicated by arrows at Figs. 9 and 10.

(2) Effect of force on configulation

The above mentioned problem is solved by varying vertical forces with other parameters fixed. The result obtained is shown in Fig. 11, and is that very truss-like frame structures are built for a light force. The amount of the substance is considerably smaller than that of the initial domain.

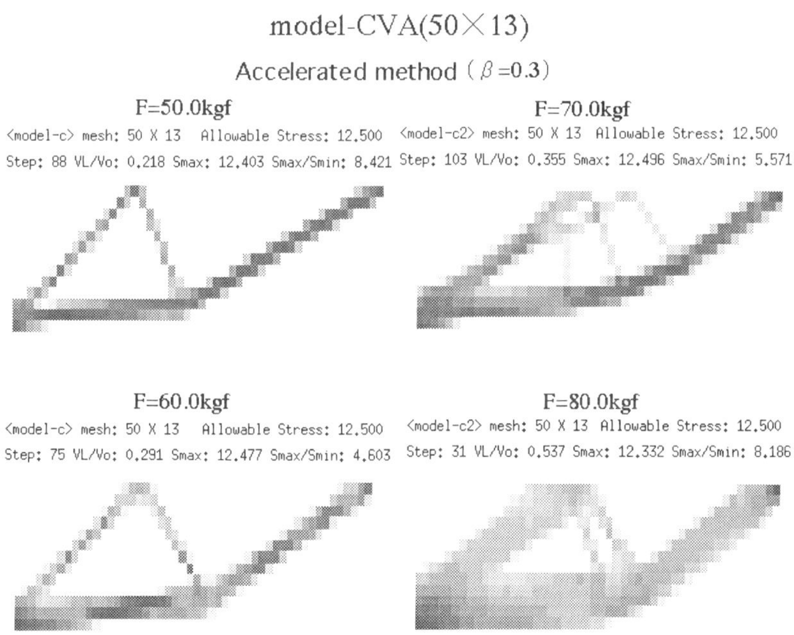

Fig. 11. Optimal configuration for different forces.

For heavier forces the amount of the substance becomes larger, the configuration diverges from truss-like structures. Curved frames appear, and more continuum-like shapes are formed. This tendency is similiar to what is obtained for various values of the volume by homogenization method.[9] The ratios V/V_0 versus forces are listed in Table 3.

Table 3. Effect of force, volume ratio and stress ratio.

	iteration	allowable stress ratio (σ_{max}/σ_{al})	volume ratio (V/V_0)	miximum stress ratio ($\sigma_{max}/\sigma_{min}$)
F=50.0kgf	167	0.992	0.218	6.832
F=60.0kgf	108	0.998	0.291	4.612
F=70.0kgf	284	0.999	0.355	4.806
F=80.0kgf	43	0.987	0.537	6.412

3.2 Michell truss

As the second example of the generalized layout problem, the authors' method is applied how fine it can reproduce Michell truss as shown by Fig. 12. Michell truss, which is solved analytically, is already known as the optimal truss structure for bending and as the most typical problem of structural layout. Fig. 13 shows the setting of nondesign domain as well as design domain divided into 64×20 isoparametric quadrilateral finite elements used in previous problem. The domains are supported rididly along the boundary of the circle. It is shown in Fig. 14 that the present method can be roughly reproduce Michell truss for the domain except the vicinity of the circular boundary, where the pattern of the substance is very different from Michell truss. It may be due to the fact that the circular boundary of the support too crudely approximates in staircase-like form. But as authors' aim set up to reproduce the truss on the whole of the domain by the present method, they are unconecerned with this fact.

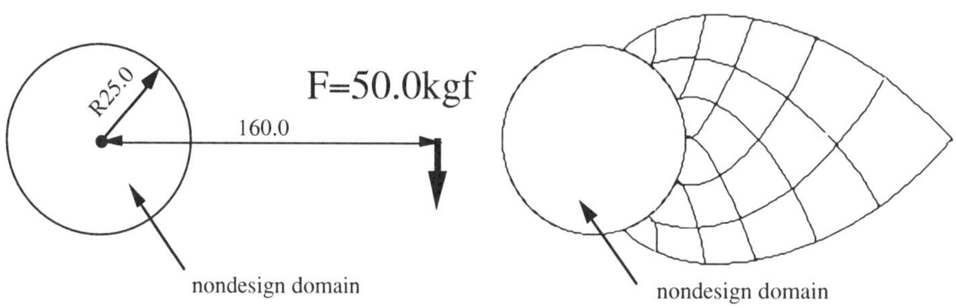

Fig. 12. Analytical solution of Michell truss.

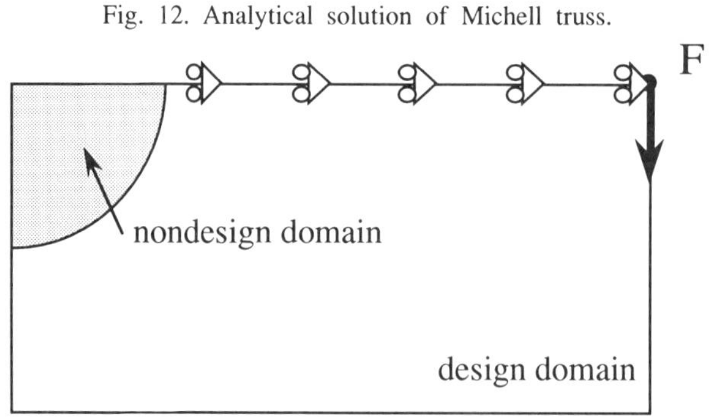

Fig. 13. Setting of the problem similar to Michell.

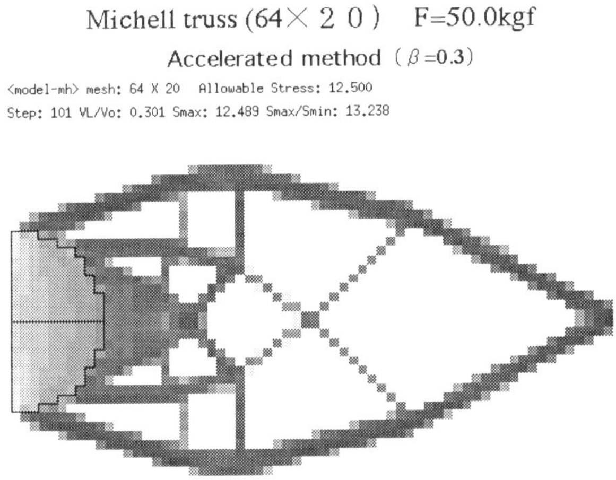

Fig. 14. Optimal configuration for Michell truss.

IV. CONCLUSION

As described above the proposed optimization method has been confirmed to be able to produce not only the optimal shape and topology but also the size of continuums, based on a kind of modified sensitivity analysis method. Also a fuzzy c-means method has been recognized to have a good influence to stablilize convergence to the optimal solution, as well as to be effective on the reduction of the number of iteration.

REFERENCES

1) Haug, E.J., Choi, K.K. and Komkov, V., Design Sensitivity Analysis of structural System, Academic Press, New York (1986).
2) Choi, K.K. and Haug, E.J., Shape design sensitivity analysis of elastic structues, J., structural Mech. vol. 11. No. 2 (1983), pp.231-269.
3) Choi, K.K. and Seong , K.K., A domain method for Shape design sensitivity analysis of built-up Structures, Comput. Meths. Appl. Mech. Engrg. vol. 57 (1986), pp. 1-16.
4) Okouchi, T., Aihara, A. and Fukutomi, Y., Optimum Shape Design of Mechanical Structures (sensitivity function to optimize two dimensional structures), Trans. of JASM. vol. 53. No. 496 (1987), pp. 2474-2480.
5) Okouchi, T., Ito, M. and Aihara, A., Lightweight Design of Mechanical Structure (Structural Weight Lightening by Creation of a Multiconnected Structure), Trans. of JASM. vol. 56. No. 525 (1990), pp. 1304-1309.
6) Cheng, K.T. and Olhoff, N., An Investigation Concerning Optimal Design of Solid Elastic Plates, Internat. J. Solids and Structures. vol. 16. No. 3 (1981), pp. 305-323.
7) Cheng, K.T. and Olhoff, N., Regularized Formulation for Optimal Design of Axisymmetric Plates, Internat. J. Solids and Structures. vol. 18. No. 2 (1982), pp. 153-170.
8) Bendsoe, M.P. and Kikuchi, N., Generating optimal topologies in structural design using a homogenization method, Comput. Meths. Appl. Mech. Engrg. vol. 71. No. 2 (1988), pp. 197-224.
9) Suzuki, K. and Kikuchi, N., A homogenization method for shape and topology optimization, Comput. Meths. Appl. Mech. Engrg. vol. 93 (1991), pp. 291-318.
10) Kohn, R.V. and Strang, G., Optimal design and relaxation of variational problems, Comm. Pure. Appl. Math. vol. 39 (1986), pp. 1-25 (Part I), pp. 139-182 (Part II), and pp. 353-337 (Part III).
11) Bezdek, J.C, Pattern Recognition with Fuzzy Objective Function Algorithms, Plenum, New York (1981).
12) Michell, A.G.M., The limits of economy of material in framed structures, Phil. Mag. vol. 6 (1904), pp. 589-597.

Minimal Surface Analysis by Using Finite Element Method

Kay ISHIHARA and Hiroshi OHMORI

Department of Architecture, Faculty of Engineering, Nagoya University, Nagoya

In the beginning of the design process of the membrane structures, it is recommended to adopt the minimal surface as the original design surface. In the numerical calculation for pursuing the objective minimal surface, we, however, can not always obtain the converged solution because of the strong nonlinearity involved in the problem. In the present paper, the new scheme for calculation of the minimal surface is proposed, by which we can obtain the solution of the minimal surface satisfying the arbitrary given subsidiary conditions.

I. INTRODUCTION

Membrane structures are essentially constructed without bending stress couples or shear stress resultants in the out-plane direction and, therefore, the realizable shape of the membrane structure cannot help being the physically restricted one. Consequently, for determination of the design shape for the membrane structure, the soap film surface is often adopted as the original surface for the initial stage of the structural design. Soap film, which is regarded as the surface in which membrane stresses in every direction are equal together, can be easily obtained by the experiment in the laboratory and often used for the confirmation of the general view of the curved surface subtended on the specified boundaries.

On the other hand, the soap film surface, or in another word, the equally tensioned curved surface, can be theoretically shown to coincide with the surface of which surface area is minimum among the other various curved surfaces spanned on the same boundaries. It is called minimal surface or minimal area surface and can be seen to coincide completely with the equally tensioned surface mentioned above.

Under these circumstances, the shape finding analysis for the objective surface being reasonably spanned on the specified geometrical and physical boundaries are conducted under two different inquiry conditions, one of which is demanding the objective surface to become the equally tensioned surface and another is based on the direct pursuit of the minimal surface. For the approach of pursuing the equally tensioned surface, the finite element method by using some kind of membrane elements has been usually adopted, where one can find the objective surface by enforcing the condition in the nonlinear process such as the membrane resultants in all direction become to be equal together at every point on the surface. This numerical technique is so simple that usual ready-made computer programs can be used only adding a little modification in the nonlinear iterative process.

This simplicity, however, cannot be overestimated taking the defect in consideration, that is, the convergence of the numerical solution can not always be obtained. Moreover, it can be said to be rather unusual that one can obtain the converged solution especially in the case of which boundary contains the sharp edges or some kind of displacement-free boundaries such as the edges reinforced by the flexible cables or the completely free boundaries.

On the other hand, the numerical approach, making use of the condition such that the objective surface coincides with the minimum area one, is usually based upon the variational principles, where the surface area functional represented by the whole coordinates or the displacements of the surface is adopted. The mathematical problem for the minimal surface subtended on the specified closed boundaries has been well-known as Plateau's problem. As in the other case of the variational problems, the closed solution of the present problem can not be obtained in almost cases except for the special cases such as the spherical surface spanned upon a circular boundary and, therefore, the numerical analysis based upon some kind of the direct method have to be carried out.

In order to obtain the minimum area surface subtended on the arbitrary given boundaries, Hinata [3] et al. have shown the effectiveness of the finite element method, and Suzuki [2] et al. have also shown the reduction technique of the degree-of-freedom of the problem and its validity from the aspect of the differential geometry of the curved surface.

In the present paper, the efficiency of the modified numerical approach for the minimum area surface for the objective solution is mainly investigated, where the subsidiary conditions of the original variational problem and the restraint condition on the coordinates of the node of the finite element are explicitly introduced into the process of the nonlinear calculation in order to improve the convergence characteristics and suppress the instabilities in the iterative process.

II. THEORY

The following variational functional which is used for the present minimal surface problem with the subsidiary condition of the specified internal volume is given as follows;

$$J(\boldsymbol{r},\lambda) = S(\boldsymbol{r}) + \lambda(V_0 - V(\boldsymbol{r})) \tag{1}$$

where $\boldsymbol{r}, S, V_0, V$ and λ represent the position vector, the surface area, the specified internal volume and Lagrange's multiplier (Gaussian mean curvature), respectively. Using the direction cosine components μ, ν and ξ, the coordinates of the i-th node can be represented as follows;

$$\begin{aligned} x_i &= \mu_i a_i + b_{i1} \\ y_i &= \nu_i a_i + b_{i2} \\ z_i &= \xi_i a_i + b_{i3} \end{aligned} \tag{2}$$

Eq.(2) implies that only the distance a_i between the present node and that of the next step in the numerical process has to be decided as shown in Fig.1, provided that the direction cosine vector is settled beforehand. Eq.(2) can be rewritten in the vector form as the following form.

$$\boldsymbol{r} = \boldsymbol{T}\boldsymbol{a} + \boldsymbol{b} \tag{3}$$

$$\begin{aligned} \boldsymbol{r} &= (\boldsymbol{r}_i\ \boldsymbol{r}_j\ \boldsymbol{r}_k)^T \\ &= (x_i\ y_i\ z_i\ x_j\ y_j\ z_j\ x_k\ y_k\ z_k)^T \end{aligned}$$

$$\boldsymbol{T} = \begin{bmatrix} \mu_i & \nu_i & \xi_i & 0 & 0 & 0 & 0 & 0 & 0 \\ 0 & 0 & 0 & \mu_j & \nu_j & \xi_j & 0 & 0 & 0 \\ 0 & 0 & 0 & 0 & 0 & 0 & \mu_k & \nu_k & \xi_k \end{bmatrix}^T$$

$$\boldsymbol{a} = (a_i\ a_j\ a_k)^T$$

$$\boldsymbol{b} = (b_{i1}\ b_{i2}\ b_{i3}\ b_{j1}\ b_{j2}\ b_{j3}\ b_{k1}\ b_{k2}\ b_{k3})^T$$

Substituting Eq.(3) into Eq.(1), we obtain the expression of the variational functional represented by the new variable a_i as follows;

$$J = J(\boldsymbol{r},\lambda) = J(\boldsymbol{r}(\boldsymbol{a}),\lambda) \tag{4}$$

Carrying out the operation of the first variation of the functional, we obtain the following equation.

$$\delta J = \frac{\partial J}{\partial \boldsymbol{a}}\delta \boldsymbol{a} + \frac{\partial J}{\partial \lambda}\delta\lambda = \frac{\partial J}{\partial \boldsymbol{r}}\frac{\partial \boldsymbol{r}}{\partial \boldsymbol{a}}\delta \boldsymbol{a} + \frac{\partial J}{\partial \lambda}\delta\lambda \tag{5}$$

Consequently, Euler's equations introduced by $\delta J = 0$ are as follows;

$$\frac{\partial J}{\partial \boldsymbol{r}}\frac{\partial \boldsymbol{r}}{\partial \boldsymbol{a}} = 0 \tag{6}$$

$$\frac{\partial J}{\partial \lambda} = 0 \tag{7}$$

In the determination of the direction cosine vector, we firstly adopted the normal vector to the present shape. Fig.2 shows the direction cosine vector which is chosen so that it coincides with the z-direction. Fig.3 depicts the direction cosine vector used in the numerical analysis, where the direction cosine vectors are chosen as the mean among the normal vectors of the contiguous elements of the referring node. As can be seen from these examples, the direction cosine vectors have to be chosen a priori rather than referring some theoretical observations.

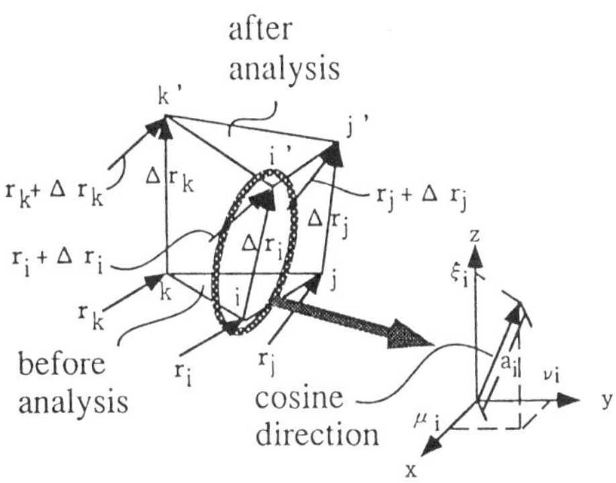

Figure 1. Direction cosine vector.

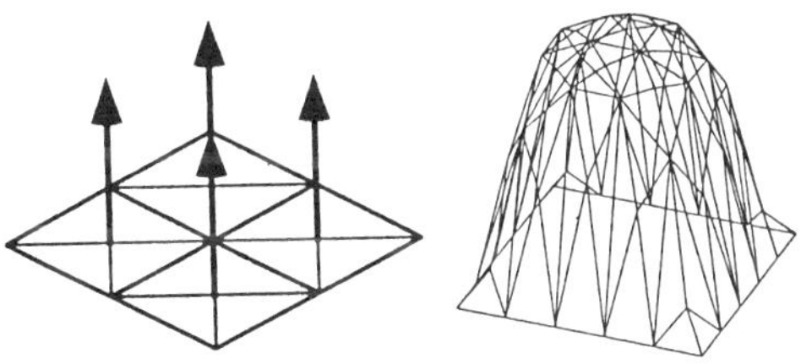

Figure 2. Direction cosine vector in z-direction and its example view.

$$\boldsymbol{r} = \frac{\boldsymbol{r}_1 + \boldsymbol{r}_2 + \boldsymbol{r}_3 + \boldsymbol{r}_4}{|\boldsymbol{r}_1 + \boldsymbol{r}_2 + \boldsymbol{r}_3 + \boldsymbol{r}_4|}$$

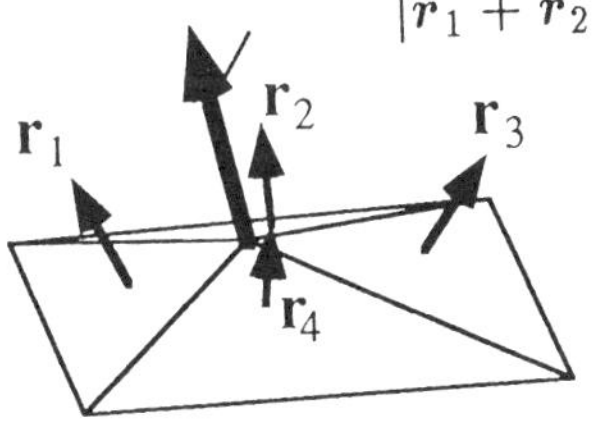

Figure 3. Normal vector.

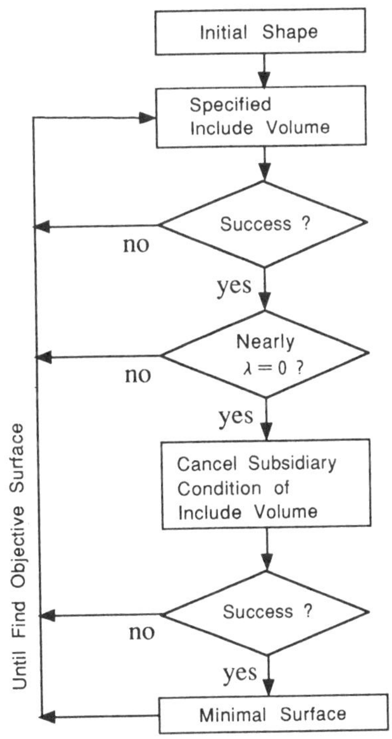

Figure 4. Flow chart of the calculating process.

III. NUMERICAL ANALYSIS

Global flow chart of the calculating process pursuing the final objective surface is as shown in Fig.4. As the example models for the numerical analysis, we adopt three different ones, that is, 1.Square boundary model, 2.Folded rectangle boundary model, and 3.Wiener frame model, respectively. Detailed explanation of each model is given below.

1. Square boundary model (Figs.5~7)

In the initial stage of the numerical analysis, it is necessary to set the initial configuration. In the present example, we can use the completely flat plane as the starting solution which is also the minimal surface shown in Fig.5, and in the continuous steps, the value of the included volume is forced to be increased little by little. Fig.6 and 7 shows the intermediate and the final state of the objective surface.

2. Folded rectangular boundary model (Figs.8~17)

The approximate configuration of the rectangular boundary model and the length of each part of the model are shown in Fig.8. Three different procedures of the numerical analysis are adopted as follows;

(a) Flat model (Figs.10~12)
Three different models are adopted as shown in Figs.10,11 and 12. Because the objective surface cat not be obtained directly, we adopt the process of volume control as shown in Fig.9.

(b) Quarter model (Figs.13 and 14)
Because of characteristic of symmetricity, we can solve only a quarter part of the original model and can obtain the objective minimal surface directly (Fig.13). The total shape, which is obtained by gathering up the solutions, is shown in Fig.14.

(c) Boundary folding model (Fig.15~17)
Folded frame in 0°, 45°, and 90° are shown in Figs.15,16 and 17.

In the present model, it can be found that the value D/H in Fig.8 depends on the intermediate procedure of the numerical analysis although the final shape has to coincide with each other.

3. Wiener frame (Figs.18~23)

The minimal surface problem of Wiener frame, which consists of two parallel circles connected with two parallel lines, has two stable solutions together with one unstable solution. In the present analysis, we adopt the direction cosine vector which is produced being fixed in the global coordinates as shown in Fig.19 instead of the normal vector mentioned above. As a result, two stable solutions and one unstable solution are confirmed to exist as mentioned above (Figs.20,21 and 22). Relation between the surface area and the internal volume and relation between the surface area and λ(Gaussian mean curvature) are shown in Figs.23, respectively. We can confirm that the value of area has the minimal and maximal value, when $\lambda=0$.

Additionally, Fig.24 shows the results of the HP shape curved surface, where both the initial configuration and the final HP surface are demonstratively shown, respectively. In the present analysis, to solve the minimal surface, we adopt the process controlled by volume in the same manner shown as Fig.9.

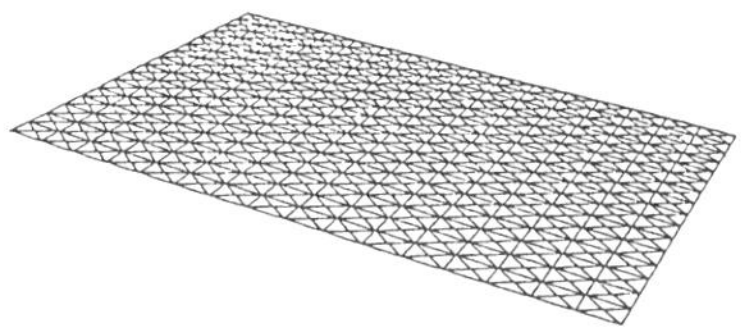

Figure 5. Initial shape (a side = 16.0).

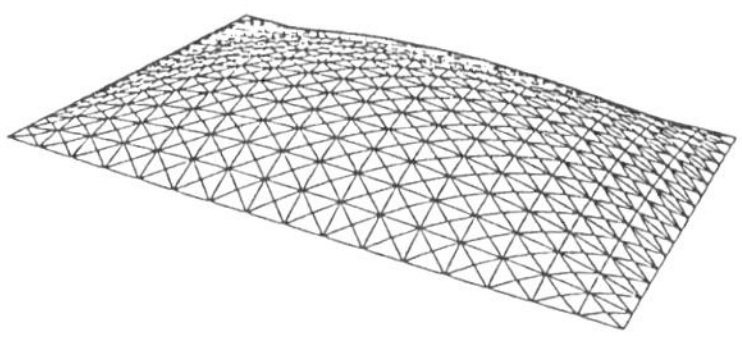

Figure 6. Internal Vol. = 2000.0 .

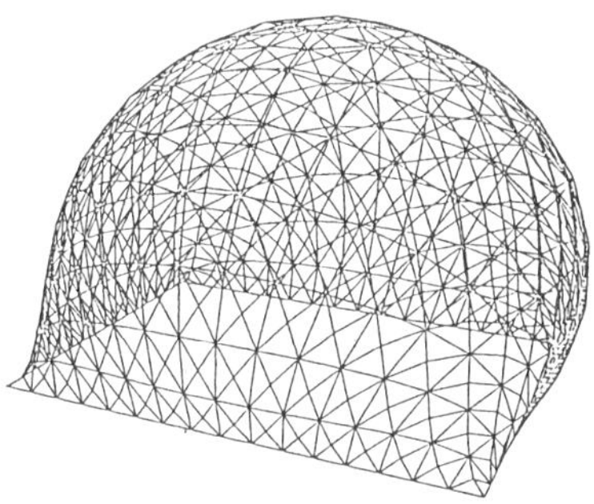

Figure 7. Internal Vol. = 20000.0 .

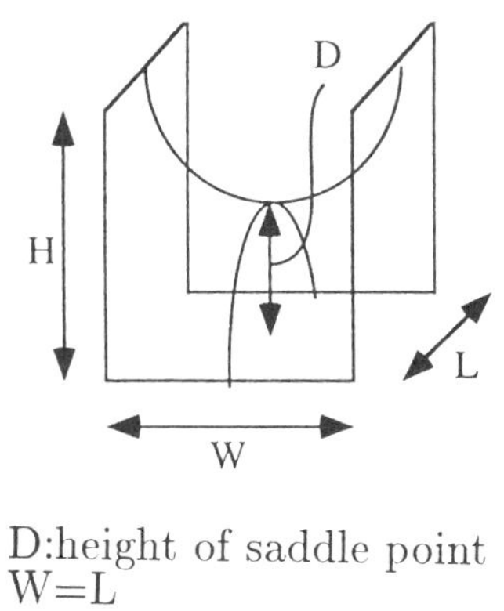

Figure 8. Details of folded rectangular boundary model.

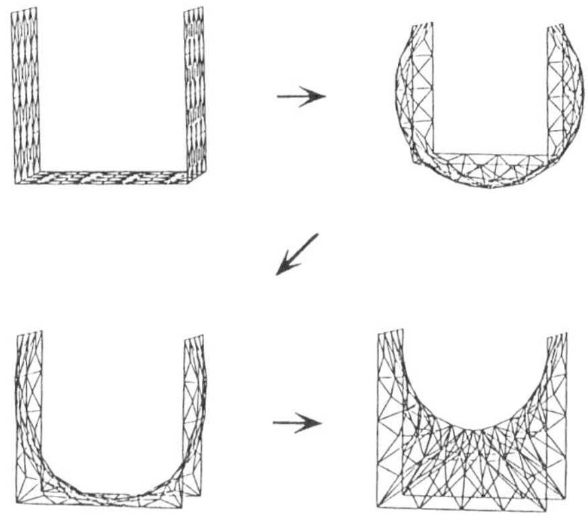

Figure 9. Process for the minimal surface.

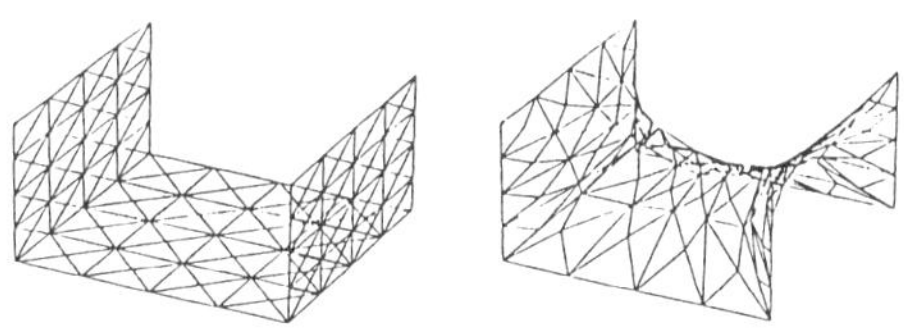

Figure 10. Initial shape and minimal surface ($\frac{H}{L} = 0.5, \frac{D}{H} = 0.5$).

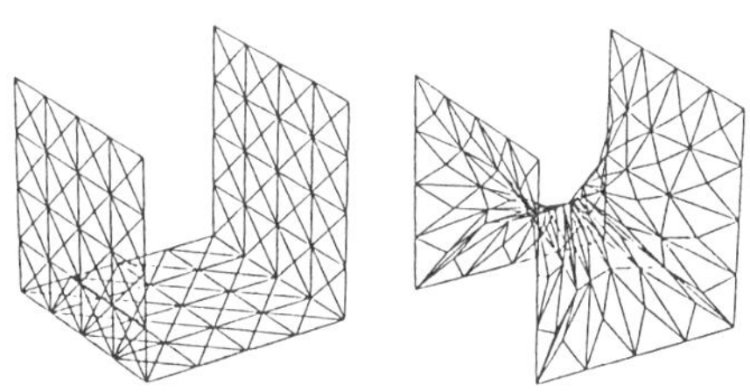

Figure 11. Initial shape and minimal surface ($\frac{H}{L} = 1.0, \frac{D}{H} = 0.42 \sim 0.47$).

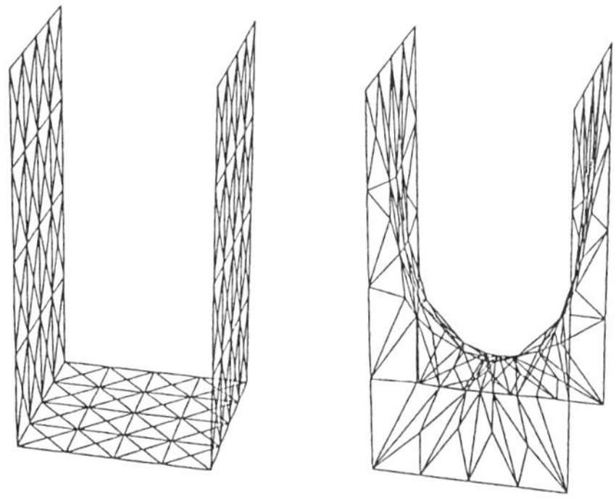

Figure 12. Initial shape and minimal surface ($\frac{H}{L} = 2.0, \frac{D}{H} = 0.23 \sim 0.25$).

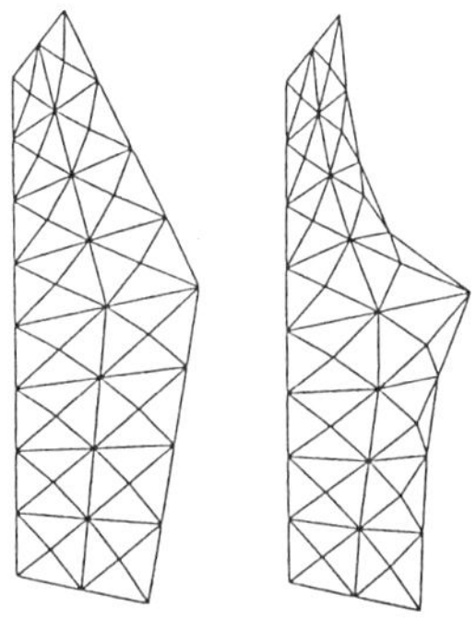

Figure 13. Quarter model.

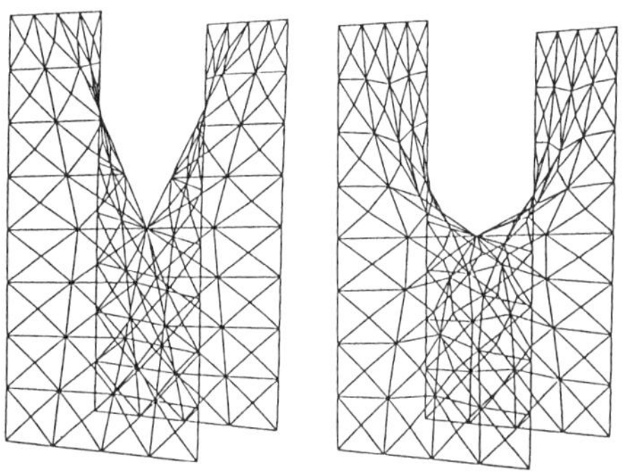

Figure 14. Total view of quarter model.

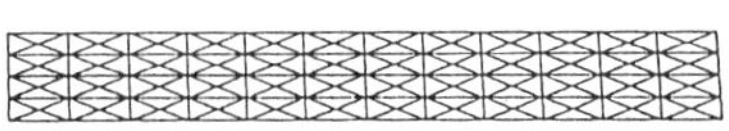

Figure 15. Boundary folding model ($\theta = 0°$).

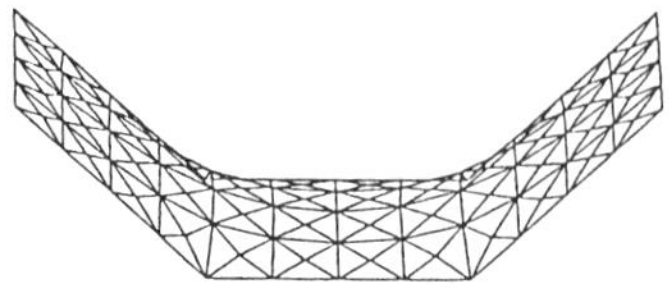

Figure 16. Boundary folding model ($\theta = 45°$).

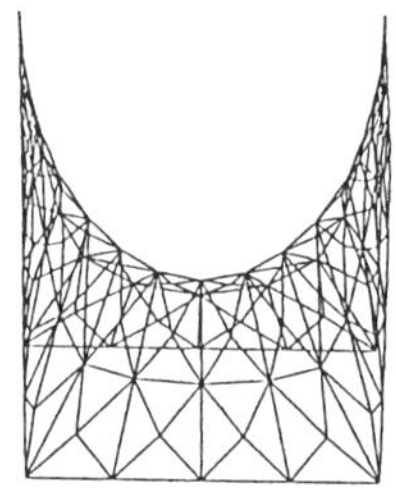

Figure 17. Boundary folding model ($\theta = 90°$).

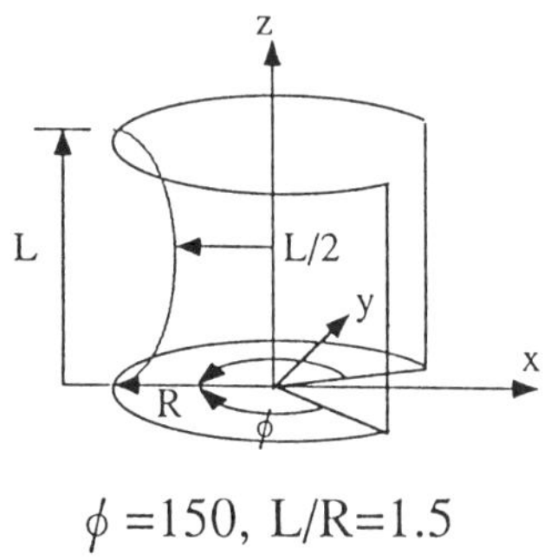

Figure 18. Wiener frame.

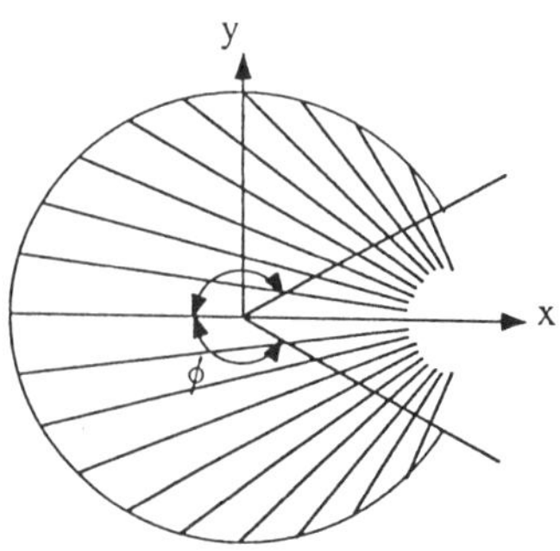

Figure 19. Direction cosine vectors in Wiener frame.

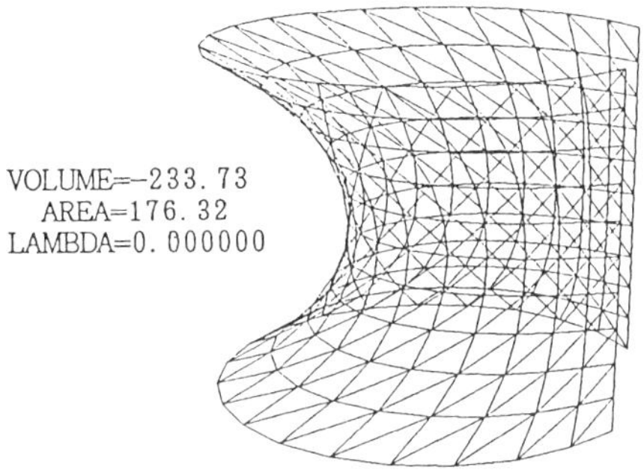

Figure 20. Stable solution (1).

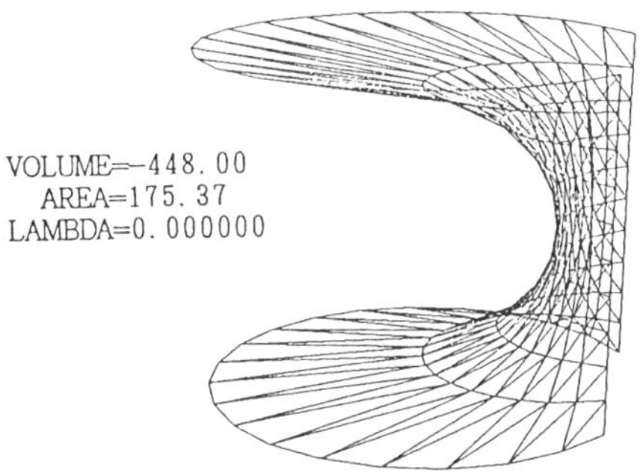

Figure 21. Stable solution (2).

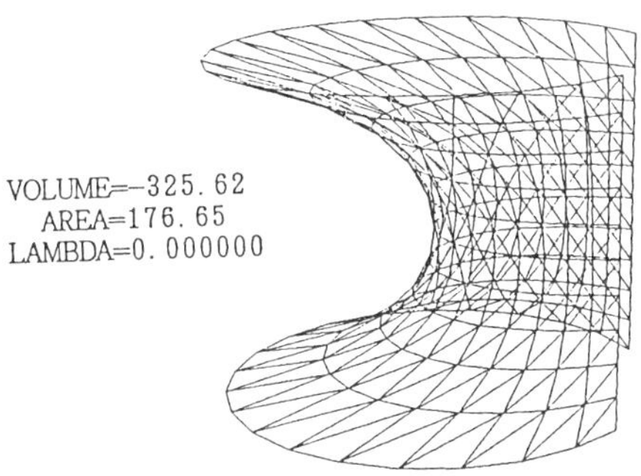

Figure 22. Unstable solution.

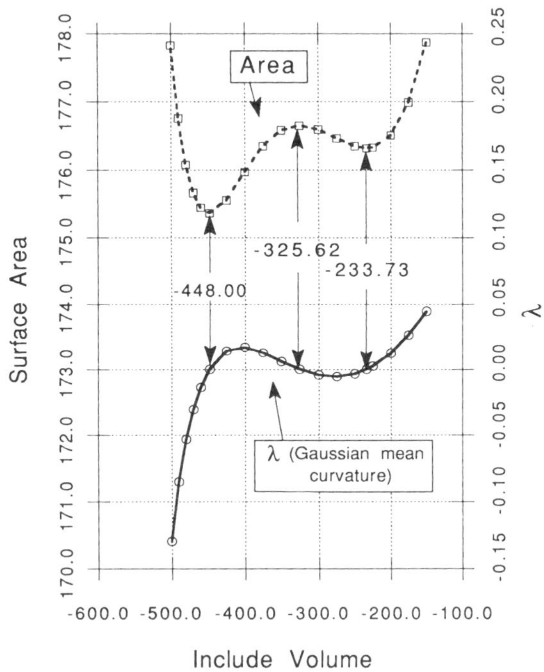

Figure 23. Correspondence between extremal value of area and Gaussian mean curvature.

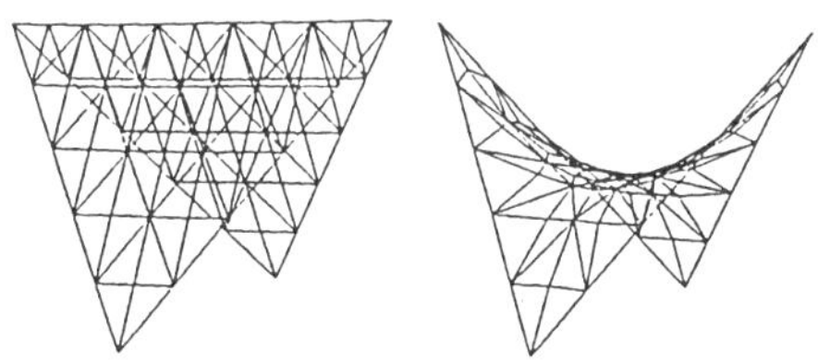

Figure 24. HP as a initial surface.

IV. MINIMUM LENGTH PROBLEM WITH SUBSIDIARY CONDITION ON EACH FINITE ELEMENT

As mentioned above, in the reduction of the degree-of-freedom of the problem, the intermediate process of the calculation has a significant influence on the final shape. In order to suppress the discrepancy, we substitute the condition of the reduction of degree-of-freedom for the newly introduced subsidiary condition by which the length or the area of each finite element comes to be equal together. As the clue, let us deal with the two-dimensional minimal length problem, where each finite element is imposed to be equal in length and area. But, we use discrete expression in following equations, in order to get element positive.

The variational functional for the segment link model shown Fig.25 is given as follows;

$$I(x_i, y_i, \lambda_s, \lambda_t) = \sum_e l_e + \lambda_s(S_0 - \sum_e S_e) + \lambda_t \sum_e (l_e - \frac{1}{m}\sum_e l_e)^2 \tag{8}$$

where l_e and S_e are shown as follows;

$$l_e = \sqrt{(x_2^e - x_1^e)^2 + (y_2^e - y_1^e)^2}$$
$$S_e = \frac{1}{2}(x_2^e - x_1^e)(y_1^e + y_2^e)$$

and $x_i, y_i, l_e, S_0, S_e, \lambda_s, \lambda_t$ and m represent x-coordinate, y-coordinate, length of an element, specified area, area of an element, Lagrange's multiplier for specified area, Lagrange's multiplier for equalizing element length and total number of elements.

Carrying out the operation of the first variation of the functional in every element, we obtain the following equation:

$$\delta x_1^e : \frac{1}{l_e}x_1^e - \frac{1}{l_e}x_2^e + \frac{y_1^e + y_2^e}{2}\lambda_s - \frac{2L_e(x_2^e - x_1^e)}{l_e}\lambda_t = 0 \tag{9}$$
$$\delta y_1^e : \frac{1}{l_e}y_1^e - \frac{1}{l_e}y_2^e + \frac{x_2^e - x_1^e}{2}\lambda_s - \frac{2L_e(y_2^e - y_1^e)}{l_e}\lambda_t = 0 \tag{10}$$
$$\delta x_2^e : -\frac{1}{l_e}x_1^e + \frac{1}{l_e}x_2^e - \frac{y_1^e + y_2^e}{2}\lambda_s + \frac{2L_e(x_2^e - x_1^e)}{l_e}\lambda_t = 0 \tag{11}$$
$$\delta y_2^e : -\frac{1}{l_e}y_1^e + \frac{1}{l_e}y_2^e - \frac{x_2^e - x_1^e}{2}\lambda_s + \frac{2L_e(y_2^e - y_1^e)}{l_e}\lambda_t = 0 \tag{12}$$
$$\delta \lambda_s : S_0 - \sum_e S_e = 0 \tag{13}$$
$$\delta \lambda_t : \sum_e (l_e - \frac{1}{m}\sum_e l_e)^2 = 0 \tag{14}$$

where

$$L_e = l_e - \frac{1}{m}\sum_e l_e$$

(x^e, y^e : Express x and y on an element)

The results of numerical calculation by means of Newton-Raphson method are shown in Figs.26 and 27. From Figs.28 and 29, it can be certified that we can obtain the final form even when the initial shape is given roughly as shown in the figure. Observing these results, it can be said that the newly imposed subsidiary condition on each element has the significant efficiency on improvement of the convergence characteristics of the numerical calculation of the present problem.

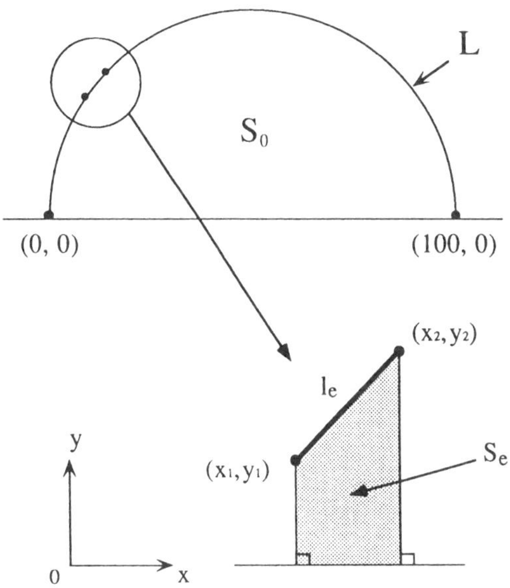

Figure 25. Segment link model.

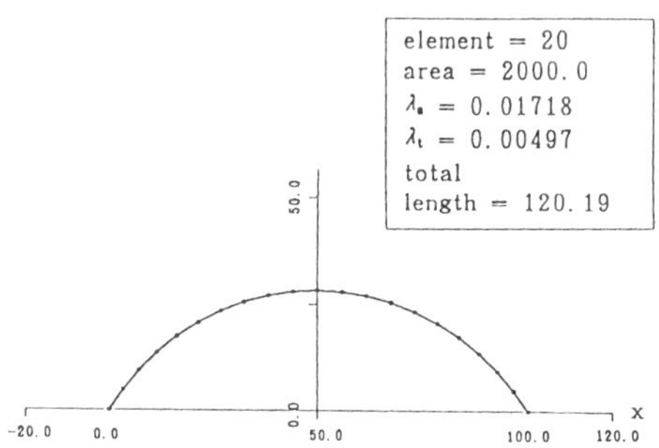

Figure 26. Minimum length problem (1).

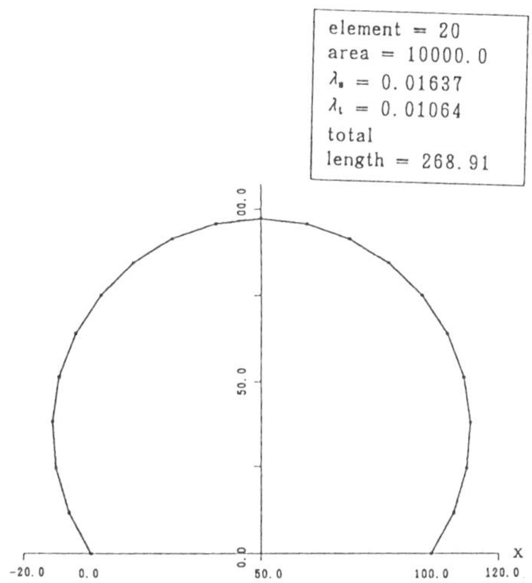

Figure 27. Minimum length problem (2).

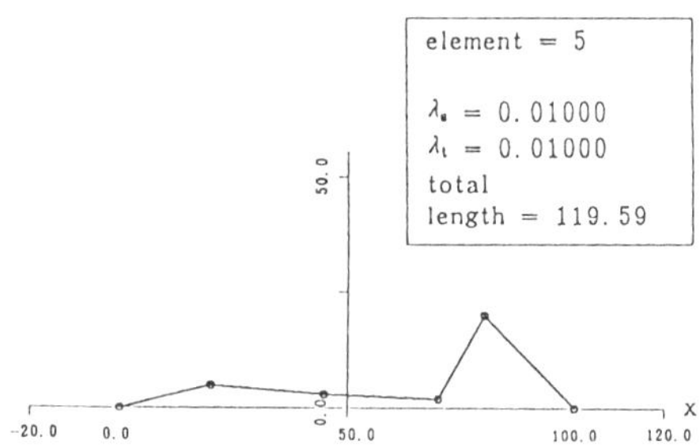

Figure 28. Initial shape for minimum length problem.

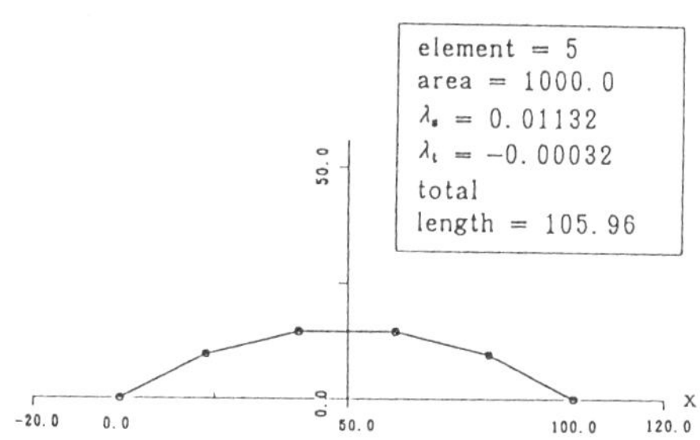

Figure 29. Find shape for minimum length problem.

V. CONCLUSION

In the present analysis, it is shown that we can easily obtain the objective surface, provided that the direction cosine is determined beforehand and only the distance of the movement of each node is taken into account as the unknowns. As shown in the numerical example through the folded rectangular boundary model to the square boundary one, the final shape of the solution of the minimal surface is clearly shown to depend on the intermediate process of the calculation, that is, the different procedures of calculation result in the different final configuration. This unreasonable result is resulted from the different restrictions imposed on the problem, which have to be released in the final stage of the calculation.

Consequently, the effect of the reduction of the degree-of-freedom of the present problem by introduction of the direction cosine concept is confirmed to be very effective for the numerical procedure of the minimal surface problem, provided that the intermediate procedure is carried out correctly, paying attention to the mesh division and to how to take the direction cosine vectors.

Additionally, the newly introduced subsidiary condition imposed on each element by which the length or the area of finite element comes to be equal together is shown to have the significant efficiency on the convergence characteristics of the numerical procedure. Investigation in the three dimensional problem, however, is left for the future research.

References

[1] Ohmori,H., Hagiwara, N.,Matsui,T. and Matsuoka,O., Numerical Analysis of Minimal Surface by Finite Element Method, Journal of Membrane Struct. Japan Institute of Membrane Structure, 1988.

[2] Suzuki,T. and Hangai,Y., Finite Element Analysis of Minimal Surface Based on The Reduction of Variables, Journal of Struct. Constr. Engng, Architectural Institute of Japan, 1991.

[3] Hinata,M.,Simasaki,M. and Kiyono,T., Numerical Solution of Plateau's Problem by Finite Element Method, Mathematics of Computation, Vol.28, 1974.

[4] Courant,R., Dirichlet's Principle, Conformal Mapping and Minimal Surface, Interscience, 1950.

II
MICROMECHANICS AND CONTACT MECHANICS

Analysis of Particle-Reinforced Composites by Means of the Equivalent Inclusion Method

Shigetoshi ARAKI*, Kenji SAITO*, and Megumu SUZUKI**

* *Department of Mechanical and System Engineering, Kyoto Institute of Technology, Matsugasaki, Sakyo-ku, Kyoto*
** *Department of Mechanical Engineering, Kanazawa Institute of Technology Ogigaoka-71, Nonoichi, Kanazawa, Ishikawa*

To study the reinforcement mechanism of particle-reinforced composites, the analysis of such composites is performed by means of the equivalent inclusion method of micromechanics. The analytical expressions to the stress in a matrix and the stress intensity factor for a matrix crack are derived by introducing the periodic distribution of the particles to the analytical model. Both the stress in a matrix and the stress intensity factor for a matrix crack increase with the decrease of the periodicity of the arrangement of particles, i.e. the volume fraction of the particles whose Young's modulus is higher than that of a matrix. It can be concluded that the increase of the volume fraction of particles may not contribute to toughen the particle-reinforced composites in contrast with the fiber-reinforced composites.

I. INTRODUCTION

The strength of the composites is said to be considerably dependent on the condition of the load-transmission from a matrix to the inclusions being introduced for reinforcing the composites. For instance, in the research of discontinuous fiber-reinforced composites, it was experimentally cleared to the composites containing only one fiber that the shear stress produced at the interface between a matrix and a fiber play an important role to the reinforcement of the composites[1).

Furthermore, it is necessary to consider the interaction effects between inclusions in addition to the above phenomenon, since the real composites contain many inclusions. For the discontinuous fiber-reinforced composites, the stress fields in the matrix region surrounded by fibers were studied by using FEM, finite difference method, and photoelasticity method, etc., for the case that seven or more fibers were in a matrix[2),3)].

On the other hand, there is a particle-reinforced composite material like a spheroidal graphite cast iron among the composites occupied an attention recently. The mechanism of the reinforcement of the particle-reinforced composites is said to be different from that of the fiber-reinforced composites. However, it seems that the detailed research to the reinforcement mechanism has not been performed to the case that many particles are dispersed in a matrix. Micromechanics is one of the powerful methods to analyze the stress distribution and the interaction effects, etc., i.e. reinforcement mechanism of such composites, taking into consideration of the effects of the microstructure[4)].

In the present research, the particle-reinforced composite is modeled by the periodic distribution of particles in a matrix, and by applying the equivalent inclusion method of micromechanics to this model, the stress distribution in a matrix is calculated. The interaction due to many particles is supposed to be comprised in the eigenstress fields caused by the eigenstrain which substitutes the particle based on the concept of the equivalent inclusion method. Furthermore, the stress intensity factor of a crack in the matrix is also evaluated by using the stress distribution obtained. The mechanism of the reinforcement of particle-reinforced composites is discussed from the standpoint of the stress distribution occurred in a matrix.

II. MODELING AND ANALYSIS

2.1 Modeling of Particles in a Matrix

The geometric modeling of the distribution of particles in a matrix is shown in Fig.1. The cross-hatched circles with radius d in the figure correspond to the circular inclusions. The distribution of the inclusions is assumed to be periodic. The periodicity of the arrangement of the particles is a and b to the x_1 and the x_2 directions, respectively. Furthermore, it is assumed that the present model is in the state of plane strain, and the direction of

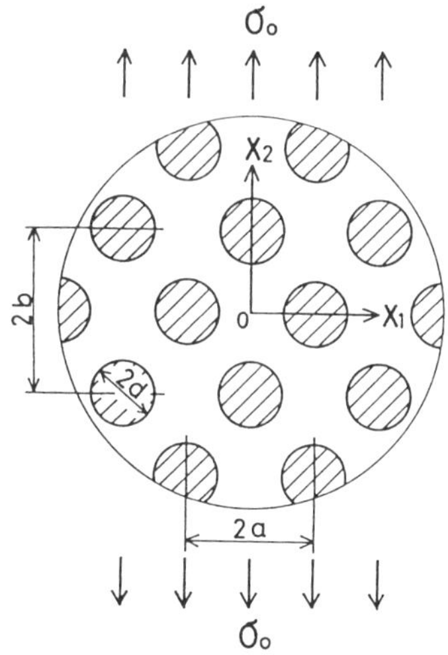

Fig.1. Modeling of particles in a matrix.

the applied stress σ_0 is parallel to the x_2 axis.

2.2 Equivalent Eigenstrain for a Particle

Under the uniform external stress, the generalized equivalent expression in a particle is

$$C^{p}_{ijkl}(\varepsilon^{o}_{kl}+S_{klmn}\varepsilon^{*p}_{mn})=C_{ijkl}(\varepsilon^{o}_{kl}+S_{klmn}\varepsilon^{*p}_{mn}-\varepsilon^{*p}_{kl}) \quad , \tag{1}$$

where

C_{ijkl}, C^{p}_{ijkl} : elastic constants of a matrix and a particle, respectively,
ε^{o}_{ij} : applied strain,
S_{ijkl} : *Eshelby*'s tensor for a particle,

and ε^{*p}_{ij} is the equivalent eigenstrain introduced in order to simulate the difference of the elastic constant of the particle from that of a matrix.

The relations between applied stress σ_0 and ε^{o}_{ij} are

$$\varepsilon^{o}_{11}=(1-\nu^2)\frac{\sigma_o}{E} \quad , \quad \varepsilon^{o}_{22}=-\nu(1+\nu)\frac{\sigma_o}{E} \quad , \tag{2}$$

where E and ν are *Young*'s modulus and *Poisson*'s ratio of a matrix, respectively. Thus, by substituting Eq. (2) into Eq. (1), the equivalent eigenstrains for a particle are obtained as follows:

$$\varepsilon^{*p}_{11}=(B_1-B_2)\frac{(1-\nu)}{2\mu}\sigma_o \quad , \quad \varepsilon^{*p}_{22}=(B_1+B_2)\frac{(1-\nu)}{2\mu}\sigma_o \quad , \tag{3}$$

where

$$B_1=\frac{(2\nu-1)\left(1+\dfrac{\Delta\lambda}{\Delta\mu}\right)}{\left(1+\dfrac{\Delta\lambda}{\Delta\mu}\right)+2(1-\nu)\dfrac{(\lambda+\mu)}{\Delta\lambda}} \quad , \quad B_2=\frac{2\Delta\mu}{(3-4\nu)\Delta\mu+4(1-\nu)\mu} \quad , \tag{4}$$

$$\Delta\nu=\nu^{p}-\nu \quad , \quad \Delta\mu=\mu^{p}-\mu \quad , \quad \Delta\lambda=\lambda^{p}-\lambda \quad ,$$

in which λ and μ are *Lame*'s constants of a matrix and the superscript p implies a particle.

2.3 Periodic Distribution of Eigenstrain

All particles are replaced by the equivalent eigenstrain given by Eq. (3). Then, expanding the periodic distribution of the equivalent eigenstrain into the double complex *Fourier* series, we have

$$\varepsilon^*_{ij}(x_1,x_2)=\sum_{m=-\infty}^{\infty}\sum_{n=-\infty}^{\infty}\bar{\varepsilon}^*_{ij}(\xi_1,\xi_2)\exp\{i(\xi_1x_1+\xi_2x_2)\}\quad, \tag{5-a}$$

$$\bar{\varepsilon}^*_{ij}(\xi_1,\xi_2)=\varepsilon^{*p}_{ij}\frac{d}{2b}\frac{\{(-1)^m+(-1)^n\}}{\sqrt{m^2+\left(\frac{a}{b}n\right)^2}}J_1\left(\frac{\pi d}{a}\sqrt{m^2+\left(\frac{a}{b}n\right)^2}\right)\quad, \tag{5-b}$$

where

$$\xi_1=\frac{\pi m}{a}\quad,\quad \xi_2=\frac{\pi n}{b}\quad, \tag{6}$$

and $J_1(x)$ is the *Bessel* function of the first kind of order one.
Furthermore, rewriting Eq.(5) in the form of the double *Fourier* cosine series, we have

$$\begin{aligned}\varepsilon^*_{ij}(x_1,x_2)=&\ \varepsilon^{*p}_{ij}\frac{d}{b}G_1(x_1)+\varepsilon^{*p}_{ij}\frac{d}{a}G_2(x_2)\\&+\varepsilon^{*p}_{ij}\frac{2d}{b}\sum_{m=1}^{\infty}\sum_{n=1}^{\infty}\frac{\{(-1)^m+(-1)^n\}}{\sqrt{m^2+\left(\frac{a}{b}n\right)^2}}J_1\left(\frac{\pi d}{a}\sqrt{m^2+\left(\frac{a}{b}n\right)^2}\right)\\&\times\cos\left(\frac{\pi m x_1}{a}\right)\cos\left(\frac{\pi n x_2}{b}\right)\quad,\end{aligned} \tag{7}$$

where

$$G_1(x_1)=\begin{cases}\frac{1}{d}\sqrt{d^2-(x_1-l_1a)^2}\ , & l_1a-d\le x_1\le l_1a+d\quad,\\ 0\ , & l_1a+d<x_1<(l_1+1)a-d\quad,\end{cases} \tag{8-a}$$

$$G_2(x_2)=\begin{cases}\frac{1}{d}\sqrt{d^2-(x_2-l_2b)^2}\ , & l_2b-d\le x_2\le l_2b+d\quad,\\ 0\ , & l_2b+d<x_2<(l_2+1)b-d\quad.\end{cases} \tag{8-b}$$

2.4 Stress Fields in a Matrix

When the eigenstrain is given in the form of the complex *Fourier* series,

$$\varepsilon^*_{ij}(x)=\sum\bar{\varepsilon}^*_{ij}(\xi)\exp(i\xi\cdot x)\quad, \tag{9}$$

the corresponding eigenstress fields are derived by *Mura*[5] as follows:

$$\sigma_{ij}(x)=C_{ijkl}\left\{\sum C_{pqmn}\bar{\varepsilon}^*_{mn}(\xi)\xi_q\xi_lN_{kp}(\xi)D^{-1}(\xi)\exp(i\xi\cdot x)-\varepsilon^*_{kl}(x)\right\}\quad, \tag{10}$$

where, for the case of the isotropic materials,

$$D(\xi)=\mu^2(\lambda+2\mu)\xi^6\ ,\quad N_{ij}(\xi)=\mu\xi^2\{(\lambda+2\mu)\delta_{ij}\xi^2-(\lambda+\mu)\xi_i\xi_j\}\ ,\quad \xi^2=\xi_k\xi_k\ , \tag{11-a}$$

$$\begin{aligned}C_{ijkl}C_{pqmn}\xi_q\xi_lN_{kp}(\xi)D^{-1}(\xi)=&(\lambda+2\mu)^{-1}\xi^{-4}\{\lambda^2\delta_{ij}\delta_{mn}\xi^4+2\lambda\mu\delta_{mn}\xi_i\xi_j\xi^2\\&+2\lambda\mu\delta_{ij}\xi_m\xi_n\xi^2+\mu(\lambda+2\mu)(\delta_{im}\xi_j\xi_n+\delta_{jm}\xi_i\xi_n\\&+\delta_{in}\xi_j\xi_m+\delta_{jn}\xi_i\xi_m)\xi^2-4\mu(\lambda+\mu)\xi_i\xi_j\xi_m\xi_n\}\end{aligned} \tag{11-b}$$

From Eqs. (3) and (5), non-zero components of the *Fourier* coefficient in Eq. (10) are $\bar{\varepsilon}^*_{11}$ and $\bar{\varepsilon}^*_{22}$. Thus, substituting Eq. (11) into Eq. (10), we have finally

$$\sigma_{11} = -\frac{2\mu}{(1-\nu)} \sum_{m=-\infty}^{\infty} \sum_{n=-\infty}^{\infty} \frac{\xi_2^2}{\xi^4} \left\{ \xi_2^2 \bar{\varepsilon}^*_{11} + \xi_1^2 \bar{\varepsilon}^*_{22} \right\} \exp(i\xi \cdot \boldsymbol{x}) \ , \tag{12-a}$$

$$\sigma_{22} = -\frac{2\mu}{(1-\nu)} \sum_{m=-\infty}^{\infty} \sum_{n=-\infty}^{\infty} \frac{\xi_1^2}{\xi^4} \left\{ \xi_2^2 \bar{\varepsilon}^*_{11} + \xi_1^2 \bar{\varepsilon}^*_{22} \right\} \exp(i\xi \cdot \boldsymbol{x}) \ , \tag{12-b}$$

$$\sigma_{12} = \frac{2\mu}{(1-\nu)} \sum_{m=-\infty}^{\infty} \sum_{n=-\infty}^{\infty} \frac{\xi_1 \xi_2}{\xi^4} \left\{ \xi_2^2 \bar{\varepsilon}^*_{11} + \xi_1^2 \bar{\varepsilon}^*_{22} \right\} \exp(i\xi \cdot \boldsymbol{x}) \ . \tag{12-c}$$

The eigenstresses caused by the periodic distribution of the eigenstrain can be derived by using Eqs. (3), (5-b), and (12). Furthermore, the total stress in the matrix can be obtained by superposing the applied stress on the eigenstress fields given by Eq. (12). Thus, we have

$$\frac{\sigma_{11}}{\sigma_o} = -\frac{d}{a}(B_1 - b_2) G_2(x_2) - \frac{2a^2 d}{b^3} \sum_{m=1}^{\infty} \sum_{n=1}^{\infty} \left[B_1 L_{mn} + B_2 K_{mn} \left\{ m^2 - \left(\frac{a}{b} n \right)^2 \right\} \right] n^2 \cdot \cos\left(\frac{\pi m x_1}{a} \right) \cos\left(\frac{\pi n x_2}{b} \right) , \tag{13-a}$$

$$\frac{\sigma_{22}}{\sigma_o} = 1 - \frac{d}{b}(B_1 - B_2) G_1(x_1) - \frac{2d}{b} \sum_{m=1}^{\infty} \sum_{n=1}^{\infty} \left[B_1 L_{mn} + B_2 K_{mn} \left\{ m^2 - \left(\frac{a}{b} n \right)^2 \right\} \right] m^2 \cdot \cos\left(\frac{\pi m x_1}{a} \right) \cos\left(\frac{\pi n x_2}{b} \right) , \tag{13-b}$$

$$\frac{\sigma_{12}}{\sigma_o} = -\frac{2ad}{b^2} \sum_{m=1}^{\infty} \sum_{n=1}^{\infty} \left[B_1 L_{mn} + B_2 K_{mn} \left\{ m^2 - \left(\frac{a}{b} n \right)^2 \right\} \right] mn \cdot \sin\left(\frac{\pi m x_1}{a} \right) \sin\left(\frac{\pi n x_2}{b} \right) , \tag{13-c}$$

where

$$L_{mn} = \frac{\{(-1)^m + (-1)^n\}}{\{m^2 + (an/b)^2\}^{3/2}} J_1 \left\{ \left(\frac{\pi d}{a} \right) \sqrt{m^2 + \left(\frac{a}{b} n \right)^2} \right\} ,$$

$$K_{mn} = \frac{\{(-1)^m + (-1)^n\}}{\{m^2 + (an/b)^2\}^{5/2}} J_1 \left\{ \left(\frac{\pi d}{a} \right) \sqrt{m^2 + \left(\frac{a}{b} n \right)^2} \right\} . \tag{14}$$

2.5 Modeling of Crack in a Matrix and Stress Intensity Factor

In this section, we consider the case that there is one crack in a matrix whose surface is perpendicular to the loading axis. The situation is shown in Fig.2. The center of the crack whose length is 2c locates in the origin of the coordinates. The particle distribution is identical to the previous model. Mode I stress intensity factor K_I for the crack tip can be calculated by using the value of the stress σ_{22} in the area occupied by the crack as follows:

$$K_I = \frac{1}{\sqrt{\pi c}} \int_{-c}^{c} \sigma_{22}|_{x_2=0} \sqrt{\frac{c+s}{c-s}} \, ds \ . \tag{15}$$

Substituting Eq. (13-b) into Eq. (15), we have

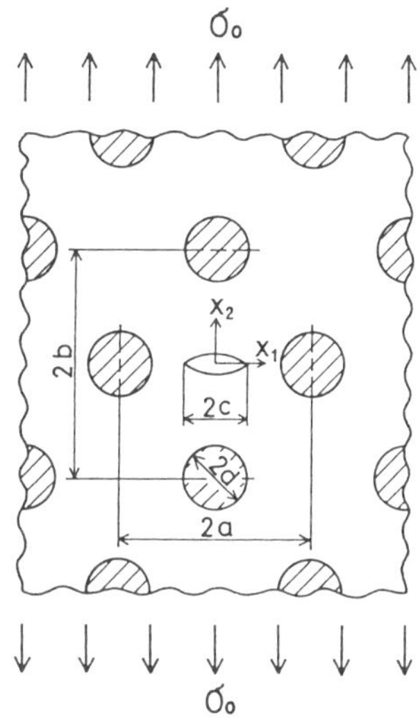

Fig.2. Modeling of crack in a matrix.

(i) $0 \le c \le d$

$$\frac{K_I}{\sigma_0\sqrt{\pi c}} = 1 - \frac{2d}{\pi b}(B_1 - B_2)\,E\left(\frac{c}{d}\right) + 2\frac{d}{b}\sum_{m=1}^{\infty}\sum_{n=1}^{\infty}\left[B_1 L_{mn} + B_2 K_{mn}\left\{m^2 - \left(\frac{a}{b}n\right)^2\right\}\right] m^2 \cdot J_0\left(\frac{\pi m c}{a}\right) , \tag{16-a}$$

(ii) $0 \le d \le c$

$$\frac{K_I}{\sigma_0\sqrt{\pi c}} = 1 - \frac{2a}{b}(B_1 - B_2)\left\{\frac{c}{a}E\left(\frac{d}{c}\right) - \left(\frac{c}{a} - \frac{d^2}{ca}\right)K\left(\frac{d}{c}\right)\right\} - 2\frac{d}{b}\sum_{m=1}^{\infty}\sum_{n=1}^{\infty}\left[B_1 L_{mn} + B_2 K_{mn}\left\{m^2 - \left(\frac{a}{b}n\right)^2\right\}\right] m^2 \cdot J_0\left(\frac{\pi m c}{a}\right) . \tag{16-b}$$

where E(x) and K(x) are the complete elliptic integral of the first and the second kind, respectively, and J_0(x) is the *Bessel* function of the first kind of order zero.

Ⅲ. NUMERICAL RESULTS AND DISCUSSIONS

3.1 Stress Distribution in a Matrix

In the present research, Alumina particle-reinforced Aluminum is used as a sample material of the composites under consideration. Elastic properties of these constitutive materials are listed in Table 1. In numerical calculations, the infinite double series in Eq. (13-a) are truncated up to m=1000 and n=2000, respectively, and the other double series up to m=2000 and n=1000.

The contours of the distribution of a maximum principal stress σ_1 normalized by the applied stress σ_0 in the matrix are shown in Fig.3. Fig.3(a) shows the present result, and Fig.3(b) is given by *Goodier*[6] to the case that there is only one circular particle in an infinite body. By comparing Fig.3(a) with Fig.3(b), it can be seen that the present result is in good agreement with *Goodier*'s one, especially near the inclusion. The difference between both results at some distances from the inclusion is considered to be caused by the influence of the adjacent inclusions. Thus, it can be confirmed from the above discussions that the stress distribution in a matrix can be exactly calculated by using the present method.

The distributions of σ_1 to various volume fractions Vp of the particles are shown in Fig.4. The shaded areas in these figures imply the domains that the maximum principal stress σ_1 is smaller than the applied stress σ_o . It can be seen from these results that the shaded area becomes smaller and the maximum value of σ_1 occurred at the upper and the lower part of the inclusion in the figure increases as Vp increases. Furthermore, the shaded area bridging over the inclusions appears only in the result of Vp=6.3%, and vanishes in the case of larger Vp. The existence of the shaded bridging area implies that some of the applied load may be supported by the particles. Thus, the increase of the volume fraction of particles may not contribute to the reinforcement of the matrix in the case of the particle-reinforced composites in contrast with the fiber-reinforced composites.

Table 1. Elastic properties of matrix and particles.

	Elastic modulus (GPa)	Poisson's ratio	Density (kg/m^3)
Matrix (Al)	70.6	0.33	2.65×10^4
Particle (Al_2O_3)	171.6	0.27	3.88×10^4

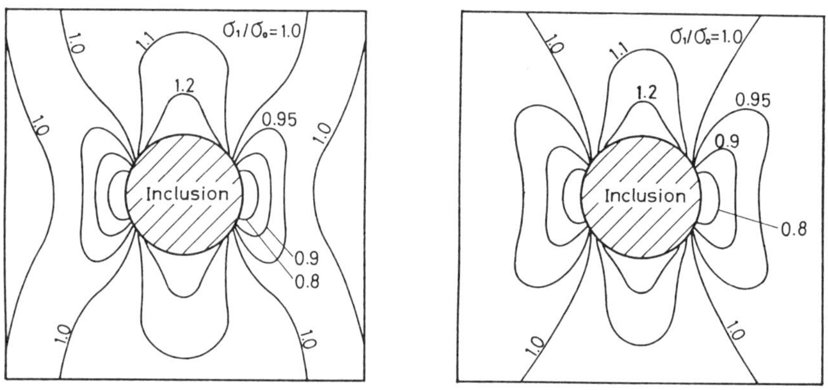

(a) present result. (b) *Goodier*'s result.
Fig.3. Distribution of maximum principal stress σ_1 normalized by the applied stress σ_o in the matrix.

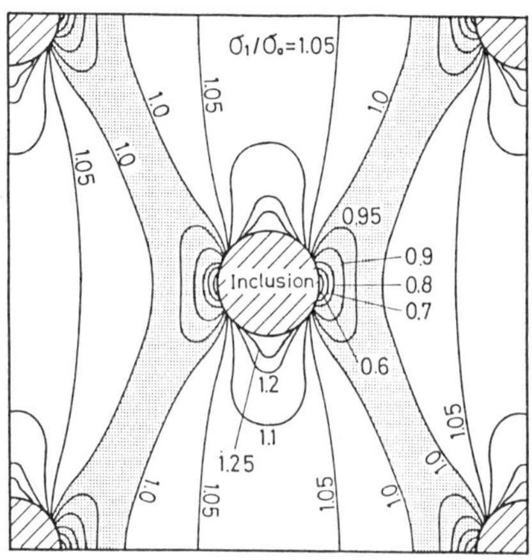

(a) Vp=6.3%.
Fig.4. Distribution of σ_1 to the volume fraction Vp of the particles.

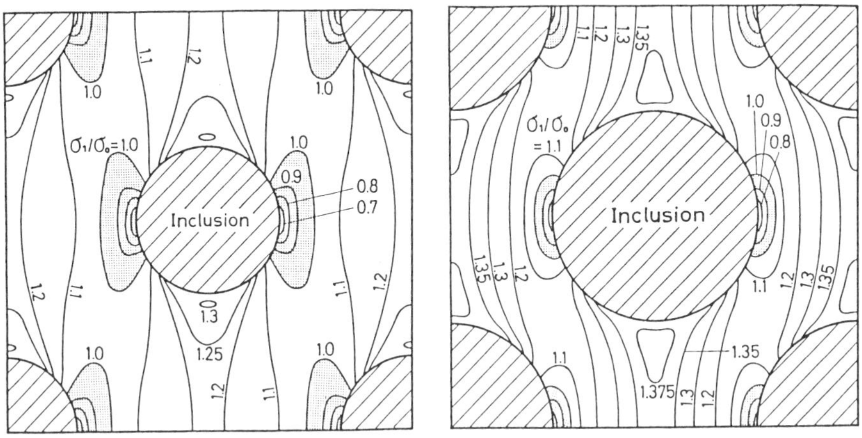

(b) Vp=19.2%. (c) Vp=39.3%.

Fig.4. Distribution of σ_1 to the volume fraction Vp of the particles.

3.2 Stress Intensity Factor for a Matrix Crack

To evaluate the stress intensity factor for a crack located between two inclusions, the model shown in Fig.5 is employed. The present results obtained and Ishida's ones calculated by using the *Laurent* expansion method[7] are shown in Table.2. By comparing both results, it can be seen that the present results are in good agreement with Ishida's ones, especially in the case of the small crack length. Thus, the validity of the present method to evaluate the stress intensity factor is confirmed from these results.

The normalized stress intensity factor $K_I / \sigma_0\sqrt{\pi c}$ of the crack surrounded by the rigid inclusions are shown to various arrangements of the particles b/a in Fig.6. The horizontal axis is taken to c/a, i.e. the crack length normalized by the period to the x_1 direction. b is the period to the x_2 direction. It is obvious from the above results that K_I increase with the decrease of b/a. The decrease of b/a can be interpreted as the increase of the volume fraction Vp of the particles. Furthermore, It is obvious from the results of the previous section that the stress in the matrix becomes higher as Vp increases. Thus, the increase of K_I with the increase of b/a is considered to be reasonable.

On the other hand, the results to the case of the circular holes are shown in Fig.7. It can be seen that the variations of K_I with b/a of the circular holes are opposite to those of the rigid particles.

By applying the concept of Fracture mechanics to both results, it is considered that the particle-reinforced composites become more brittle when such composites contain many rigid particles, while more tough in the case of the composites containing many voids.

K_I versus the elastic constant of the particle is shown in Fig.8. The horizontal axis is taken to E/E^p and E^p/E in the left and the right side of the figure, respectively. It is obvious from the results in Fig.8 that K_I in the range of $b/a \leq 4/3$ decreases, while in $b/a \geq 2$ increase as Young's modulus of the particle decreases. Thus, it can be seen that the particle-reinforced composites may be toughened when the composites contain many particles like a void, or when the volume fraction of the harder inclusions in the composites becomes lower.

Table 2. Stress intensity factor for a crack located between two inclusions.

	c/a		0.1	0.2	0.3	0.4	0.5
$\frac{K_I}{\sigma_0\sqrt{\pi c}}$ (d/a =0.2)	Rigid inclusons	Present result	0.979	0.979	0.978	0.969	0.959
		Ishida's result	0.979	0.979	0.978	0.967	0.946
	Circular holes	Present result	1.05	1.05	1.06	1.07	1.09
		Ishida's result	1.05	1.05	1.06	1.07	1.10

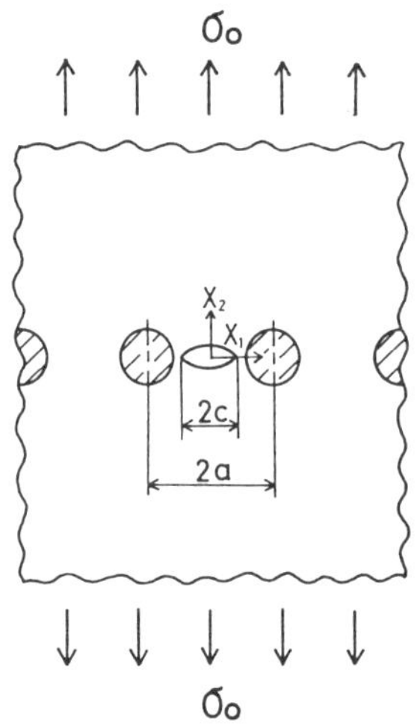

Fig.5. Modeling for a crack located between two particles.

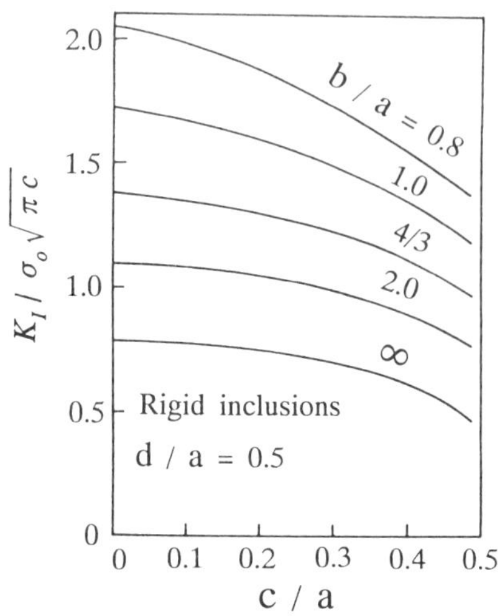

Fig.6. Variation of stress intensity factor to arrangements of rigid particles.

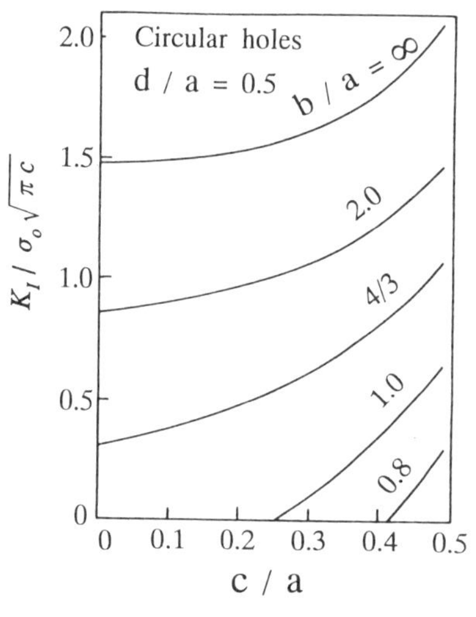

Fig.7. Variation of stress intensity factor to arrangements of circular holes.

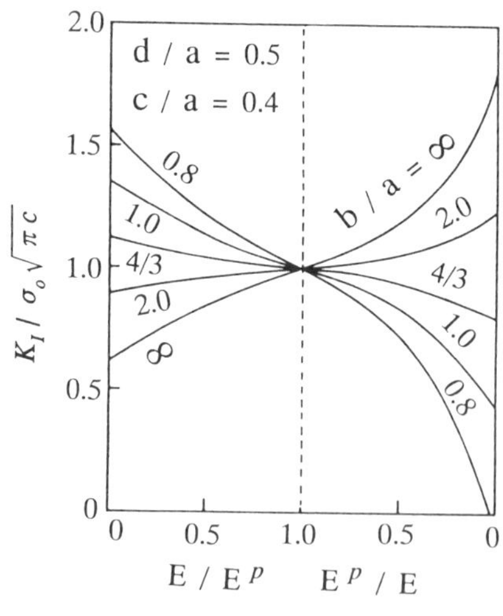

Fig.8. Stress intensity factor versus *Young*'s modulus of the particles.

IV. CONCLUSION

The analysis of particle-reinforced composites is performed by means of the equivalent inclusion method of micromechanics. The analytical expressions to the stress in a matrix and the stress intensity factor for a matrix crack can be derived by introducing the periodic distribution of the particles to the analytical model.

It can be seen that both the stress in a matrix and stress intensity factor for a matrix crack increase with the decrease of the periodicity of the arrangement of particles, i.e. the volume fraction of the particles, in the case of such composites containing many particles with higher Young's modulus than that of a matrix. Thus, it can be concluded that the increase of the volume fraction of particles may not contribute to toughen the particle-reinforced composites in contrast with the fiber-reinforced composites.

Acknowledgement

The authors would like to express their appreciations to Mr. N. Matsuda, the technical official in K.I.T.

REFERENCES

1) Tyson.W.R. and Davies.G.J., *A photoelastic study of the shear stresses associated with the transfer of stress during fibre reinforcement*, Brit. J.Appl. Phys. 16 (1965), pp.199–205.
2) Chen.P.E. and Lavengood.R.E., *Stress fields around multiple inclusions*, Structure, Solid Mechanics and Engineering Design, 1 (The Proc. the Southampton 1969 Civil Engineering Materials Conf.), pp.75–86.
3) Tolf.G., *Mechanical behavior of a short fiber composites*, Proc. 4th Int. Conf. on Mechanical Behavior of Materials, 1 (1983), pp.587–593.
4) Saito.K., Yamazaki.H., and Suzuki.M., *Analysis of short fiber reinforced composites by means of the modified equivalent inclusion method*, Microstructure and mechanical behavior of materials, 1 (Chameleon press ltd., London,1985), pp.199–208.
5) Mura.T., *Micromechanics of Defects in Solids*, (Martinus Nijhoff, Harge, 1982).
6) Goodier.J.N., *Concentration of stress around spherical and cylindrical inclusions and flaws*, J. Appl. Mec. 55 (1933), pp.39.
7) Ishida,M., *Mechanics of Fracture*, vol.1 (Noodhoff Int. Pus.,1973), pp.56.

Behaviors of an Edge Dislocation Both Inside and Outside a Circular Inclusion Near a Free Surface

Tatsuro MORITA*, Kenji SAITO*, and Shyunsuke SHIOYA**

* *Department of Mechanical and System Engineering, Kyoto Institute of Technology, Matsugasaki, Sakyo-ku, Kyoto*

** *Department of Mechanical Engineering, the Defence Academy, Hashirimizu, Yokosuka, Kanagawa*

The elastic interaction between an edge dislocation and an inclusion near a free surface is treated by using the exact elasticity solution. The results show that both the ratio of shear modulus of the inclusion to that of the matrix and the position of the inclusion affect strongly the behaviors of the dislocation. Their influences become to be strong as the difference of the shear modulus increases and as the inclusion approaches to the surface. Moreover, when the shear modulus of the inclusion is higher than that of the substrate, the matrix could escape easily from the surface since the dislocation is forced toward the free surface all over the slip plane due to the elastic interaction.

I. INTRODUCTION

It has been widely accepted that inclusions play an important role among some factors which determine mechanical properties of metals because of their interactions with dislocations. There are two kinds of the interactions between a dislocation and an inclusion. The first is the elastic interaction of a dislocation with eigenstress fields caused by a misfit of the atomic structure between the inclusion and the matrix[1], and the second is the elastic interaction arising from a difference of their elastic constants. The second case analysis to a cylindrical inclusion was made by J. Dundurs and T. Mura[2], and they showed that for certain combinations of elastic constants an edge dislocation possesses a stable equilibrium position in the matrix at a finite distance from the interface.

If there exists a free surface near an inclusion, the behavior of a dislocation near an inclusion changes considerably. The authors analyzed the behaviors of an edge dislocation in the neighborhood of an inclusion both near a free surface[3,4] and a curved free surface representing a void[5,6] as a two-dimensional elasticity problem in bipolar coordinates. As a consequence, it can be clarified that the edge dislocation which has only a stable equilibrium position near an inclusion possesses an unstable equilibrium position, depending upon the combination of the modulus of rigidity of the inclusion and the matrix when the free surface exists. An analysis of such problem is important to explain the fact that inclusions near the free surface provide sites for fatigue crack initiations in the high-strength steel.

Therefore, in the present study, the behaviors of an edge dislocation both inside and outside the circular inclusion near a free surface are treated exactly by using Airy stress function expressed in bipolar coordinates. The exact solution is constructed by superposing the following three solutions, i.e., (i) a solution to the edge dislocation in a semi-infinite solid, (ii) a solution to the region of the matrix, and (iii) a solution to the region of the circular inclusion.

II. ANALYSIS

2.1 Bipolar Coordinates

Because of the geometrical nature of the problem, bipolar coordinates are used in the analysis. The following relations exist between bipolar coordinates (α, β) and complex coordinates (x, y) as shown in Fig.1

$$z=a\coth\frac{\xi}{2}, \tag{1a}$$

where $z=x+iy$ and $\xi=\alpha+i\beta$ ($-\infty<\alpha<\infty$, $-\pi\leqq\beta\leqq\pi$), and a is a positive real number. From Eq. (1a), x and y are represented by α and β as follows.

$$x = J\sinh\alpha, \quad y = -J\sin\beta \,, \tag{1b}$$

where $J = 1/h = a/(\cosh\alpha - \cos\beta)$.

For many purposes, Airy stress function provides the most convenient representation of a solution in plane elasticity. In bipolar coordinates, Airy stress function $h\chi$ is derived from the solutions of the following partial differential equation[7].

$$\left(\frac{\partial^4}{\partial\alpha^4} + 2\frac{\partial^4}{\partial\alpha^2\partial\beta^2} + \frac{\partial^4}{\partial\beta^4} - 2\frac{\partial^2}{\partial\alpha^2} + 2\frac{\partial^2}{\partial\beta^2} + 1\right) h\chi = 0 \tag{2a}$$

The displacement function hQ is determined from the relation between $h\chi$ and hQ.

$$\frac{\partial^2}{\partial\alpha\partial\beta} hQ = \left(\frac{\partial^2}{\partial\alpha^2} - \frac{\partial^2}{\partial\beta^2} - 1\right) h\chi \tag{2b}$$

Stress components $\sigma_\alpha, \sigma_\beta, \tau_{\alpha\beta}$ and displacement components u_α, u_β in bipolar coordinates are obtained by $h\chi$ and hQ as follows.

$$\begin{aligned} a\sigma_\alpha &= \left\{(\cosh\alpha - \cos\beta)\frac{\partial^2}{\partial\beta^2} - \sinh\alpha\frac{\partial}{\partial\alpha} - \sin\beta\frac{\partial}{\partial\beta} + \cosh\alpha\right\} h\chi \,, \\ a\sigma_\beta &= \left\{(\cosh\alpha - \cos\beta)\frac{\partial^2}{\partial\alpha^2} - \sinh\alpha\frac{\partial}{\partial\alpha} - \sin\beta\frac{\partial}{\partial\beta} + \cos\beta\right\} h\chi \,, \\ a\tau_{\alpha\beta} &= -(\cosh\alpha - \cos\beta)\frac{\partial^2}{\partial\alpha\partial\beta} h\chi \,, \end{aligned} \tag{3a}$$

$$\begin{aligned} 2Gu_\alpha &= \frac{\kappa-1}{2}\left(\frac{\partial}{\partial\alpha} - \frac{\sinh\alpha}{\cosh\alpha - \cos\beta}\right) h\chi - \frac{\kappa+1}{4}\left(\frac{\partial}{\partial\beta} - \frac{\sin\beta}{\cosh\alpha - \cos\beta}\right) hQ \,, \\ 2Gu_\beta &= \frac{\kappa-1}{2}\left(\frac{\partial}{\partial\beta} - \frac{\sin\beta}{\cosh\alpha - \cos\beta}\right) h\chi + \frac{\kappa+1}{4}\left(\frac{\partial}{\partial\alpha} - \frac{\sinh\alpha}{\cosh\alpha - \cos\beta}\right) hQ \,, \end{aligned} \tag{3b}$$

where G is shear modulus, ν is Poisson's ratio, $\kappa = 3 - 4\nu$ for plane strain, and $\kappa = (3-\nu)/(1+\nu)$ for plane stress.

2.2 Elasticity Solutions of an Edge Dislocation in a Semi-Infinite Solid

Figure 2 shows an edge dislocation of Burgers vector b_x parallel to the x direction situated at $(\alpha_0, \pm\pi)$ in a semi-infinite solid. By considering its image dislocation at $(-\alpha_0, \pm\pi)$, Airy stress function $\hat{\chi}$ for this

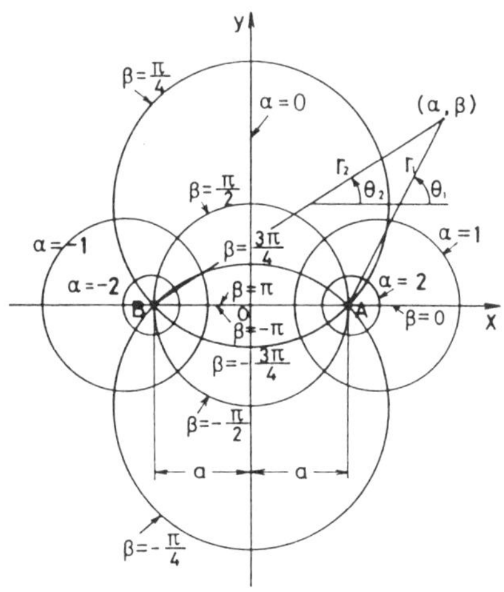

Fig.1. Bipolar coordinates.

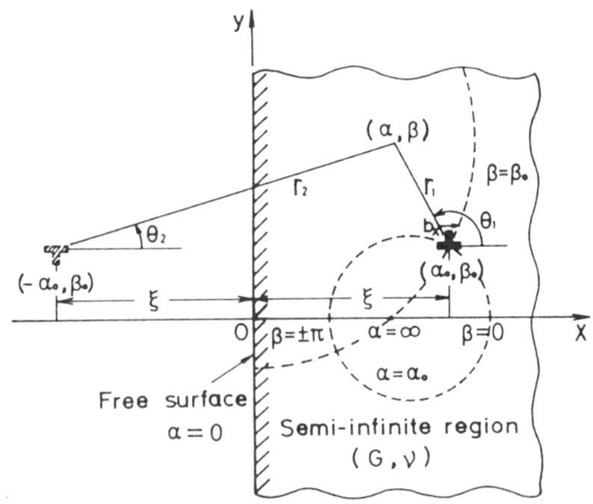

Fig.2. An edge dislocation existing in a semi-infinte solid.

dislocation is given by two coordinates (r_1, θ_1) and (r_2, θ_2) as follows.

$$\hat{\chi}=-\frac{2Gb_x}{\pi(\kappa+1)}\{r_1\sin\theta_1\ln r_1-r_2\sin\theta_2\ln r_2+\xi(\sin2\theta_2-2\xi\frac{\sin\theta_2}{r_2})\}\,, \tag{4}$$

where the second term gives the contribution from the image dislocation, and the third term represents the elasticity solution to make a plane $x=0$ be a free surface. Making a transformation of coordinates and inserting $\pm\pi$ into β, the stress function $h\hat{\chi}$ is given below in bipolar coordinates.

$$h\hat{\chi}=-\frac{2Gb_x}{\pi(\kappa+1)}\frac{\sin\beta}{2}\{\ln\frac{\cosh(\alpha-\alpha_0)+\cos\beta}{\cosh(\alpha+\alpha_0)+\cos\beta}+\frac{2\sinh\alpha\,\sinh\alpha_0}{\cosh(\alpha+\alpha_0)+\cos\beta}\} \tag{5}$$

Next, Eq. (5) is expanded to Fourier series with respect to β. Fourier expansions inside and outside a circle $\alpha=\alpha_0$ are given as follows.

(i) Inside the circle $\alpha=\alpha_0$ $(\alpha>\alpha_0)$;

$$\begin{aligned} h\hat{\chi}&=-\frac{2Gb_x}{\pi(\kappa+1)}\{q_1\sin\beta+\sum_{n=1}^{\infty}\hat{\phi}_n(\alpha)\sin n\beta\}\,,\\ h\hat{Q}&=-\frac{2Gb_x}{\pi(\kappa+1)}\sum_{n=1}^{\infty}\hat{\psi}_n(\alpha)\cos n\beta\,, \end{aligned} \tag{6a}$$

where

$$\begin{aligned} &\hat{\phi}_1(\alpha)=p_1e^{-2\alpha},\ \hat{\phi}_n(\alpha)=p_ne^{-(n+1)\alpha}+q_ne^{-(n-1)\alpha}\quad(n\geq2)\,,\\ &\hat{\psi}_1(\alpha)=2p_1e^{-2\alpha},\ \hat{\psi}_n(\alpha)=2\{p_ne^{-(n+1)\alpha}+q_ne^{-(n-1)\alpha}\}\quad(n\geq2)\,,\\ &p_n=(-1)^{n+1}\{\frac{\sinh(n+1)\alpha_0}{n+1}-e^{-n\alpha_0}\sinh\alpha_0\}\quad(n\geq1)\,,\\ &q_1=-\alpha_0+e^{-\alpha_0}\sinh\alpha_0\,,\\ &q_n=(-1)^{n+1}\{-\frac{\sinh(n-1)\alpha_0}{n-1}+e^{-n\alpha_0}\sinh\alpha_0\}\quad(n\geq2)\,. \end{aligned} \tag{6b}$$

(ii) Outside the circle $\alpha=\alpha_0$ $(0<\alpha<\alpha_0)$;

$$\begin{aligned} h\hat{\chi}&=-\frac{2Gb_x}{\pi(\kappa+1)}\{s_1\alpha\sin\beta+q_1\sin\beta+\sum_{n=1}^{\infty}\hat{\phi}_n(\alpha)\sin\beta\}\,,\\ h\hat{Q}&=-\frac{2Gb_x}{\pi(\kappa+1)}\{2s_1\beta\sin\beta+\sum_{n-1}^{\infty}\hat{\psi}_n(\alpha)\cos n\beta\}\,, \end{aligned} \tag{7a}$$

where

$$\begin{aligned} &\hat{\phi}_1(\alpha)=p_1\cosh2\alpha+r_1\sinh2\alpha,\ \hat{\psi}_1(\alpha)=-2(p_1\sinh2\alpha+r_1\cosh2\alpha)\,,\\ &\hat{\phi}_n(\alpha)=p_n\cosh(n+1)\alpha+q_n\cosh(n-1)\alpha+r_n\sinh(n+1)\alpha+s_n\sinh(n-1)\alpha\quad(n\geq2)\,,\\ &\hat{\psi}_n(\alpha)=-2\{p_n\sinh(n+1)\alpha+q_n\sinh(n-1)\alpha+r_n\cosh(n+1)\alpha+s_n\cosh(n-1)\alpha\}\quad(n\geq2)\,,\\ &p_n=(-1)^ne^{-n\alpha_0}\sinh\alpha_0\quad(n\geq1)\,,\\ &q_n=(-1)^ne^{-n\alpha_0}\sinh\alpha_0\quad(n\geq1)\,,\\ &r_n=(-1)^{n+1}\{\frac{e^{-(n+1)\alpha_0}}{n+1}+e^{-n\alpha_0}\sinh\alpha_0\}\quad(n\geq1)\,,\\ &s_1=-1,\ s_n=(-1)^n\{\frac{e^{-(n-1)\alpha_0}}{n-1}+e^{-n\alpha_0}\sinh\alpha_0\}\quad(n\geq2)\,. \end{aligned} \tag{7b}$$

In the above expressions, $q_1\sin\beta$ gives rigid body displacements, and $s_1\alpha\sin\beta$ gives multivalued parts of displacements which are equal to Burgers vector b_x of the dislocation. In the case that the dislocation exists inside the inclusion $(\alpha_0>\alpha_1)$, it should be noted that shear modulus $\bar{G}$ and Poisson's ratio $\bar{\nu}$ of the inclusion are used in Eqs. (6) and (7) instead of G and ν of the matrix.

2.3 Elasticity Solutions both inside and outside a Circular Inclusion

Outside a circular inclusion $(0<\alpha<\alpha_1)$ shown in Fig.3, both the stress function $h\tilde{\chi}$ and the displacement function $h\tilde{Q}$ are derived in series forms by the general solutions of Eq. (2) corresponding to the previous results in the section 2.2.

$$h\tilde{\chi}=-\frac{2Gb_x}{\pi(\kappa+1)}\{d_1\sin\beta+b_1\sin\beta+\sum_{n=1}^{\infty}\tilde{\psi}_n(\alpha)\sin n\beta\},$$
$$h\tilde{Q}=-\frac{2Gb_x}{\pi(\kappa+1)}\{2d_1\beta\sin\beta+\sum_{n=1}^{\infty}\tilde{\phi}_n(\alpha)\cos n\beta\}, \tag{8a}$$

where

$$\tilde{\phi}_1(\alpha)=a_1\cosh2\alpha+c_1\sinh2\alpha,\quad \tilde{\psi}_1(\alpha)=-2(a_1\sinh2\alpha+c_1\cosh2\alpha),$$
$$\tilde{\phi}_n(\alpha)=a_n\cosh(n+1)\alpha+b_n\cosh(n-1)\alpha+c_n\sinh(n+1)\alpha+d_n\sinh(n-1)\alpha\quad(n\geq2), \tag{8b}$$
$$\tilde{\psi}_n(\alpha)=-2\{a_n\sinh(n+1)\alpha+b_n\sinh(n-1)\alpha+c_n\cosh(n+1)\alpha+d_n\cosh(n-1)\alpha\}\quad(n\geq2).$$

There are four kinds of constants a_n, b_n, c_n, d_n in Eqs. (8a) and (8b), and the following relations exist between them in order to make σ_α and $\tau_{\alpha\beta}$ be zero on the free surface $\alpha=0$.

$$a_n+b_n=0\ (n\geq2),\quad (n+1)c_n+(n-1)d_n=0\ (n\geq2) \tag{9}$$

Inside the circular inclusion $(\alpha>\alpha_1)$ shown in Fig.4, both the stress function $h\bar{\chi}$ and the displacement function $h\bar{Q}$ are also derived in the same way.

$$h\bar{\chi}=-\frac{2\bar{G}b_x}{\pi(\bar{\kappa}+1)}\{f_1\sin\beta+\sum_{n=1}^{\infty}\bar{\phi}_n(\alpha)\sin n\beta\},$$
$$h\bar{Q}=-\frac{2\bar{G}b_x}{\pi(\bar{\kappa}+1)}\sum_{n=1}^{\infty}\bar{\psi}_n(\alpha)\cos n\beta, \tag{10a}$$

where

$$\bar{\phi}_1(\alpha)=e_1\cosh2\alpha+g_1\sinh2\alpha,\quad \bar{\psi}_1(\alpha)=-2(e_1\sinh2\alpha+g_1\cosh2\alpha),$$
$$\bar{\phi}_n(\alpha)=e_n\cosh(n+1)\alpha+f_n\cosh(n-1)\alpha+g_n\sinh(n+1)\alpha+h_n\sinh(n-1)\alpha\quad(n\geq2), \tag{10b}$$
$$\bar{\psi}_n(\alpha)=-2\{e_n\sinh(n+1)\alpha+f_n\sinh(n-1)\alpha+g_n\cosh(n+1)\alpha+h_n\cosh(n-1)\alpha\}\quad(n\geq2).$$

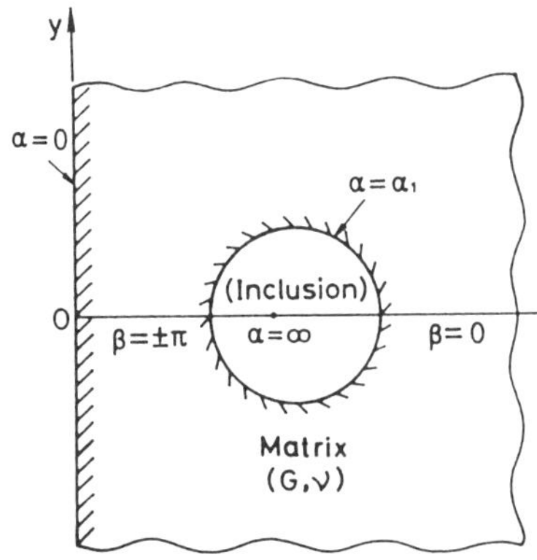

Fig.3. The domain outside a circular inclusion near a free surface.

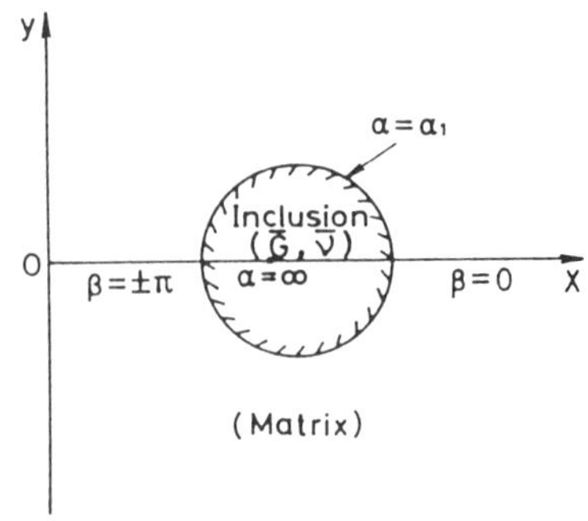

Fig.4. The domain inside a cicular inclusion.

In these expressions, $\bar{G}$ is shear modulus of the inclusion and $\bar{\nu}$ is its Poisson's ratio, and e_n, f_n, g_n, h_n are unknown constants. However, they must satisfy the following relations since both stress and displacement components become to be finite values at the pole of the coordinate ($\alpha \to \infty$).

$$f_1=0, \quad e_n+g_n=0 \ (n\geq 1), \quad f_n+h_n=0 \ (n\geq 2) \tag{11}$$

All the kinds of unknown constants can be determined by the boundary conditions on the interface between an inclusion and a matrix together with Eqs. (9) and (11). The details of the boundary conditions will be mentioned in the next section 2.4.

2.4 Determination of Unknown Constants

In Eqs. (8) and (10), there are eight kinds of unknown constants (a_n–h_n) in all. Assuming a perfect bonding between the matrix and the inclusion, the constants are determined by the following two requirements regarding the regularity of both the displacement and stress fields on the interface together with Eqs. (9) and (11).

(I) The case that the dislocation exists between the free surface and the interface (Fig.5a) ;

$$\hat{\sigma}_\alpha+\tilde{\sigma}_\alpha=\bar{\sigma}_\alpha, \quad \hat{\tau}_{\alpha\beta}+\tilde{\tau}_{\alpha\beta}=\bar{\tau}_{\alpha\beta}, \qquad \hat{u}_\alpha+\tilde{u}_\alpha=\bar{u}_\alpha, \quad \hat{u}_\beta+\tilde{u}_\beta=\bar{u}_\beta . \tag{12}$$

(II) The case that the dislocation exists inside the inclusion (Fig.5b) ;

$$\hat{\sigma}_\alpha+\bar{\sigma}_\alpha=\tilde{\sigma}_\alpha, \quad \hat{\tau}_{\alpha\beta}+\bar{\tau}_{\alpha\beta}=\tilde{\tau}_{\alpha\beta}, \qquad \hat{u}_\alpha+\bar{u}_\alpha=\tilde{u}_\alpha, \quad \hat{u}_\beta+\bar{u}_\beta=\tilde{u}_\beta . \tag{13}$$

Since all the coefficients of the each stress functions (cf. Eqs. (6), (7), (8) and (10)) are derived by solving Eqs. (12) and (13) together with Eqs. (9) and (11), stresses acting on the dislocation which gives the elastic interaction for the above two cases, (I) and (II), can be obtained by superposing the each stress components except their singular terms.

III. RESULTS AND DISCUSSIONS

In order to know the behaviors of the dislocation both inside and outside the inclusion, the force F_x acting on the dislocation is the best indicator to them, and is calculated by the Peach–Koehler formula[8] expressed in bipolar coordinates. The force F_x acting on the dislocation situated at (α_0, β_0) is given by the following equation.

$$F_x=\frac{b_x}{\cosh\alpha_0-\cos\beta_0}[(\cosh\alpha_0-\cos\beta)\sinh\alpha_0\cos\beta_0\{\sigma^*_\alpha(\alpha_0,\beta_0)-\sigma^*_\beta(\alpha_0,\beta_0)\} + \{\sinh^2\alpha_0\sin^2\beta_0-(1-\cos\alpha_0\cos\beta_0)^2\}\tau^*_{\alpha\beta}(\alpha_0,\beta_0)] \tag{14}$$

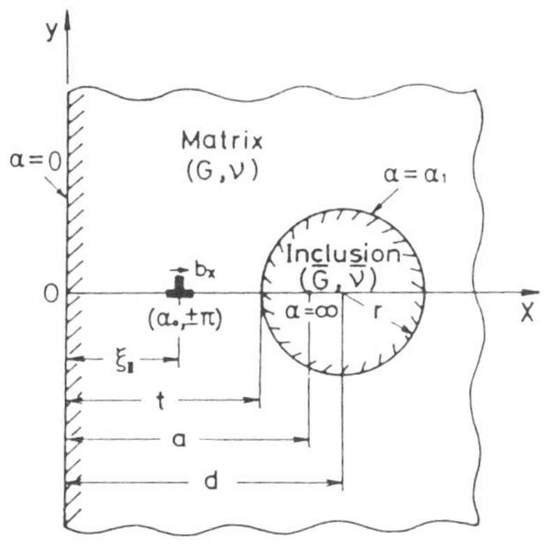

Fig.5(a). An edge dislocation existing between a free surface and an interface.

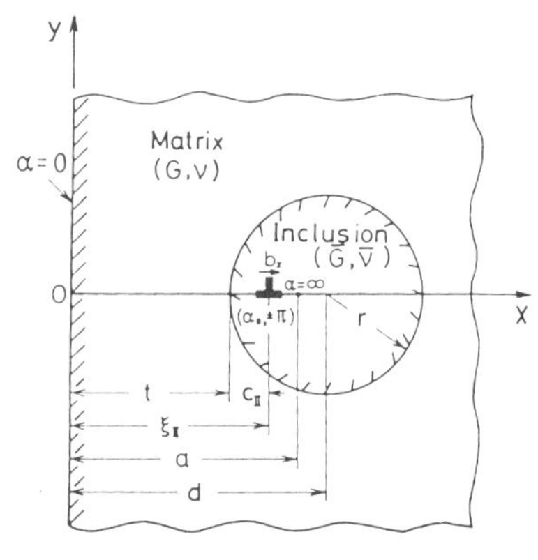

Fig.5(b). An edge dislocation existing inside a circular inclusion.

where σ_α^*, σ_β^*, $\tau_{\alpha\beta}^*$ represent the stress components arising from the elastic interaction, and they are derived by the results in the section 2. In the case that the dislocation exists on a slip plane $\beta=\pm\pi$, F_x is obtained by inserting $\beta_0=\pm\pi$ into Eq. (14).

$$F_x=-b_x\tau^*_{\alpha\beta}(\alpha_0,\pm\pi) \tag{15}$$

The force acting on a dislocation is calculated numerically by using Eq. (15) with the following non-dimensional parameters.

$$\rho=\frac{r}{d},\quad \Gamma=\frac{G}{\bar{G}},\quad \epsilon=\frac{\xi}{t}\ (0<\xi<1),\quad \eta=\frac{c}{2r}\ (0<\eta<1)$$

The results are summarized in Fig.6 together with the ones in the infinite solid given by J. Dundurs and T. Mura [2] for reference.

In this figure, it is clear that when the shear modulus of the inclusion is higher than that of the matrix (Γ =0.2, 0.5), the dislocation both inside and outside the inclusion is forced toward the free surface all over the slip plane. From this aspects, it is supposed that an initiation of slip step on the surface of metals due to motion of dislocations under an alternative applied stress could be promoted by presence of the inclusion near the free surface.

In contrast to this, when the shear modulus of the inclusion is less than that of the matrix (Γ = 2, 5), the behaviors of a dislocation are complex. Inside the inclusion, a stable equilibrium position exists and the dislocation is trapped at this position even if dislocations could move inside the inclusion. Between the interface and the free surface, an unstable equilibrium position (F_x= 0) is present, and the dislocation is attracted to the surface when it locates at the right side of this position, and is strikingly forced toward the interface when at the left side.

In order to assess mechanical properties of real industrial materials, it is important to know behaviors of dislocations existing between the interface and the free surface. Then, influences of each parameter $\Gamma(=G/\bar{G})$, $\nu/\bar{\nu}$ and $\rho(=r/d)$ on the behavior of dislocations are discussed below.

Firstly, the influence of the parameter Γ is shown in Fig.7. This figure clearly indicates that the value of Γ affects extremely the aspects of the acting force. By paying attention to a domain near the inclusion in this figure, it is understood that the acting force decreases with decreasing Γ, and that when Γ becomes to be less than unity, the dislocation is forced toward the surface all over the slip plane. From this fact, it is considered that the dislocation could escape more easily from the surface according as the shear modulus of the inclusion increases comparing with that of the matrix.

Secondly, Fig.8 shows the relations between the parameter $\nu/\bar{\nu}$ and the acting force. At each value of Γ shown in this figure, the curves of the acting force are almost the same irrespective of the value of $\nu/\bar{\nu}$, so that it is considered that the parameter $\nu/\bar{\nu}$ gives almost the same effect on the behaviors of the dislocation.

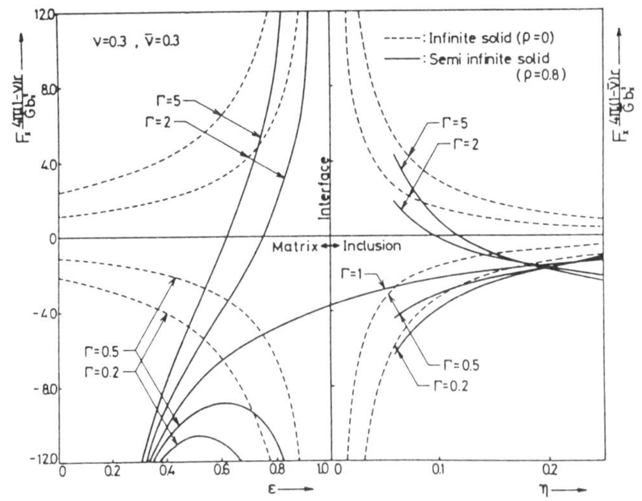

Fig.6. The force acting on an edge dislocation both inside and outside a circular inclusion.

Moreover, influence of ρ on aspects of the acting force is shown in Fig.9. At each value of Γ shown in this figure, the magnitude of the acting force increases as the value of ρ decreases although the shape of the curvature is the same irrespective of the value of ρ. This fact means that the elastic interactions of the dislocation with the inclusion become to be strong as the inclusion approaches to the free surface.

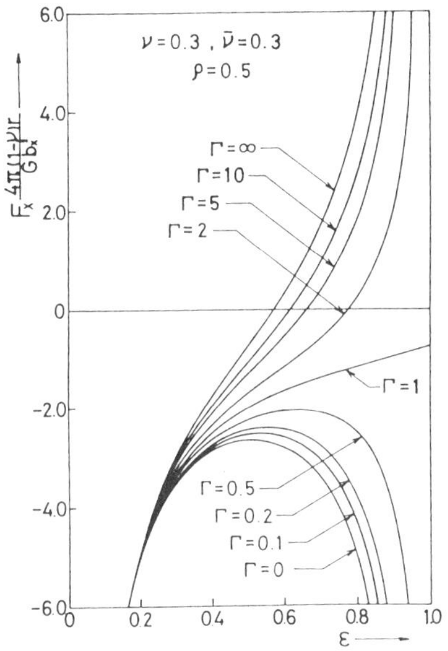

Fig.7. The influemce of the shear modulus ratio Γ on the aspects of the acting force.

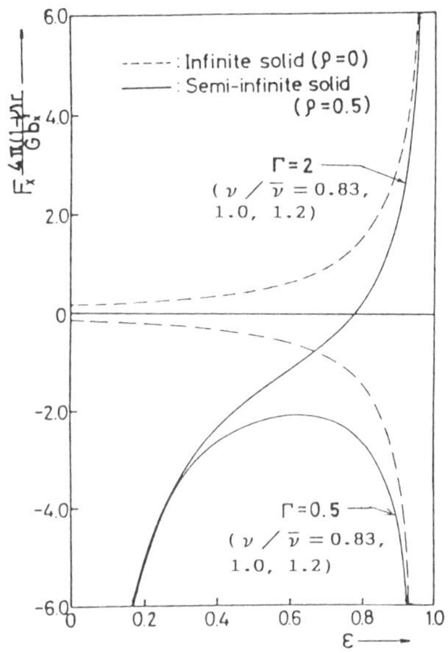

Fig.8. The relation between $\nu/\bar{\nu}$ and the aspects of the acting force.

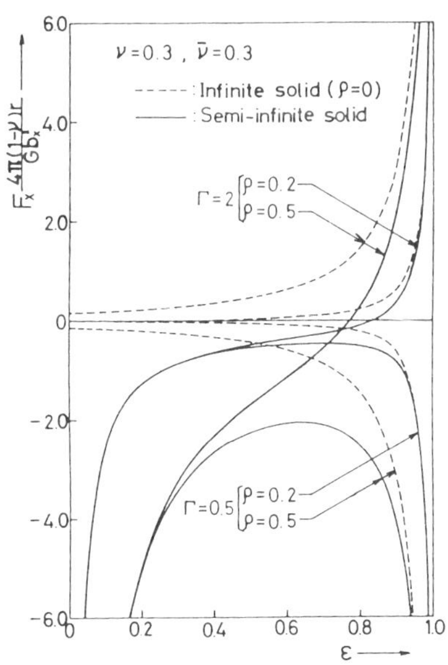

Fig.9. The influence of the position of a cicular inclusion on the aspects of the acting force.

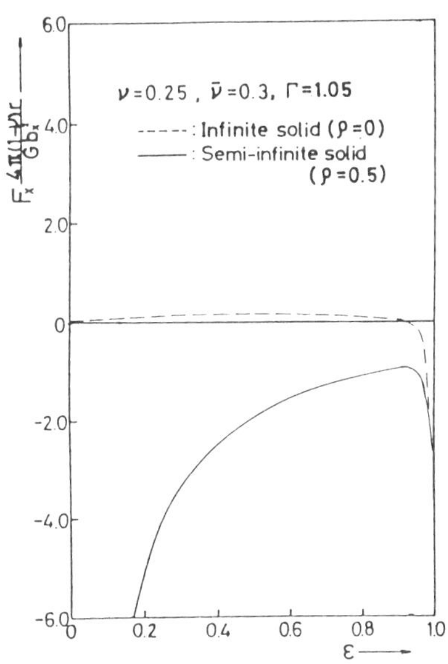

Fig.10. The disappearance of the stable equilibrium position due to the presence of the free surface.

Finally, the influence of the free surface on the interaction between the dislocation and the inclusion is discussed by comparing the results in the semi-infinite solid obtained in this study with those to the infinite solid given by J. Dundurs and T. Mura[2]. By paying attention to the cases in Fig.6, it is understood that the free surface affects strikingly the aspects of the acting force although the aspects near the interface are similar. Especially, when the difference of the elastic constants is small, the free surface could affect strongly the aspects of the acting force. For example, even if the combination of elastic constants are such that the stable equilibrium position exists in the infinite solid, this position disappears in the semi-infinite solid as shown in Fig.10. This phenomenon is certainly developed by the presence of the free surface.

IV. CONCLUSIONS

In the present study, elastic interactions of the edge dislocation with the circular inclusion near the free surface were treated exactly by using the elasticity solution represented by the bipolar coordinates. Especially, the behaviors of the dislocation both inside and outside the inclusion and influence of the each parameters ($G/\bar{G}$, $\nu/\bar{\nu}$ and r/d) on the behavior of the dislocation were investigated. The conclusions in this study are summarized as follows.

(1) For the case that the shear modulus of the inclusion is higher than that of the matrix, the dislocation could escape easily from the free surface under an applied stress since the dislocation both inside and outside the inclusion is forced toward the free surface all over the slip plane due to the elastic interaction. In contrast to this, for the case that the shear modulus of the inclusion is less than that of the matrix, both stable and unstable equilibrium positions of the dislocation exist on the slip plane and the behaviors of the dislocation depend on its position.

(2) Both the ratio of shear modulus of the inclusion to that of the matrix and the position of the inclusion affect strongly the behaviors of the dislocation. Their influences become to be strong as the difference of the shear modulus increases and as the inclusion approaches to the surface. Comparing with them, it is considered that the difference of Poisson's ratio gives little effect to the elastic interaction.

(3) The free surface affects strikingly the behavior of the edge dislocation, so that the stable equilibrium position disappears in the semi-infinite solid even for the combination of elastic constants for which the stable equilibrium position exists in the infinite solid.

REFERENCES

1) Cottrell, A. H., Report of conference on strength of solids, Phys. Soc., (1948), p.30.
2) Dundurs, J., and Mura, T., Interaction between an edge dislocation and a circular inclusion, J. Mech. Phys. Solids, 12 (1964), p.177.
3) Matsuoka, A., Shioya, S., and Saito, K., The elastic interaction of an edge dislocation with a circular inclusion in a semi-infinite medium (1st report), Trans. JSME, Series 1 (1976), 42–362, p.3051.
4) Saito, K., Matsuoka, A., and Shioya, S., The elastic interaction of an edge dislocation with a circular inclusion in a semi-infinite medium (2nd report), Trans. JSME, Series 1 (1977), 43–371, p.2492.
5) Yoshioka, K., Shioya, S., and Saito, K., The interaction of an edge dislocation with a circular hole and a circular inclusion in an infinite body (1st report), Trans. JSME, Series A (1986), 52–474, p.308.
6) Yoshioka, K., Shioya, S., and Saito, K., The interaction of an edge dislocation with a circular hole and a circular inclusion in an infinite body (2nd report), Trans. JSME, Series A (1986), 52–474, p.507.
7) Jeffery, G. B., Plane stress and plane strain in bipolar coordinates, Phil. Trans. Roy. Soc. Lond., A–211 (1921), p.265.
8) Peach, M., and Köehler, J. S., The forces exerted on dislocations and the stress field produced by them, Phy. Rev., 80 (1950), p.436.

An Asymmetric Contact Problem of a Transversely Isotropic Layer Subjected to Moment by an Annular Rigid Punch

Makoto SAKAMOTO*, Qian-Fu ZHU*, Toshiaki HARA**, Toshikazu SHIBUYA***, and Takashi KOIZUMI***

* *Department of Mechanical Engineering, Niigata College of Technology, Niigata*
** *Department of Mechanical Engineering, Niigata University, Niigata*
*** *Department of Mechanical Engineering, Tokyo Institute of Technology, Tokyo*

An exact solution is presented for an asymmetric contact problem of a transversely isotropic layer subjected to moment by an annular rigid punch, with the layer resting on a rigid foundation. The three-part mixed boundary-value problem is reduced to a solution of an infinite system of simultaneous equations using the method of expanding the normal stress under the punch to an appropriate series. Numerical results for the contact stress and the surface displacement in two examples of transversely isotropic materials as well as in an isotropic one are presented.

I. INTRODUCTION

The mixed boundary-value problem of a transversely isotropic half-space has been considered by Elliott[1], Deutsch[2], Svekl0[3], Guidera et al.[4], Pouyet and Lataillade[5], and Fabrikant[6] for the axisymmetric punch problem,by Deutsch[7] and Erguven[8] for the case of torsion,and by Fabrikant[9)-11] for the asymmetric punch problem. However, the punch problem of a transversely isotropic layer has been dealt with in relatively few publications. England[12] and Tsai[13] solved the axisymmetric problem of the contact between a spherical rigid punch and a transversely isotropic layer, utilizing the method of Hankel transforms. The authors presented analytical solutions for axisymmetric stress field in a transversely isotropic layer indented by an annular punch[14] and a circular one[15]. Furthermore, the only asymmetric problem[16], where the layer was subjected to a slightly inclined indentation by a circular punch, was solved recently by the authors.

In this paper, we consider an asymmetric contact problem of a transversely isotropic layer subjected to a moment by an annular rigid punch. The problem is reduced to a solution of an infinite simultaneous equations using the method of expressing the normal stress under the punch as an appropriate series. Numerical results are obtained clarifying the effect of material anisotropy on the stress field.

II. FORMULATION OF THE PROBLEM AND ITS SOLUTION

We denote the cylindrical coordinate of a point by (r,θ,z), where the z-axis is parallel to the axis of symmetry of the transversely isotropic solid. The asymmetric torsionless displacements and stresses can be represented in terms of two potential functions ψ_1 and ψ_2 as follows[17]:

$$\left.\begin{aligned} & u_r=\frac{\partial\psi_i}{\partial r}, \quad v_\theta=\frac{\partial\psi_i}{r\,\partial\theta}, \quad w_z=k_i\frac{\partial\psi_i}{\partial z} \\ & \frac{\sigma_r}{c_{44}}=-(1+k_i)\{\beta_i(\frac{\partial\psi_i}{r\,\partial r}+\frac{\partial^2\psi_i}{r^2\partial\theta^2})+\frac{\partial^2\psi_i}{\partial z^2}\} \\ & \frac{\sigma_\theta}{c_{44}}=-(1+k_i)(\beta_i\frac{\partial^2\psi_i}{\partial r^2}+\frac{\partial^2\psi_i}{\partial z^2}) \\ & \frac{\sigma_z}{c_{44}}=\mu_i(1+k_i)\frac{\partial^2\psi_i}{\partial z^2}, \quad \frac{\tau_{\theta z}}{c_{44}}=(1+k_i)\frac{\partial^2\psi_i}{r\,\partial\theta\,\partial z} \end{aligned}\right\} \quad (i=1,\ 2) \qquad (1)$$

$$\frac{\tau_{rz}}{c_{44}} = (1+k_i)\frac{\partial^2\psi_i}{\partial r\,\partial z}, \quad \frac{\tau_{r\theta}}{c_{44}} = \beta_i(1+k_i)\frac{\partial}{\partial r}\left(\frac{\partial\psi_i}{r\,\partial\theta}\right)$$

where

$$\left(\frac{\partial^2}{\partial r^2} + \frac{\partial}{r\,\partial r} + \frac{\partial^2}{r^2\partial\theta^2} + \mu_i\frac{\partial^2}{\partial z^2}\right)\psi_i = 0, \quad (i=1,2) \tag{2}$$

$$k_i = (c_{11}\mu_i - c_{44})/(c_{13}+c_{44}), \quad (i=1,2) \tag{3}$$

$$\beta_i = (c_{11}-c_{12})/\{c_{44}(1+k_i)\}, \quad (i=1,2) \tag{4}$$

and $\mu_i(i=1,2)$ are the roots of the following characteristic equation:

$$c_{11}c_{44}\mu^2 + \{c_{13}(c_{13}+2c_{44}) - c_{11}c_{33}\}\mu + c_{33}c_{44} = 0 \tag{5}$$

As shown in Fig.1, we consider a transversely isotropic, elastic layer $0 \le z \le h$ of infinite extent resting on a frictionless, rigid foundation at z=0. The layer is subjected to a slightly inclined indentation by a smooth annular rigid punch. The solution of the problem can be obtained by considering two cases; one is the axisymmetric problem of the punch subjected to a normal force P, the other is the asymmetric problem under a moment M. The former problem has already been solved by the authors[13]. In this paper, therefore, we analyze the latter case.

The punch is acted on by the moment M about the y–axis, let r_i and r_0 be respectively the inner and outer radii of the punch and α be the rotational angle. The normal deformation ε_m on the surface is assumed to be sufficiently small and of the form $\varepsilon_m = \alpha r_0$. The boundary conditions on the layer are

$$\left.\begin{array}{ll} \text{(i)} & (\sigma_z)_{z=h} = 0, \quad (0 \leqq r < r_i,\ r_0 < r) \\ \text{(ii)} & (w_z)_{z=h} = -(r/r_0)\varepsilon_m\cos\theta, \quad (r_i \leqq r \leqq r_0) \\ \text{(iii)} & (\tau_{rz})_{z=h} = (\tau_{\theta z})_{z=h} = 0, \quad (0 \leqq r) \\ \text{(iv)} & (w_z)_{z=0} = (\tau_{rz})_{z=0} = (\tau_{\theta z})_{z=0} = 0, \quad (0 \leqq r) \\ \text{(v)} & \text{All stress components vanish at infinity.} \end{array}\right\} \tag{6}$$

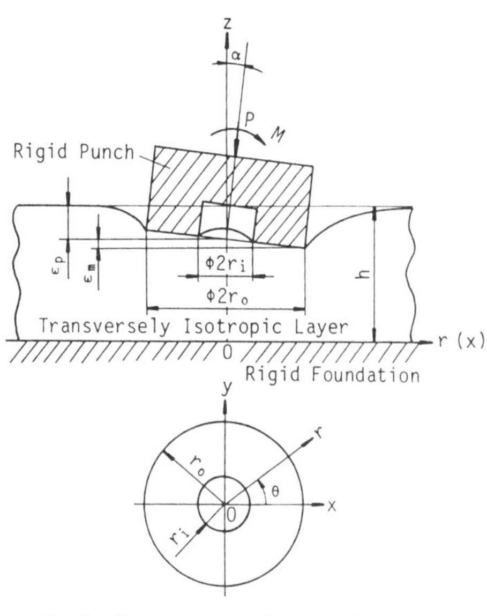

Fig.1. Geometry of the problem.

Choosing the functions ψ_i so as to satisfy the conditions (iv) and (v), we set

$$\psi_i = \cos\theta \int_0^\infty A_i(\lambda)\, J_1(\lambda r) \cosh \lambda z_i \, d\lambda, \quad (i=1,\ 2) \tag{7}$$

where $A_i(\lambda)$ are unknown functions, $J_n(x)$ is the Bessel function of the first kind of order n, and

$$z_i = z/\sqrt{\mu_i}, \quad (i=1,2) \tag{8}$$

From Eqs.(1) and (7), the conditions (iii) leads to

$$A_2(\lambda) = -\eta \sqrt{\mu_2/\mu_1}\, A_1(\lambda) \sinh \lambda h_1 / \sinh \lambda h_2 \tag{9}$$

where

$$\eta = (1+k_1)/(1+k_2) \tag{10}$$

$$h_i = h/\sqrt{\mu_i}, \quad (i=1,\ 2) \tag{11}$$

From Eq.(9), the conditions (i) and (ii) yield the following dual integral equations:

$$\frac{(\sigma_z)_{z=h}}{\cos\theta} = c_{44}(1+k_1) \int_0^\infty \lambda^2 p(\lambda)\, A_1(\lambda)\, J_1(\lambda r)\, d\lambda = 0, \quad (0 \le r < r_i,\ r_o < r) \tag{12}$$

$$\frac{(w_z)_{z=h}}{\cos\theta} = -\frac{1-k_1}{\sqrt{\mu_1}} \int_0^\infty \lambda A_1(\lambda)\, J_1(\lambda r) \sinh \lambda h_1 \, d\lambda = -(r/r_o)\varepsilon_m, \quad (r_i \le r \le r_o) \tag{13}$$

where

$$p(\lambda) = \cosh \lambda h_1 - \sqrt{\mu_2/\mu_1} \sinh \lambda h_1 \cdot \coth \lambda h_2 \tag{14}$$

We can assume that the stress $(\sigma_z)_{z=h}$ in the present problem has the singularity of the form $\{(r^2-r_i^2)(r_0^2-r^2)\}^{-1/2}$. Then, the stress $(\sigma_z)_{z=h}$ can be expressed by the Fourier series as follows:

$$\frac{(\sigma_z)_{z=h}}{\cos\theta} = -\frac{2c_{44} r}{\pi\sqrt{(r^2-r_i^2)(r_o^2-r^2)}} \sum_{n=0}^{\infty} c_n' \cos n\phi$$

$$= -\frac{c_{44} r}{\pi r_c b \sin\phi} \sum_{n=0}^{\infty} c_n' \cos n\phi, \quad (r_i < r < r_o) \tag{15}$$

where

$$\phi = \cos^{-1}\{(r_c^2 + b^2 - r^2)/(2r_c b)\}$$
$$r_c = (r_o + r_i)/2, \quad b = (r_o - r_i)/2 \tag{16}$$

and c_n' $(n=0,1,\cdots)$ are unknown coefficients.

Here, putting

$$M_n(\lambda) = \partial\{J_n(\lambda r_c)\, J_n(\lambda b)\}/\partial\lambda, \quad (n=0,1,\cdots) \tag{17}$$

and utilizing the following integral formula for the Bessel function[18];

$$\int_0^\infty \lambda J_1(\lambda r)\, M_n(\lambda)\, d\lambda = \begin{cases} -\dfrac{r}{\pi r_c b}\dfrac{\cos n\phi}{\sin\phi}, & (r_i < r < r_o) \\ 0, & (0 \le r < r_i,\ r_o < r) \end{cases} \tag{18}$$

we find that the Hankel inversion of Eqs.(12) and (15) yields

$$\lambda(1+k_1)\, p(\lambda)\, A_1(\lambda) = \sum_{n=0}^{\infty} c_n' M_n(\lambda) \tag{19}$$

Substituting Eq.(19) into Eq.(13) and making use of the Gegenbauer's formula, we obtain

$$\sum_{n=0}^{\infty} c_n \int_0^{\infty} q(\lambda)\, M_n(\lambda) \sum_{m=0}^{\infty} (2-\delta_{0m})\, M_m(\lambda) \cos m\phi \, d\lambda$$
$$= 1-\frac{2r_c b}{r_c{}^2+b^2}\cos\phi, \quad (0\leq\phi\leq\pi) \tag{20}$$

where

$$c_n = -\frac{r_0 c_n'}{\varepsilon_m\sqrt{\mu_1}\,\xi(r_c{}^2+b^2)}, \quad (n=0,1,\cdots) \tag{21}$$

$$\xi = (1+k_1)/(1-k_1) \tag{22}$$

$$q(\lambda) = \sinh \lambda h_1/p(\lambda) \tag{23}$$

and δ_{0m} is the Kronecker's delta.

Since Eq.(20) must hold for arbitrary values of ϕ, we obtain the following infinite simultaneous equations for the determination of the coefficients c_n:

$$\sum_{n=0}^{\infty} c_n \int_0^{\infty} q(\lambda)\, M_m(\lambda)\, M_n(\lambda)\, d\lambda = \delta_{0m} - \frac{r_c b}{r_c{}^2+b^2}\delta_{1m},$$
$$(m=0,1,\cdots) \tag{24}$$

From Eqs.(19) and (21), the function $A_1(\lambda)$ can be rewritten as

$$\lambda A_1(\lambda) = -\frac{\varepsilon_m\sqrt{\mu_1}(r_c{}^2+b^2)}{r_0(1-k_1)\,p(\lambda)} \sum_{n=0}^{\infty} c_n M_n(\lambda) \tag{25}$$

For a large value of λ, we obtain

$$q(\lambda) \longrightarrow \frac{\sqrt{\mu_1}}{\sqrt{\mu_1}-\sqrt{\mu_2}} = \gamma \tag{26}$$

Therefore, the normal displacement $(w_z)_{z=h}$ may be expressed as

$$\frac{(w_z)_{z=h}}{\varepsilon_m \cos\theta} = \frac{r_c{}^2+b^2}{r_0} \sum_{n=0}^{\infty} c_n \Big[\int_0^{\infty} \{q(\lambda)-\gamma\}\, M_n(\lambda)\, J_1(\lambda r)$$
$$\times d\lambda - \frac{\gamma r}{2}(I_0^n - I_2^n)\Big], \quad (0\leqq r<r_i,\ r_0<r) \tag{27}$$

where

$$I_j^n = \int_0^{\infty} J_j(\lambda r)\, J_n(\lambda r_c)\, J_n(\lambda b)\, d\lambda \tag{28}$$

and the infinite integral of Eq.(28) can be calculated by the method in ref.19).

From Eqs.(12) and (18), the normal stress $(\sigma_z)_{z=h}$ can be rewritten as

$$\frac{(\sigma_z)_{z=h}}{\cos\theta} = \frac{c_{44}\varepsilon_m\sqrt{\mu_1}\,\xi r(r_c{}^2+b^2)}{\pi r_0 r_c b \sin\phi} \sum_{n=0}^{\infty} c_n \cos n\phi,$$
$$(r_i<r<r_0) \tag{29}$$

The total moment M required to maintain a slight inclination of the punch is

$$M = -\int_0^{2\pi}\int_{r_i}^{r_0} r^2 (\sigma_z)_{z=h} \cos\theta \, dr\, d\theta$$
$$= -\pi c_{44}\varepsilon_m\sqrt{\mu_1}\,\xi(r_c{}^2+b^2)^2\{c_0 - r_c b c_1/(r_c{}^2+b^2)\}/r_0 \tag{30}$$

In the particular case of $h \to \infty$, Eq.(23) gives $q(\lambda) \to \gamma$; hence Eq.(24) can be rewritten as

$$\sum_{n=0}^{\infty} x_n \int_0^{\infty} M_m(\lambda)\, M_n(\lambda)\, d\lambda = \delta_{0m} - \frac{r_c b}{r_c^2 + b^2}\, \delta_{1m}, \quad (m=0,1,\cdots) \tag{31}$$

where

$$x_n = -\frac{r_0 c_n'}{\varepsilon_m(\sqrt{\mu_1} - \sqrt{\mu_2})\xi(r_c^2 + b^2)} \tag{32}$$

Then, we obtain

$$\frac{(w_z)_{z=h}}{\varepsilon_m \cos\theta} = \frac{r(r_c^2+b^2)}{2r_0} \sum_{n=0}^{\infty} x_n\, (I_0^n - I_2^n), \quad (0 \leqq r < r_i,\ r_0 < r) \tag{33}$$

$$\frac{r_0^3(\sigma_z)_{z=h}}{M\cos\theta} = \frac{r_0^3}{\pi^2 r_c b(r_c^2+b^2)\sin\phi} \times \frac{r}{(x_0 - r_c b x_1)/(r_c^2+b^2)} \sum_{n=0}^{\infty} x_n \cos n\phi, \quad (r_i < r < r_0) \tag{34}$$

$$M = \pi c_{44}\varepsilon_m(\sqrt{\mu_1} - \sqrt{\mu_2})\xi(r_c^2+b^2)^2 \times \{x_0 - r_c b x_1/(r_c^2+b^2)\}/r_0 \tag{35}$$

Equations (31), (33) and (34) coincide with the results of the problem for an isotropic half–space[18]. The contact stress $(\sigma_z)_{z=h}$, under the action of the normal force P and the moment M as shown in Fig.1, can be obtained by superposing Eq.(29) on the axisymmetric stress field[14], i.e.,

$$(\sigma_z)_{z=h} = \frac{c_{44}\sqrt{\mu_1}\xi\varepsilon_p}{\pi r_c b \sin\phi} \sum_{n=0}^{\infty} \{a_n + \frac{\varepsilon_m}{\varepsilon_p} \frac{r(r_c^2+b^2)}{r_0} c_n \cos\theta\} \times \cos n\phi, \quad (r_i < r < r_0) \tag{36}$$

where the coefficients a_n are solutions of the infinite simultaneous equations in ref.19).

III. NUMERICAL RESULTS AND DISCUSSION

The parameters μ_i and k_i are related to the engineering constants as follows:

$$\left.\begin{aligned} \mu_i &= (\chi \pm \sqrt{\chi^2 - 4\omega})/2 \\ k_i &= \frac{(\zeta - \nu'^2)\mu_i - \rho\{(1-\nu)\zeta - 2\nu'^2\}/2}{(1+\nu)\nu'\zeta + \rho\{(1-\nu)\zeta - 2\nu'^2\}/2} \end{aligned}\right\} (i=1,2) \tag{37}$$

where

$$\left.\begin{aligned} \zeta &= E'/E, \quad \rho = G'/G \\ \chi &= 2\zeta(1+\nu)(\zeta - \nu'\rho)/\{\rho(\zeta - \nu'^2)\} \\ \omega &= \zeta^2(1-\nu^2)/(\zeta - \nu'^2) \end{aligned}\right\} \tag{38}$$

and E, G and ν are Young's modulus, shear modulus and Poisson's ratio in the isotropic plane, respectively. The symbol " ' " denotes these constants along the principal direction of anisotropy.

Numerical calculations are performed for E–glass/epoxy composite[20] and single crystals of magnesium[21]. These elastic constants are listed in Table 1. In Tables 2 and 3, we show the coefficients c_n. Since the convergence of c_n becomes slow as the value of r_i/r_0 decreases, it had been checked in advance that 15 terms of c_n are sufficient to evaluate the stress in the case of $r_i/r_0=0.25$. Following results are compared with those of an isotropic layer[19] which are independent of Poisson's ratio.

Table 1. Elastic constants.

Material	ζ	ρ	ν	ν'
E-glass/Epoxy V_f=0.63*	3.44	1.01	0.37	0.24
Cadmium	0.35	0.45	0.12	0.25

*fiber volume fraction

Table 2. Values of coefficients c_n (r_i/r_o=0.5).

n	E/E			Cd		
	h/r_o=1.5	h/r_o=2	h/r_o=3	h/r_o=1.5	h/r_o=2	h/r_o=3
0	1.95333505	1.74424732	1.58316600	0.72490759	0.70758551	0.69704394
1	-0.87566072	-0.87272353	-0.85140166	-0.39542811	-0.38929372	-0.38458746
2	-0.19235715	-0.14720009	-0.11783668	-0.05240038	-0.05013389	-0.04903343
3	-0.04655795	-0.03474704	-0.02742885	-0.01219549	-0.01170871	-0.01147431
4	-0.01176370	-0.00922428	-0.00749094	-0.00334979	-0.00322644	-0.00316491
5	-0.00334490	-0.00268065	-0.00220804	-0.00099006	-0.00095482	-0.00093695
6	-0.00102107	-0.00082093	-0.00068015	-0.00030537	-0.00029470	-0.00028924
7	-0.00031736	-0.00025508	-0.00021177	-0.00009513	-0.00009183	-0.00009013
8	-0.00011003	-0.00008965	-0.00007542	-0.00003398	-0.00003286	-0.00003227
9	-0.00001895	-0.00001379	-0.00001037	-0.00000455	-0.00000433	-0.00000423

Table 3. Values of coefficients c_n (h/r_o=1.5).

n	E/E			Cd		
	r_i/r_o=0.25	r_i/r_o=0.5	r_i/r_o=0.75	r_i/r_o=0.25	r_i/r_o=0.5	r_i/r_o=0.75
0	2.51680327	1.95333505	1.27794612	0.91721386	0.72490759	0.49315308
1	-1.30234242	-0.87566072	-0.41183528	-0.57861763	-0.39542811	-0.18831378
2	-0.40999535	-0.19235715	-0.04295180	-0.10714526	-0.05240038	-0.01285996
3	-0.14816280	-0.04655795	-0.00499293	-0.03822053	-0.01219549	-0.00141581
4	-0.06098277	-0.01176370	-0.00058192	-0.01694420	-0.00334979	-0.00018164
5	-0.02919728	-0.00334490	-0.00007694	-0.00833442	-0.00099006	-0.00002584
6	-0.01508534	-0.00102107	-0.00003415	-0.00434016	-0.00030537	-0.00001281
7	-0.00809904	-0.00031736		-0.00234008	-0.00009513	
8	-0.00446005	-0.00011003		-0.00129288	-0.00003398	
9	-0.00248315	-0.00001895		-0.00071992	-0.00000455	
10	-0.00141686			-0.00041290		
11	-0.00076946			-0.00022082		
12	-0.00046247			-0.00013687		
13	-0.00014842			-0.00003537		

The radial distributions of the dimensionless contact stress $(\overline{\sigma}_z)_{z=h}=r_0^3(\sigma_z)_{z=h}/M\cos\theta$ are shown in Figs.2 to 4. The minimum value of $(\overline{\sigma}_z)_{z=h}$ occurs at the vicinity of the inner edge of the punch and magnitude of the value is in the order of E/E>Iso.>Cd. In the limit case as $h/r_0\rightarrow\infty$, the curve of $(\overline{\sigma}_z)_{z=h}$ agrees with that of an isotropic half-space.

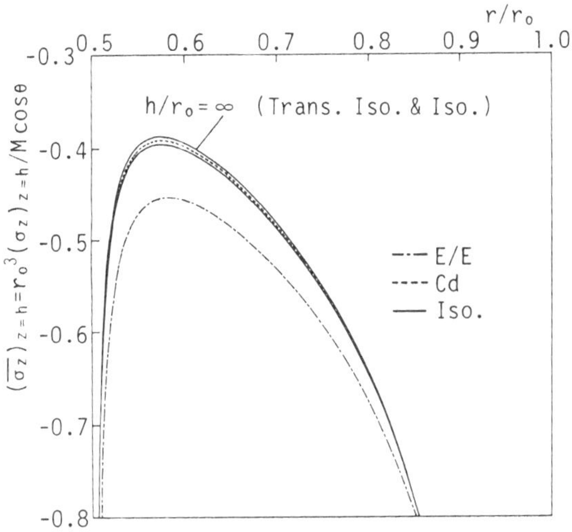

Fig.2. Distribution of $(\overline{\sigma}_z)_{z=h}$ (r_i/r_0=0.5, h/r_0=1.5).

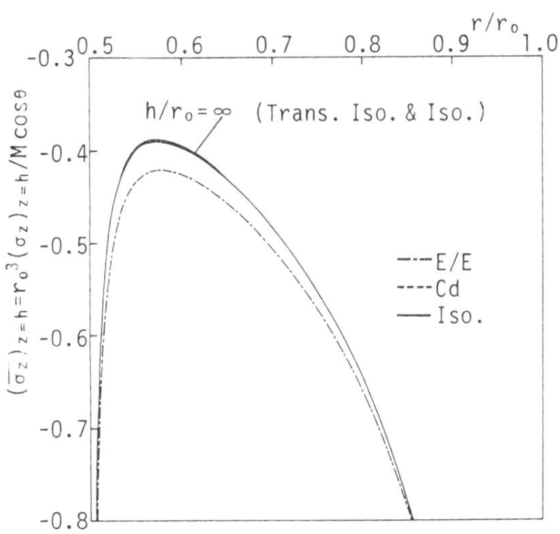

Fig.3. Distribution of $(\overline{\sigma}_z)_{z=h}$ (r_i/r_0=0.5, h/r_0=2).

Figures 5 and 6 show the radial distributions of the dimensionless displacement $(\overline{w}_z)_{z=h}=(w_z)_{z=h}/\varepsilon_m\cos\theta$. It is noted that the slope of $(\overline{w}_z)_{z=h}$ approaches infinity as $r\rightarrow r_i-0$ and $r\rightarrow r_0+0$. The magnitude of $(\overline{w}_z)_{z=h}$ for $0<r<r_i$ and $r_0<r$ becomes lower with decreasing h/r_0.

Figures 7 and 8 show the variations of the dimensionless tilting moment $\overline{M}=(1-\nu)M/G\varepsilon_m r_0^3$ with h/r_0 and r_i/r_0, respectively. It is seen from Fig.7 that $\overline{M}$ decreases with increasing h/r_0. Figure 8 shows that $\overline{M}$ decreases with increasing r_i/r_0 and the decreasing rate of M for E/E is more significant than that for other materials.

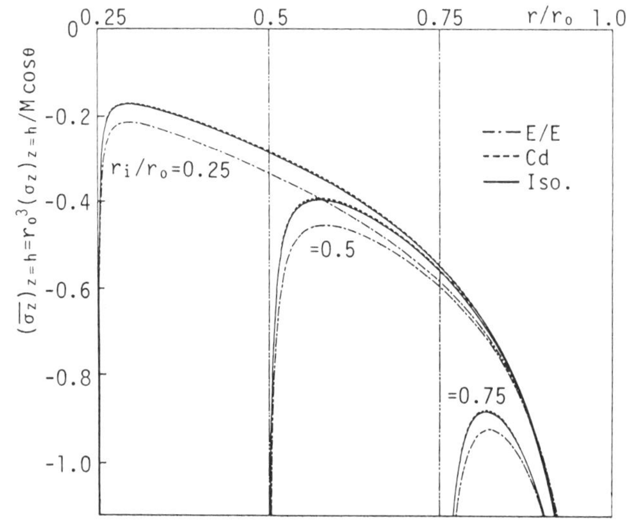

Fig.4. Distribution of $(\overline{\sigma}_z)_{z=h}$ $(h/r_0=1.5)$.

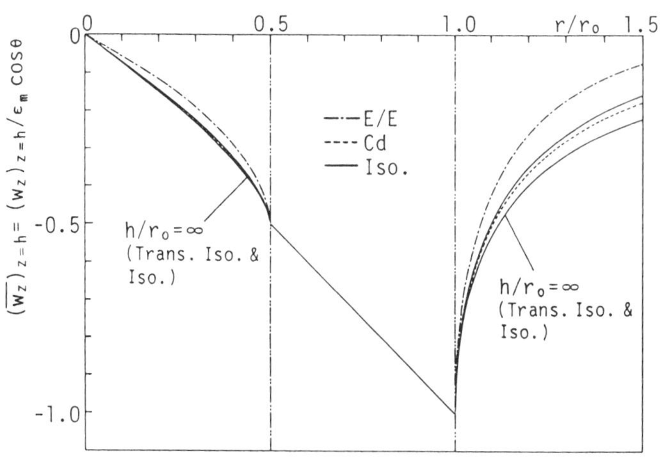

Fig.5. Distribution of $(\overline{w}_z)_{z=h}$ $(r_i/r_0=0.5,\ h/r_0=1.5)$.

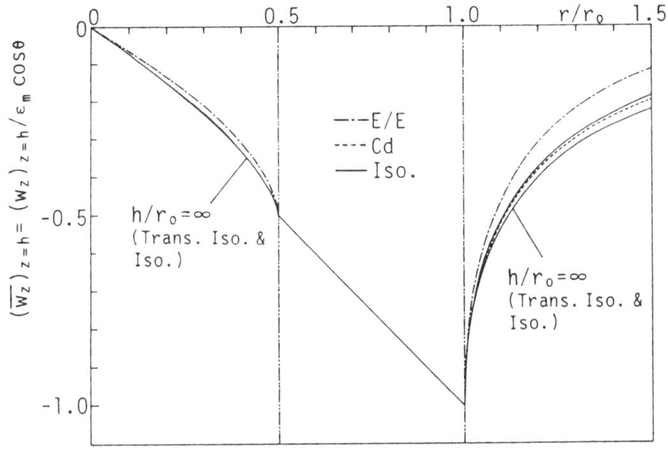

Fig.6. Distribution of $(\overline{w}_z)_{z=h}$ $(r_i/r_0=0.5,\ h/r_0=2)$.

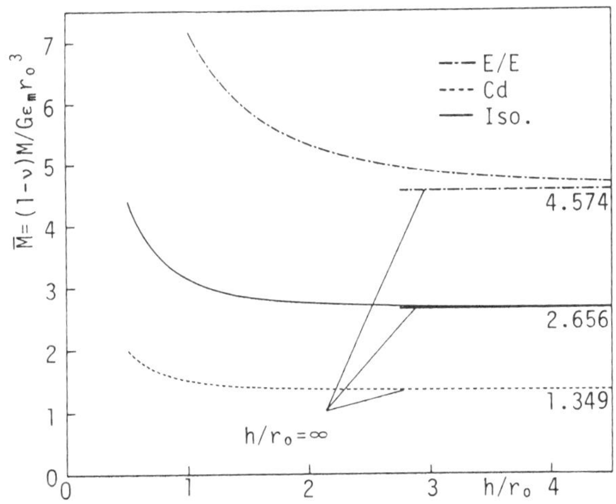

Fig.7. Variation of $\overline{M}$ with h/r_o $(r_i/r_0=0.5)$.

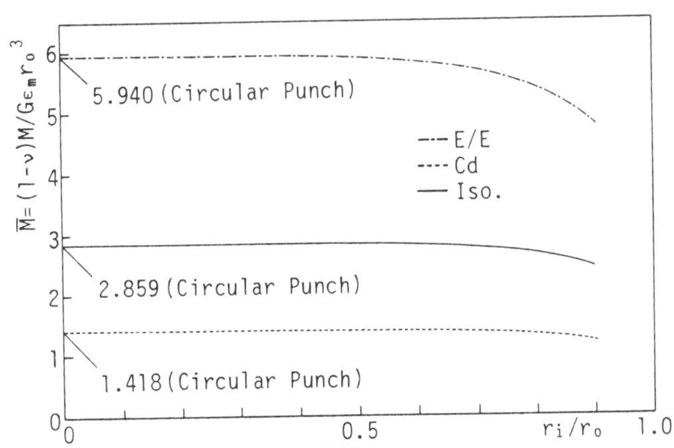

Fig.8. Variation of $\overline{M}$ with r_i/r_0 $(h/r_o=1.5)$.

Finally, by utilizing the axisymmetric stress field[14], we consider that the layer is subjected to the slightly inclined indentation by the annular punch. It is clear that $\varepsilon_m/\varepsilon_P$ must be chosen so that the face of the punch does not separate from the layer. To prevent separation from the surface, the following inequality must hold for $\varepsilon_m/\varepsilon_P$:

$$\varepsilon_m/\varepsilon_p \leqq (\varepsilon_m/\varepsilon_p)_{max.} \tag{39}$$

where $(\varepsilon_m/\varepsilon_P)_{max.}$ satisfies the following condition:

$$\lim_{r \to r_0 - 0} (\sigma_z)_{z=h,\,\theta=\pi} = 0 \tag{40}$$

Figures 9 and 10 show the variations of $(\varepsilon_m/\varepsilon_P)_{max.}$ with h/r_0 and r_i/r_0, respectively. Figures 11 and 12 show respectively the dimensionless stress $(\sigma_z^*)_{z=h}=r_0^2(\sigma_z)_{z=h}/P$ and displacement $(w_z^*)_{z=h}=(w_z)_{z=h}/\varepsilon_P$ for $r_i/r_0=0.5$, $h/r_0=1.5$, and $\varepsilon_m/\varepsilon_P=0.5$. In the limit case as $h/r_0 \to \infty$(a transversely isotropic half-space), the curves $(\sigma_z^*)_{z=h}$ and $(w_z^*)_{z=h}$ agree with those of the isotropic half-space.

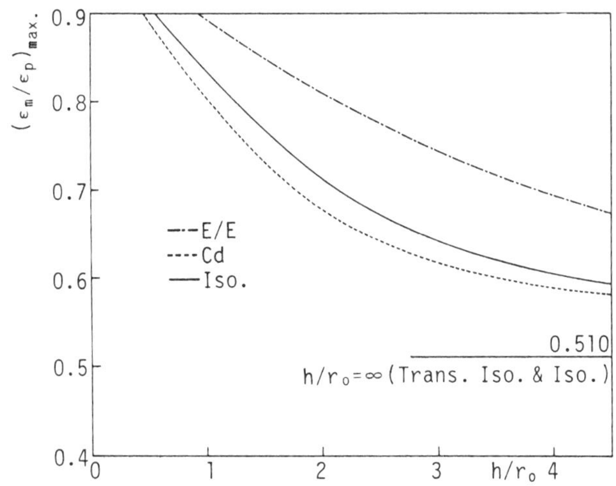

Fig.9. Variation of $(\varepsilon_m/\varepsilon_P)_{max.}$ with h/r_o $(r_i/r_0=0.5)$.

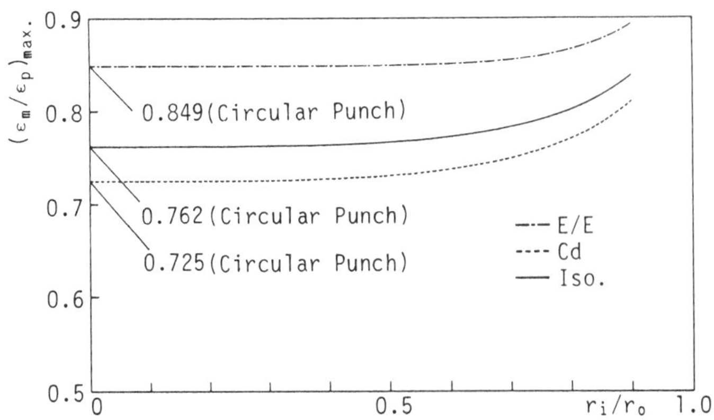

Fig.10. Variation of $(\varepsilon_m/\varepsilon_P)_{max.}$ with r_i/r_0 $(h/r_0=1.5)$.

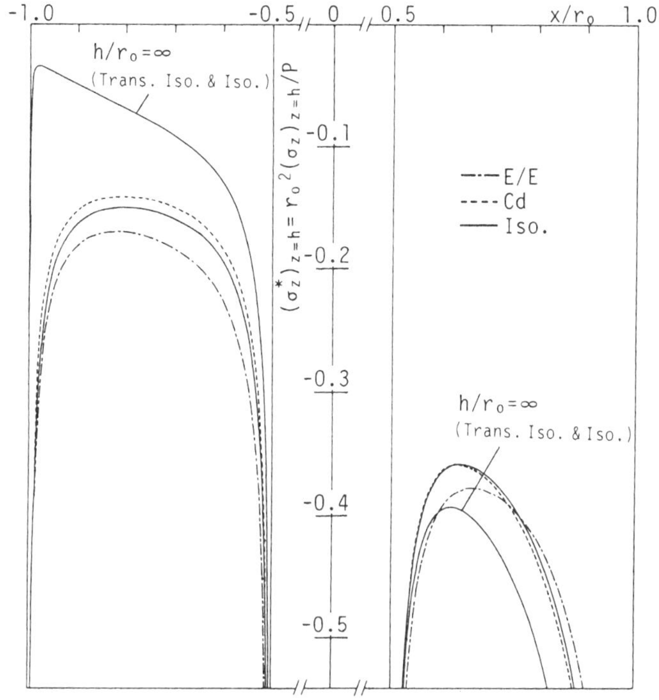

Fig.11. Distribution of $(\sigma_z^*)_{z=h}$ (r_i/r_0=0.5, h/r_0=1.5, $\varepsilon_m/\varepsilon_P$=0.5).

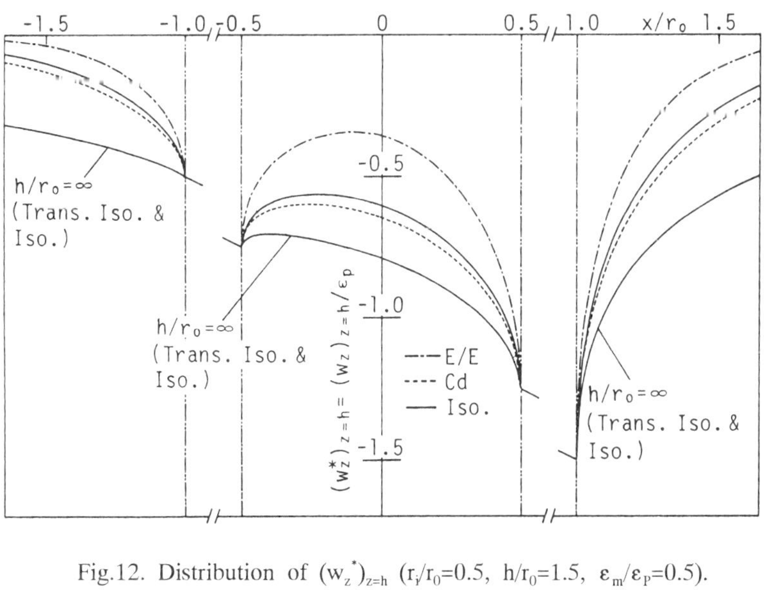

Fig.12. Distribution of $(w_z^*)_{z=h}$ (r_i/r_0=0.5, h/r_0=1.5, $\varepsilon_m/\varepsilon_P$=0.5).

REFERENCES

1) Elliott,H.A., Axial symmetric stress distributions in aeolotropic hexagonal crystals, the problem of the plane and related problems, Proc.Camb.Phil.Soc., 45(1949), pp.621–630.
2) Deutsch,E., Indentation of a transversely isotropic half–space by a cylindrical, conical and paraboloidal punch, Acad.Pol.Sci.Class.Quart., 11(1963), pp.311–319.
3) Sveklo,V.A., The action of a stamp on an elastic anisotropic half–space, J.Appl.Math.Mech., 34(1970), pp.165–171.
4) Guidera,J.T., Lardner,R.W. and Tupholme,G.E., The indentation of a half–space of hexagonal elastic material by a circular punch of arbitrary end–profile J.Eng.Math., 12(1978), pp.77–82.
5) Pouyet,J. and Lataillade,J.L., Elastic strain of a transversely isotropic half–space caused by a rigid punch, Lett. Appl.Eng.Sci., 17(1979), pp.765–772.
6) Fabrikant,V.I., Four types of exact solutions to the problem of an axisymmetric punch bonded to a transversely isotropic half–space, Int.J.Eng.Sci., 24(1986), pp.785–801.
7) Deutsch,E., Torsion of a transversely isotropic semi–infinite body, Arch.Mech.,15(1963), pp.517–533.
8) Erguven,M.E., Torsional loading of an elastic transversely isotropic nonhomogeneous semi–infinite solid, Acta Mech., 61(1986), pp.165–174.
9) Fabrikant,V.I., Effect of shearing force and tilting moment on a cylindrical punch attached to a transversely isotropic half–space, J.Appl.Math.Mech., 35(1971), pp.147–151.
10) Fabrikant,V.I., Inclined circular punch bonded to transversely isotropic half–space, J.Eng.Mech., 116(1990), pp.1599–1617.
11) Fabrikant,V.I., Nonaxisymmetric annular punch problem, Trans.ASME, J.Appl.Mech., 58(1991), pp.947–953.
12) England,A.H., A punch problem for a transversely isotropic layer, Proc.Camb.Phil.Soc., 58(1962), pp.539–547.
13) Tsai,Y.M., Symmetrical contact problem of a thick transversely isotropic plate, J.Elasticity, 16(1986), pp.179–188.
14) Hara,T., Sakamoto,M., Shibuya,T. and Koizumi,T., An axisymmetric contact problem of a transversely isotropic layer indented by an annular punch, JSME Int.J.Ser.I, 32(1989), pp.94–100.
15) Sakamoto.M., Hara,T., Shibuya,T. and Koizumi,T., Indentation by a circular rigid punch of a transversely isotropic layer on a rigid foundation, JSME Int.J.Ser.I, 34(1991), pp.130–134.
16) Sakamoto,M., Hara,T., Shibuya,T. and Koizumi,T., An asymmetric contact problem of a transversely isotropic layer subjected to moment by a circular rigid punch, Trans.JSME, (in Japanese), 57A(1991), pp.3016–3022.
17) Elliott,H.A., Three–dimensional stress distributions in hexagonal aeolotropic crystals, Proc.Camb.Phil.Soc., 44(1948), pp.522–533.
18) Hara,T.,Shibuya,T., Koizumi,T. and Nakahara,I., An asymmetric mixed boundary value problem of the elastic half–space subjected to moment by an annular rigid punch, Bull.JSME, 21(1978), pp.566–571.
19) Hara,T., Shibuya,T., Koizumi,T. and Nakahara,I., An asymmetric mixed boundary value problem of the thick elastic plate: moment applied to the annular rigid stamp, Theo.Appl.Mech., 25(1976),pp.505–512.
20) Kasano,H., Tsuchiya,H., Matsumoto,H. and Nakahara,I., Stresses and deformations in a transversely isotropic hollow cylinder under a ring of radial load, Bull.JSME,25(1982), pp.143–148.
21) Huntington,H.B., The elastic constant of crystals, Solid State Physics 7,(Academic Press, 1958), pp.213.

Torsion of a Transversely Isotropic Cylinder with an External Crack

Takao AKIYAMA*, Toshiaki HARA**, Toshikazu SHIBUYA***, and Takashi KOIZUMI***

* *Department of Mechanical System Engineering, Yamagata University, Yonezawa, Yamagata*
** *Department of Mechanosystem Engineering, Niigata University, Niigata*
*** *Department of Mechanical Production Engineering, Tokyo Institute of Technology, Tokyo*

The Paper deals with the axisymmetric torsion problem for a transversely isotropic cylinder with an external crack. Expressing the stress component along the crack plane as an appropriate series, the problem is reduced to the solution of an infinite system of simultaneous equations. Displacement and stress distributions around the crack tip are shown graphically for various magnitudes of the crack radius. It is shown that the stress intensity factor is independent of transverse isotropy under constant torque.

I. INTRODUCTION

In resent years, we have been making use of the fiber reinforced composites such as FRP for many kinds of machines and structures. Composites reinforced in one direction can particulary be assumed to be transversely isotropic elasticity.

On the other hand, from the viewpoint of the strength design for various turning shafts, elastic solutions for the cylinders with cracks under torsion are basic and significant in engineering. In such researches, many investigators[1)-4)] have dealt with isotropic cylinders with cracks subjected to torsion, and a few[5)] analyzed the corresponding problems for the case of transverse isotropy. Then, in this study, we analyze the transversely isotropic infinite cylinder, as composites reinforced in one direction, with an external crack subjected to torque. By expressing the tangent stress along the crack plane as an appropriate series with unknown coefficients, the problem is reduced to the solution of an infinite system of algebraic equations. Numerical results are presented in order to illustrate displacement and stress distributions around the crack tip and the stress intensity factor for various magnitudes of the crack radius and the effect of transverse isotropy on the displacement and stress fields.

II. PROBLEM FORMURATION

In this problem, a circular cylindrical coodinate system (r, θ, z) is employed and the z-axis is chosen as the principal axis of transverse isotropy. If (u_r, v_θ, w_z) denotes the displacement vector and $(\sigma_r, \sigma_\theta, \sigma_z, \tau_{rz}, \tau_{\theta z}, \tau_{r\theta})$ the stress tensor, a general solution of the axisymmetric equilibrium equations with torsion can be expressed by the stress function φ_3 [6)] as follows:

$$v_\theta = -\partial\varphi_3/\partial r, \quad u_r = w_z = 0,$$

$$\tau_{\theta z} = G'\,\partial^2\varphi_3/(\partial r\,\partial z), \quad \tau_{r\theta} = G'\,\partial^2\varphi_3/\partial z^2 + 2G\,\partial\varphi_3/(r\,\partial r), \tag{1}$$

$$\sigma_r = \sigma_\theta = \sigma_z = \tau_{rz} = 0,$$

where G and G' are the shear modulus of the isotropic plane ($r\theta$-plane) and the one of the orthotropic axis (z-axis), respectively. The stress function satisfies the following equation.

$$\{\partial^2/\partial r^2+\partial/(r\partial r)+\partial^2/\partial z_3^2\}\varphi_3=0, \tag{2}$$

where $z_3=z/\mu_3^{1/2}$ and $\mu_3=G'/G$.

We consider an infinite transversely isotropic cylinder of radius r_o with an external crack of radius r_c subjected to torque T, as shown in Fig.1. Since the displasement and stress fields are symmetry on the plane containing the crack, the boundary conditions of this problem are taken as

$$\begin{aligned}
&(\mathrm{i})\ (v_\theta)_{z=0}=0,\ (0\leq r\leq r_c),\\
&(\mathrm{ii})\ (\tau_{\theta z})_{z=0}=0,\ (r_c<r\leq r_o),\\
&(\mathrm{iii})\ (\tau_{r\theta})_{r=r_o}=0,\ (0\leq z<\infty),\\
&(\mathrm{iv})\ \text{Torque}=T\ \text{at}\ z\to\infty.
\end{aligned} \tag{3}$$

To satisy the stress field of the cylinder at infinity, we choose the function φ_3 as

$$\varphi_3=A_0(r^2-2z_3^2)+B_0z_3(2z_3^2-3r^2)+\sum_{m=1}^{\infty}A_mJ_0(\xi_m r)\exp(-\xi_m z_3), \tag{4}$$

where $J_n(x)$ is a Bessel funtion of the first kind in order n , and A_m and B_0 are unknown coefficients of m. From Eqs.(1) and (4) we obtain the components of displacement and stress.

$$\begin{aligned}
v_\theta&=-2A_0r+6B_0rz^3+\sum_{m=1}^{\infty}\xi_mA_mJ_1(\xi_m r)\exp(-\xi_m z_3),\\
\tau_{\theta z}&=6(GG')^{1/2}B_0r-(GG')^{1/2}\sum_{m=1}^{\infty}\xi_m^2A_mJ_1(\xi_m r)\exp(-\xi_m z_3),\\
\tau_{r\theta}&=-G\sum_{m=1}^{\infty}\xi_m^2A_mJ_2(\xi_m r)\exp(-\xi_m z_3).
\end{aligned} \tag{5}$$

It is immediately found that the boudary condition (iii) will be satisfy for all $z\geqq 0$ if $\xi_m(0<\xi_1<\xi_2\cdots)$ being positive roots of equation

$$J_2(\xi_m r_o)=0. \tag{6}$$

The applied torque T from the boundary condition (iv) is given by

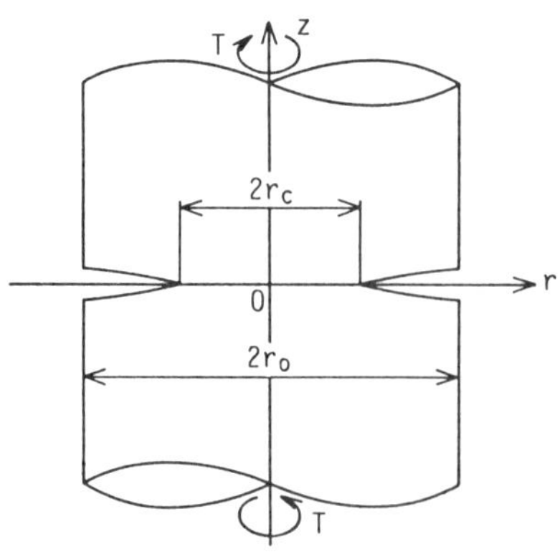

Fig.1. Geometry of the problem.

$$T=2\pi\int_0^{r_o} r^2(\tau_{\theta z})_{z\to\infty}dr=3\pi(GG')^{1/2}B_o r_o^4, \tag{7}$$

so that

$$B_o=T/\{3\pi(GG')^{1/2}r_o^4\}. \tag{8}$$

The remaining two boundary conditions (i) and (ii) may be written in the form

$$(v_\theta)_{z=0}=-2A_o r+\sum_{m=1}^{\infty}\xi_m A_m J_1(\xi_m r)=0,\quad (0\leq r\leq r_c), \tag{9}$$

$$(\tau_{\theta z})_{z=0}=2Tr/(\pi r_o^4)-(GG')^{1/2}\sum_{m=1}^{\infty}\xi_m^2 A_m J_1(\xi_m r)=0,\quad (r_c<r\leq r_o), \tag{10}$$

from Eq.(8). The problem is now reduced to the determination of A_m in the dual series equations (9) and (10). To solve the Eqs.(9) and (10), we use the analytical method described in Ref.(3).

Taking the results by Sih[7] and Kumar[8] into consideration, we can find that the stress $(\tau_{\theta z})_{z=0}$ has the singularity at the crack tip ($r=r_c$) and is a finite and continious function of r in $0\leq r<r_c$. Therefore, the stress may be expressed in the form of the suitable series with unknown coefficiants α_n by

$$(\tau_{\theta z})_{z=0}=(r_c^2-r^2)^{-1/2}\sum_{n=0}^{\infty}(-1)^{n+1}\alpha_n T_{2n+1}(r/r_c)H(r_c-r),\quad (0\leq r\leq r_o), \tag{11}$$

where $T_n(x)$ and $H(x)$ are Chebyshev polynomial of the first kind in order n and Heaviside's step function, respectively. Making a comparison between Eqs.(10) and (11), we immediately get the following identical equation:

$$\begin{aligned}&2Tr/(\pi r_o^4)-(GG')^{1/2}\sum_{m=1}^{\infty}\xi_m^2 A_m J_1(\xi_m r)\\&=(r_c^2-r^2)^{-1/2}\sum_{n=0}^{\infty}(-1)^{n+1}\alpha_n T_{2n+1}(r/r_c)H(r_c-r),\quad (0\leq r\leq r_o).\end{aligned} \tag{12}$$

Expanding the right hand side of Eq.(12) in $0\leq r\leq r_o$ with respect to $J_1(\xi_m r)$ under the condition of Eq.(6) and equating the coefficiants of $J_1(\xi_m r)$ on both sides, we obtain

$$T=4\pi r_c^2\sum_{n=0}^{\infty}\alpha_n/\{(2n-1)(2n+1)(2n+3)\}, \tag{13}$$

$$A_m=\{r_c^2(GG')^{-1/2}/2\}\sum_{n=0}^{\infty}\alpha_n P(n,m), \tag{14}$$

where

$$P(n,m)=\{j_n(z_m)n_{n-1}(z_m)-j_{n+1}(z_m)n_n(z_m)\}/\{\xi_m r_o^2 J_1^2(\xi_m r_o)\},\quad z_m=\xi_m r_c/2, \tag{15}$$

and $j_n(x)$ and $n_n(x)$ are spherical Bessel functions of the first and second kinds in order n, respectively. It is shown that Eq.(14) for arbitrary coefficients α_n satisfies the boundary condition for $(\tau_{\theta z})_{z=0}$. Then, to determine the coefficients α_n, we use Eq.(9). Expanding r and $J_1(\xi_m r)$ in $0\leq r\leq r_c$ with a Chebyshev polynomial, we get[9]

$$r=r_c T_1(r/r_c),$$

$$J_1(\xi_m r)=2\sum_{p=0}^{\infty}(-1)^p J_p(z_m)J_{p+1}(z_m)T_{2p+1}(r/r_c),\quad (0\leq r\leq r_c). \tag{16}$$

Substituting Eqs.(16) into Eq.(9), we get

$$\sum_{p=0}^{\infty}(-1)^p T_{2p+1}(r/r_c)\sum_{m=1}^{\infty}\xi_m A_m J_p(z_m)J_{p+1}(z_m)=A_0 r_c T_1(r/r_c),\quad (0\leq r\leq r_c). \tag{17}$$

Since Eq.(17) must hold for arbitrary value of r in $0\leq r\leq r_c$, we obtain

$$\sum_{m=1}^{\infty}\xi_m A_m J_p(z_m)J_{p+1}(z_m)=A_0 r_c \delta_{0p},\quad (p=0,1,2,\cdots), \tag{18}$$

where δ_{ij} denotes a Kronecker delta. Then the addition of the p-th equation to the (p+1)-th one in the Eqs.(18) leads to the following equations

$$\sum_{m=1}^{\infty}A_m J_{p+1}^2(z_m)=A_0 r_c^2\delta_{0p}/4,\quad (p=0,1,2,\cdots). \tag{19}$$

Substituting Eq.(14) into Eqs.(19) and interchanging the order of summations, we obtain an infinite system of simultaneous equations

$$\sum_{n=0}^{\infty}\alpha_n D(n,p)=A_0(GG')^{1/2}\delta_{0p},\quad (p=0,1,2,\cdots), \tag{20}$$

where

$$D(n,p)=2\sum_{m=1}^{\infty}J_{p+1}^2(z_m)P(n,m). \tag{21}$$

Once $\alpha_n/\{A_0(GG')^{1/2}\}$ are determined by Eqs.(20), the coefficients A_0, A_m and the displacement and stresses at any point may be calculated by Eqs.(13),(14),(5) and (11) in order.

Defining the stress intensity factor K_{III} in mode III by

$$K_{III}=\lim_{r\to r_c-0}\{2\pi(r_c-r)\}^{1/2}(\tau_{\theta z})_{z=0}, \tag{22}$$

we see, by using Eq.(11), that

$$K_{III}=(\pi/r_c)^{1/2}\sum_{n=0}^{\infty}(-1)^{n+1}\alpha_n. \tag{23}$$

It is shown that the stress intensity factor is independent of transverse isotropy under constant torque by Eqs.(13) and (23), and the results in the case of $\mu_3=1$ agree with those in the case of the isotrpic cylinder[3].

The increase of a twist angle due to the presence of a crack, $\Delta\theta$ is given as

$$\Delta\theta=(v_\theta-v_\theta^0)_{z\to\infty}/r=-2A_0. \tag{24}$$

III. NUMERICAL RESULTS

To determine $\alpha_n/\{A_0(GG')^{1/2}\}$ from Eqs.(20), we calculated Eq.(21) by the same procedure given in Ref.(3). The coefficient numbers N_0 of $\alpha_n/\{A_0(GG')^{1/2}\}$ for parameters r_c/r_0 were determined from the acceptable convergence of the values of stress and the stress intensity factor. Table 1 shows the values of coefficients $\alpha_n/\{A_0(GG')^{1/2}\}$ for r_c/r_0=0.1, 0.5 and 0.9. N_0 becomes larger with an increase of r_c/r_0, and for example, it is observed that N_0=10 for r_c/r_0=0.9. In this section, we mention the nondimensional displacement and stresses defined by $r_o\tau_\infty/G$ and τ_∞, respectively, where $\tau_\infty=2T/(\pi r_o^3)$, and the various nondimensional quantities are shown with a symbol "*" for convenience.

Figs.2 to 4 show the distributions of v_θ^*, $\tau_{\theta z}^*$ and $\tau_{r\theta}^*$ for $r_c/r_o=0.5$ and $\mu_3=0.5$, 1

Table 1. Values of $\alpha_n/\{A_o(GG')^{1/2}\}$.

$x\alpha_n/A_0$	$r_c/r_o=0.1$	$r_c/r_o=0.5$	$r_c/r_o=0.9$
0	0.5103434	3.3185981	60.1110248
1	-0.0000049	0.0375936	17.7171569
2	0.0000113	-0.0005031	- 1.8121818
3	-0.0000131	0.0000060	0.3247854
4	0.0000124	0.0000050	- 0.0679361
5			0.0150914
6			- 0.0033005
7			0.0005489
8			0.0000658
9			- 0.0001396

$x=2/(GG')^{1/2}$

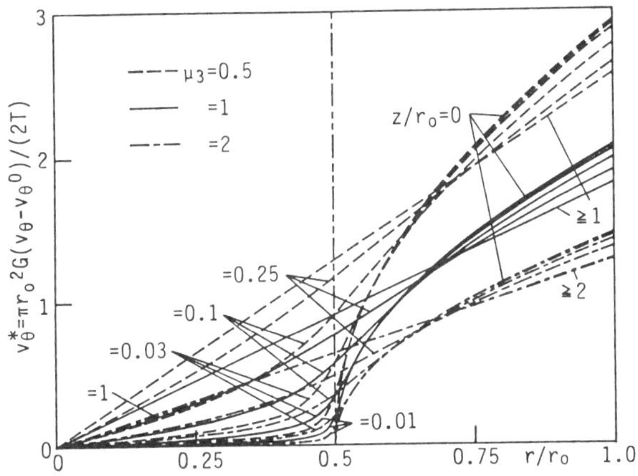

Fig.2. The distribution of v_θ^* for $r_c/r_o=0.5$ and $\mu_3=0.5$, 1 and 2.

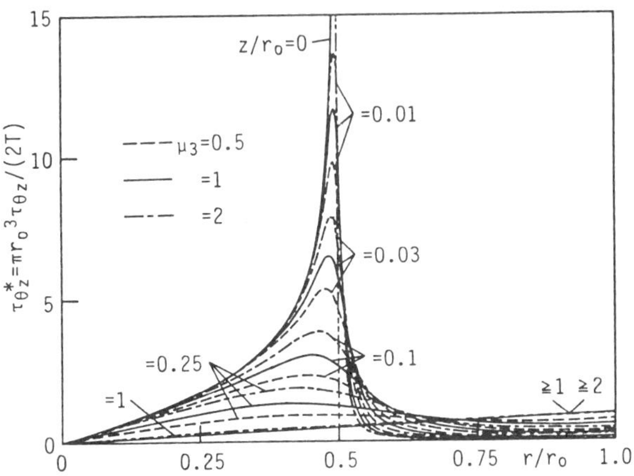

Fig.3. The distribution of $\tau_{\theta z}^*$ for $r_c/r_o=0.5$ and $\mu_3=0.5$, 1 and 2.

and 2, respectively, where $v_\theta^0 = 2Trz/(\pi G' r_o^4)$. It can be seen from Fig. 2 that $(v_\theta^*)_{\mu_3=0.5} > (v_\theta^*)_{\mu_3=1} > (v_\theta^*)_{\mu_3=2}$ for any value of z/r_o. From Figs. 3 and 4, we find that both

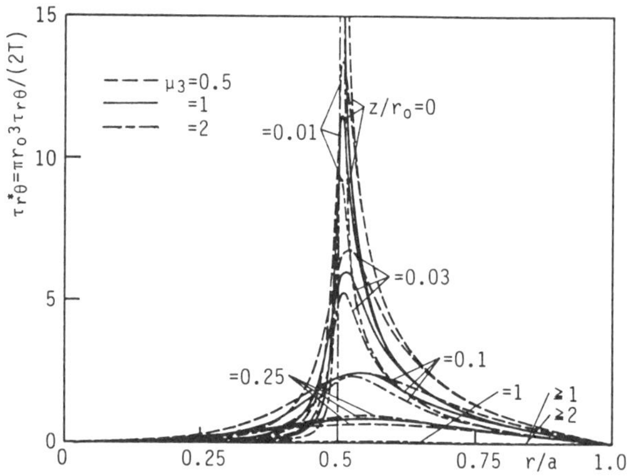

Fig. 4. The distribution of $\tau_{r\theta}^*$ for $r_c/r_o=0.5$ and $\mu_3=0.5$, 1 and 2.

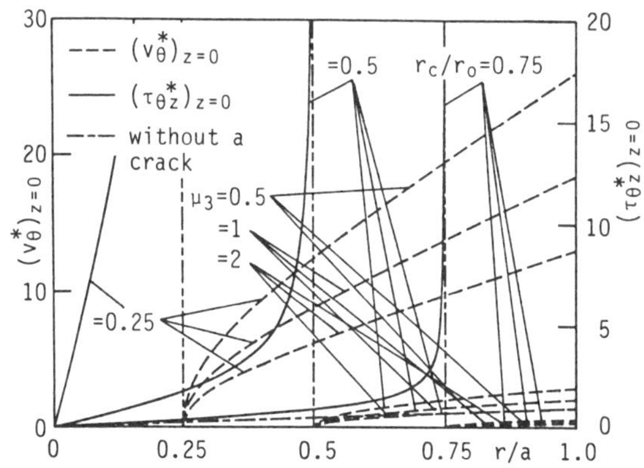

Fig. 5. The variation of $(v_\theta^*)_{z=0}$ and $(\tau_{\theta z}^*)_{z=0}$ with r_c/r_o and μ_3.

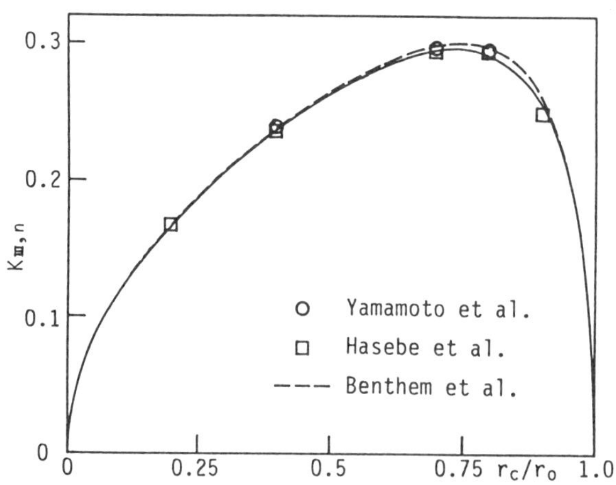

Fig. 6. The variation of $K_{III,n}^*$ with r_c/r_o.

stresses $\tau_{\theta z}^*$ and $\tau_{r\theta}^*$ for each range of μ_3 distribute smoothly and tend to the results without the crack as z/r_o increases.

Fig.5 shows that the distributions of $(v_\theta^*)_{z=0}$ and $(\tau_{\theta z}^*)_{z=0}$ for different values of r_c/r_o and μ_3. It is seen that the magnitude of $(v_\theta^*)_{z=0}$ becomes greater with decreasing r_c/r_o and μ_3. $(\tau_{\theta z}^*)_{z=0}$ for each r_c/r_o is independent of μ_3 and has a singularity as $r \to r_c^-$, and the effect of a crack tip on $(\tau_{\theta z}^*)_{z=0}$ is locally significant with increasing r_c/r_o.

The variations of nondimensional stress intensity factors $K_{III,x}^*$ $(=K_{III}/[\tau_x(\pi r_o)^{1/2}])$, $x=\infty, n$ with r_c/r_o are shown in Figs.6 and 7, respectively, where $\tau_n = 2T/(\pi r_c^3)$. In Fig.6, marks ○ and □ denote the numerical results by Yamamoto et al.[1] and Hasebe et al.[2], respectively, and a dashed line is the approximate result by Benthem et al.[10]. It is observed from Figs.6 and 7 that $K_{III,x}^*$ is independent of μ_3, so that present results agree with those in the case of isotropy. From Fig.6, the present result agrees well with the results discribed in Ref.(1), (2) and (10). It is seen from Fig.7 that $K_{III,\infty}^*$ increases rapidly as r_c/r_o decreases.

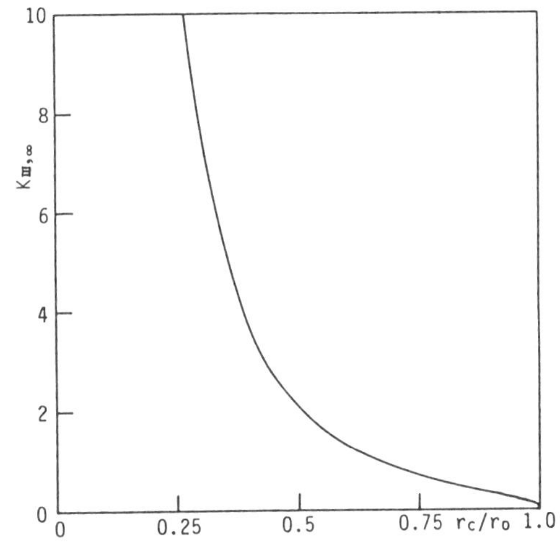

Fig.7. The variation of $K_{III,\infty}^*$ with r_c/r_o.

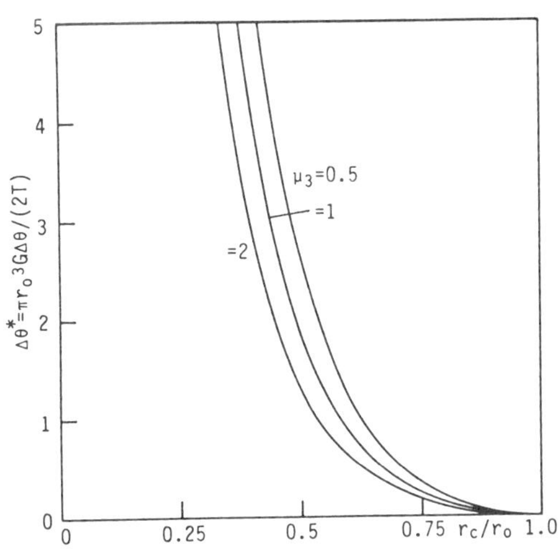

Fig.8. The variation of $\Delta\theta^*$ with r_c/r_o and μ_3.

Fig.8 present $\Delta\theta^*$ $(=\Delta\theta G/\tau_\infty)$ for a range of values of r_c/r_o and μ_3. $\Delta\theta^*$ becomes larger with decreasing r_c/r_o and μ_3.

IV. CONCLUSIONS

Within a three-dimensional theory of elasticity, we have studied the torsion problem of a transversely isotropic cylinder containing an external crack. The analytical solutions are given for the displacement and stresses around the crack tip and the stress intensity factor. Numerical results are also presented for some cases.

REFERENCES

1) Yamamoto, Y. and Sumi, Y., Stress Intensity Factors of a Twistedround Bar with a Circumferential Crack, Int. J. Fract., 10(1974), pp. 269-271.
2) Hasebe, N. and Kuse, K., Torsion of a Cylindrical Bar with Ring-shaped Edge Crack by the Mapping Function and the Finite Difference Method, Theo. Appl. Mech., Vol. 27(1977), pp. 37-46.
3) Shibuya, T., Koizumi, T. and Okuya, T., The Axisymmetric Stress Field in an Infinite Solid Cylinder with an External Crack under Torsion, Bull. of JSME, Vol. 22, No. 170(1979), pp. 1049-1052.
4) Uchiyama, T., Tsuchida, E., Nakahara, I. and Kodama, M., Trans. Japan Soc. Mech. Engng, Vol. 47, No. 422, A(1981), pp. 1012-1018. (In Japanese)
5) Zhang, X. S., The General Solution of a Finite Cylinder with a Concentric Penny-Shaped Crack under Torsion, Eng. Fract. Mech., Vol. 31, No. 5(1988), pp. 872.
6) Chen, W. T., Stress Concentration Around Spheroidal Inclusions and Cavities in a Transversely Isotropic Material under Pure Shear, J. Appl. Mech., 37(1970), pp. 85-92.
7) Sih, G. C., Mechanics of Fracture, 2, Noordhoff, (1975), pp. 7-8.
8) Kumar, M., J. Math. Phy. Sci., Vol. 14, No. 6(1980), pp. 619.
9) Shibuya, T., Koizumi, T. and Okuya, T., Torsion of a Two-Part Cylindrical Rod with a Circumferential Crack, Theo. Appl. Mech., Vol. 28(1978), pp. 187-194.
10) Benthem, J. P. and Koitor, W. T., Asymptotic Approximations to Crack Problems, Method of Analysis and Solutions of Crack Problems, 1, Noordhoff, (1973), pp. 131-178.

The Elastic Fields of Disclination Dipoles and Their Applications

Kenji SAITO*, Shigetoshi ARAKI*, and Mikio HAYASHI**

* *Department of Mechanical and System Engineering, Kyoto Institute of Technology, Matsugasaki, Sakyo-ku, Kyoto*
** *Technical Research and Development Institute, J.D.A., Kurihama, Yokosuka, Kanagawa*

Stress and displacement fields of a disclination dipole are derived by differentiating a field of a straight disclination as a source function. Applications of the disclination dipole fields to the source potential of the indirect boundary element method for solving *St.Venant* torsion problem are examined. In order to confirm the validity of the present method thus developed, the torsional stress concentration factors for the angle and the square tube are analyzed.

I. INTRODUCTION

It can be said that the concepts of disclinations and dislocations were introduced into theory of solids at the same time[1]. *Volterra* had considered six types of linear imperfections characteristic of deformed solid, later they were called *Volterra* dislocations. Nowadays they usually subdivided into translational dislocation (or simply dislocations) and rotational dislocations (or disclinations).

In the present paper, the elastic fields of a straight disclination are first treated as the plastic strain problem of the elastic solid. The elastic fields of a straight disclination thus obtained have the character that some of them have a singularity $\sigma \sim \ln r$ as $r \to 0$ and diverges as $\ln r$ when $r \to \infty$, so that an individual straight disclination can not exist because too much elastic energy is stored in a crystal. The logarithmic divergence of the long-range stress of a straight disclination can be disposed of by joining straight disclination in dipole and other multipole configuration. Therefore, the elastic fields of a infinitesimal disclination dipole are derived by differentiating the field of a straight disclination as the source function.

Next, as the applications of the fields of a disclination dipole, the torsion of a prismatic bar, i.e. *St.Venant* torsion problem is considered. *St.Venant* torsion problem of a prismatic bar with a geometrically complex configuration can hardly be solved analytically, so that numerical treatments such as FEM are suitable. However, FEM analysis of *St.Venant* problem becomes the so-called hybrid formulation and complex as compared with the displacement formulation.

In the present paper, the elastic fields of a disclination dipole are used as a source potential of the indirect boundary element method for solving *St.Venant* torsion problem, and validity of the present method are confirmed by some practical examples.

II. DISCLINATION DIPOLE FIELDS

Total displacement u^T_i in an isotropic elastic body containing plastic strain e^P_{ij} can be derived by the following volume integral with respect to plastic strain e^P_{ij}[2].

$$u_i^T(\boldsymbol{r}) = \frac{1}{8\pi} \int \left\{ 2R_{,jkk}\, e_{ij}^P(\boldsymbol{r}') - \frac{1}{1-\nu} R_{,ijk}\, e_{jk}^P(\boldsymbol{r}') + \frac{\nu}{1-\nu} R_{,ijj}\, e_{kk}^P(\boldsymbol{r}') \right\} dV' \quad , \tag{1}$$

where $\boldsymbol{R} = \boldsymbol{r} - \boldsymbol{r}'$ is a relative position vector.
For a straight disclination line L given in Fig.1, plastic distortion β^*_{kl} is given by

$$\beta^*_{2l} = \varepsilon_{lqr}\, \Omega_q X_r H(-X_1)\, \delta(X_2) \quad , \tag{2}$$

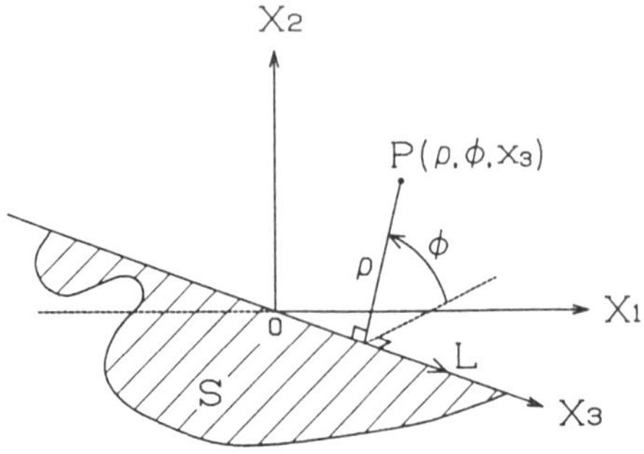

Fig.1. Straight disclination line and coordinates.

where, Ω_q , ε_{lqr} , δ (x) and H (x) are *Frank* vector of a disclination, permutation tensor, *Dirac* delta function and *Heaviside* step function, respectively. The relation $\varepsilon^p_{kl} = \beta^*_{(kl)}$ holds between a plastic strain ε^p_{kl} and a plastic distortion β^*_{kl}, where the symmetric part of a tensor T_{ij} is denoted by $T_{(ij)} = (1/2)(T_{ij} + T_{ji})$.

Substituting Eq. (2) into Eq. (1) and integrating, we get

$$u_1^T = -\frac{\Omega_1 X_3}{4\pi(1-\nu)}\left\{(1-2\nu)\ln\rho + \frac{X_2^2}{\rho^2}\right\} + \Omega_2 X_3\left\{\frac{\phi}{2\pi} + \frac{X_1 X_2}{4\pi(1-\nu)\rho^2}\right\} - \Omega_3\left\{\frac{X_2\phi}{2\pi} - \frac{1-2\nu}{4\pi(1-\nu)}X_1(\ln\rho - 1)\right\} \quad , \tag{3}$$

where

$$\rho = (X_1^2 + X_2^2)^{1/2} \quad , \quad \phi = \tan^{-1}(X_2/X_1) + \pi H(-X_1)\{H(X_2) - H(-X_2)\} \quad . \tag{4}$$

Using the relations $e^T_{kl} = u^T_{(l,k)}$, $e_{kl} = u^T_{(l,k)} - e^p_{kl}$, and *Hooke*'s law , we get the stress fields for a straight disclination shown in Fig.1.

$$\sigma_{11} = -\frac{G\Omega_1 X_3}{2\pi(1-\nu)}\left\{\frac{X_1}{\rho^2} - 2\frac{X_1 X_2^2}{\rho^4}\right\} - \frac{G\Omega_2 X_3}{2\pi(1-\nu)}\left\{\frac{X_2}{\rho^2} + 2\frac{X_1^2 X_2}{\rho^4}\right\} + \frac{G\Omega_3}{2\pi(1-\nu)}\left\{\ln\rho + \frac{X_2^2}{\rho^2} + \frac{\nu}{1-2\nu}\right\} \quad , \tag{5-a}$$

$$\sigma_{22} = -\frac{G\Omega_1 X_3}{2\pi(1-\nu)}\left\{\frac{X_1}{\rho^2} + 2\frac{X_1 X_2^2}{\rho^4}\right\} - \frac{G\Omega_2 X_3}{2\pi(1-\nu)}\left\{\frac{X_2}{\rho^2} - 2\frac{X_1^2 X_2}{\rho^4}\right\} + \frac{G\Omega_3}{2\pi(1-\nu)}\left\{\ln\rho + \frac{X_1^2}{\rho^2} + \frac{\nu}{1-2\nu}\right\} \quad , \tag{5-b}$$

$$\sigma_{33} = -\frac{G\nu X_3}{\pi(1-\nu)\rho^2}(\Omega_1 X_1 + \Omega_2 X_2) + \frac{G\Omega_3}{2\pi(1-\nu)}\left\{2\nu\ln\rho + \frac{\nu}{1-2\nu}\right\} \quad , \tag{5-c}$$

$$\sigma_{12} = -\frac{G\Omega_1 X_3}{2\pi(1-\nu)}\left\{\frac{X_2}{\rho^2} - 2\frac{X_1^2 X_2}{\rho^4}\right\} + \frac{G\Omega_2 X_3}{2\pi(1-\nu)}\left\{\frac{x_1}{\rho^2} - 2\frac{X_1 X_2^2}{\rho^4}\right\} - \frac{G\Omega_3 X_1 X_2}{2\pi(1-\nu)\rho^2} \quad , \tag{5-d}$$

$$\sigma_{23} = \frac{G\Omega_1 X_1 X_2}{2\pi(1-\nu)\rho^2} - \frac{G\Omega_2}{2\pi(1-\nu)}\left\{(1-2\nu)\ln\rho + \frac{X_1^2}{2\rho^2}\right\} \quad , \tag{5-e}$$

$$\sigma_{31} = -\frac{G\Omega_1}{2\pi(1-\nu)}\left\{(1-2\nu)\ln\rho + \frac{X_2^2}{2\rho^2}\right\} + \frac{G\Omega_2 X_1 X_2}{2\pi(1-\nu)\rho^2} \quad . \tag{5-f}$$

It is found from Eq. (5) that the stress fields of the straight disclination line diverge at infinity, and that elastic strain energy of the medium involving it is energetically unstable. Therefore, practical usage of the stress and displacement fields of a straight disclination is hardly possible as compared with those of a straight dislocation.

Consequently, if we consider a disclination dipole instead of a straight disclination, its field may converge even at infinity. Field quantities for infinitesimal disclination dipole $S^D(\boldsymbol{r})$ can be defined as

$$S^D(\boldsymbol{r}) = -\xi_i\, S_{,i}(\boldsymbol{r}) \quad , \tag{6}$$

where $S(\boldsymbol{r})$ is called the source function.

Applying Eq. (3) to Eq. (6), we can get the displacement fields for an infinitesimal disclination dipole, among which only u^D_1 is described here[3).

$$\begin{aligned} u_1^D = {} & \frac{M_{11}X_3}{4\pi(1-\nu)}\left\{(1-2\nu)\frac{X_1}{\rho^2} - 2\frac{X_1X_2^2}{\rho^4}\right\} - \frac{M_{22}X_3}{4\pi(1-\nu)}\left\{\frac{(3-2\nu)X_1}{\rho^2} - 2\frac{X_1X_2^2}{\rho^4}\right\} \\ & + \frac{(M_{12}+M_{21})X_3}{4\pi(1-\nu)}\left\{\frac{(1-2\nu)X_2}{\rho^2} + 2\frac{X_1^2X_2}{\rho^4}\right\} \\ & -(M_{23}-M_{32})\left\{\frac{\phi}{2\pi} + \frac{X_1X_2}{4\pi(1-\nu)\rho^2}\right\} - \frac{M_{31}-M_{13}}{4\pi(1-\nu)}\left\{(1-2\nu)\ln\rho + \frac{X_2^2}{\rho^2}\right\} , \end{aligned} \tag{7}$$

where $M_{gm} = \Omega_g \xi_m$ is used for the intensity of the disclination dipole.

Ⅲ. APPLICATIONS OF DISCLINATION DIPOLE FIELDS TO *ST.VENANT* TORSION PROBLEM

The so-called *St.Venant* torsion problem, i.e., torsion of a prismatic bar is shown in Fig.2. The displacement u_1, u_2 and u_3 in the x_1 , x_2 and x_3 directions, respectively are given by the following expressions.

$$u_1 = -\omega X_2 X_3 \quad , \quad u_2 = \omega X_1 X_3 \quad , \quad u_3 = \omega\,\phi(X_1, X_2) \quad . \tag{8}$$

where ω is a specific angle of torsion and ϕ is a warping of the cross section in the x_3 direction.

Movement of the point which locates on the $x_1 - x_2$ plane must satisfy the following condition of the rigid body rotation,

$$X_1 u_1 + X_2 u_2 = 0 \tag{9}$$

In order to apply the disclination dipole fields to *St.Venant* torsion problem, the displacement fields of the disclination dipole must satisfy the above condition, and from this condition, the following relations can be obtained regarding the intensities of the disclination dipole.

$$M_{11} = M_{22} = 0 \quad , \quad M_{12} = -M_{21} \quad , \quad M_{23} = M_{32} \quad , \quad M_{31} = M_{13} \quad . \tag{10}$$

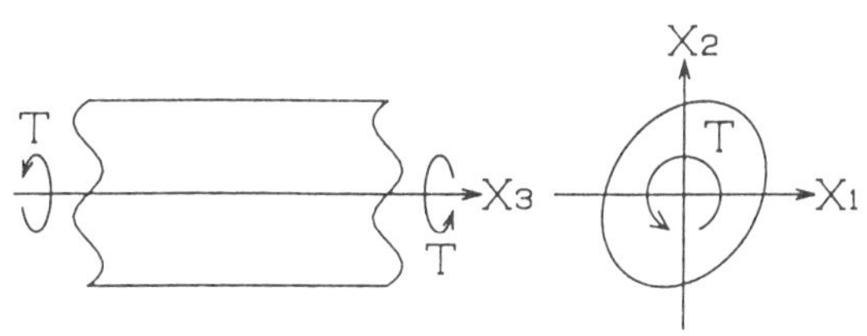

Fig.2. *St.Venant* torsion problem.

Therefore, the following displacement and stress fields which can be applied to the solution of *St.Venant* torsion problem can be derived.

$$u_3^D = M_{21}\frac{\phi}{\pi} \quad , \quad u_1^D = u_2^D = 0 \quad ,$$
$$\sigma_{23}^D = \frac{GM_{21}X_1}{\pi\rho^2} \quad , \quad \sigma_{31}^D = -\frac{GM_{21}X_2}{\pi\rho^2} \quad , \quad \sigma_{11}^D = \sigma_{22}^D = \sigma_{33}^D = \sigma_{12}^D = 0 \quad . \tag{11}$$

In the above expressions, only M_{21} component of the intensities of the disclination dipole appears, so that it can be seen that the disclination dipole which is suitable to the solution of *St.Venant* torsion problem is a twist disclination dipole.

IV. INDIRECT BOUNDARY ELEMENT METHOD USING TWIST DISCLINATION DIPOLE FIELDS

Figure 3 shows that the prismatic bar having the cross section bounded by the solid line L is twisted with the specific angle of torsion ω . The solid line L represents the real boundary of the body, the domain D, which is to be analyzed, while the dotted line L^* is the fictitious boundary on which disclination dipoles are distributed.

If we distribute the disclination dipoles on the real boundary, the degree of accuracy of the numerical solution degenerates due to the singularity of the disclination dipole over the real boundary on which in most cases the maximum stress may occur. From the above reason, we establish a fictitious boundary outside the real one, and the influence of the singularity of the dipole is intended to be shut up inside the domain D^* between the both boundaries.

When the disclination dipole of intensity M_{21} is located at an arbitrary point $Z = Z_d$ on the fictitious boundary L^* in a complex plane as shown in Fig.3, the stress and displacement fields at an arbitrary point Z can be expressed from Eq. (11) in a complex form as follows:

$$u_z = -\frac{iM_{21}}{2\pi}\left(\ln\frac{Z-Z_d}{\overline{Z}-\overline{Z}_d} - \ln\frac{Z_0-Z_d}{\overline{Z}_0-\overline{Z}_d}\right) \quad , \tag{12}$$

$$\sigma_{zx} = \frac{iGM_{21}}{2\pi}\left(\frac{1}{\overline{Z}-\overline{Z}_d} - \frac{1}{Z-Z_d}\right) \quad , \quad \sigma_{zy} = \frac{GM_{21}}{2\pi}\left(\frac{1}{\overline{Z}-\overline{Z}_d} - \frac{1}{Z-Z_d}\right) \quad , \tag{13}$$

where $i = \sqrt{-1}$, Z_0 denotes the reference point of the displacement, and

$$\phi_d = \frac{1}{2i}\ln\frac{Z-Z_d}{\overline{Z}-\overline{Z}_d} \quad . \tag{14}$$

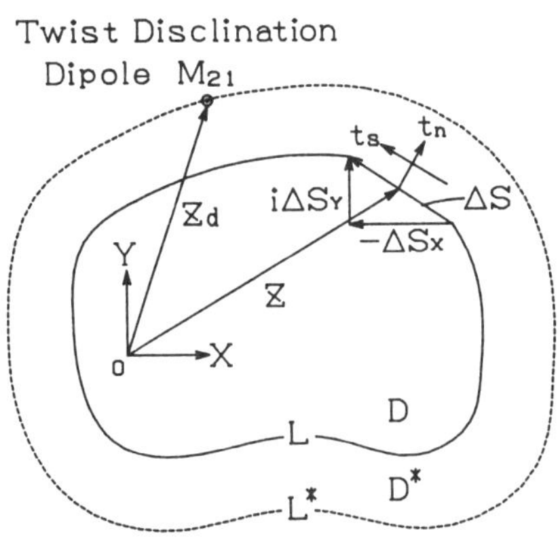

Fig.3. Actual boundary L, fictitious boundary L^* and twist disclination dipole.

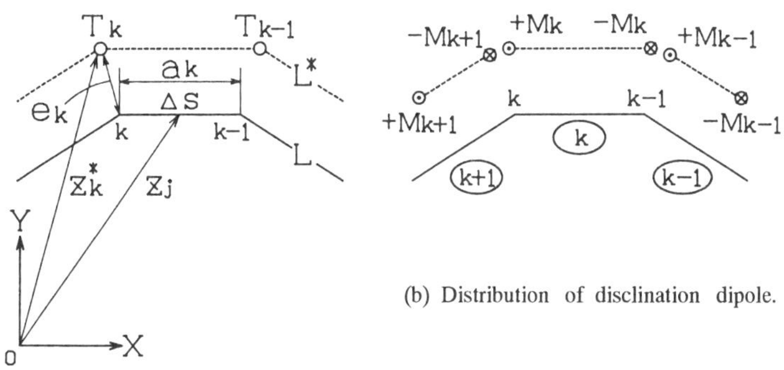

(b) Distribution of disclination dipole.

(a) Element subdivisions.

Fig.4 Distributions of disclination dipole.

The normal stress t_n acting over the line element $\Delta S = \Delta S_x + i\Delta S_y$ along the real boundary can be obtained by the transformation of stress as follows:

$$t_n = -\frac{1}{2|\Delta S|}\left\{ i\,\sigma_{zx}(\Delta S - \overline{\Delta S}) + \sigma_{zy}(\Delta S + \overline{\Delta S})\right\} \quad , \tag{15}$$

where $\overline{\Delta S}$ is the conjugate of ΔS and $|\Delta S|$ is the absolute value of ΔS.

Next, we divide the real boundary into N linear boundary elements, and correspondingly the fictitious one into N linear elements. At both ends of the each element, i.e. node, we distribute the equal and opposite intensity of the twist disclination dipole which forms the dipole pair with each other as shown in Fig.4(a). Here, the coefficient of fictitious-boundary distance α_k is introduced, which is defined as the ratio of the distance between the real and the fictitious boundary e_k to the length a_k of the K th element of the real boundary as shown in Fig.4(a). Therefore, the intensity M_k of the twist disclination dipole which exists at the node K of the K th element (subscript denotes the number of the node) is given by

$$M_k = T_k - T_{k+1} \quad , \tag{16}$$

where T_k denotes the intensity of the dipole pair at the K th element as shown in Fig.4(b).

If we divide the real boundary into N elements and distribute the disclination dipoles of the intensity M_k (k = 1~N) at the corresponding fictitious boundary nodes of the point $Z = Z_k$, then the stress and displacement at the middle point $Z = Z_j$ on the real boundary can be obtained by the superposition of the contributions due to all the dipoles, and thus

$$\tau_n^j = \sum_{k=1}^{N} M_k V(Z_j, Z_k^*) \quad , \tag{17}$$

$$u_z^j = \sum_{k=1}^{N} M_k W(Z_j, Z_k^*) \quad , \tag{18}$$

where

$$V(Z_j, Z_k^*) = -\frac{G}{2\pi|\Delta S_j|}\left\{\left(\frac{\Delta S_j}{Z_j - Z_k^*} + \frac{\overline{\Delta S_j}}{\overline{Z_j - Z_K^*}}\right) - \left(\frac{\Delta S_j}{Z_j - Z_{k-1}^*} + \frac{\overline{\Delta S_j}}{\overline{Z_j - Z_{k-1}^*}}\right)\right\} , \tag{19}$$

$$W(Z_j, Z_k^*) = -\frac{i}{2\pi}\left(\ln\frac{Z_j - Z_k^*}{Z_j - Z_{k-1}^*} - \ln\frac{Z_o - Z_k^*}{Z_o - z_{k-1}^*} - \ln\frac{\overline{Z_j - Z_k^*}}{\overline{Z_j - Z_{k-1}^*}} + \ln\frac{\overline{Z_o - Z_k^*}}{\overline{Z_o - Z_{k-1}^*}}\right) . \tag{20}$$

As the fundamental solution of *St.Venant* torsion problem, the solution to the torsion of an infinite body is employed, and the normal stress of the differential element ΔS in Fig.3 can be expressed as

$$\sigma_n = -\frac{G\omega}{2|\Delta S|}(Z\overline{\Delta S} + \overline{Z}\Delta S) \quad , \tag{21}$$

where ω is a specific angle of torsion.

The boundary conditions for *St.Venant* torsion problem are such that the normal stress is zero on the lateral surface of the prismatic bar and the warping of the cross section is zero along the geometrically symmetric line of the cross section. When linear elements having the element number from 1 to M exist on the contour and those with the element number from M+1 to N locate on the geometrically symmetric line of the cross section, the boundary condition for the torsion of the prismatic bar can be expressed in a matrix form as the residual problem of the torsion of an infinite body as follows.

$$\begin{bmatrix} V(Z_1,Z_1^*) & \cdots & V(Z_1,Z_N^*) \\ \cdot & \cdots & \cdot \\ \cdot & \cdots & \cdot \\ V(Z_M,Z_1^*) & \cdots & V(Z_M,Z_N^*) \\ W(Z_{M+1},Z_1^*) & \cdots & W(Z_{M+1},Z_N^*) \\ \cdot & \cdots & \cdot \\ \cdot & \cdots & \cdot \\ W(Z_N,Z_1^*) & \cdots & W(Z_N,Z_N^*) \end{bmatrix} \begin{Bmatrix} M1 \\ \cdot \\ \cdot \\ M_M \\ M_{M+1} \\ \cdot \\ \cdot \\ M_N \end{Bmatrix} = \begin{bmatrix} -\sigma_n(Z_1) \\ \cdot \\ \cdot \\ -\sigma_n(Z_M) \\ 0 \\ \cdot \\ \cdot \\ 0 \end{bmatrix} \tag{22}$$

Once all the intensities of the disclination dipoles can be determined by solving the above linear simultaneous algebraic equations with respect to M_k , then the stress and displacement fields in the prismatic bar can be determined by the superposition of all the dipoles with the intensities thus determined.

V. PRACTICAL EXAMPLES

5.1 Torsion of the Angle

As a practical application of the indirect boundary element method developed above, torsion problems of the angle and the square tube are treated in this section, and the torsional stress concentration factors for them are clarified by this method.

The geometry of the angle to be analyzed is shown in Fig.5. Figure 6 is one example of the element subdivisions and the fictitious boundary denoted by the two dotted line, in which the coefficient of fictitious boundary distance α is taken to α = 3.

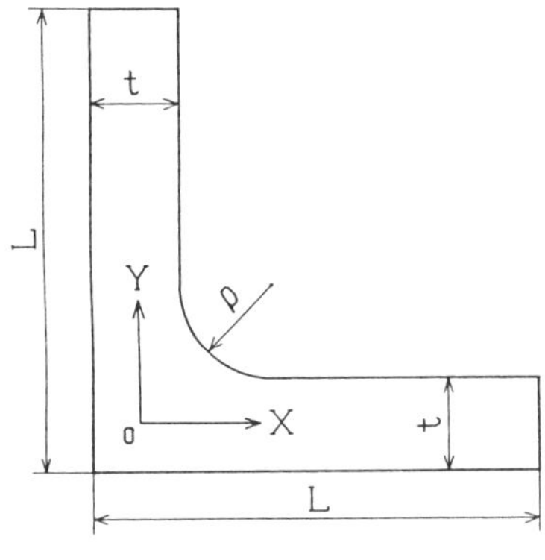

Fig.5. Dimension of the angle.

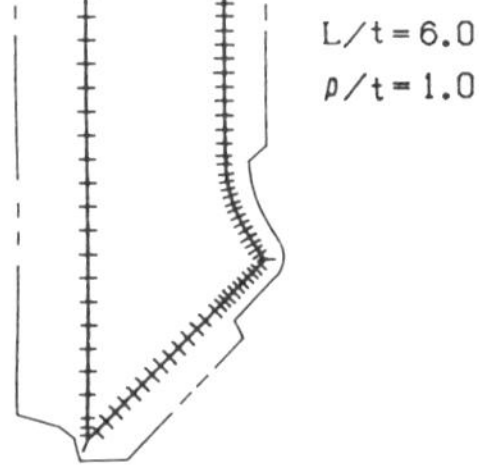

Fig.6. Element subdivisions of the angle.

The torsional stress concentration factors for the angle are shown in Table 1 and in Fig.7. The stress concentration factor K is defined as $K = \tau_{max} / \tau_o$, where τ_{max} is the maximum resultant shear stress and τ_o is the nominal stress which is given by $\tau_o = G\omega t$.

The results indicated by *Kishida* in Table 1 are those due to *Kishida*[4] by boundary-integral method. It can be seen from the figure that the present results are in good agreement with those due to *Huth*[5] , which is the most reliable results ever existed.

5.2 Torsion of the Square Tube

The geometry of the square tube to be analyzed is shown in Fig.8, and one example of the element subdivisions is also given in Fig.9.
The coefficient of fictitious-boundary distance α is also taken to $\alpha = 3$ in this case. Figure 10 shows the torsional stress concentration factors K for the square tube, in which the nominal stress is taken to $\tau_o = G\omega\ (L+t)/2$.

Table 1. Stress concentration factor.

ρ/t \ L/t	3.0	4.0	5.0	6.0	Kishida
0.3	1.646	1.649	1.650	1.650	1.647
1.0	1.498	1.502	1.503	1.504	1.500
1.5	1.578	1.583	1.583	1.583	1.580
2.0	1.684	1.693	1.694	1.694	1.691

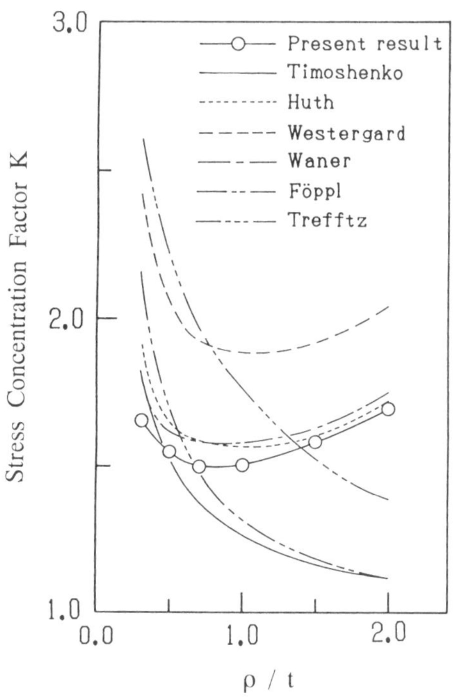

Fig.7. Torsional stress intensity factor in the angle.

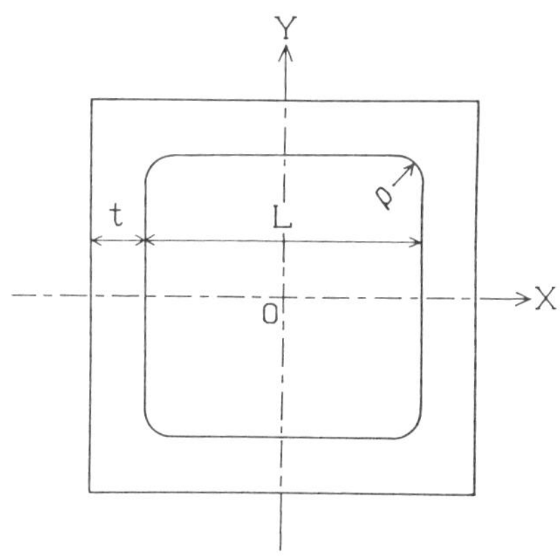

Fig.8. Dimension of the square tube.

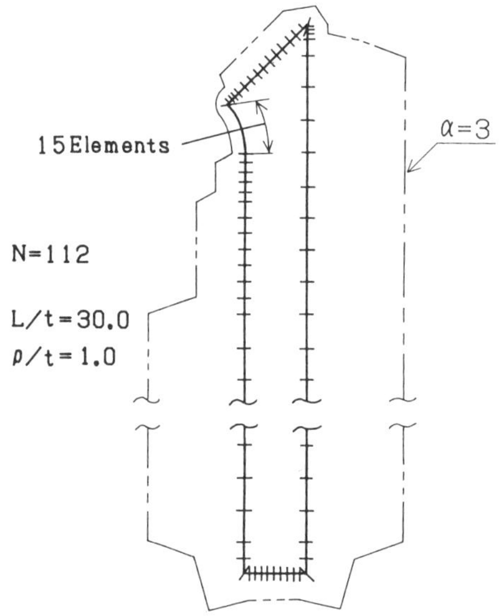

Fig.9. Element subdivision of the square tube.

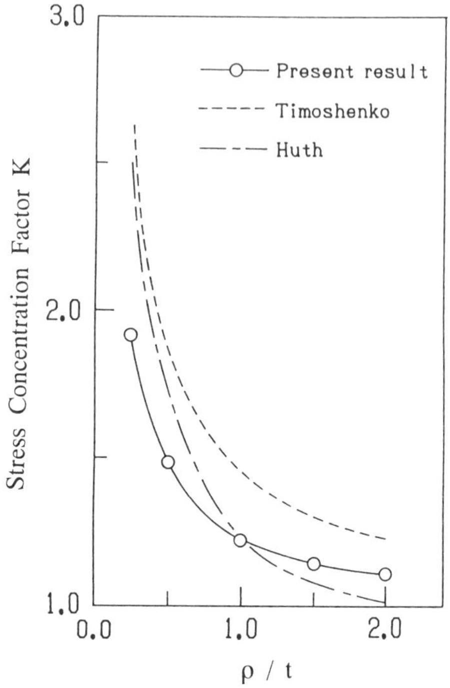

Fig.10. Torsional stress intensity factor in the square tube.

VI. CONCLUSIONS

In the present investigation, the elastic fields of a disclination dipole are obtained, and their applications to the indirect boundary element method for solving *St.Venant* torsion problem are discussed. Validity of the present method is examined by some practical examples.

Acknowledgement

The authors would like to express their appreciations to Mr. N. Matsuda, the technical official in K.I.T.

REFERENCES

1) Romanov,A.E. and Vladimirov,V.I., *Disclination in crystalline solids*, in Dislocations in Solids, Vol.9, (Editor F.R.N.Nabarro, North-Holland, 1992), p.191.
2) deWit, R., *Theory of disclinations*, IV *straight disclinations*, J.Research of N.B.S'.-A, Phy. and Chem., 77a-5(1973), p.607.
3) Hayashi, M.,Saito,K., and Kunieda,Y., *Disclinations and their applications to stress analysis (1st report)*, Trans.JSME, Series A, Vol.58, No.550(1992), p.895 (in Japanese).
4) Kishida,M., Hanzawa,H., Yanaka,T., and Honma,K., *Applications of fictitious boundary method-boundary integral method to torsion problem*, Trans.JSME, series A, Vol.46, No.408(1980), p.901 (in Japanese).
5) Huth,J.H., *Torsional stress concentration in angle and square tube fields*, J.Appl.Mech., 17(1950), p.388.

Analysis of Impact Problem Considered Friction on Contact Surface

Jinwook KIM*, Takuzo IWATSUBO**, Yoshihiro TOMITA**, and Shozo KAWAMURA**

* *Graduate School of Science and Technology, Kobe University, Kobe, Hyogo*
** *Department of Mechanical Engineering, Kobe University, Kobe, Hyogo*

The contact problem is analyzed by the implicit method or the explicit method. Although the implicit method is preferable for greater efficiency for the numerical stability and accuracy of solution, it has some practical problem of the calculation time for iteration. The other hand, the explicit method is superior for calculation time, but it is very difficult to consider the friction term at the contact surface. So the algorithm considered the friction force in explicit method is required. In this paper the simplified explicit method, considering the equivalent contact frictional force is proposed. As numerical example, the two dimensional axisymmetric model is considered and the stress wave propagation is analyzed. As a result, the proposed explicit method have the advantage of saving the calculation time and it is applicable for the case of small friction coefficient. And it is shown that the fracture behaviour can be pre-estimated roughly.

I. INTRODUCTION

Many studies of the dynamic contact problem using the finite element method have been performed[1)~4)]. The time integration algorithms used in nonlinear dynamic analysis are commonly classified into two groups of the implicit method and the explicit method. Although the implicit method is preferable for greater efficiency for the numerical stability and accuracy of solution, it has some practical problem with respect to the calculation time for iteration and to the larger computer memory. The other hand, the explicit method in which the frictional force on contact surfaces is not considered, has the advantage of saving the calculation time, but it may have the problem of the accuracy[1)]. Recently, the explicit method in which the frictional force is considered has been presented and it is well known as DYNA3D[5)]. In DYNA3D, for search the contact points, the iterative calculation using the penalty method is performed. The interface stiffness is chosen to be approximately the same order of magnitude as the stiffness of the iterface. In practice, the interface stiffness may be larger according to the penalty size, so the penalty size is limited by stability considerations[2),5)]. So by scaling up the interface stiffness and scaling down the time step size, the iterative calculating is need to obtain the contact point on contact surface.

In the present method, the contact problem can be analyzed without iterative calculation for the case of small friction coefficint. As numerical example, the two dimensional axisymmetric model is considered and the stress wave propagation is analyzed. The results are compared with the ones obtained by using the implicit method[4)], which is based on the incremental finite element method, and the ones obtained by using the explicit method in which the frictional force is not considered. And the accuracy of solution, the calculating time and the applicability of the presented method are discussed. Moreover the experimental results of fracture of the brittle materials are simulated by using the presented method, and it is shown that the fracture behaviour can be pre-estimated roughly.

II. FORMULATION OF IMPACT PROBLEM

2.1 Principle of the Virtual Work for the States of Contact and Separation

Two elastic bodies A and B in contact are considered as shown in Fig.1. The domain of interest β (β=A, B) has two boundaries; the boundary tractions and displacements on S_σ and S_u are respectively prescribed. The contact states are classified into (1) opening state(co; where the two bodies are separated), (2) sticking state(cst; where the two bodies are in contact and slipping doesn't occur), (3) slipping state(csl; where the two bodies are in contact and slipping occurs). It is necessary to satisfy the conditions which are the moment equilibrium and the continuum of displacement, and the Coulumb's friction on contact states, respectively. The contact conditions are shown in Table 1. The contact problem as shown in Fig.1 is formulated by using the principle of the virtual work. And using the functional Π, the variational formulation may be regarded as a generalization of the principle of virtual displacement as follows.

$$\delta\Pi = \sum_{\beta=A,B} (\delta E^\beta - \delta T^\beta - \delta W^\beta - \delta W_c^\beta) = 0 \tag{1}$$

In the above equation, δE^β, δT^β,δW^β, δW_c^β are the works of the internal elastic, the inertia, the traction boundary force and contact force, respectively. Then the virtual work equation can be obtained by invoking the stationarity of Π with respect to the displacements and the contact condition of time step.

$$\begin{aligned}
&\int_{\Omega_A} [\delta\epsilon_N]^T \sigma_N d\Omega_A - \int_{\Omega_A} [\delta\mathbf{U}_N]^T (\,\mathbf{b}_N - \rho_N \ddot{\mathbf{U}}_N)\, d\Omega_A - \int_{S_A} [\delta\mathbf{U}_N]^T \mathbf{t}_N dS_A \\
&+ \int_{\Omega_B} [\delta\epsilon_N]^T \sigma_N d\Omega_B - \int_{\Omega_B} [\delta\mathbf{U}_N]^T (\,\mathbf{b}_N - \rho_N \ddot{\mathbf{U}}_N)\, d\Omega_B - \int_{S_B} [\delta\mathbf{U}_N]^T \mathbf{t}_N dS_B \\
&- \int_{cst} [\delta\mathbf{U}_{Nc}]^T \mathbf{T}_{c(N-1)} dS - \int_{csl} [\delta\mathbf{U}_{Nt}]^T \mathbf{T}_{t(N-1)} dS \\
&- \int_{csl} [\delta\mathbf{U}_{Nz}]^T \mathbf{T}_{z(N-1)} dS - \int_{co} [\delta\mathbf{U}_{Nc}]^T \mathbf{T}_{c(N-1)} dS = \mathbf{0}
\end{aligned} \tag{2}$$

where c means z and t. The subscription z and t are the normal and tangential direction respectively. $\delta\mathbf{U}_N$ and $\delta\epsilon_N$ are the vector of virtual displacement and the associated virtual strains, respectively. $\mathbf{b}_N$, $\mathbf{t}_N$ are respectively the vector of applied body forces and the vector of surface tractions. σ_N, ρ_N and Ω_β are the vector of stresses, the mass density and the domain of interest. $\mathbf{T}_{c(N-1)}$ is the vector of equivalent contact forces of contact surface tractions with respect to N time step, and two dot refer to the differentiation with respect to time, N-1 and N are the time step for any incremental time Δt.

2.2 Formulation by the Finite Element Method

Using the relation of the displacement and strain at the nodes, we can derive the equilibrium equation by discretisation of Eq.(2). In the displacements and strains of all nodal points, we obtain for each node

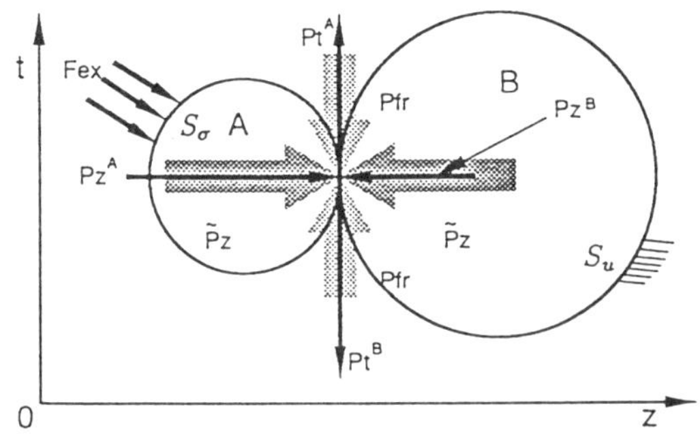

Fig.1. Equivalent contact forces on contact surface.

Table 1. Contact conditions.

States	Contact conditions
Opening	$\mathbf{P}_z^A \geq 0, \mathbf{u}_z^A - \mathbf{u}_z^B < 0$
Sticking	$\mathbf{u}_c^A = \mathbf{u}_c^B, \dot{\mathbf{u}}_c^A = \dot{\mathbf{u}}_c^B, \ddot{\mathbf{u}}_c^A = \ddot{\mathbf{u}}_c^B$
Slipping	$\lvert\mathbf{P}_t^\beta\rvert \geq \mathbf{P}_{fr}, \mathbf{u}_z^A = \mathbf{u}_z^B, \dot{\mathbf{u}}_z^A = \dot{\mathbf{u}}_z^B, \ddot{\mathbf{u}}_z^A = \ddot{\mathbf{u}}_z^B$

i the following equilibrium equation[9].

$$\{\mathbf{f}_{Ii}\}_N - \{\mathbf{f}_{Bi}\}_N + \{\mathbf{P}_{Ii}\}_N - \{\mathbf{f}_{Ti}\}_N - \{\mathbf{T}_{ci}\}_{N-1} = \mathbf{0} \tag{3}$$

where the inertia forces and the nodal forces are respectively.

$$\{\mathbf{f}_{Ii}\}_N = \sum_{j=1}^{m}[\mathbf{M}_{ij}]_N\{\ddot{\mathbf{d}}_j\}_N, \qquad \{\mathbf{P}_{Ii}\}_N = \int_\Omega \mathbf{B}_i{}^T\sigma_N d\Omega \tag{4}$$

The applied body forces and the traction boundary forces are respectively

$$\{\mathbf{f}_{Bi}\}_N = \int_\Omega \mathbf{N}_i{}^T\mathbf{b}_N d\Omega, \qquad \{\mathbf{f}_{Ti}\}_N = \int_S \mathbf{N}_i{}^T\mathbf{t}_N dS \tag{5}$$

The equivalent contact force, $\{\mathbf{T}_{ci}\}_{N-1}$ at N time step is obtained from the nodal force at N-1 time step according to the contact states.

2.3 Contact Condition and Judgement of Contact Force

The equivalent normal contact forces, $\tilde{\mathbf{P}}_z$, and the equivalent frictional forces, $\mathbf{P}_{fr}$ are respectively evaluated by the force of the same magnitude and opposite direction as shown in Fig.1. Considering with the frictional force on the contact surface and it is judged and estimated by the contact states at N time step, as follows in Table 2.

Table 2. Judgement and estimation of the equivalent contact force.

<table>
<tr><td>States</td><td colspan="2">Contact (Sticking or Slipping)</td><td>Opening</td></tr>
<tr><td rowspan="2">Judgement</td><td colspan="2">$\mathbf{P}_z^A < 0$</td><td rowspan="2">$\mathbf{P}_z^A \geq 0$</td></tr>
<tr><td>$\mathbf{P}_z^A \cdot \mathbf{P}_z^B < 0$</td><td>$\mathbf{P}_z^A \cdot \mathbf{P}_z^B \geq 0$</td></tr>
<tr><td rowspan="2">Result</td><td>$\tilde{\mathbf{P}}_z = \frac{1}{2}(\lvert\mathbf{P}_z^A\rvert + \lvert\mathbf{P}_z^B\rvert)$</td><td>$\tilde{\mathbf{P}}_z = \frac{1}{2}(\lvert\mathbf{P}_z^A - \mathbf{P}_z^B\rvert)$</td><td>$\tilde{\mathbf{P}}_z = 0$</td></tr>
<tr><td colspan="3">$\mathbf{P}_{fr} = \mu\tilde{\mathbf{P}}_z$</td></tr>
</table>

In Table 2, μ is the frictional coefficient. Based on the above arguments and procedures, the contact states are judged. If simultaneously $|\mathbf{P}_t^\beta| > \mathbf{P}_{fr}$ on contact surface, it is in slipping state, if not, it is in sticking state. The update equivalent contact force of the normal and tangential at N+1 time step is approximated as follows. In the case of the open state, of course, because $\tilde{\mathbf{P}}_z$=0, the update nodal force is the same value that of the nodal force at N time step. The update tangential and normal force are respectively evaluated according to compare the tangential nodal force and equivalent frictional force at N time step as follows.

$$\mathbf{T}_t^{\beta(N+1)} = \frac{1}{2}(\mathbf{P}_t^{A(N)} + \mathbf{P}_t^{B(N)}) \rightleftharpoons (sticking) \tag{6}$$

$$\mathbf{T}_t^{\beta(N+1)} = \mathbf{P}_t^{\beta(N)} - sgn(\mathbf{P}_{fr}, \mathbf{P}_t^{\beta(N)})|\phi(\overline{v})| \rightleftharpoons (slipping) \tag{7}$$

$$\mathbf{T}_z^{\beta(N+1)} = \frac{1}{2}(\mathbf{P}_z^{A(N)} + \mathbf{P}_z^{B(N)} - sgn(\tilde{\mathbf{P}}_z, \mathbf{P}_z^{A(N)}) - sgn(\tilde{\mathbf{P}}_z, \mathbf{P}_z^{B(N)})) \rightleftharpoons (contacting) \tag{8}$$

$$\phi(\overline{v}) = tanh(\xi(\dot{\mathbf{u}}_t^{A(N)} - \dot{\mathbf{u}}_t^{B(N)})), 0 < \xi \leq 1 \tag{9}$$

where $\phi(\overline{v})$ is the hyperbolic function expressed as follows in order to relax the numerical unstable problem due to a rapid change of slipping direction. This form may be used to obtain the stable solution of the nonlinear frictional system. In case of changing from the slipping state to the sticking state, however, if the contact state at the previous instant time step is in the slipping state, the update normal and tangential forces are exceptionally approximated as follows.

$$\mathbf{T}_t^{\beta(N+1)} = \mathbf{P}_t^{\beta(N)} \tag{10}$$

And the equivalent normal contact forces, $\tilde{\mathbf{P}}_z$=0. But not the open state in this instant, the update normal force is evaluated as the one of a half of the Eq.(8). This procedure is introduced to avoid the numerical instability and to progress to the next incremental time step smoothly when the contact state changes suddenly to the sticking state from the slipping state. And the compatibility of velocities and accelerations should be satisfied at the instant time step of the impact between two contact points. The update velocities and accelerations on the contact points are formulated to the same type as the approach of references[4),10)]. And also, corresponding nodal masses on contact surface are calculated by the approach of Wilkins[6),7)].

III. NUMERICAL ANALYSIS AND DISCUSSION

The Newmark-β method is employed to solve the Eq.(3) and the parameters β and γ of the Newmark-β method are $\beta = 0, \gamma = 1/2$. Incremental time, Δt, is taken to be $0.054 \mu sec$. The analytical model is shown in Fig.2. It is two dimensional axisymmetric model and the mechanical properties are shown at Table 3. And the length and the diameter of the striker are $8mm$ and $10mm$ respectively, its impact velocity is $30m/sec$. Numerical calculations are performed for the case that the Coulomb's frictional coefficients between A and B, μ, is 0.0, 0.1 and 0.3.

Table 3. Mechanical and physical properties of materials.

No.	Metal	Young's modulus (GPa)	Poisson's ratio	Density (g/cm^3)
A	S45C	207	0.3	7.86
B	Cemented Carbides	626.8	0.207	14.9

For various frictional coefficient, μ=0.1, 0.3, the variations of the principal stress in the 5th element shown in Fig.2 on contact surface are shown in Fig.3. In the case of μ=0.1, the numerical results obtained by using the presented explicit method agree with those of the implicit method[4)]. However, in the case of μ=0.3, the results of the presented explicit method disagree with the results of the implicit method according to incremental time. Next the accuracy of the presented explicit method is checked. The relative errors among the implicit method, the explicit method without frictional force and the presented explicit method are compared. The root mean square errors (RMS) of the principal stress on the contact surface are shown in Fig.4, which obtained during the contact state until to the separation state. The RMS error is defined as follows.

$$RMS = \sqrt{\frac{\sum_{i=1}^{N} \left(\frac{\sigma_{\mathbf{imp}i}}{\sigma_o} - \frac{\sigma_{\mathbf{exp}i}}{\sigma_o}\right)^2}{N}} \tag{11}$$

Under the small number of the friction coefficient, the errors of the presented explicit method are relatively smaller than those of the explicit method without friction. And the calculation time(KUBOTA COMPUTER TITAN3000) of the presented explicit method is checked. In case of $\mu = 0.1$ and the same incremental time step, it is required about $26sec$ for the presented explicit method, but for the implicit method it is required about $1119sec$. The presented explicit algorithm in which the frictional force and separation is considered, have the advantage of saving the calculation time, but it is necessary for better accuracy to improve and study.

IV. EXPERIMENTAL RESULTS AND SIMULATIONS

A series of impactive fracture test with applying the split-Hopkinson pressure-bar method which using a truncated cone tool was conducted. The diameter and length of the steel striker are $25mm$ and $100mm$ respectively. The impact velocity of the striker was about $30m/sec$. The material of the test specimen is the brittle material, i.e cemented carbides. The experimental results are shown in Fig.5 and Fig.6. Fig.5 is the fracture type of the $15mm$ thickness specimen, and Fig.6 is the fracture type of $5mm$ thickness specimen. Most of the specimen cracked in the similar fracture type.

Theoretical impact analysis was performed as follows. The calculation model of multi-contact bodies is shown in Fig.7. The properties of materials are shown in Table 3. The material of A and C is steel, and

B is cemented carbides. In Fig.7 the frictional coefficients on contact surface are set as $\mu_1 = \mu_2 = 0.2$ and

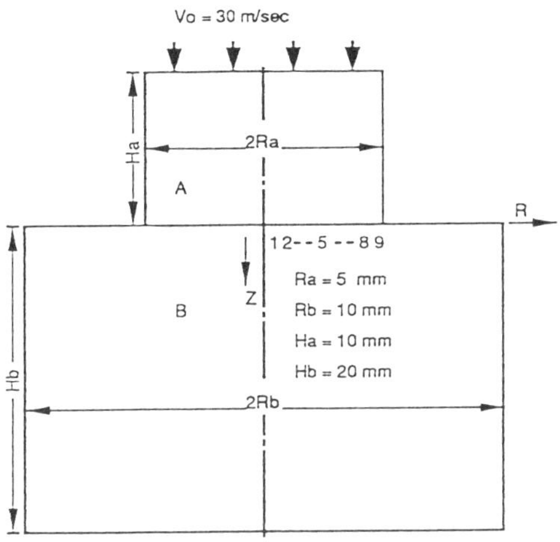

Fig.2. Calculation model of the finite element method.

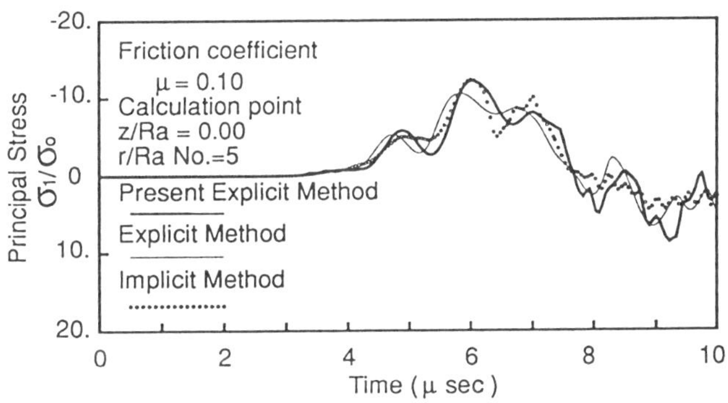

(a) $\mu = 0.1$

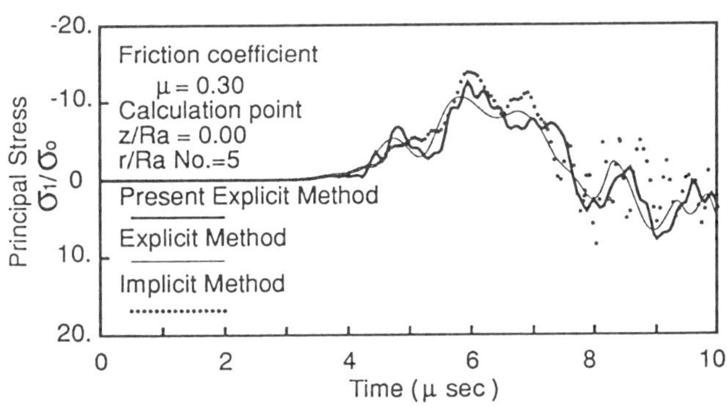

(b) $\mu = 0.3$

Fig.3. The time history of principal stress with friction.

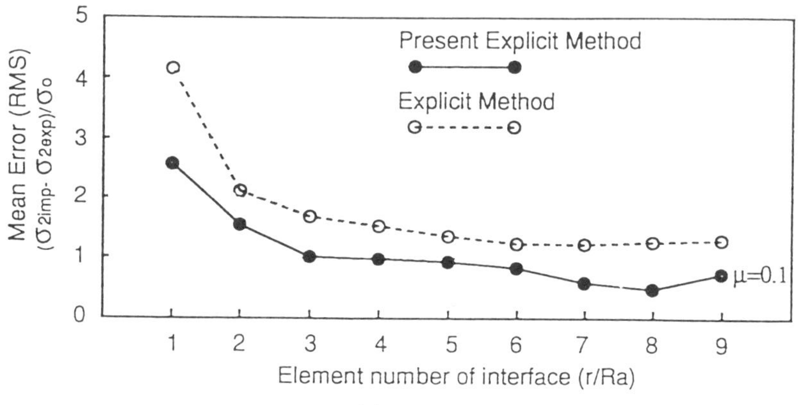

(a) $\mu = 0.1$

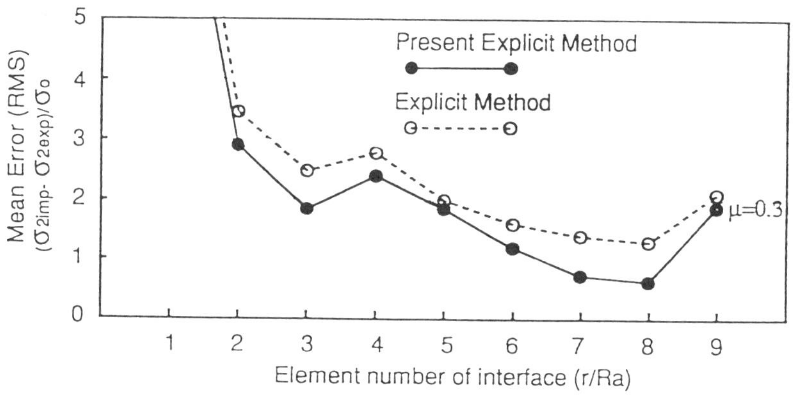

(b) $\mu = 0.3$

Fig.4. Comparison of the mean error of the principal stress.

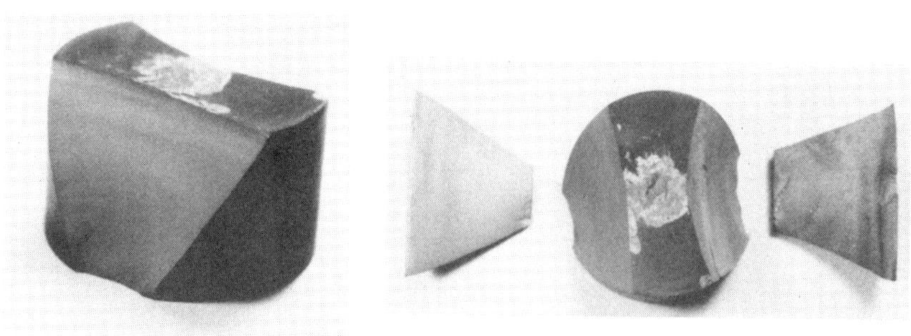

Fig.5. Fractured specimen (thickness $15mm$).

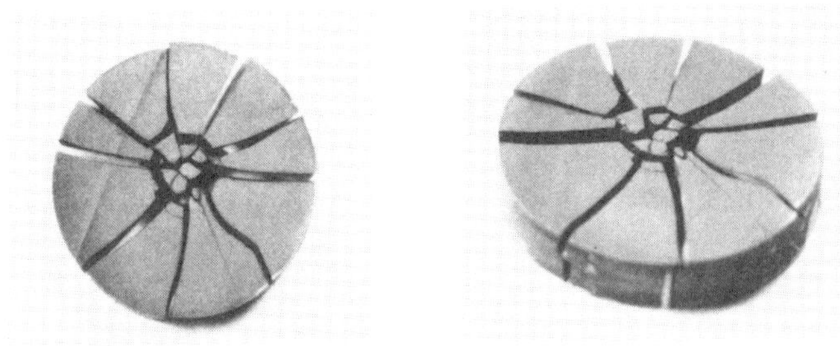

Fig.6. Fractured specimen (thickness $5mm$).

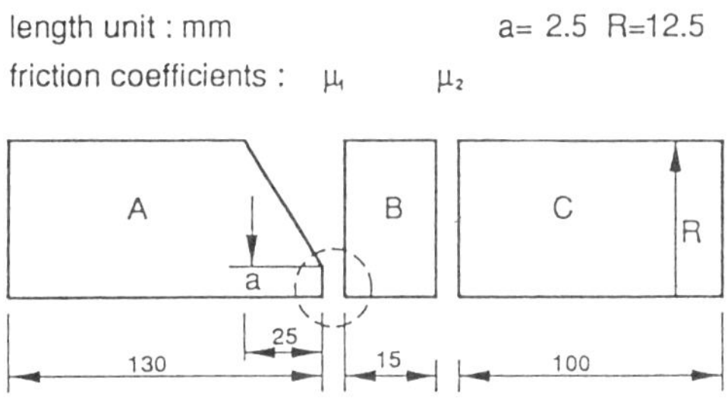

Fig.7. Calculation model of multi contact bodies.

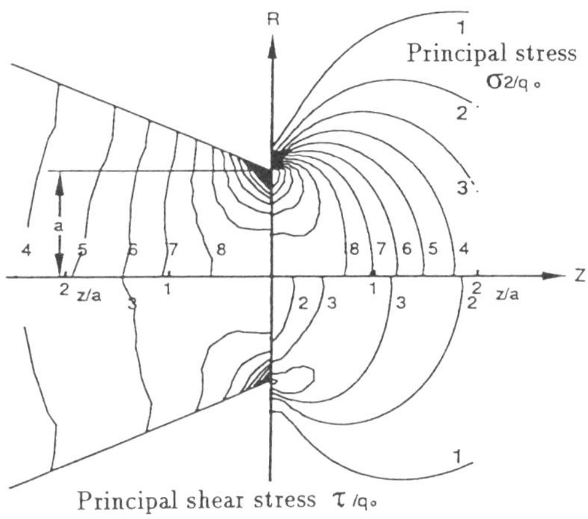

(a) Principal stress σ_2 and principal shear stress τ (30.4 μsec)

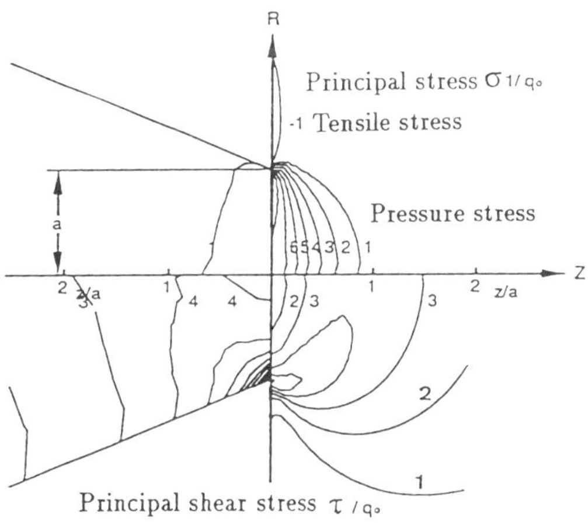

(b) Principal stress σ_1 and principal shear stress τ (31.6 μsec)

Fig.8. Distribution of principal stress and principal shear stress (thickness $15mm$).

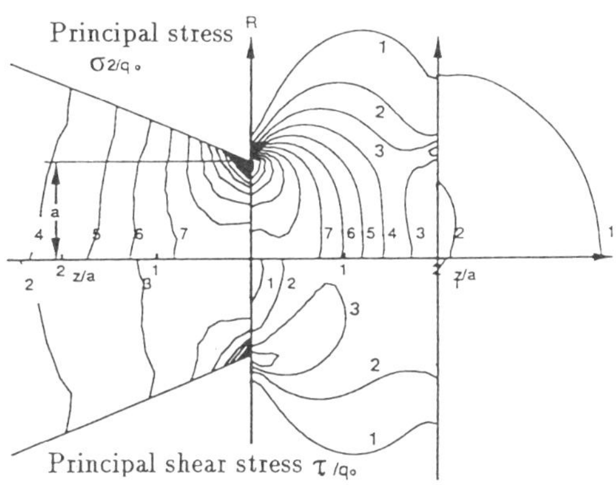

Fig.9. Principal stress σ_2 and principal shear stress τ (30.4 μsec, thickness $5mm$).

incremental time step, $\Delta t = 0.024\mu sec$. In case of $15mm$ thickness spesimen, the distribution of principal stress and principal shear stress are shown in Fig.8. Fig.8(a) is the results of σ_2/q_o ($q_o = 608MPa$) and τ/q_o at $30.4\mu sec$ and Fig.8(b) is the results of σ_1/q_o and τ/q_o at $31.6\mu sec$. From Fig.8(a), it is shown that the stress σ_2 is concentrated at about $2.5mm$ diameter and the stress τ is concentrated at about $2.6mm$ diameter. From Fig.8(b), it is shown the stress σ_1 is concentrated at about $2.7mm$ diameter. It can be considered that the initial crack will occur when the maximum principal stress σ_1 and maximum principal shear stress τ exceed the fracture strength of the material[8),11)]. So it is considered the initial crack will occur at about $2.6mm$ or $2.7mm$ diameter. And it is considered that, after the initial crack occurs, the crack will propagate obliquely in the radial direction and the specimen will fracture as shown in Fig.5.

In case of $5mm$ thickness specimen, the distribution of principal stress σ_2 and principal shear stress τ are shown in Fig.9. Comparing Fig.8 and Fig.9, the distribution of the principal stress and principal shear stress is different but it is similar that the stress is concentrated at about $2.5mm$ diameter. And the noteworthy finding is the concentration of the stress on the contact surface between B and C at about $2.5mm$ diameter. We can consider that initial crack occurs at the edge on the contact surface between A and B and there is a possibility that the next crack occurs on the contact surface between B and C. As shown above, it is shown that fracture behaviour can be pre-estimated roughly by using the numerical results.

V. CONCLUSIONS

The simplified explicit method considering the frictional force is proposed. The proposed method is based on the consideration of equivalent contact force under the small friction condition , and the contact problem can be analyzed without iterative calculation. As a numerical example, the two dimensional axisymmetric model is considered and the stress wave propagation is analyzed. The root mean square errors (RMS) of the principal stress on the contact surface normalized by the result of implicit method are used and the results are compared with the conventional method. Then, the proposed explicit method have the advantage in the accuracy and the calculation time. And it is shown that the fracture behaviour can be pre-estimated roughly.

REFERENCES

1) Hughes, T. J. R., Taylor, R. L. and Sackman, J. L., "A Finite Element Method for a Class of Contact-Impact Problems ", Comp. Meth. App. Mech. Eng. 8(1976), pp.249-276.

2) Hallquist, J. O., "Sliding Interfaces with Contact-Impact in Large-Scale Lagrangian Computations", Comput. Meth. App. Mech. Eng., 51(1985), pp.107-137.

3) Asano, N., "A Penalty Function Type of Virtual Work Principle for Impact Contact Problems of Two Bodies ", JSME, A, 52-475(1986), pp.773-780(in Japanese).

4) Yagawa, G., Kanto, Y. and Ando Y., "Analysis of dynamic Contact Problem Using a Penalty Function ", JSME, A, 49-448(1983), pp.1581-1589(in Japanese).

5) Hallquist, J. O., "Three-Dimensional Lagrangian Computations with Sliding Interfaces Using Finite Element of Finite Difference Techniques ", DYNA3D Course Notes(1987), p.226.

6) Wilkins, M. L., "Calculation of Elastic-Plastic Flow ", Meth. Comp. Phys., Vol. 3, Academic Press, New York(1964), pp.211-263.

7) Wilkins, M. L., "Calculation of Elastic-Plastic Flow ", UCRL-7322(1969), Rev. 1, Lawrence Livermore National Lab., Univ. of California.

8) Liaw, B. M., Kobayashi, A. S. and Emery, A. F., "Theoretical Model of Impact Damage in Structural Ceramics ", J. Amer. Cera. Soci., Vol. 67, No. 8(1984), pp.544-548.

9) Owen, D. R. T. and Hinton E., "Finite Elements in Plasticity: Theory and Practice ", Pineridge Press Limited(1980), pp.378-379.

10) Chen, W. H., Tsai, P., "Finite Element Analysis of Elastodynamic Sliding Contact Problems with Friction ", Comp. Struc. Vol. 22(1986), pp.925-938.

11) Teramae, T. and Hamada, S., "Failure of Ceramic Circular Plates Impacted with Steel Balls ", Symmp. JSME, No. 900-50(1990), pp.359-364(in Japanese).

III
FRACTURE MECHANICS AND COMPOSITE MATERIALS

Calculation of Effective Elastic Moduli of Composite Materials Reinforced by Long Fibers

Nozomu WATARI and Nobunori OSHIMA

College of Industrial Technology, Nihon University, Chiba

It can be proved that the composite materials reinforced by long fibers are elastic bodys of the tetragonal system. In this case, we have six independent elastic coefficients. Plain deformation problems are considered in the present paper, where appear three of six elastic coefficients. It is aimed to estimate these three elastic coefficients and Poisson's ratio by applying the boundary element method (BEM). We assume that fibers are circular in the cross section and are located regularly at square lattice points in a plane. The result by the present method is compared with the existing calculated data and experiments.

I. INTRODUCTION

It is the aim of the present paper to calculate the effective elastic moduli and Poisson's ratio of composite materials, which are reinforced by continuous parallel long fibers. It can be proved that the composite materials are elastic bodys of the tetragonal system of which axis of the fiber has 4-valued symmetry. In the case of materials of the tetragonal system, there are six independent effective elastic coefficients[1].

If we assume that the fibers are circular in the cross section and are located regularly at square lattice points in xy-plain, we can take a square domain containing a fiber and the surrounding matrix(continuous medium) to be called a "cell" forming a unit domain of the whole field. Boundary conditions of the cell are specified by the strains. Plain deformation problem will be considered in this paper, where the displacement takes place in xy-plain perpendicular to the parallel continuous fibers. In this case, three of six elastic coefficients make appearance. These three coefficients are calculated by applying the boundary element method (BEM)[2]. Elasticity problems of the cell are considered for three assigned type of deformation of extension, contraction and shear.

The boundary element method is extremely efficient in the present problem, since it can be performed simply with informations on nodal points of the cell boundary and on the fiber-matrix interface. There is no need to analyze the interior field of the cell. The present method to be used is based on the constant element approximation. By making use of physical symmetry of the problem, the number of unknown nodal tractions p and nodal displacements u over the cell boundary and fiber surface can be considerably reduced. This reduction provides great saving in computational effort.

Many works have been done theoretically to estimate these coefficients in terms of physical properties and composition of constituent materials. In the case of transversal isotropy, overall elastic coefficients can be calculated by the cell model method which is very efficient in contrast to its simplicity.[3),4)] The result by the present method will be compared with the existing calculated data[3),6)] and experiments.[7)]

II. THE STRESS - STRAIN RELATIONS IN TETRAGONAL SYSTEM

The stress-strain relations in tetragonal system are described by the following equations:

$$\sigma_{xx} = (\lambda_1 + 2\mu_1)\varepsilon_{xx} + \lambda_1\varepsilon_{yy} + \lambda_2\varepsilon_{zz}, \tag{1}$$

$$\sigma_{yy} = \lambda_1\varepsilon_{xx} + (\lambda_1 + 2\mu_1)\varepsilon_{yy} + \lambda_2\varepsilon_{zz}, \tag{2}$$

$$\sigma_{zz} = \lambda_1\varepsilon_{xx} + \lambda_2\varepsilon_{yy} + (\lambda_3 + 2\mu_1)\varepsilon_{zz}, \tag{3}$$

$$\sigma_{xy} = 2\mu_2\varepsilon_{xy}, \tag{4}$$

$$\sigma_{yz} = 2\mu_3\varepsilon_{yz}, \tag{5}$$

$$\sigma_{xz} = 2\mu_3\varepsilon_{xz}, \tag{6}$$

where σ_{xx} , ... , σ_{xz} are stresses, ε_{xx} , ... , ε_{xz} are strains and λ_1, λ_2, λ_3, μ_1, μ_2 and μ_3 are elastic coefficients. Strains defined by

$$\varepsilon_{xx} = \frac{\partial u}{\partial x},\ \varepsilon_{yy} = \frac{\partial v}{\partial y},\ \varepsilon_{zz} = \frac{\partial w}{\partial z}\ , \tag{7}$$

$$\varepsilon_{xy} = \frac{1}{2}\left(\frac{\partial v}{\partial x} + \frac{\partial u}{\partial y}\right), \tag{8}$$

$$\varepsilon_{yz} = \frac{1}{2}\left(\frac{\partial w}{\partial y} + \frac{\partial v}{\partial z}\right), \tag{9}$$

$$\varepsilon_{zx} = \frac{1}{2}\left(\frac{\partial w}{\partial x} + \frac{\partial u}{\partial z}\right), \tag{10}$$

where u, v, and w are infinitesimal displacements in the direction of x-, y- and z-axis,respectively.

We assume that the displacement takes place in the plane perpendicular to the parallel long fibers. Therefore, plane deformation problem is considered in the present paper. In this case, the stress-strain relations reduce as follows:

$$\sigma_{xx} = (\lambda_1 + 2\mu_1)\varepsilon_{xx} + \lambda_1\varepsilon_{yy}\ , \tag{11}$$

$$\sigma_{yy} = \lambda_1\varepsilon_{xx} + (\lambda_1 + 2\mu_1)\varepsilon_{yy}\ , \tag{12}$$

$$\sigma_{xy} = 2\,\mu_2\varepsilon_{xy}\,. \tag{13}$$

Three elastic coefficients: λ_1 , μ_1 and μ_2 are calculated by applying the boundary element method in this paper. The equations of equilibrium for constituent materials are given by

$$\mu\ \Delta u + (\lambda + \mu)\partial\varepsilon/\partial x = 0\,, \tag{14}$$

$$\mu\ \Delta v + (\lambda + \mu)\partial\varepsilon/\partial y = 0\,, \tag{15}$$

$$\varepsilon = \partial u/\partial x + \partial v/\partial y\,, \tag{16}$$

where λ and μ are elastic coefficients of respective constituent materials, Δ being the two-dimensional Laplacian operator.

III. BOUNDARY CONDITIONS ON THE CELL AND DETERMINATION OF ELASTIC COEFFICIENTS

3.1 Expansion of the Cell in xy-plain

Let $2R$ be the distance between axes of two neighboring fibers and a the radius of a cylindrical fiber. In this case, if we take a square domain with $2R$ around a fiber, it forms a unit domain of the whole fields. We shall call the unit domain a "cell". $BB_1B_2B_3B$ in Fig.1 denotes the cell. The effective elastic moduli will be formulated as a function of the volume fraction z_d , which is given by

$$z_d = \frac{\pi\, a^2}{4\,R^2}\,. \tag{17}$$

We first consider the deformation case where the cell is subjected to the expansion in xy-plain as shown in Fig.1. We can express this deformation case as follows:

$$\bar{\varepsilon}_{xx} = X\ ,\ \bar{\varepsilon}_{yy} = X\,, \tag{18$_{1,2}$}$$

$$\bar{\varepsilon}_{xy} = 0\,, \tag{19}$$

where $\bar{\varepsilon}_{xx}$ and $\bar{\varepsilon}_{yy}$ denote overall strain and $X = \partial\,\bar{u}\,/\,\partial\,x$,where $\bar{u}$ is the overall displacement.

Considering the Eq $(18)_{1,2}$ in this case, we can obtain overall stresses $\bar{\sigma}_{xx}$ and $\bar{\sigma}_{yy}$ from the Eqs(11) and (12) as follows:

$$\bar{\sigma}_{xx} = 2\,(\lambda_1 + \mu_1)\,X\,,\ \bar{\sigma}_{yy} = 2\,(\lambda_1 + \mu_1)\,X\,. \tag{20$_{1,2}$}$$

From Eq$(20)_{1,2}$, we have elastic coefficient $\lambda_1 + \mu_1$ as follows:

$$\lambda_1 + \mu_1 = \bar{\sigma}_{xx}\,/\,2\,X = \bar{\sigma}_{yy}\,/\,2\,X\,. \tag{21}$$

The boundary conditions for this case are given by

$$u = RX \ , \ \sigma_{xy} = 0 \qquad \text{(on } BB_3\text{)}, \qquad (22)_1$$

$$\sigma_{yx} = 0 \ , \ v = RX \qquad \text{(on } BB_1\text{)}, \qquad (22)_2$$

$$u = -RX \ , \ \sigma_{xy} = 0 \qquad \text{(on } B_1B_2\text{)}, \qquad (22)_3$$

$$\sigma_{yx} = 0 \ , \ v = -RX \qquad \text{(on } B_2B_3\text{)}. \qquad (22)_4$$

The overall stresses $\bar{\sigma}_{xx}$ and $\bar{\sigma}_{yy}$ are obtained by

$$\bar{\sigma}_{xx} = \int_A^B \sigma_{xx}\, dy / R \, , \qquad (23)_1$$

$$\bar{\sigma}_{yy} = \int_C^B \sigma_{yy}\, dx / R \, . \qquad (23)_2$$

3.2 Extension and Contraction of the Cell in xy-plain

In this section, we shall consider the deformation case where the cell is subjected to the extension to the x-axis and the contraction of the same magnitude to the y-axis as shown in Fig.2. We can express this deformation case as follows:

$$\bar{\varepsilon}_{xx} = X \ , \ \bar{\varepsilon}_{yy} = -X \, , \qquad (24)_{1,2}$$

$$\bar{\varepsilon}_{xy} = 0. \qquad (25)$$

Considering the Eq$(24)_{1,2}$ in this case, we can obtain elastic coefficient μ_1 from the Eqs(11) and (12) as follows:

$$\mu_1 = \bar{\sigma}_{xx} / 2X = -\bar{\sigma}_{yy} / 2X \, . \qquad (26)$$

The boundary conditions for this case are given by

$$u = RX \ , \ \sigma_{xy} = 0 \qquad \text{(on } BB_3\text{)}, \qquad (27)_1$$

$$\sigma_{yx} = 0 \ , \ v = -RX \qquad \text{(on } BB_1\text{)}, \qquad (27)_2$$

$$u = -RX \ , \ \sigma_{xy} = 0 \qquad \text{(on } B_1B_2\text{)}, \qquad (27)_3$$

$$\sigma_{yx} = 0 \ , \ v = RX \qquad \text{(on } B_2B_3\text{)}. \qquad (27)_4$$

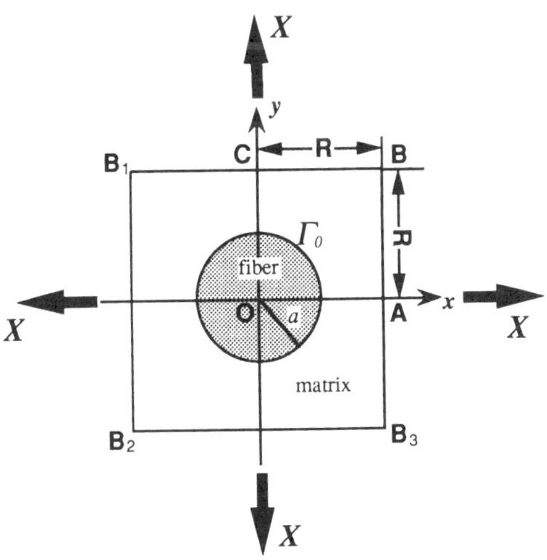

Fig.1. Illustration of the cell $BB_1B_2B_3B$ subjected to the expansion in xy-plain.

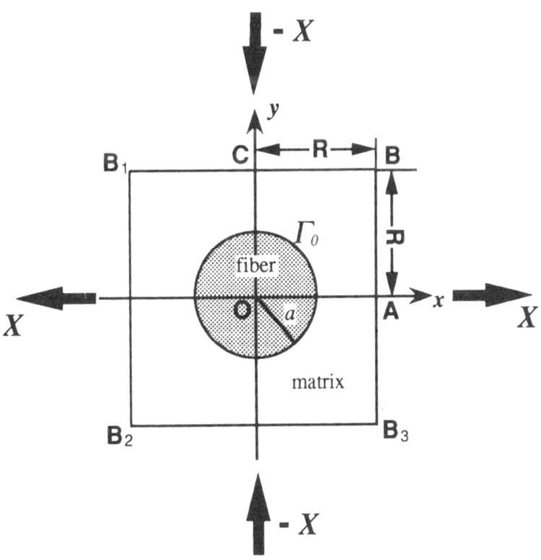

Fig.2. Illustration of the cell $BB_1B_2B_3B$ subjected to the extension to the x-axis and contraction of the same magnitude to the y-axis.

3.3 Shear of the Cell in xy-plain

Finally, we shall consider the deformation case where the cell is subjected to the shear as shown in Fig.3. In this case, we have

$$\bar{\varepsilon}_{xy} = X, \tag{28}$$

$$\bar{\varepsilon}_{xx} = \bar{\varepsilon}_{yy} = 0. \tag{29}$$

From Eqs(28) and (13), it follows that

$$\mu_2 = \bar{\sigma}_{xy} / 2X. \tag{30}$$

The overall stress $\bar{\sigma}_{xy}$ is obtained by

$$\bar{\sigma}_{xy} = \int_A^B \sigma_{xy}\, dy / R. \tag{31}$$

The boundary conditions for this case are given by

$$\sigma_{xx} = 0\ ,\ v = RX \qquad \text{(on } BB_3\text{)}, \tag{32$_1$}$$

$$u = RX\ ,\ \sigma_{yy} = 0 \qquad \text{(on } BB_1\text{)}, \tag{32$_2$}$$

$$\sigma_{xx} = 0\ ,\ v = -RX \qquad \text{(on } B_1B_2\text{)}, \tag{32$_3$}$$

$$u = -RX\ ,\ \sigma_{yy} = 0 \qquad \text{(on } B_2B_3\text{)}. \tag{32$_4$}$$

IV. FORMULATION BY THE BOUNDARY ELEMENT METHOD

Plane deformation problem of composite materials composed of two constituent materials $\Omega^{(1)}$(matrix) and $\Omega^{(2)}$(fiber) is solved by applying the boundary element method. The method to be used is based on the constant element approximation as shown in Fig.4. There are four variables associated with each boundary mid-nodes i: two displacement components u_{ix} and u_{iy} and two traction components p_{ix} and p_{iy}. Let M be the number of boundary elements on AB, BC and Γ_0. The total number of mid-nodes is $12M$ ($M = 3$ in Fig.4). By the symmetry of the problem, it is enough to consider only a quarter of all nodes and we can reduce the unknowns. In the deformation case of 3.1 and 3.2, we have the symmetry of displacement components as follows:

$$u_{i_2 x} = -u_{ix},\ u_{i_3 x} = -u_{ix},\ u_{i_4 x} = u_{ix}, \tag{33}$$

$$u_{i_2 y} = u_{iy},\ u_{i_3 y} = -u_{iy},\ u_{i_4 y} = -u_{iy}. \qquad (i = 1, 2, \dots, 3M)$$

In the case of 3.3, we have

$$u_{i_2 x} = u_{ix},\ u_{i_3 x} = -u_{ix},\ u_{i_4 x} = -u_{ix}, \tag{34}$$

$$u_{i_2 y} = -u_{iy},\ u_{i_3 y} = -u_{iy},\ u_{i_4 y} = u_{iy}.$$

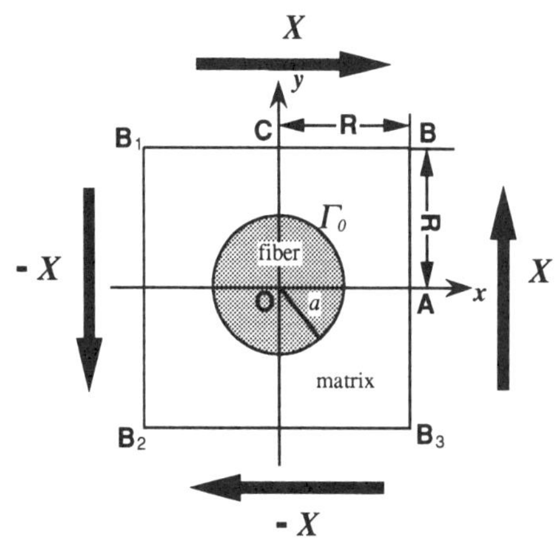

Fig.3. Illustration of the cell $BB_1B_2B_3B$ which is subjected to the shear.

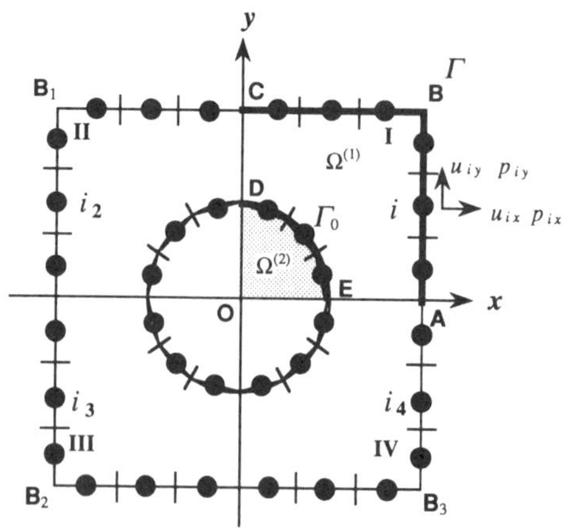

Fig.4. Discretization and mid-nodes of the cell boundary Γ and the interface Γ_0 by the constant element approximation.

where i, i_2, i_3 and i_4 denote the corresponding number of mid-nodes belong to the first (I) through the fourth (IV) quadrant,respectively. Similarly, relations in Eqs(33) and (34) are also valid for p_{ix} and p_{iy}. From the relations in Eqs (33) or (34), we have the following equation in the matrix domain $\Omega^{(1)}$

$$\sum_m \boldsymbol{H}_m^{(1)} \boldsymbol{u}_m^{(1)} + \sum_n \boldsymbol{H}_n^{(1)} \boldsymbol{u}_n^{(1)} = \sum_m \boldsymbol{G}_m^{(1)} \boldsymbol{p}_m^{(1)} + \sum_n \boldsymbol{G}_n^{(1)} \boldsymbol{p}_n^{(1)} , \tag{35}$$

where $1 \leqq m \leqq 2M$ (on AB and BC), $2M+1 \leqq n \leqq 3M$ (on Γ_0).
In the fiber domain $\Omega^{(2)}$, we have

$$\sum_n \boldsymbol{H}_n^{(2)} \boldsymbol{u}_n^{(2)} = \sum_n \boldsymbol{G}_n^{(2)} \boldsymbol{p}_n^{(2)} , \tag{36}$$

where $\boldsymbol{H}_j^{(k)}$ and $\boldsymbol{G}_j^{(k)}$ (k =1, 2) are $6M \times 2$ matrices and j denotes suffices m or n. $\boldsymbol{u}_j^{(k)}$ and $\boldsymbol{p}_j^{(k)}$ are 2-colum vector.

$$\boldsymbol{H}_j^{(k)} = \left[\boldsymbol{H}_{ij}^{(k)}\right], \; \boldsymbol{G}_j^{(k)} = \left[\boldsymbol{G}_{ij}^{(k)}\right], \tag{37}$$

$$\boldsymbol{u}_j^{(k)} = \begin{bmatrix} u_{jx}^{(k)} \\ u_{jy}^{(k)} \end{bmatrix}, \; \boldsymbol{p}_j^{(k)} = \begin{bmatrix} p_{jx}^{(k)} \\ p_{jy}^{(k)} \end{bmatrix}, \tag{38}$$

where $\boldsymbol{H}_j^{(k)}, \boldsymbol{G}_j^{(k)}$ are 2×2 matrices.

$$\left[\boldsymbol{H}_{ij}^{(k)}\right] = \begin{bmatrix} H_{11}^{(k)} & H_{12}^{(k)} \\ H_{21}^{(k)} & H_{22}^{(k)} \end{bmatrix}_{ij}, \; \left[\boldsymbol{G}_{ij}^{(k)}\right] = \begin{bmatrix} G_{11}^{(k)} & G_{12}^{(k)} \\ G_{21}^{(k)} & G_{22}^{(k)} \end{bmatrix}_{ij}. \tag{39}$$

($i, j = 1, 2, \ldots, 3M$)

For the case of 3.1 and 3.2

$$\left[\boldsymbol{H}_{ij}^{(k)}\right] = \begin{bmatrix} h_{11} & h_{12} \\ h_{21} & h_{22} \end{bmatrix}_{ij_1}^{(k)} - \begin{bmatrix} h_{11} & -h_{12} \\ h_{21} & -h_{22} \end{bmatrix}_{ij_2}^{(k)} - \begin{bmatrix} h_{11} & h_{12} \\ h_{21} & h_{22} \end{bmatrix}_{ij_3}^{(k)} + \begin{bmatrix} h_{11} & -h_{12} \\ h_{21} & -h_{22} \end{bmatrix}_{ij_4}^{(k)}, \tag{40}$$

$$\left[\boldsymbol{G}_{ij}^{(k)}\right] = \begin{bmatrix} g_{11} & g_{12} \\ g_{21} & g_{22} \end{bmatrix}_{ij_1}^{(k)} - \begin{bmatrix} g_{11} & -g_{12} \\ g_{21} & -g_{22} \end{bmatrix}_{ij_2}^{(k)} - \begin{bmatrix} g_{11} & g_{12} \\ g_{21} & g_{22} \end{bmatrix}_{ij_3}^{(k)} + \begin{bmatrix} g_{11} & -g_{12} \\ g_{21} & -g_{22} \end{bmatrix}_{ij_4}^{(k)}. \tag{41}$$

For the case of 3.3

$$\left[\boldsymbol{H}_{ij}^{(k)}\right] = \begin{bmatrix} h_{11} & h_{12} \\ h_{21} & h_{22} \end{bmatrix}_{ij_1}^{(k)} + \begin{bmatrix} h_{11} & -h_{12} \\ h_{21} & -h_{22} \end{bmatrix}_{ij_2}^{(k)} - \begin{bmatrix} h_{11} & h_{12} \\ h_{21} & h_{22} \end{bmatrix}_{ij_3}^{(k)} - \begin{bmatrix} h_{11} & -h_{12} \\ h_{21} & -h_{22} \end{bmatrix}_{ij_4}^{(k)}, \tag{42}$$

$$\left[\boldsymbol{G}_{ij}^{(k)}\right] = \begin{bmatrix} g_{11} & g_{12} \\ g_{21} & g_{22} \end{bmatrix}_{ij_1}^{(k)} + \begin{bmatrix} g_{11} & g_{12} \\ g_{21} & g_{22} \end{bmatrix}_{ij_2}^{(k)} - \begin{bmatrix} g_{11} & g_{12} \\ g_{21} & g_{22} \end{bmatrix}_{ij_3}^{(k)} - \begin{bmatrix} g_{11} & -g_{12} \\ g_{21} & -g_{22} \end{bmatrix}_{ij_4}^{(k)}. \tag{43}$$

where j_1 (= j), j_2, j_3 and j_4 denote the corresponding number of the element belong to the first (I) through the fourth (IV) quadrant,respectively. The integral over the element ' j_s' (s = 1,2,3,4) is carried out as follows:

$$\begin{bmatrix} h_{11} & h_{12} \\ h_{21} & h_{22} \end{bmatrix}_{ij_s}^{(k)} = \frac{1}{2} \boldsymbol{E}_{ij_s} + \int_{\Gamma_{j_s}} \begin{bmatrix} p_{11}^* & p_{12}^* \\ p_{21}^* & p_{22}^* \end{bmatrix} d\Gamma , \tag{44}$$

$$\begin{bmatrix} g_{11} & g_{12} \\ g_{21} & g_{22} \end{bmatrix}_{ij_s}^{(k)} = \int_{\Gamma_{j_s}} \begin{bmatrix} u_{11}^* & u_{12}^* \\ u_{21}^* & u_{22}^* \end{bmatrix} d\Gamma . \tag{45}$$

$$\boldsymbol{E}_{ij_s} = \begin{bmatrix} 0 & 0 \\ 0 & 0 \end{bmatrix} \quad (i \neq j_s), \tag{46}$$

$$= \begin{bmatrix} 1 & 0 \\ 0 & 1 \end{bmatrix} \quad (i = j_s), \tag{47}$$

where p^*_{lk}, u^*_{lk} (l,k =1,2) are the fundamental solution for two-dimensional isotopic materials[2]. The integrals in Eqs.(44) and (45) are calculated by using the 4-point Gauss quadrature formulae[2].

On the interface Γ_0, the following conditions are demanded for three cases

$$\boldsymbol{u}_n^{(1)} = \boldsymbol{u}_n^{(2)}, \tag{48}$$

$$\boldsymbol{p}_n^{(1)} = -\boldsymbol{p}_n^{(2)}. \tag{49}$$

From Eqs.(36) and conditions in Eqs.(48) and (49), it follows that

$$\boldsymbol{p}_n^{(1)} = -\boldsymbol{p}_n^{(2)} = -\sum_n [\boldsymbol{K}_{nn}]\, \boldsymbol{u}_n^{(1)} \quad (2M+1 \leqq n \leqq 3M), \tag{50}$$

where

$$[\boldsymbol{K}_{nn}] = \left[\boldsymbol{K}_{nn}^{(2)}\right]^{-1}\left[\boldsymbol{H}_{nn}^{(2)}\right]. \tag{51}$$

In the case of 3.1 and 3.2, substituting Eq.(50) into Eq.(35), considering the boundary conditions in Eqs$(22)_{1,2}$ and $(27)_{1,2}$,respectively and transposing all unknowns to the left hand side, the following system of equation is obtained.

$$\sum_{m=1}^{M}\begin{bmatrix} -G_{11}^{(1)} & H_{12}^{(1)} \\ -G_{21}^{(1)} & H_{22}^{(1)} \end{bmatrix}_{im}\begin{bmatrix} p_{mx}^{(1)} \\ u_{my}^{(1)} \end{bmatrix} + \sum_{m=M+1}^{2M}\begin{bmatrix} H_{11}^{(1)} & -G_{12}^{(1)} \\ H_{21}^{(1)} & -G_{22}^{(1)} \end{bmatrix}_{im}\begin{bmatrix} u_{mx}^{(1)} \\ p_{my}^{(1)} \end{bmatrix}$$
$$+\sum_{n=2M+1}^{3M}\begin{bmatrix} F_{11}^{(1)} & F_{12}^{(1)} \\ F_{21}^{(1)} & F_{22}^{(1)} \end{bmatrix}_{in}\begin{bmatrix} u_{mx}^{(1)} \\ u_{my}^{(1)} \end{bmatrix} = -\sum_{m=1}^{M}\begin{bmatrix} H_{11}^{(1)} \\ H_{21}^{(1)} \end{bmatrix}_{im} RX \mathop{-}_{+} \sum_{m=M+1}^{2M}\begin{bmatrix} H_{12}^{(1)} \\ H_{22}^{(1)} \end{bmatrix}_{im} RX, \tag{52}$$

where in the last term, $-$ is for 3.1 and $+$ for 3.2,

$$\begin{bmatrix} F_{11}^{(1)} & F_{12}^{(1)} \\ F_{21}^{(1)} & F_{22}^{(1)} \end{bmatrix}_{in} = \begin{bmatrix} H_{11}^{(k)} & H_{12}^{(k)} \\ H_{21}^{(k)} & H_{22}^{(k)} \end{bmatrix}_{in} + \sum_{n=2M+1}^{3M}\begin{bmatrix} G_{11}^{(1)} & G_{12}^{(1)} \\ G_{21}^{(1)} & G_{22}^{(1)} \end{bmatrix}_{in}\begin{bmatrix} K_{11}^{(1)} & K_{12}^{(1)} \\ K_{21}^{(1)} & K_{22}^{(1)} \end{bmatrix}_{nn}. \tag{53}$$

Equation (52) corresponds a mid-node 'i' ($1 \leqq i \leqq 3M$). We have $6M$ unknowns in Eq.(52). By solving Eq.(52), $\boldsymbol{p}_{mx}^{(1)}$ and $\boldsymbol{p}_{my}^{(1)}$ are obtained. The overall stresses $\bar{\sigma}_{xx}$ and $\bar{\sigma}_{yy}$ are calculated by as follows:

$$\bar{\sigma}_{xx} = \frac{1}{M}\sum_{m=1}^{M} p_{mx}^{(1)}, \quad \bar{\sigma}_{yy} = \frac{1}{M}\sum_{m=M+1}^{2M} p_{my}^{(1)}. \tag{54}$$

In the case of 3.3, it follows that

$$\sum_{m=1}^{M}\begin{bmatrix} H_{11}^{(1)} & -G_{12}^{(1)} \\ H_{21}^{(1)} & -G_{22}^{(1)} \end{bmatrix}_{im}\begin{bmatrix} u_{mx}^{(1)} \\ p_{my}^{(1)} \end{bmatrix} + \sum_{m=M+1}^{2M}\begin{bmatrix} -G_{11}^{(1)} & H_{12}^{(1)} \\ -G_{21}^{(1)} & H_{22}^{(1)} \end{bmatrix}_{im}\begin{bmatrix} p_{mx}^{(1)} \\ u_{my}^{(1)} \end{bmatrix}$$
$$+\sum_{n=2M+1}^{3M}\begin{bmatrix} F_{11}^{(1)} & F_{12}^{(1)} \\ F_{21}^{(1)} & F_{22}^{(1)} \end{bmatrix}_{in}\begin{bmatrix} u_{nx}^{(1)} \\ u_{ny}^{(1)} \end{bmatrix} = -\sum_{m=1}^{M}\begin{bmatrix} H_{12}^{(1)} \\ H_{22}^{(1)} \end{bmatrix}_{im} RX - \sum_{m=M+1}^{2M}\begin{bmatrix} H_{11}^{(1)} \\ H_{21}^{(1)} \end{bmatrix}_{im} RX. \tag{55}$$

The overall stresses $\bar{\sigma}_{xy}$ are calculated by as follows:

$$\bar{\sigma}_{xy} = \frac{1}{M}\sum_{m=1}^{M} p_{my}^{(1)}. \tag{56}$$

All the coefficient matrices of these equations (52) and (55), if $z_d = 0.1 \times K$, vary with the increase of K. However, the coefficient matrices $\boldsymbol{H}_n^{(k)}$ and $\boldsymbol{G}_n^{(k)}$ corresponding to Γ_0 are simply calculated as follows:

$$\left(\boldsymbol{H}_n^{(k)}\right)_K = \left(\boldsymbol{H}_n^{(k)}\right)_{K=1}, \quad \left(\boldsymbol{G}_n^{(k)}\right)_K = \sqrt{K} \times \left(\boldsymbol{G}_n^{(k)}\right)_{K=1}, \tag{57}$$

where $k = 1, 2$ and $(\quad)_{K=1}$ denotes the value at $K = 1$. [5),6)]

V. NUMERICAL CALCULATION AND CONCLUSION

From the stress-strain relations in Eqs.(11) through (13), strains ε_{xx}, ε_{yy} and ε_{xy} are represented by stresses σ_{xx}, σ_{yy} and σ_{xy} as follows:

$$\varepsilon_{xx} = \frac{1}{E_{eff}} (\sigma_{xx} - \nu\, \sigma_{yy}), \tag{58$_1$}$$

$$\varepsilon_{yy} = \frac{1}{E_{eff}} (-\nu\, \sigma_{xx} + \sigma_{yy}), \tag{58$_2$}$$

$$\varepsilon_{xy} = \frac{1}{2\mu_2}\, \sigma_{xy}, \tag{58$_3$}$$

where E_{eff} and ν denote the effective Young's modulus and Poisson's ratio of composite materials, respectively. and E_{eff} and ν are related to the elastic coefficients λ_1 and μ_1 as follows:

$$E_{eff} = 4\mu_1 \frac{\lambda_1 + \mu_1}{\lambda_1 + 2\mu_1}, \tag{59$_1$}$$

$$\nu = \frac{\lambda_1}{\lambda_1 + 2\mu_1}. \tag{59$_2$}$$

The elastic coefficients $\lambda_1 + \mu_1$ and μ_1 are calculated by the Eqs.(21) and (26). E_{eff} and ν are determined by Eqs.$(59)_{1,2}$.

In the Ref(6), tensile strain is assigned in direction of x-axis as shown in Fig.5 and $\sigma_{yx} = \sigma_{yy} = 0$ are assumed on BB_1 and B_2B_3 instead of $(18)_2$ in the case of 3.1. The effective modulus E'_{eff} and Poisson's ratio ν' are calculated by BEM as follows:

$$E'_{eff} = \frac{\overline{\sigma}_{xx}}{X} = \int_A^B \sigma_{xx}\, dy \Big/ RX, \tag{60}$$

$$\nu' = -\frac{\overline{\varepsilon}_{yy}}{\varepsilon_{xx}} = -\frac{\int_C^B \varepsilon_{yy} dx \Big/ R}{\varepsilon_{xx}} \tag{61}$$

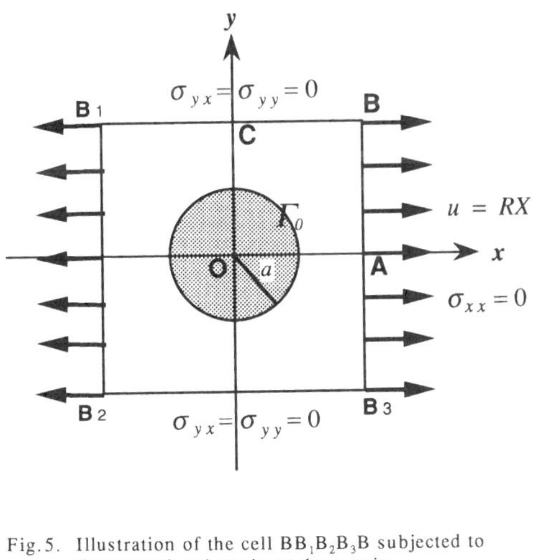

Fig.5. Illustration of the cell $BB_1B_2B_3B$ subjected to the extensional strain to the x-axis.

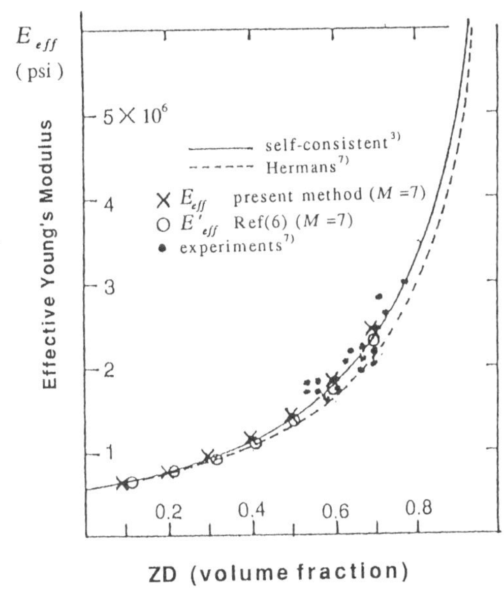

Fig.6. The effective Young's modulus E_{eff} vs. the volume fraction z_d. M denotes the number of boundary elements on each side of a quarter of the cell.

Let E_d and ν_d be the Young's modulus and Poisson's ratio of the fiber respectively. Similarly, let E_c and ν_c be the Young's modulus and Poisson's ratio of the matrix respectively. The numerical results by the present method and Ref(6) (Eqs.(60) and (61)) are plotted in Fig.6 and Fig.7. We take $M=7$, $E_d = 10.6\times10^6$ psi (7.31×10^{10} Pa), $\nu_d = 0.22$ for glass fiber and $E_c = 0.5\times10^6$ psi (0.34×10^{10} Pa), $\nu_c = 0.35$ for epoxi resin matrix.

The solid line is the result obtained by the method of self-consistent cell model[3),4)]. The broken line is the result obtained by J.J.Hermans[7)]. We notice that the values of E_{eff} by the present method are greater slightly than the values of E'_{eff} in Ref(6) as shown in Fig.6. It may be because the cell is constrained more severely in the present paper by the boundary conditions for three assigned type of deformation of extension,contraction and shear,respectively.

We notice that Poisson's ratio ν and ν' of composite materials are greater than Poisson's ratio of the matrix ($\nu_c = 0.35$) and that $\nu < \nu'$ as shown in Fig.7. For the isotropic materials, it is shown theoretically that Poisson's ratio of plane deformation problems is greater than three dimensional Poisson's ratio. Similar results are obtained in Ref(8) for the case of composite materials with the hexagonal array of fibers. The numerical results of μ_1 and μ_2 by the present method are plotted in Fig.8. It is noted that $\mu_1 > \mu_2$.

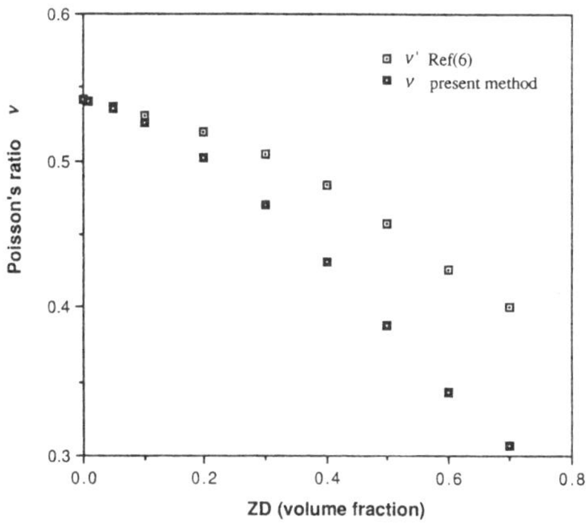

Fig.7. Poisson's ratio ν vs. the volume fraction z_d .

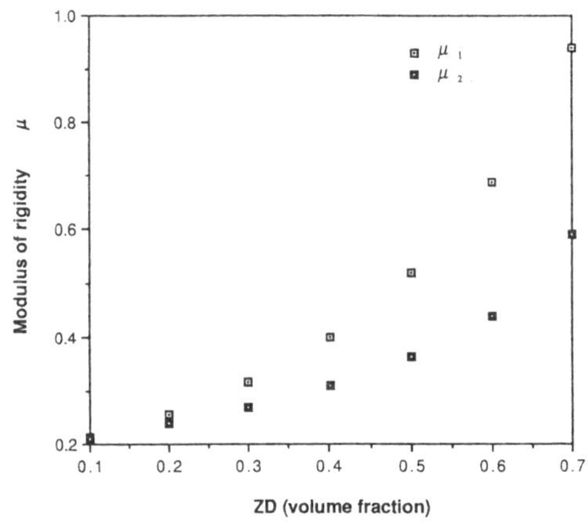

Fig.8. Moduli of rigidity μ_1 and μ_2 vs. the volume fraction z_d .

REFERENCES

1) Max Planck:Einführung in die Theoretische Physik (Translation by K.Terazawa).
2) Brebbia,C,A., The Boundary Element Method for Engineers. Pentech Press,1984, pp.139-176.
3) Oshima,N.,Calculation of Effective Elastic Moduli of Transversally Isotropic Composite Materials. Trans.JSME,48,434,1982,pp.1253-1260(in Japanese).
4) Oshima,N.,Method for Estimating Effective Elastic Moduli of Dispersely Reinforced Composite Materials.Proceedings of the Fourth International Conference on Composite Materials,ICCM-IV (Tokyo JAPAN),1982, pp.1123-1128.
5) Oshima,N.and Watari,N., The Calculation of the Effective Thermal Conductivity of Media with Regularly Dispersed Parallel Cylinders.JSME International Journal,SeriesII,Vol.32, No.2, 1989, pp.225-230.
6) Oshima,N.and Watari,N.,Calculation of Effective Elastic Moduli of Composite Materials with Dispersed Parallel Fibers.Theoret.Appl.Mech.,41(University of Tokyo Press,1992),pp.181-188.
7) Hermans,J,J., The Elastic Properties of Fiber Reinforced Materials When the Fibers are Aligned. Konik.Neder.Akad.Waten.,Proc.,B70,1967,pp.1-9.
8) Wan-Lee Yin and Zhizhong Fei.,Boundary-Element Analysis of Micromechanical Models for Computing the Elastic Moduli of a Unidirectional Composite.Proceedings of the International Symposium on Composite Materials and Structures(Beijing,CHINA), 1992,pp.382-389.

Load-Carrying Capacity of Beam-Columns under Compression and End-Moments

Masahiro DOGAKI*, Izumi MASUDA**, and Hiroshi YONEZAWA***

* *Department of Civil Engineering, Kansai University, Osaka*

** *Formerly Postgraduate Student at Kansai University, Osaka*

*** *Emeritus Professor of Kansai University, Osaka*

This paper is aimed at developing a finite difference analysis for predicting the non-linear behaviour and load-carrying capacity at which the hot-rolled beam-columns with I-shaped cross-section occur flexural-torsional buckling. Both the geometric and material non-linearities are taken into account. The beam-columns are subjected to both the compression and equal end-moments about the minor and major principal axes. The initial crookednesses and twisting, and residual stress are considered. A set of non-linear algebraic equations is precisely solved by the modified Newton-Raphson procedure with incremental approach, in which some effective acceleration techniques are employed to improve convergence characteristics.

The comprehensive parametric computation is carried out to examine the influences of slenderness parameter, initial imperfections and combination of applied loadings on the non-linear behaviour and load-carrying capacity of beam-columns. The interaction curves of load-carrying capacity are presented for the beam-columns under various combination of compression and biaxial end-moments.

I. INTRODUCTION

Almost members in steel frame structures are beam-columns subjected to compression, biaxial bending moments and torsion on the ends simultaneously. They may respond in a variety of fashions to the imposed loads on frames, depending upon the geometry of the cross-section, end-restraints, initial crookednesses and twisting, residual stress, and material properties. A lot of theoretical and experimental investigations have been extensively carried out[1),2)] to understand the non-linear behaviour and load-carrying capacity of beam-columns.

The objective of this paper is to develop a finite difference procedure for predicting the load-carrying capacity for the flexural-torsional buckling of doubly symmetrical hot-rolled beam-columns with I-shaped cross-section under a combination of compression and bending moments about both the minor and major principal axes equally acting on the ends. The geometric and material non-linear phenomenon of beam-columns is clarified theoretically. Initial crookednesses and twisting, and residual stress are also taken into account. A set of non-linear algebraic equations is precisely solved by the modified Newton-Raphson procedure with incremental loading.

II. METHOD OF ANALYSIS

Consider an inelastic flexural-torsional deformation of hot-rolled steel beam-columns with I-shaped cross-section under a combination of compression and end-moments about both the minor and major principal axes equally acting on both the ends. The doubly symmetrical I-shaped beam-columns are of flanges of width b_f and thickness t_f and a web of depth b_w and thickness t_w as shown in Fig.1. Initial twisting is introduced as well as initial crookednesses in both the vertical and lateral directions. These imperfections are assumed to be a half wave of sinusoidal series in the longitudinal direction with the amplitudes v_m, w_m

and ϕ_m. It is also assumed that longitudinal residual stress induced in a hot-rolled steel beam-column during cooling after rolling distributes over the cross-section as shown in Fig.1, in which σ_m is the absolute and maximum value of the compressive and tensile residual stresses.

The following assumputions are introduced in this paper: (1) the material is elastic-perfectly-plastic without strain hardening; (2) no local plate buckling of flanges and web occurs before the overall flexural-torsional buckling of members; and (3) transverse shear deformation is ignored.

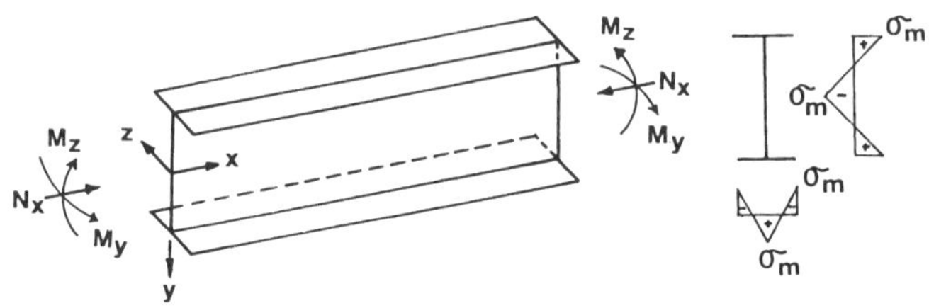

Fig.1. A doubly symmetrical hot-rolled steel beam-column with initial crookednesses and twisting, and with residual stress under a combination of compression and biaxial end-moments in both the directions.

2.1 Equations of Equilibrium

According to the Lin's or Massonnet's treatment[3),4)] for elasto-plastic behaviour of steel members, a set of the equations of equilibrium of stress resultants in the x-, y- and z-directions and with respect to twist at an elasto-plastic state for the flexural-torsional buckling of beam-columns are given as follows:

$$N_{x,x} = 0, \tag{1.1}$$

$$M_{z,xx} + N_x\bar{v}_{,xx} - M_{y,x}\bar{\phi}_{,x} - M_y\bar{\phi}_{,xx} = 0, \tag{1.2}$$

$$M_{y,xx} + N_x\bar{w}_{,xx} + M_{z,x}\bar{\phi}_{,x} + M_z\bar{\phi}_{,xx} = 0, \tag{1.3}$$

$$M_{x,x} + (I_{cp}/A)N_x\bar{\phi}_{,xx} + M_{z,x}\bar{w}_{,x} + M_z\bar{w}_{,xx} - M_{y,x}\bar{v}_{,x} - M_y\bar{v}_{,xx} = 0 \tag{1.4}$$

in which $\bar{v}$ and $\bar{w}$ are the total out-of-plane displacements in the vertical and lateral directions given by a sum of the initial crookednesses v_0 and w_0 and additional deflections v and w due to applied loading respectively; $\bar{\phi}$ is the total angle of twist about the x-axis given by a sum of the initial twist ϕ_0 and additional one ϕ due to applied loading; N_x is the axial force; M_y and M_z are the bending moments about the minor and major principal axes respectively; M_x is the torsional moment about the x-axis; I_{cp} is the sectional moment of inertia; A is the area of I-shaped cross-section; and the subscript preceding the comma describes the ordinary differentiation with respect to it.

According to the same way as Lin's and Massonnet's method on the plasticity of material, the above-mentioned stress resultants at elasto-plastic state are given as follows:

$$N_x = EA[u_{,x}+(\bar{v}_{,x}{}^2 - v_{0,x}{}^2)/2+(\bar{w}_{,x}{}^2 - w_{0,x}{}^2)/2] + EI_{cp}(\bar{\phi}_{,x}{}^2 - \phi_{0,x}{}^2)/2-N_x{}^p \tag{2.1}$$

$$M_y = -EI_y(\bar{w}_{,xx} - w_{0,xx} + \bar{v}_{,x}\bar{\phi}_{,x} - v_{0,x}\phi_{0,x}) - M_y{}^p \tag{2.2}$$

$$M_z = -EI_z(\bar{v}_{,xx} - v_{0,xx} - \bar{w}_{,x}\bar{\phi}_{,x} - w_{0,x}\phi_{0,x}) - M_z{}^p \tag{2.3}$$

$$M_x = GJ(\bar{\phi}_{,x} - \phi_{0,x}) - EI_w(\bar{\phi}_{,xxx} - \phi_{0,xxx}) - M_x{}^p \tag{2.4}$$

$$M_w = -EI_w(\bar{\phi}_{,xx} - \phi_{0,xx}) - M_w{}^p \tag{2.5}$$

in which u is the displacement in the x-direction; EI_y and EI_z are the flexural rigidities of member about its minor and major principal axes respectively, calculated by using the original cross-section; GJ and EI_w are the St. Venant's torsional constant and warping constant respectively; and the stress resultants with the superscript p are the fictitious ones being introduced to compensate on an actual stress field in elasto-plastic region, and calculated by summing up a surplus stress over the cross-section which is defined as the difference between a virtual stress σ_e obtained by assuming the beam-column to be purely elastic even if it behaves within elasto-plastic region and the yield stress σ_y as follows:

$$N_x^p = \Sigma \{\sigma_b + (\sigma^e + \sigma_r - \sigma_b)(\sigma_{eq} - \sigma_y)/\sigma_{eq}\} \Delta A \qquad (3.1)$$

$$M_y^p = \Sigma \{\sigma_b + (\sigma^e + \sigma_r - \sigma_b)(\sigma_{eq} - \sigma_y)/\sigma_{eq}\} z \Delta A \qquad (3.2)$$

$$M_z^p = \Sigma \{\sigma_b + (\sigma^e + \sigma_r - \sigma_b)(\sigma_{eq} - \sigma_y)/\sigma_{eq}\} y \Delta A \qquad (3.3)$$

$$M_x^p = \Sigma \{\tau_{xzb} + (\tau_{xz}^e - \tau_{xzb})(\sigma_{eq} - \sigma_y)/\sigma_{eq}\} y \Delta A$$
$$- \Sigma \{\tau_{xyb} + (\tau_{xy}^e - \tau_{xyb})(\sigma_{eq} - \sigma_y)/\sigma_{eq}\} z \Delta A - M_w^p{}_{,x} \qquad (3.4)$$

$$M_w^p = \Sigma \{\sigma_b + (\sigma^e + \sigma_r - \sigma_b)(\sigma_{eq} - \sigma_y)/\sigma_{eq}\} \omega_n \Delta A \qquad (3.5)$$

in which the stresses with the subscript b are the fictitious values determined at the previous loading stage; σ_r is the residual stress; ΔA is the area of elements on any cross-section of beam-columns; ω_n is the normalized unit warping with respect to the centroid; and the stress σ_{eq} is the von Mises' equivalent stress given by

$$\sigma_{eq} = \{ (\sigma^e + \sigma_r - \sigma_b)^2 + 3(\tau_{ij}^e - \tau_{ijb})^2 \}^{1/2} \qquad (4)$$

in which the shearing stress τ_{xy} is used instead of τ_{ij} when the yielding of material is checked at the web and τ_{xz} at the flanges.

2.2 Boundary Conditions

The beam-column is assumed to be simply supported at both the ends, i.e., free with respect to rotation and warping. It is also assumed that the member is subjected to combined compression and biaxial end-moments. The corresponding boundary condition at the end of x=0 is given as follows:

$$u=0,\ \bar{v}=v_0,\ \bar{w}=w_0,\ \bar{\phi}=\phi_0,\ \bar{v}_{,x} - v_{0,x} = \theta_{z1},\ \bar{w}_{,x} - w_{0,x} = \theta_{y1},\ M_w = 0 \qquad (5)$$

and at the end of $x=\ell$

$$u=-u_1,\ \bar{v}=v_0,\ \bar{w}=w_0,\ \bar{\phi}=\phi_0,\ M_z = M_{z0},\ M_y = M_{y0},\ M_w = 0 \qquad (6)$$

in which u_1 is the compulsorily applied displacement in the x-direction; θ_{z1} and θ_{y1} are the compulsorily applied rotations about the major and minor principal axes at the end of x=0 respectively; and M_{z0} and M_{y0} are the bending moments about the major and minor principal axes at the end of x=0 respectively.

2.3 Numerical Procedure

The flexural-torsional buckling of beam-columns can be clarified by precisely solving Eqs.(1) together with the boundary conditions (5) and (6) for an incremental load. The numerical analysis is carried out by the finite difference method. To express the fundamental equations and boundary conditions into finite difference form, the beam-column is divided into the number of subdivisions n_x in the x-direction. Eqs.(1) for the displacements u, $\bar{v}$, $\bar{w}$, and $\bar{\phi}$ are rewritten in terms of the central finite difference expression at the inner nodal points. Boundary conditions (5) and (6) are expressed in terms of the central finite difference at the end of beam-columns by using the fictitious nodal points out of the member. As a result of expressing the equations in terms of the finite difference, a set of non-linear algebraic equations for the unknown incremental displacements is obtained as follows:

$$[K]\{\Delta d\} = \{R\} \tag{7}$$

in which [K] is a matrix consisting of constant coefficients relating to the dimensions of beam-columns; $\{\Delta d\}$ is a row vector of unknown incremental displacements; and $\{R\}$ is a row vector of residual quantities calculated by substituting displacements obtained at each iteration into the equilibrium equations and boundary conditions.

To obtain the solution of Eq.(7), the modified Newton-Raphson method is adopted. Acceleration techniques such as Jacobi method and Aitken method are used to improve convergence characteristics. Incremental displacements can be obtained by solving Eq.(7) with a residual quantity by the method of triangular decomposition. It is checked whether a new iterative solution satisfies the equilibrium equations and boundary conditions.

III. NUMERICAL RESULTS AND CONSIDERATIONS

The elasto-plastic flexural-torsional buckling of beam-columns with large deflection is numerically investigated under combined compression and biaxial end-moments. The material adopted is of mild steel with a yield stress σ_y = 235MPa(2,400kgf/cm^2) and modulus of elasticity E=206GPa(2.1$\times 10^6$ kgf/cm^2).

3.1 Number of Subdivisions in Longitudinal Direction

The number of subdivisions n_x in the longitudinal direction should be carefully decided, so as to enable the load-carrying capacity to be predicted accurately. The relationship between the number of subdivisions n_x and non-dimensionalized load-carrying capacity N_u/N_p is given in Table 1 for the beam-column initially deformed in the lateral direction without residual stress, in which the modified slenderness ratio $\overline{\lambda}_y$ is selected as $\sqrt{N_p/N_{cr}}=1$ and the magnitude of initial crookedness in the lateral direction is assumed to be $w_m = \ell/1000$. The beam-column is subjected a combination of compression and biaxial end-moments for the case of the ratio of non-dimensionalized bending moment about the minor axis to non-dimensionalized compressive force $(M_y/M_{py})/(N_x/N_p)=1$, and ratio of non-dimensionalized bending moment about the major axis to non-dimensionalized compressive force $(M_z/M_{pz})/(N_x/N_p)=1$, in which N_{cr}, N_p, M_{py} and M_{pz} are the elastic Euler buckling strength, full plastic axial force, and full plastic moments about the minor and major axes respectively. The cross-section is divided into small elements of n_y =20 and n_z =20 along the middle plane of web and flanges respectively.

It can be found from this table that the finite difference solution converges the value N_u/N_p =0.242 as the number of subdivisions n_x in the longitudinal direction is more than n_x =14. There are also the relative differences between the finite difference solution and converged value N_u/N_p =0.242 in this table. Although the present finite difference procedure can estimate the acculate load-carrying capacity even if the number of subdivisions is less than n_x =10, the number of subdivisions n_x =20 is adopted so as to trace the elasto-plastic large deformation and to estimate the load-carrying capacity accurately.

Table 1. Relationship between number of subdivisions and load-carrying capacity.

Number of subdivision n_x	Load-carrying capacity N_u/N_p	Relative error (%)
6	0.237	2.0
10	0.241	0.6
14	0.242	0
16	0.242	0
20	0.242	0
24	0.242	0
30	0.242	0
Converged value	0.242	

3.2 Number of Subdivisions over Cross-Section

The load-carrying capacity of beam-columns is given in Table 2 for various values of the number of subdivisions $n_y = n_z$ over the cross-section, in which the beam-column initially deformed in the lateral direction with $w_m = \ell/1000$ and with no residual stress is analysed with the modified slenderness ratio of compressed column $\bar{\lambda}_y = 1$ in combined loading of the ratio of non-dimensionalized bending moment about the minor axis to non-dimensionalized compressive force $(M_y/M_{py})/(N_x/N_p)=1$, and ratio of non-dimensionalized bending moment about the major axis to non-dimensionalized compressive force $(M_z/M_{pz})/(N_x/N_p)=1$. The number of subdivisions in the x-direction is fixed as $n_x = 20$. It may be recognized from this table that the appropriate number of subdivisions over the cross-section is $n_y = 14$ and $n_z = 14$ along the middle plane of web and flanges respectively, to obtain the accurate non-linear solution.

Table 2. Relationship between number of subdivisions over the cross-section and load-carrying capacity.

Number of subdivision $n_x = n_y$	Load-carrying capacity N_u/N_p	Relative error (%)
4	0.257	5.9
6	0.249	2.6
8	0.245	1.2
10	0.244	0.5
14	0.242	0
18	0.242	0
20	0.242	0
Converged value	0.242	

3.3 Comparison with Other Results

To compare the present numerical solution with appropriate other result, the column is analysed under compression. The relationship between the non-dimensionalized load-carrying capacity N_u/N_p for the lateral buckling of compressed column and modified slenderness ratio $\bar{\lambda}_y$ is shown by the solid lines in Fig.2. The initially deformed column with $w_m = \ell/1000$ is analysed for the various magnitudes of residual stress. The b-curve of ECCS column strength is also drawn in this figure. It may be seen that the ECCS b-curve is very good agreement with the present curve of the column with $w_m = \ell/1000$ and $\sigma_m = 0.3\sigma_y$.

Figure 3 shows the relationship between the non-dimensionalized load-carrying capacity M_z/M_{pz} and buckling parameter $\lambda_b = \sqrt{M_{pz}/M_{cr}}$ for the flexural-torsional buckling of beams under pure bending about the major axis, in which the solid and broken lines indicate the present solution and finite element solution by Nishimura and Komatsu respectively, and the dotted lines describe the ECCS beam strength curve for the system parameter of n=1.5 and 2.5. The initial imperfections are assumed as the initial crookedness $w_m = \ell/2000$ in the lateral direction and residual stress $\sigma_m = 0.3\sigma_y$. It can be seen that the finite difference solution accurately coincides with the finite element solution by Nishimura and Komatsu.

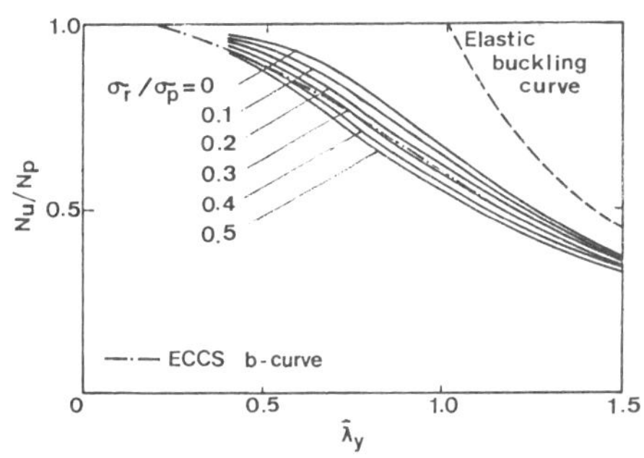

Fig.2. Ultimate strength curves for compressed columns with initial crookedness.

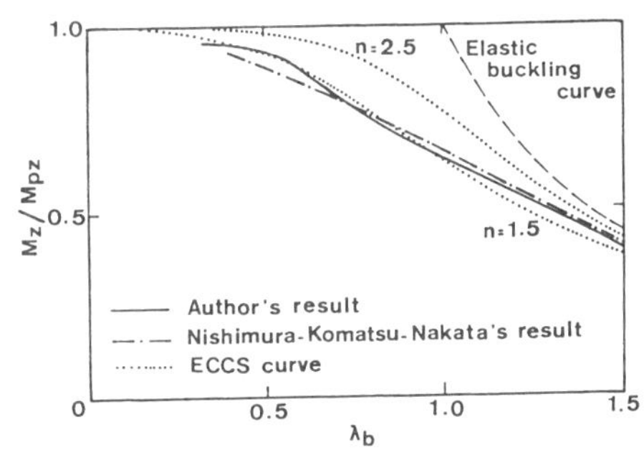

Fig.3. Ultimate strength curves for beams with initial crookedness under equal end-moments.

3.4 Columns under Compression

To examine the effect of initial twisting on the load-carrying capacity of compressed columns, the non-dimensionalized load-carrying capacity N_u/N_p is plotted against the modified slenderness of column in Fig.4, in which the broken lines indicate the results for the column with initial twisting. On the other hand, the solid lines mean the results for the columns with initial crookedness in the lateral direction being the same value as that for the initial twisting. The column with no residual stress is analysed. It is found from this figure that the load-carrying capacity of columns with initial twisting almost coincides with the elastic buckling strength as the modified slenderness is larger than $\bar{\lambda}_y=1.3$. The initial twisting does not have almost any influence on the load-carrying capacity of columns in the case of $\bar{\lambda}_y \geq 1.3$. On the other hand, it is found that initial crookedness in the lateral direction significantly affects the load-carrying capacity of compressed columns in comparison with the columns with the initial twisting about the centroid when the slenderness parameter is less than $\bar{\lambda}_y=1$, i.e., the stocky columns.

Figure 5 shows the spread of yielding zone over the cross-section and along the longitudinal direction against the compressive force for the columns with the modified slenderness ratio of $\bar{\lambda}_y=0.4$, 1 and 1.5, in which the compressed columns are analysed with initial crookedness $w_m=\ell/1000$ and without residual stress. It is obvious that for the stocky column the plastic zone widely spreads both in the x-direction and over the cross-section.

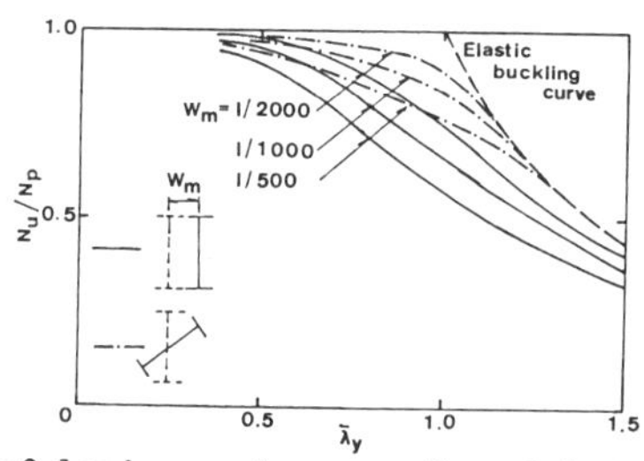

Fig.4. Comparison of load-carrying capacity of initially twisted columns with that of initially deformed columns in the lateral direction.

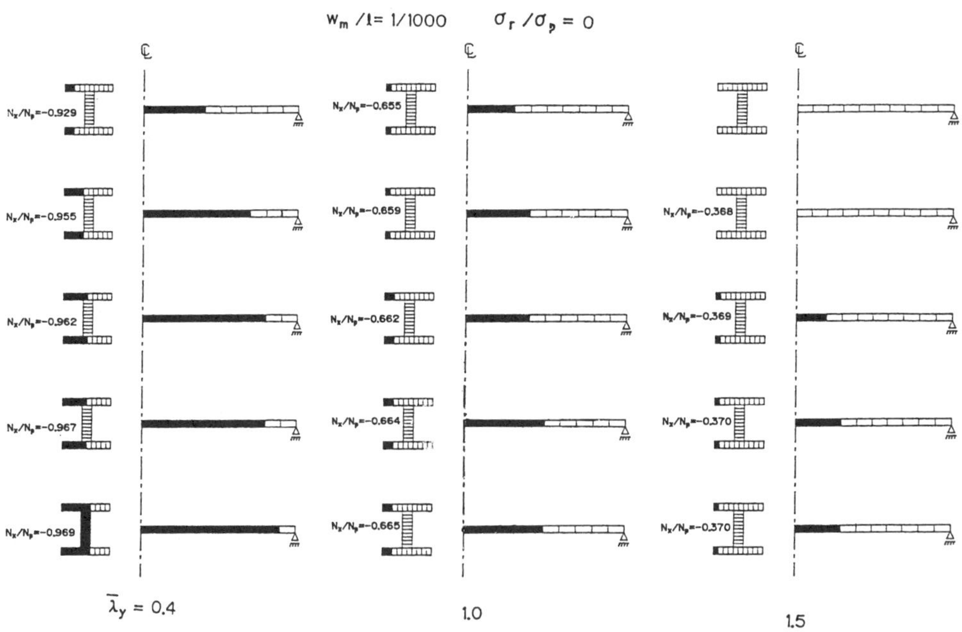

Fig.5. Spread of yielding zone over the cross-section and along the x-direction for compressed columns.

3.5 Beams under Pure Bending

The load-carrying capacity of beams under pure bending is plotted against the buckling parameter λ_b in Fig.6, in which the broken and solid lines indicate the ultimate strength curves for the beams with initial twisting about the centroid and for the beams with initial crookedness in the lateral direction respectively. It is found that the ultimate strength curves for the beams with initial twisting are almost similar to those for the beams with initial crookedness in the lateral direction. The load-carrying capacities of beams with initial twisting are smaller than those for the beams with initial crookedness.

Figure 7 shows the spread of yielding zone over the cross-section and along the longitudinal direction against the bending moment for the beams with the buckling parameter of λ_b =0.6, 1 and 1.5, in which the beams under bending are analysed with initial crookedness $w_m = \ell/1000$ and with no residual stress.

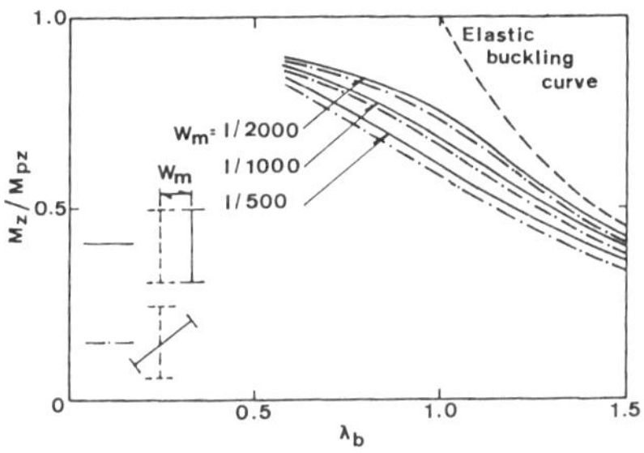

Fig.6. Comparison of ultimate bending strength of initially twisted beams with that of initially deformed beams in the lateral direction.

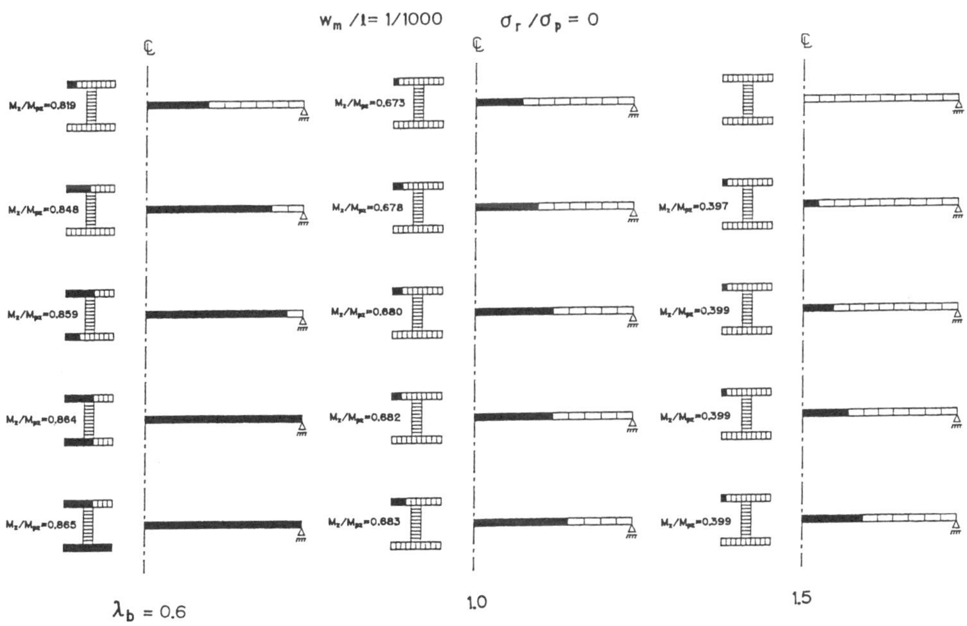

Fig.7. Spread of yielding zone over the cross-section and along the x-direction for beams in pure bending.

3.6 Interaction Curves for Beam-Columns under Compression and End-Moments about Major Principal Axis

The interaction curve of load-carrying capacity for the beam-columns with the modified slenderness ratio $\bar{\lambda}_y$ =1 under various combination of compression and equal end-moments about the major principal axis is shown in Fig.8, in which the symbols ○ and ● indicate the finite difference and finite element solutions for the beam-columns with initial imperfections of $w_m = \ell/2000$ and $\sigma_m = 0.2\sigma_y$ presented by the authors and Nishhimura respectively. The symbol × means the author's result for the beam-column with initial imperfections of $w_m = \ell/1000$ and $\sigma_m = 0.3\sigma_y$. Also, the solid line is the interaction curve proposed by Nishimura, and the broken and dotted lines are ones presented in the Japanese Specification for Highway Bridges and EUROCODE3 respectively. The present finite difference results illustrated by the symbol ○ agree well with the Nishimura's finite element results. It seems that the interaction curve for the Japanese Specification of Highway Bridges is conservative in comparison with the author's and Nishimura's results.

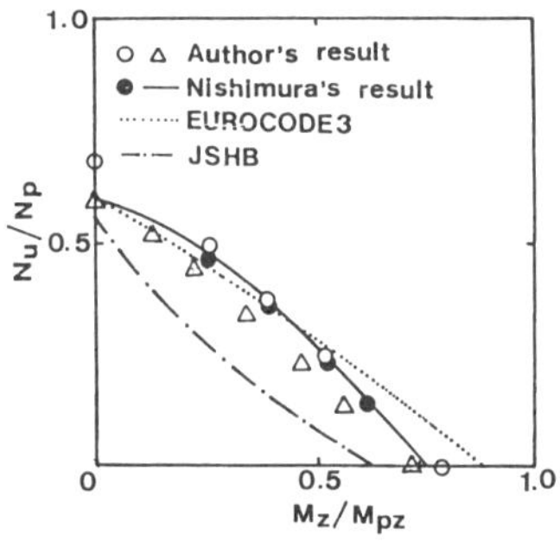

Fig.8. Interaction curves of load-carrying capacity for beam-columns under combined compression and end-moments about major principal axis.

3.7 Interaction Curves for Beam-Columns under Various Combination of Compression and End-Moments

Figure 9 shows the interaction curves for the beam-columns under combination of compression and end-moments about both the minor and major principal axes, in which the beam-columns with the modified slenderness ratio of $\bar{\lambda}_y$ =0.5, 0.75 and 1 are analysed for the initial crookedness $w_m = \ell/1000$ and residual stress $\sigma_m = 0.3\sigma_y$ under constant combination of compression and end-moments about the major principal axis, i.e., $(M_z/M_{pz})/(N_x/N_p)=1$. In this figure, the solid and broken lines indicate the present solution and value calculated by the Japanese Specification for Highway Bridges respectively. It is obvious from this figure that the flexural-torsional buckling strengths estimated by the JSHB is lower than those predicted by the elasto-plastic large deflection analysis. It seems that the present JSHB's design rule is conservative one for the beam-colunns under combined loading.

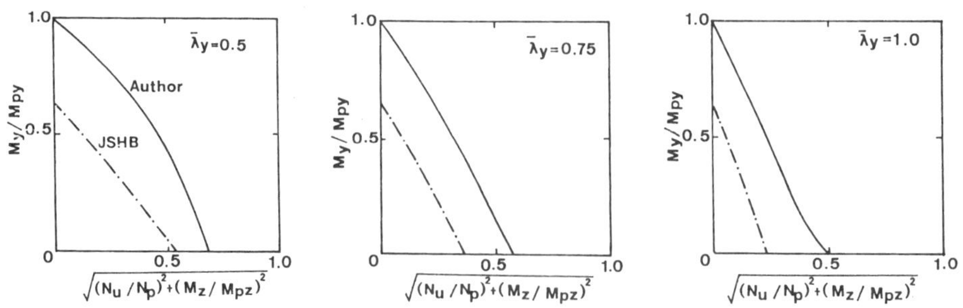

Fig.9. Interaction curves of load-carrying capacity for beam-columns under combination of compression and end-moments about both the minor and major principal axes.

IV. CONCLUSIONS

The inelastic flexural-torsional buckling of simply supported hot-rolled beam-columns with initial imperfections is analytically investigated under a combination of compression and biaxial end-moments about the minor and major principal axes, taking into account the geometric and material non-linearities. The finite difference approach is developed for numerically predicting the flexural-torsional buckling strength of beam-columns. A system of non-linear algebraic equations with incremental form is precisely solved by the modified Newton-Raphson technique with incremental loading sequence.

The effects of the slenderness parameter, initial crookednesses and twisting, residual stress, and a combination of compression and biaxial end-moments on the elasto-plastic large deflection of beam-columns are numerically clarified.

REFERENCES

1) Narayanan, R.(ed.) : Beams and Beam Columns - Stability and Strength -, Applied Science Publishers, 1983.
2) Fukumoto, Y.(ed.) : Guidelines for Stability Design of Steel Structures, Japan Society of Civil Engineers, Oct., 1987.
3) Lin, T. H. : Theory of Inelastic Structures, John Wiley & Sons, 1968.
4) Massonnet, Ch. : General theory of elastic-plastic membrane-plates, Engineering Plasticity, eds. by Heyman, J. and Leckie, F. A., Cambridge University Press, pp.443-471, 1968.

Studies on Connecting Structure between Steel Shell and Steel Reinforced Concrete

Akihiro MORIMOTO*, Hisashi MORIKAWA**, Noboru ITO**, and Yukio ABE***

* *Department of Electrical and Mechanical Engineering, Kinki University, Osaka*
** *Kanagawa Constructing Bureau, Metropolitan Expressway Public Corporation, Kanagawa*
*** *Construction Technology Department, Sumitomo Metal Industries, Ltd., Ibaragi*

The main tower of the Tsurumi Fairway Bridge, a long span cable-stayed bridge in Yokohama, consists of the steel shell structure for the upper part and the steel reinforced concrete structure for the lower part. To connect these two parts, a new type structure, using short deformed-bars inserted to holes of the steel shell walls, have been developed on the basis of experimental study.

The basic shear tests and the structural model test were carried out. It was recognized that the proposed connecting structure had a enough load carrying capacity and deformation characteristics. Besides, the estimating method of shear strength and deformation characteristics were obtained.

As a result, it has been confirmed that the proposed connecting structure is able to be put to practical use.

I. INTRODUCTION

As a connecting device in the mixed structure that consists of steel and concrete, the following methods have been proposed and utilized.

① Prestressing method to connect steel beam and concrete beam for the composite cable-stayed bridge.[1]

② Inserted deformed-bar method to connect steel pipe sheet piles and concrete footing for the foundation structure.[2]

The main tower of the Tsurumi Fairway Bridge shown in Fig.1 and Fig.2,[3] the subject of this study, consists of the steel shell structure for the upper part and

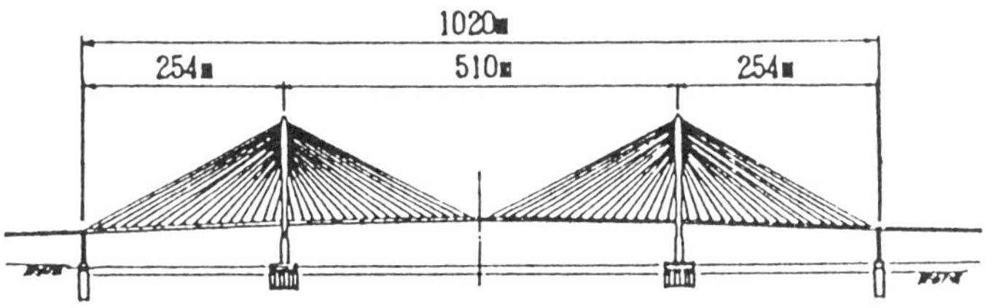

Fig. 1. Tsurumi Fairway Bridge.

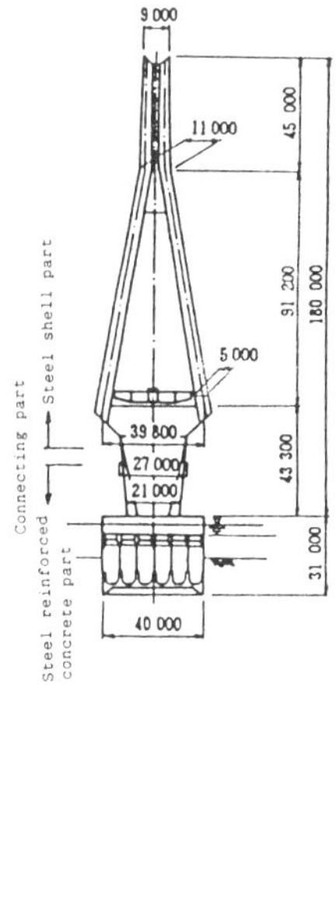

Fig. 2. Main tower.

the steel reinforced concrete structure for the lower part considering the reduction in dead weight, durability of the lower part of the tower and aesthetical effects. Therefore, it was considered that the above mentioned method ② was preferable as the connecting structure to prevent the peeling of concrete from steel shell and for the simplicity of construction.

To clarify the load carrying mechanism of this connecting structure, the basic shear tests for the inserted deformed-bars and the feasible connecting structure model test were carried out. Then the applicability of the proposed new connecting structure was discussed.

II. EXPERIMENTS

2.1 Basic Shear Tests

The tests were carried out to investigate the load carrying capacity, the deformation characteristics and the failure mode of the connecting structure shown in Fig.3. The test specimens are 1/2 scale model of a part of actual connecting structure . Four specimens with various ratios of inserted deformed-bars were tested. These test specimens are named A Specimen. The types of these specimens are shown in Table 1 and the configurations are shown in Fig.4. The ratio of inserted deformed-bars is the percentage of the shear area (two faces shear) of inserted deformed-bars to the contact area between steel plate and concrete.

Loading set-up is shown in Fig.5. The test specimens were loaded vertically in order to get the two faces shear for the deformed-bars inserted to the steel shell. The relative displacement between steel plate and concrete was measured by dialgage type displacementmeter at the point shown in Fig.4 and the strains of the inserted deformed-bar were measured by strain gages.

2.2 Structural Model Test

The test was conducted to clarify the load carrying mechanism of connecting structure, based on the results of the basic shear tests.

The test specimen is 1/2.5 scale model of actual connecting structure and has the connecting part of 2m length with 1.06% inserted deformed-bars. This specimen shown in Fig.6 is named B Specimen. The specimen was loaded vertically in the same way as the basic shear test. During the test, the relative displacements between steel plate and concrete were measured by dialgage type displacementmeters at several points shown in Fig.6. Furthermore the strains of the inserted deformed-bars and steel shell were measured by strain gages.

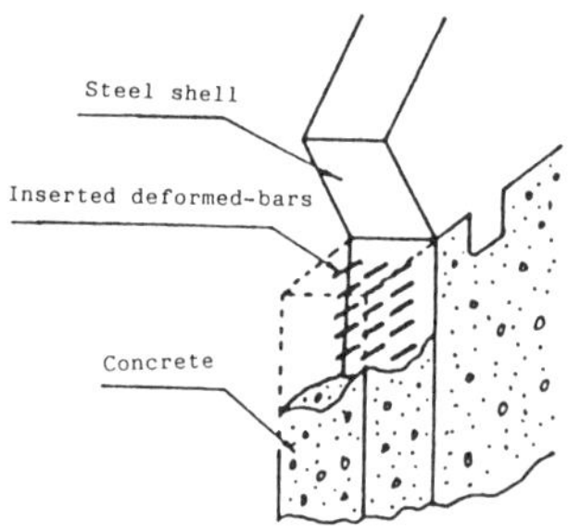

Fig.3. Use of inserted deformed-bars.

Table 1. Types of specimen (A Specimen).

No.	Connecting structure	Number of inserted deformed-bars	Percentage of inserted deformed-bars
A 1	Inserted deformed-bars (D13)	0	0
A 2		36	1.45
A 3		20	0.80
A 4		16	0.64

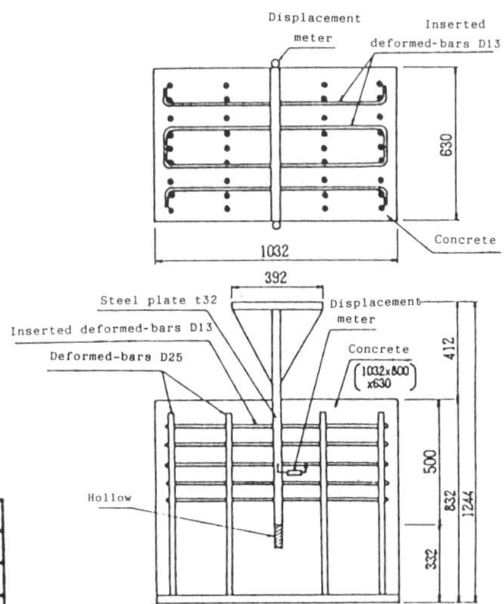

Fig.4. Specimen configuration (A Specimen).

III. PROPERTIES OF MATERIAL

The mechanical properties of several types of steel, deformed-bars and concrete are shown in Table 2.

IV. EXPERIMENTAL RESULTS AND DISCUSSION

4.1 Basic Shear Test

Relationship between the load P or shear stress τ, which is defined as P divided by the contact area between steel plate and concrete, and the relative displacement between steel plate and concrete is shown in Fig.7. Figure 8 shows the strain distribution of deformed-bars at the yield load Py (defined as the load where the load-relative displacement relation becomes nonlinear) and at 0.9Pmax. The following findings are obtained from these figures.

① As to A1 specimen without inserted deformed-bars, the bond strength τ_0 between steel plate and concrete is 3.6kgf/cm2. Beyond this bond strength τ_0, the bond stress is about 1.0kgf/cm2 regardless of relative displacement.

② Relationship between shear stress and relative displacement for A2~A4 specimens with inserted deformed-bars consists of first slope after the bond strength τ_0 and second slope after the yield strength τ_y as shown in Fig.9. Furthermore, the first slope, second slope, yield strength and maximum strength depend on the ratio of inserted deformed-bars.

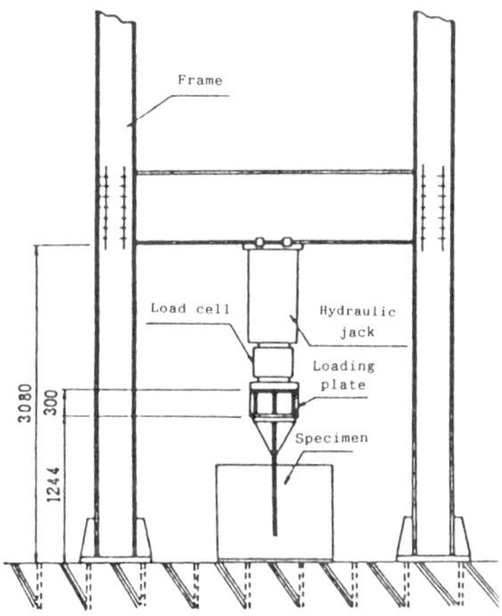

Fig.5.Loading set-up (A Specimen).

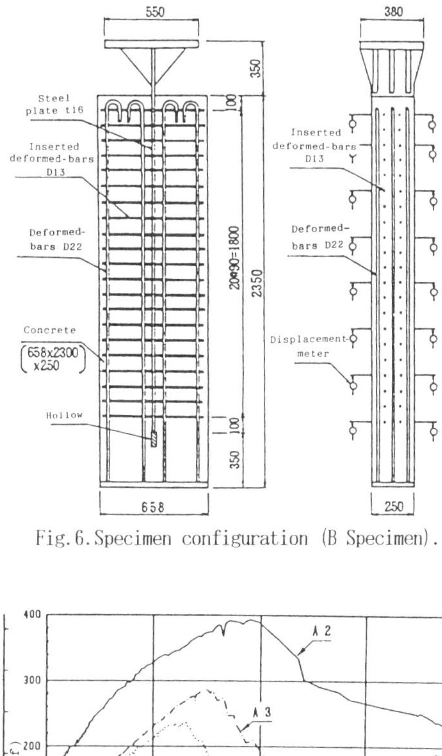

Fig.6.Specimen configuration (B Specimen).

Table 2.Mechanical properties of materials.

Materials	Dimension	Yield strength σ_y (kgf/cm²)	Tensile strength σ_B (kgf/cm²)	True tensile strength σ_T (kgf/cm²)	Elongation $e\ell$ (%)	Specimen
Deformed-bar	D13	3770	5190	9910	16.4	A
		3510	5060	8210	22.7	B
Steel plate	t16	3540	5060	9640	27.3	B
	t32	3330	5060	10100	32.0	A
Material		Compressive strength Fc (kgf/cm²)	Tensile strength Ft (kgf/cm²)	Slump (cm)	Maximum size of coarse aggregate (mm)	Specimen
Concrete		326.3	19.2	11.0	15	A
		318.0	22.1	11.5	15	B

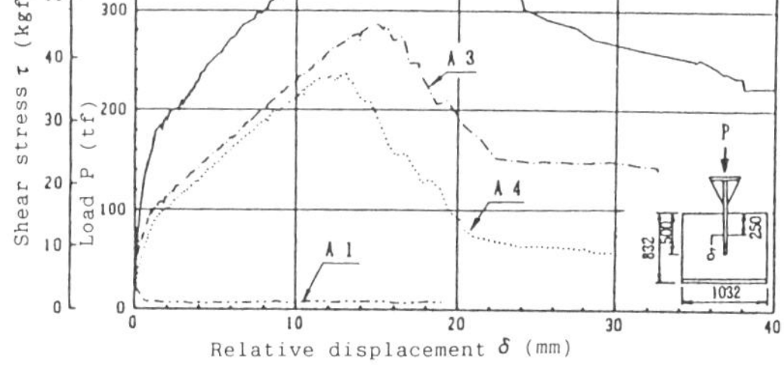

Fig.7.Load displacement relationship (A Specimen).

③ The axial strain of inserted deformed-bars at the shear plane is 400μ at Py and 1500μ at 0.9Pmax. On the other hand, the bending strain is 2000μ at Py. It is therefore presumed that extreme fiber stress of inserted deformed-bars reaches yield stress at Py and inserted deformed-bars are in full plastic range at Pmax.

4.2 Structural Model Test

Figure 10 illustrates the relationship between load P and relative displacement δ at the shear plane. Figure 11 shows the shear stress distributions (the stress transferred to the concrete through the inserted deformed-bars from the steel plate ; dσ/dy) at 120tf, which corresponds the calculated load for the model to reach allowable compressive load for design seismic effect. The following findings are obtained from these figures.

① The relative displacement δ_1 at the upper part of test specimen reaches yielding point at 140tf. The safety factor for the yield load is 1.2 (140/120) against the allowable compressive load for design seismic effect.

② The relative displacement is mostly less than 1mm at 120 tf. Therefore this displacement is considered harmless for practical use.

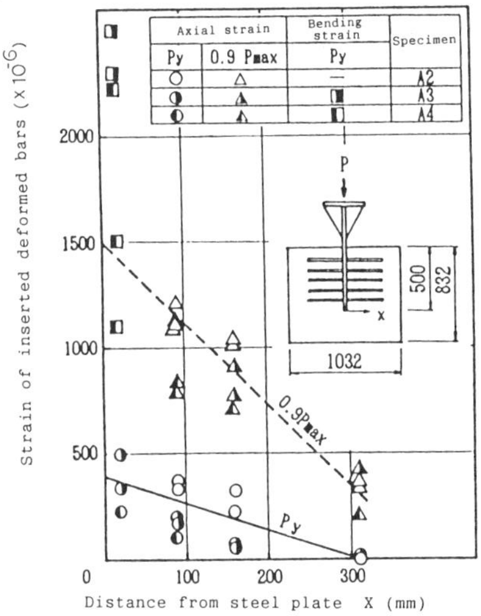

Fig. 8. Strain distributions of inserted deformed-bars.

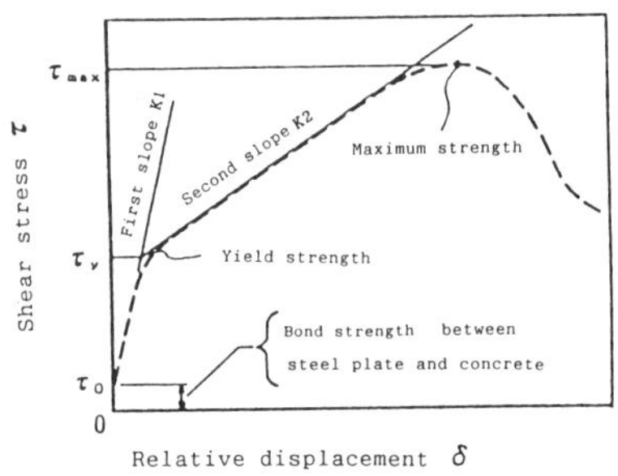

Fig. 9. Schematic relationship between shear stress and relative displacement.

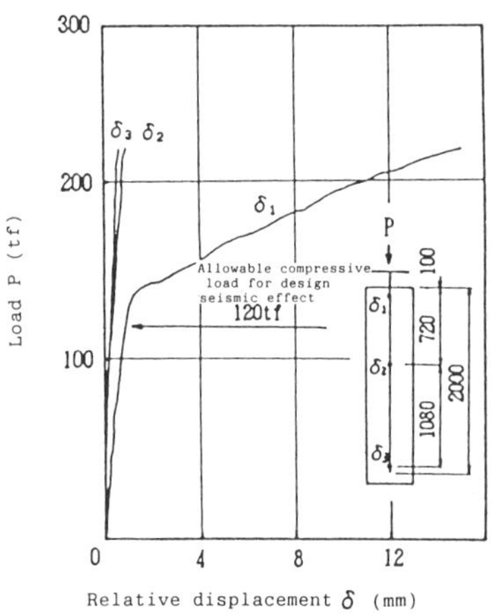

Fig. 10. Load displacement relationship (B Specimen).

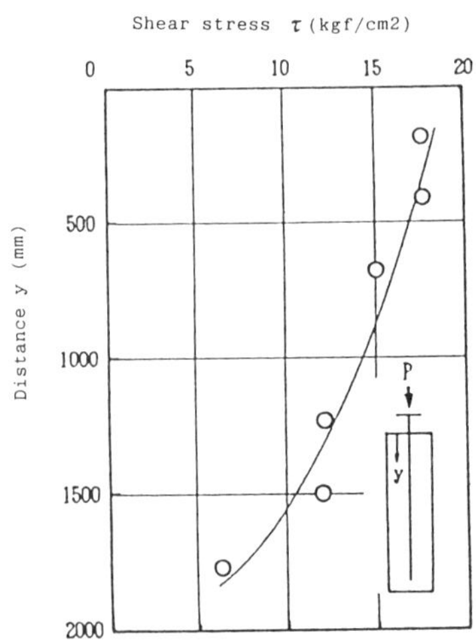

Fig. 11. Shear stress distribution (B Specimen).

③ As the shear stress for the lower part of specimen does not become lower than the bond strength τ_0 between steel plate and concrete at 120tf for 1.06% inserted deformed-bar ratio, the required connecting length is considered to be more than 2m for this specimen (Actual connecting length would be more than 5m).

Considering ① and ③, it is required for the actual connecting structure to provide a bearing plate to disperse the load in addition to the inserted deformed-bars.

V. ESTIMATION FOR SHEAR STRENGTH AND DEFORMATION CHARACTERISTICS

5.1 Estimation for Shear Strength

Based on the shear basic test results, the yield shear stress and the maximum shear stress are able to be estimated from the inserted deformed-bar ratio assuming as follows.

ⓐ The shear strength consists of the bond strength between steel plate and concrete and the shear strength of inserted deformed-bars.

ⓑ The yielding and breaking of inserted deformed-bars are obeyed to Von Mises criteria.

ⓒ The tensile strength of inserted deformed-bar is adopted as the true tensile strength in Table 2 because of a little reduction of area at the time of shear breaking.

Based on the above assumptions, the shear strength estimation is as follows.

$$\tau_y = \tau_0 + p\sqrt{(\sigma_y^2 - \sigma_1^2)/3}\ /100 = 3.6 + 18.85p \qquad (1)$$

$$\tau_{max} = \tau_0 + p\sqrt{(\sigma_T^2 - \sigma_2^2)/3}\ /100 = 3.6 + 52.91p \qquad (2)$$

τ_y ; Shear strength at the yielding
τ_{max} ; Shear strength at the breaking
τ_0 ; Bond strength between steel plate and concrete (3.6kgf/cm2)
p ; Percentage of inserted deformed-bars
σ_y ; Tensile yield strength of inserted deformed-bars
σ_T ; True tensile strength of inserted deformed-bars
σ_1 ; Axial stress at the yielding for inserted deformed-bars($\int \sigma_1^2 dA/A = \sigma_y^2/4$)
σ_2 ; Axial stress at the breaking for inserted deformed-bars($\int \sigma_2^2 dA/A = \sigma_y^2$)

Estimation results using Eq.1 and Eq.2 are shown in Fig.12. From these results the proposed estimating equations are thought to have enough accuracy.

5.2 Estimation of Deformation Characteristics

As mentioned before (4.1 ②), the deformation characteristics consist of first slope K1 after τ_0 and second slope K2 after τ_y. Figure 13 shows the relationship

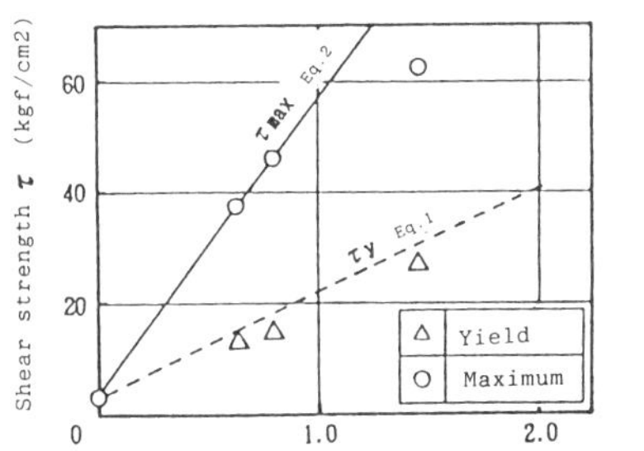

Fig. 12. Relationship between shear strength and percentage of inserted deformed-bars.

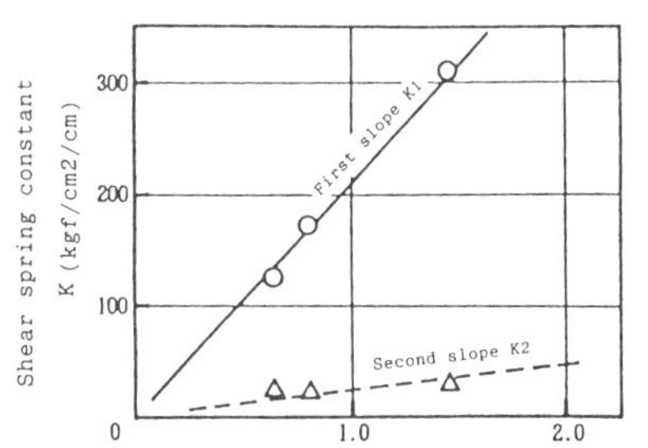

Fig. 13. Relationship between shear spring constant and percentage of inserted deformed-bars.

between these slopes and the ratio of inserted deformed-bars. It is confirmed that the estimation of shear spring constant,K1 and K2, are possible to obtain from the ratio of inserted deformed-bars.

$$K_1 = 210P \quad : \quad (\tau_0 \leqq \tau \leqq \tau_y) \qquad (3)$$

$$K_2 = 24P \quad : \quad (\tau_y \leqq \tau \leqq \tau_{max}) \qquad (4)$$

VI.CONCLUSION

The basic shear tests and the structural model test were conducted to design the connecting structure using inserted deformed-bars on the mixed structure consisted of steel shell and steel reinforced concrete.

The proposed connecting structure had a sufficient load carrying capacity and deformation characteristics. Besides, the estimation method of shear strength and deformation characteristics was proposed and confirmed. As a result, it was recognized that the bearing plate would be required for the actual connecting structure in addition to the inserted deformed-bars.

Acknowledgement

The authors wish to appreciate the helpful discussions with the members of the committee for the investigation of the Tsurumi Fairway Bridge design and construction chaired by Dr. Manabu Ito.

REFERENCES (All in Japanese)

1)Mori,K.,Hoashi,H. and Kimura,K.; Report on two experiments for the design of transition elements of the superstructure on Ikuchi Bridge, Technical report of Honshu-Shikoku Bridge Authority, Vol.13 No.49 1989,pp.48-52.

2)Abiko,T.,Kujirai,H.,Izumi,K. and Morimoto,A.; Studies on the joint structure between footing concrete and steel pipe sheet pile wall, Proceedings of JSCE, 390/V-8 1988,pp.47-56.

3)Iriyama,K.; The foundation construction for the Tsurumi Fairway Bridge of Metropolitan Expressway Bay Shore Route, Mechanization of construction, 1990, pp.11-18.

On Static and Buckling Analyses by Using Composite Degree of Two-Layered Composite Plates

Koichi SATO

Department of Civil Engineering, Hokkaido University, Sapporo, Hokkaido

This paper offers a set of practical evaluations on composite degree designed to enable static and buckling systematic analyses of partial(incomplete) composite plates consisting of two thin layers of isotropic materials (i.e., concrete slabs and steel plates). The paper also analyses analyzes the relationship between lateral deflections and critical loads among complete composite plates, partial composite plates, and individual plates which do not interact with regard to simply supported rectangular composite plates by practical evaluations on composite degree of two-layered composite plates defined by α, in which $\alpha=1$ signifies complete composite plates, $0<\alpha<1$ signifies partial composite plates, and $\alpha=0$ signifies individual plates which do not interact. Lateral deflections and critical loads can be derived by the use of practical evaluations on composite degree without much difficulty. The paper also submits the essential features of the lateral deflection and critical load characteristics of simply supported rectangular partial composite plates suitable for design purpose.

I. INTRODUCTION

In recent years bridge-slab technology has increasingly utilized hybrid structures composed of such different materials as concrete slabs and steel plates. A concrete composite steel deck plate (called here a "composite plate") consists of a concrete slab reinforced on its underside by a relatively thin flat steel plate(Fig.1). Although a composite plate offers the combined advantages of both

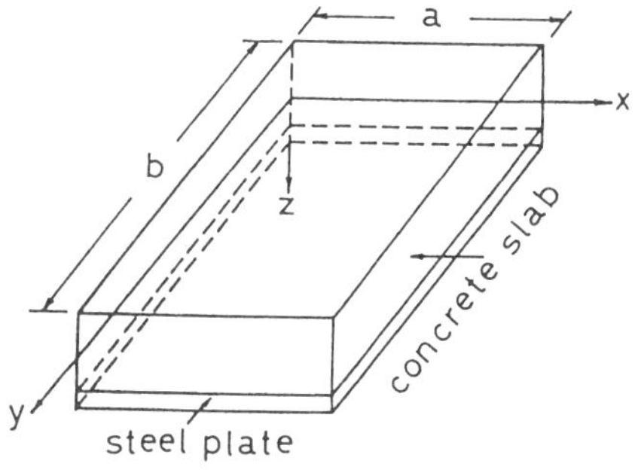

Fig.1. Concrete composite steel deck plate.

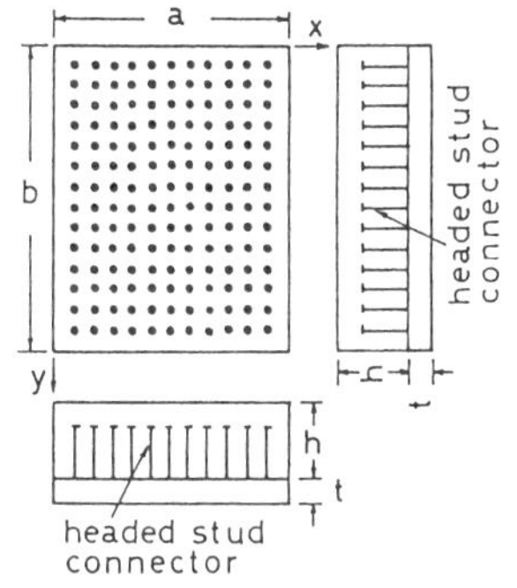

Fig.2. Composite plate.

concrete and steel, we cannot usually rely upon a natural bond between the materials. Headed stud connectors, which are welded to the steel plate and cast in the concrete, are therefore used to make the entire structure act as a composite plate(Fig.2). Headed stud connectors, however, are deformed by horizontal shear force. Complete interaction may be therefore impossible. If the parts of plates composed of two materials are not interconnected, each material acts separately, while an actual composite plate is intermediate between a complete composite plate and individual plates which do not interact. The same is true of an actual composite girder;i.e., a composite plate acts basically in a manner similar to a composite girder. A theory of partial composite plates seems to be given on p.564 of Ref.1). In that paper, however, it appears that the thickness of the steel plate is small compared with that of concrete slab, and is of a negligible order of magnitude. But, if the steel plates are thick, we cannot neglect to take their thickness into consideration.

The theory offered in this present paper can be applied to the formulation of a system of equations which govern the elastic bending and buckling of partial composite plates consisting of two layers of isotropic materials. The essential features of lateral deflection and elastic buckling load characteristics of partial composite plates for practical puroses have never been yet sufficiently understood.

The main objectives of this paper are as follows:

1) To offer a set of practical evaluations on composite degree designed to enable static and buckling systematic analyses of partial composite plates consisting of concrete slabs and steel plates.
2) To analyze the relationship between lateral deflections and critical buckling loads among complete composite plates, partial composite plates, and individual plates which do not ineract by practical evaluations on composite degree.
3) To submit a diagram that represents the essential features of the lateral deflection and elastic buckling load characteristics of simply supported rectangular partial composite plates suitable for design purpose.

II. GENERAL CONCEPTS AND ASSUMPTIONS

The plate theory of composite plates used in this paper is based on the small deflection theory, generally attributed to Kirchhoff and Love[2]. If the concrete part of the section is transformed into an equivalent steel plate (Fig.3), the sectional properties of composite plates can be estimated in Eq.(1)

$$A_c = h\ ,\quad A_s = t\ ,\quad A_v = A_s + \frac{A_c}{\bar{n}}\ ,$$

$$s = \frac{h+t}{2}\ ,\quad s_c = \frac{A_s}{A_v}\, s\ ,\quad s_s = \frac{A_c}{\bar{n}\, A_v}\, s\ ,$$

$$I_c = \frac{h^3}{12}\ ,\quad I_s = \frac{t^3}{12}\ ,\quad I_v = I_s + \frac{I_c}{\bar{n}} + A_v s_c s_s\ ,$$

$$D_v = \bar{E}_s I_v\ ,\quad \bar{n} = \frac{\bar{E}_s}{\bar{E}_c}\ ,\quad \bar{E}_c = \frac{E_c}{1-\nu_c^2}\ ,\quad \bar{E}_s = \frac{E_s}{1-\nu_s^2} \qquad (1)$$

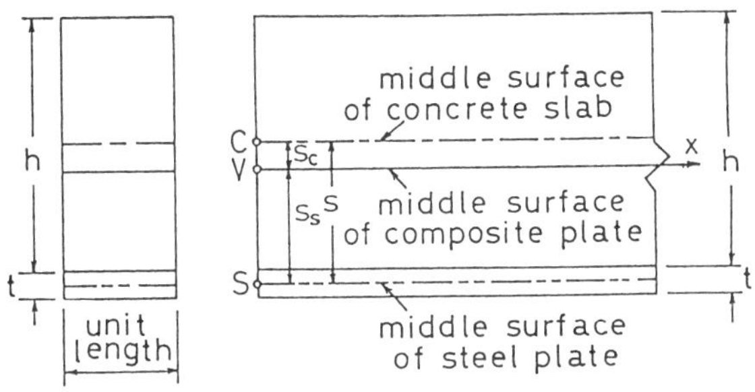

Fig.3. Cross section of composite plate.

Notations are explained in APPENDIX I. In the theory of plates it is customary to deal with internal forces and moments per unit length of the middle surface, and it should always be kept in mind that all sectional properties are defined per unit length.

III. FUNDAMENTAL DIFFERENTIAL EQUATIONS OF COMPOSITE PLATE

Unless otherwise stated, the sectional properties of composite plates defined by Eq.(1) are used throughout this paper. And it should be kept in mind that interchanging the two variables, X and Y, is valid.

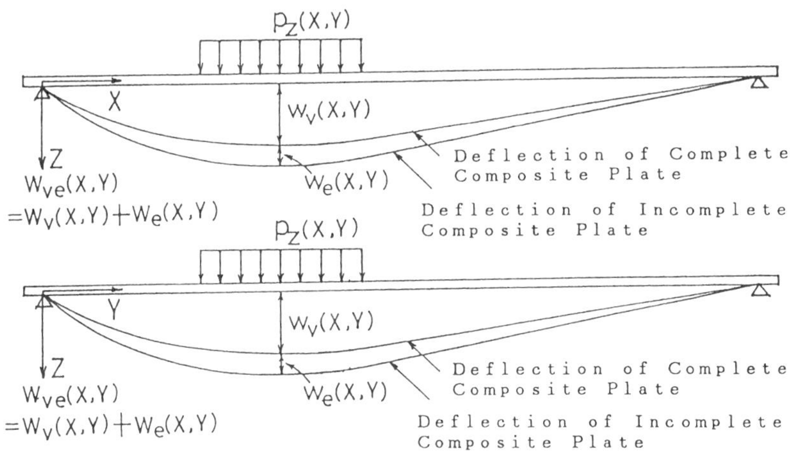

Fig.4. Deflections of composite plate.

The governing differential equation of complete composite plates expressed in terms of lateral deflection, $w_v(x,y)$ (Fig.4), of the middle surface is given by Eq.(2)[2),3)]:

$$\nabla^2\nabla^2 w_v(x,y)=\frac{p_z(x,y)}{D_v} \tag{2}$$

in which

$$\nabla^2 = \frac{\partial^2}{\partial x^2} + \frac{\partial^2}{\partial y^2} \qquad (3)$$

is the two-dimensional Laplacian operator, D_v represents flexural rigidity of the complete composite plate, and $p_z(x,y)$ means the live load intensity. If reinforcing bars are arranged, the section of concrete of a slab subjected to tensile stress is assumed to be effective in analysis. The bending moment, $M_{vx}(x, y)$ (Fig.5), which acts on the middle surface of complete composite plate in the X direction, is given by the following formula expressed in terms of lateral deflection, $w_v(x,y)$:

$$M_{vx}(x,y) = -D_v\left(\frac{\partial^2 w_v(x,y)}{\partial x^2} + \nu\frac{\partial^2 w_v(x,y)}{\partial y^2}\right) \qquad (4)$$

in which ν represents Poisson's ratio of complete composite plate. This bending moment, $M_{vx}(x,y)$, is distributed into $M_{cx}(x,y)$, $M_{sx}(x,y)$, and $N_{vx}(x,y)$ (Fig.5).

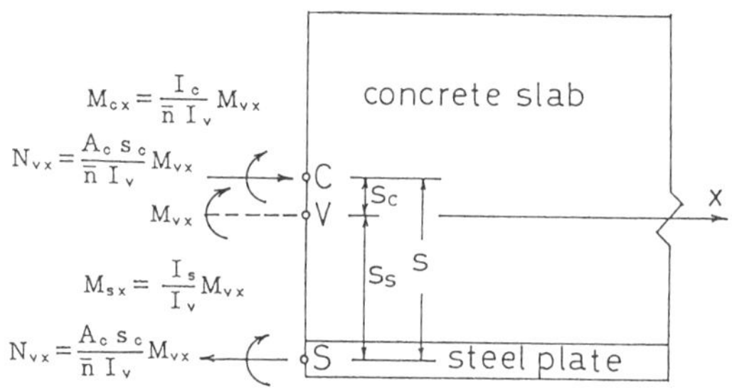

Fig.5. Distribution of bending moment Mvx(x,y)
(Complete composite plate).

$M_{cx}(x,y)$ and $M_{sx}(x,y)$ signify bending moments acting on the middle surface of a concrete slab and a steel plate, respectively; $N_{vx}(x,y)$ signifies in-plane force acting on the middle surface of a concrete slab and a steel plate. It is evident that the following relationship exists:

$$M_{vx}(x,y) = M_{cx}(x,y) + M_{sx}(x,y) + s \cdot N_{vx}(x,y) \qquad (5)$$

And we should mention that occasionally it might be an advantage to introduce what is called the moment-sum in a form as follows:

$$M(x,y) = \frac{M_{vx}(x,y) + M_{vy}(x,y)}{1+\nu} = -D_v \nabla^2 w_v(x,y) \qquad (6)$$

The introduction of this moment-sum permits us to split the governing fourth-order differential equation of a complete composite plate into two second-order differential equations. Thus, we obtain

$$\nabla^2 M(x,y) = \frac{\partial^2 M(x,y)}{\partial x^2} + \frac{\partial^2 M(x,y)}{\partial y^2} = -p_z(x,y) \tag{7}$$

and

$$\nabla^2 w_v(x,y) = \frac{\partial^2 w_v(x,y)}{\partial x^2} + \frac{\partial^2 w_v(x,y)}{\partial y^2} = -\frac{M(x,y)}{D_v} \tag{8}$$

IV. DERIVATION OF THE FUNDAMENTAL DIFFERENTIAL EQUATIONS OF PARTIAL COMPOSITE PLATE

In the case of partial composite plates, we can introduce simplifying assumptions which are practically the same as those assumed in analyzing partial composite girders[4]: (1) A continuous imperfect connection exists between the two separate materials, i.e., the shear connection between the concrete slab and steel plate is assumed to be continuous in all directions; (2) Headed stud connectors are deformed by a horizontal shear; (3) The distribution of strains throughout the depth of the concrete slab and steel plate follows Hooke's law, for the purpose of simplicity; (4) For the purpose of simplicity, it is assumed that the concrete slab and steel plate deflect equal amounts.

Let us extend the concepts used in setting up the differential equation of a partial composite girder[5] to those of a partial composite plate. If done, the procedures involved in setting up the differential equation of a partial composite plate subjected to lateral loads will be as follows:

First Step:

Let us consider composite plates with complete interaction. The bending moment, $M_{vex}(x,y)$, which acts on the middle surface of partial composite plates in the X direction, is obtained by the following equation, expressed in terms of lateral deflections, $w_{ve}(x,y)$ (Fig.4):

$$\frac{M_{vex}(x,y)}{D_v} = -\left(\frac{\partial^2 w_{ve}(x,y)}{\partial x^2} + \nu \frac{\partial^2 w_{ve}(x,y)}{\partial y^2}\right) \tag{9}$$

This bending moment, $M_{vex}(x,y)$, is distributed into $M_{cex}(x,y)$, $M_{sex}(x,y)$, and $N_{ex}(x,y)$; i.e.,

$$N_{ex}(x,y) = \frac{A_c s_c}{\bar{n} I_v} M_{vex}(x,y) \tag{10}$$

$$M_{cex}(x,y) = \frac{I_c}{\bar{n} I_v} M_{vex}(x,y) \tag{11}$$

$$M_{sex}(x,y) = \frac{I_s}{I_v} M_{vex}(x,y) \tag{12}$$

It is evident that the following relationship exists:

$$M_{vex}(x,y) = M_{cex}(x,y) + M_{sex}(x,y) + s \cdot N_{ex}(x,y) \tag{13}$$

Second Step:

Let us direct in-plane compressive forces $N_{2x}(x,y)$ in the X direction on the middle surface of a concrete slab, and direct in-plane tensile forces $N_{2x}(x,y)$ in the X direction on the middle surface of a steel plate. If the concrete slab and steel plate are not interconnected, the curvatures of the deflected middle surfaces of the concrete slab and steel plate remain unchanged. Each material behaves separately, without any interaction, and this causes an abrupt change in the strain distribution on the plane of the contact surfaces. According to the definition of strain, $\varepsilon_x(x,y)$, we can write as follows:

$$\varepsilon_x(x,y) = \left\{\frac{1}{\bar{E}_c A_c} + \frac{1}{\bar{E}_s A_s}\right\} \cdot N_{2x}(x,y) = \frac{\bar{n} \cdot s}{\bar{E}_s A_c s_c} N_{2x}(x,y) \tag{14}$$

The bending moments due to $N_{2x}(x,y)$ can be calculated from Eq.(15).

$$M_{2x}(x,y) = s \cdot N_{2x}(x,y) \tag{15}$$

Consequently, the total bending moments, $M_{vx}(x,y)$, acting on the middle surface of a composite plate will be as follows:

$$M_{vx}(x,y) = M_{vex}(x,y) + M_{2x}(x,y) \tag{16}$$

And the total in-plane forces, $N_{vex}(x,y)$, acting on the middle surfaces of a concrete slab and steel plate will be as follows:

$$N_{vex}(x,y) = N_{ex}(x,y) + N_{2x}(x,y) \tag{17}$$

Substitution of Eqs.(10),(11),(12),(13),(15) and (17) into (16) yields

$$M_{vx}(x,y) = \frac{\bar{n} I_s + I_c}{\bar{n} I_v} M_{vex}(x,y) + (s_c + s_s) \cdot N_{vex}(x,y) \tag{18}$$

Fig.6 shows that bending moment, $M_{vx}(x,y)$, which acts on the middle surface of composite plate in the X direction, is distributed into $M_{cex}(x,y)$, $M_{sex}(x,y)$, and $N_{vex}(x,y)$.

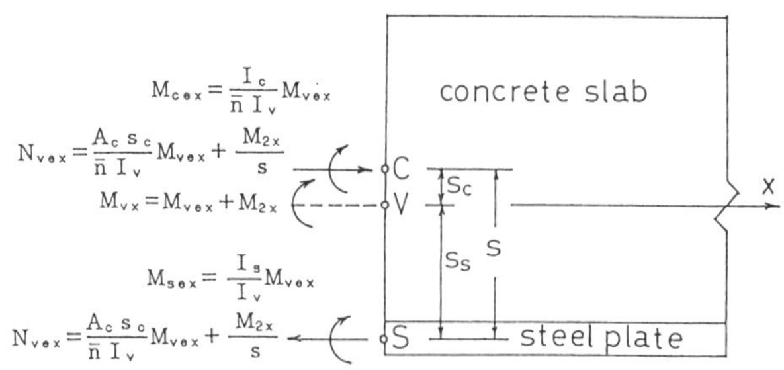

Fig.6. Distribution of bending moment Mvx(x,y) (Partial composite plate).

Third Step:

We have pointed out that headed stud connectors are deformed by a horizontal shear. If we assume that the amount of deformation (slip) permitted by the shear connection is directly proportional to the horizontal shear, we can define the amount of slip by using a modulus, i.e., the spring constant of the headed stud connector will be as follows:

$$H_{vex}(x,y) = K\,\delta_x(x,y) \tag{19}$$

in which $\delta_x(x,y)$ represents the amount of slip of the headed stud connector in the X direction; K denotes the spring constant of the headed stud connector and can be obtained by a Push-Out Test described in Ref.4); and $H_{vex}(x,y)$ signifies horizontal shears in the X direction. As the force $H_{vex}(x,y)$ which acts on headed stud connectors are equal to the increment of in-plane compressive force acting on the middle surface of concrete slab, we can derive Eq.(20).

$$\frac{\partial N_{vex}(x,y)}{\partial x} = H_{vex}(x,y) \tag{20}$$

And since the amount of the slip is produced by an addition of normal strains, we can write

$$\frac{\partial \delta_x(x,y)}{\partial x} = \varepsilon_x(x,y) \tag{21}$$

Eqs.(9)-(21) express fundamental relationships regarding any consideration of deformations of headed stud connectors.

Fourth Step:

The governing differential equation of partial composite plates expressed in terms of lateral deflection, $w_{ve}(x,y)$, is obtained by the following procedure (Fig.4). After derivative of Eq.(20) with respect to x, the use of Eqs.(19), (21) and (14) gives

$$\frac{\partial^2 N_{vex}(x,y)}{\partial x^2} = \frac{K\bar{n}}{\bar{E}_s A_c}\,\frac{s}{s_c}\,N_{2x}(x,y) \tag{22}$$

The use of Eqs.(18),(6) and (3) gives

$$(1+\nu)\cdot\nabla^2 M(x,y) = \frac{\bar{n}\,I_s + I_c}{\bar{n}\,I_v}\,\nabla^2(M_{vex}(x,y) + M_{vey}(x,y)) + s\cdot\nabla^2(N_{vex}(x,y) + N_{vey}(x,y)) \tag{23}$$

Using Eqs.(9) and (3) gives

$$\nabla^2(M_{vex}(x,y) + M_{vey}(x,y)) = -(1+\nu)\cdot D_v\cdot\nabla^2\nabla^2 w_{ve}(x,y) \tag{24}$$

The use of Eqs.(22),(3),(15),(16),(4) and (9) gives

$$s\cdot\nabla^2(N_{vex}(x,y)+N_{vey}(x,y)) = \frac{K\bar{n}}{\bar{E}_s A_c}\frac{s}{s_c}\cdot\{-(1+\nu)\cdot D_v\}\cdot(\nabla^2 w_v(x,y)-\nabla^2 w_{ve}(x,y)) \qquad (25)$$

After substituting Eqs.(24) and (25) into Eq.(23), by using Eq.(7), the governing differential equation of composite plate with partial interaction expressed in terms of lateral deflection, $w_{ve}(x,y)$, is obtained by the following form:

$$D_v\nabla^4 w_{ve}(x,y)-D_v\kappa^2\nabla^2 w_{ve}(x,y)=-D_v\kappa^2\nabla^2 w_v(x,y)+\frac{\bar{n}I_v}{\bar{n}I_s+I_c}p_z(x,y) \qquad (26)$$

in which

$$\kappa^2=\frac{\bar{n}I_v}{\bar{n}I_s+I_c}\frac{K\bar{n}}{\bar{E}_s A_c}\frac{s}{s_c} \qquad (27)$$

Application of Eq.(3) to Eq.(26), the use of Eq.(2) yields

$$D_v\nabla^6 w_{ve}(x,y)-D_v\kappa^2\nabla^4 w_{ve}(x,y)=-\kappa^2 p_z(x,y)+\frac{\bar{n}I_v}{\bar{n}I_s+I_c}\nabla^2 p_z(x,y) \qquad (28)$$

Either Eq.(26) or Eq.(28), both of which can be derived from this paper, will be the governing differential equation of partial composite plates expressed in terms of lateral deflection, $w_{ve}(x,y)$. Let us consider a condition similar to that of a partial composite girder. $w_v(x,y)$ represents the lateral deflection of a complete composite plate and must satisfy Eq.(2); i.e., Eq.(30a), and $w_{ve}(x,y)$ represents lateral deflection of a partial composite plate and must satisfy Eq.(26). If $w_e(x,y)$ is defined as the difference between $w_{ve}(x,y)$ and $w_v(x,y)$, we are able to obtain the governing differential equation expressed in terms of lateral deflection, $w_e(x,y)$, by adjudging the difference between Eq.(26) and Eq.(2). Therefore, the differential equation expressed in terms of $w_e(x,y)$ is given in the following form:

$$D_v\nabla^4 w_e(x,y)-D_v\kappa^2\nabla^2 w_e(x,y)=\frac{A_c s_c s}{\bar{n}I_s+I_c}p_z(x,y) \qquad (29)$$

Consequently, Eq.(26) consists of the two following differential equations; i.e., Eq.(30a) and Eq.(30b):

$$\begin{cases}\nabla^4 w_v(x,y)=\dfrac{p_z(x,y)}{D_v} & (30a)\\[2ex] \nabla^4 w_e(x,y)-\kappa^2\nabla^2 w_e(x,y)=\dfrac{p_z(x,y)}{D_e} & (30b)\end{cases}$$

in which

$$D_v = \bar{E}_s I_v \quad , \quad D_e = D_v \frac{\bar{n} I_s + I_c}{A_c s_c s} \tag{30c}$$

The sectional properties expressed in Eq.(30) have already been explained in Eq.(1). D_v defined in Eq.(30a) is identical with that of the two layered plates given on p.391 of Ref.2) and p.5 of Ref.6). Again, κ^2 is defined by Eq.(27), or $\kappa^2 = H/D_e$ and

$$H = K\left(\frac{\bar{n} I_v}{A_c s_c}\right)^2 \tag{31}$$

It is interesting to note that Eq.(30b) has a form similar to that of the partial differential equation with respect to the plate under lateral loads and an in-plane force($H = D_e \cdot \kappa^2$). The partial composite plate problem is considered solved if suitable expressions for the lateral deflections, $w_v(x,y)$ and $w_e(x,y)$, are found which simultaneously satisfy the differential equations,i.e., Eqs.(30a) and (30b) of equilibrium and the boundary conditions. The lateral deflection, $w_{ve}(x,y)$, of the middle surface of partial composite plates can be obtained by the following equations.

$$w_{ve}(x,y) = w_v(x,y) + w_e(x,y) \backsim \frac{1}{D_v} + \frac{1}{D_e} \tag{32}$$

V. DIFFERENTIAL EQUATIONS OF ELASTIC BUCKLING OF PARTIAL COMPOSITE PLATES

All loads considered in the preceding section have been static. In engineering practice, however, partial composite plate problems often involve consideration of elastic buckling, by assuming that there are compressive forces acting in the middle plane of the plate. Eq.(30a),in terms of deflection of complete composite plate, is the same equation as that derived by assuming that the plate is bent by only lateral loads ($p_z(x,y)$). Eq.(30b) is the the same equation as that derived by assuming that in addition to lateral loads ($p_z(x,y)$) there are tensile forces($H = D_e \cdot \kappa^2$) acting in the middle plane of the plate in two perpendicular directions. Let us investigate a flat rectangular composite plate subjected to the action in-plane forces p_{vex} and p_{vey} applied at the boundaries of the middle surface, as shown in Fig.7. The following differential equation, instead of Eq.(30a), should be used in determining the deflection surface.

$$\nabla^4 w_v(x,y) = \frac{1}{D_v}\left[p_z(x,y) - p_{vx}\frac{\partial^2 w_v(x,y)}{\partial x^2} - p_{vy}\frac{\partial^2 w_v(x,y)}{\partial y^2}\right] \tag{33a}$$

By the same reasoning as in the case of Eq.(30a), we obtain,instead of Eq.(30b),

$$\nabla^4 w_e(x,y) - \kappa^2 \nabla^2 w_e(x,y) = \frac{1}{D_e}\left[p_z(x,y) - p_{ex}\frac{\partial^2 w_e(x,y)}{\partial x^2} - p_{ey}\frac{\partial^2 w_e(x,y)}{\partial y^2}\right] \tag{33b}$$

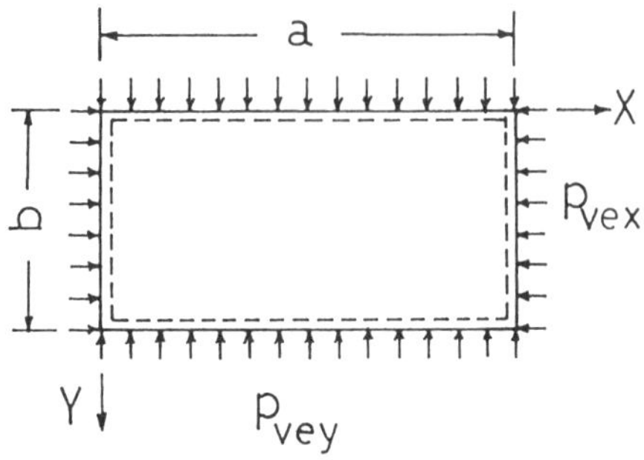

Fig.7. Simply supported rectangular composite plate uniformly compressed in two directions.

VI. ANALYSIS OF SIMPLY SUPPORTED RECTANGULAR COMPOSITE PLATES

Let us determine the lateral deflections of a partial composite plate for a simply supported rectangular composite plate subjected to lateral loads $p_z(x,y)$ and subjected to the action in-plane forces p_{vex} and p_{vey}. Since the deflection and the bending moment along the boundary are zero, the formulation of this type of boundary condition involves statements concerning displacements and forces. For simply supported rectangular plates, Navier's solution offers considerable mathematical advantages, since the solution of the governing fourth-order partial differential equation is reduced to the solution of an algebraic equation. The boundary conditions of rectangular plates, for which Navier's solution is applicable, are

$$(w_v)_{x=0,\,x=a}=0\ ,\qquad (M_{vx})_{x=0,\,x=a}=0$$
$$(w_v)_{y=0,\,y=b}=0\ ,\qquad (M_{vy})_{y=0,\,y=b}=0 \tag{34a}$$

and

$$(w_e)_{x=0,\,x=a}=0\ ,\qquad (M_{eex})_{x=0,\,x=a}=0$$
$$(w_e)_{y=0,\,y=b}=0\ ,\qquad (M_{eey})_{y=0,\,y=b}=0 \tag{34b}$$

which represent simply supported edge conditions at all edges. The solutions of the governing differential equations of the plate (Eq.(33)) subjected to a lateral loading are obtained by Navier's method as follows:

(1) The deflections are expressed by a double sine series,

$$w_v=\sum_{m=1}^{\infty}\sum_{n=1}^{\infty}{}_vW_{mn}\cdot\sin\frac{m\pi x}{a}\cdot\sin\frac{n\pi y}{b} \tag{35a}$$

$$w_e=\sum_{m=1}^{\infty}\sum_{n=1}^{\infty}{}_eW_{mn}\cdot\sin\frac{m\pi x}{a}\cdot\sin\frac{n\pi y}{b} \tag{35b}$$

which satisfy all the above-stated boundary conditions. In Eq.(35) the coefficients of expansion ${}_vW_{mn}$ and ${}_eW_{mn}$ are unknown.

(2) The lateral load $p_z(x,y)$ may also be expanded into a double sine series:

$$p_z(x,y)=\sum_{m=1}^{\infty}\sum_{n=1}^{\infty} P_{mn}\cdot\sin\frac{m\pi x}{a}\cdot\sin\frac{n\pi y}{b} \quad (36)$$

The coefficients P_{mn} of the double Fourier expansion of the load can be calculated:

$$P_{mn}=\frac{4}{ab}\int_0^a\int_0^b p_z(x,y)\cdot\sin\frac{m\pi x}{a}\cdot\sin\frac{n\pi y}{b}\,dxdy \quad (37)$$

(3) If we substitute Eq.(35) and Eq.(36) into the governing differential equations ;i.e., Eq.(33a) and Eq.(33b), we obtain algebraic equations from which the unknown ${}_vW_{mn}$ and ${}_eW_{mn}$ can be readily calculated. If $p_z(x,y)$ is equal to fully uniformly distributed load p_z in Eq.(36) and Eq.(37), ${}_vW_{mn}$ and ${}_eW_{mn}$ in Eq.(35a) and Eq.(35b) can be obtained as follows :

$${}_vW_{mn}=\frac{16p_z}{D_v\pi^6\cdot m\cdot n\cdot(A-B)} \quad (38)$$

and

$${}_eW_{mn}=\frac{16p_z}{D_e\pi^6\cdot m\cdot n\cdot(A-D+C)} \quad (39)$$

in which

$$A=\left(\frac{m^2}{a^2}+\frac{n^2}{b^2}\right)^2 \quad (40)$$

$$B=\frac{p_{vx}\,m^2}{\pi^2D_va^2}+\frac{p_{vy}\,n^2}{\pi^2D_vb^2},\qquad D=\frac{p_{ex}\,m^2}{\pi^2D_ea^2}+\frac{p_{ey}\,n^2}{\pi^2D_eb^2} \quad (41)$$

and

$$C=\frac{\kappa^2}{\pi^2}\left(\frac{m^2}{a^2}+\frac{n^2}{b^2}\right) \quad (42)$$

6.1 α_{11}, β_{11} and γ_{11} in Static Analysis

If we put $p_{vx}=p_{vy}=0$ in Eq.(38) and substitute ${}_vW_{mn}$ in Eq.(38) into w_v in Eq.(35a), we obtain an analytical solution, w_v, for the deflection in Eq.(35a). In a similar way, we can also obtain w_e for the deflection in Eq.(35b). Upon the introduction of symbols and the use of Eq.(1), the ratio, $w_e(0.5a,0.5b)/w_v(0.5a,0.5b)$, becomes

$$\gamma_{11}=\frac{w_e(0.5a,0.5b)}{w_v(0.5a,0.5b)}=\frac{{}_eW_{11}}{{}_vW_{11}}=\frac{D_v}{D_e}\beta_{11}=\frac{A_cs_cs}{\bar{n}I_s+I_c}\cdot\beta_{11} \quad (43)$$

in which

$$\beta_{11} = \frac{1}{1 + \dfrac{\kappa^2}{\pi^2 \mu_{11}^2}} = 1 - \alpha_{11} \tag{44}$$

$$\mu_{11}^2 = \frac{1}{a^2} + \frac{1}{b^2} = \frac{1}{a^2}\left(1 + \frac{a^2}{b^2}\right) \tag{45}$$

We can then obtain the lateral deflection, w_{ve}, of partial composite plates

$$w_{ve} = w_v + w_e = w_v \cdot (1 + \gamma_{11}) \tag{46}$$

Substituting γ_{11} in Eq.(43) into w_{ve} in Eq.(46) becomes

$$w_{ve} = w_v \cdot \frac{(I_{cs} + \beta_{11} \cdot I_{csv})}{I_{cs}} \tag{47}$$

D_v or D_e is the flexural rigidity of the strip of a plate of unit length. It is obvious that the lateral deflections vary inversely as D_v or D_e.

6.2 α_{11}, β_{11} and γ_{11} in Buckling Analysis

The critical value of the forces p_{vx} and p_{vy} in Eq.(38) is obtained from the condition

$$\frac{(p_{vx})_{cr}\ m^2}{\pi^2 D_v a^2} + \frac{(p_{vy})_{cr}\ n^2}{\pi^2 D_v b^2} = \left(\frac{m^2}{a^2} + \frac{n^2}{b^2}\right)^2 \tag{48}$$

where m and n are chosen so as to make p_{vx} and p_{vy} a minimum for any given value of the ratio p_{vx}/p_{vy}. In the case of a plate submitted to the action of a uniform pressure p_v in the middle plane we have $p_{vx} = p_{vy} = p_v$.

$$(p_v)_{cr} = \pi^2 D_v \left(\frac{m^2}{a^2} + \frac{n^2}{b^2}\right) \tag{49}$$

The critical value of the forces p_{ex} and p_{ey} in Eq.(39) is obtained in a similar way.

$$\frac{(p_{ex})_{cr}\ m^2}{\pi^2 D_e a^2} + \frac{(p_{ey})_{cr}\ n^2}{\pi^2 D_e b^2} = \left(\frac{m^2}{a^2} + \frac{n^2}{b^2}\right)^2 + \frac{\kappa^2}{\pi^2}\left(\frac{m^2}{a^2} + \frac{n^2}{b^2}\right) \tag{50}$$

In the case of a plate submitted to the action of a uniform pressure p_e in the middle plane we have $p_{ex} = p_{ey} = p_e$.

$$(p_e)_{cr} = \pi^2 D_e \left(\frac{m^2}{a^2} + \frac{n^2}{b^2} + \frac{\kappa^2}{\pi^2}\right) \tag{51}$$

$(p_v)_{cr}$ or $(p_{ve})_{cr}$ is the compressive load for a strip of the plate of unit length, and D_v or D_e is the flexural rigidity of the strip. It is obvious that the critical load varies directly as D_v or D_e. Upon the introduction of symbols and the use of Eq.(1), the ratio, $(p_v)_{cr}/(p_e)_{cr}$, becomes

$$\gamma_{11} = \frac{(p_v)_{cr}}{(p_e)_{cr}} = \frac{D_v}{D_e} \cdot \beta_{11} = \frac{A_c s_c s}{\bar{n} I_s + I_c} \cdot \beta_{11} = \frac{A_c s_c s}{\bar{n} I_s + I_c} \qquad (52)$$

in which

$$\beta_{11} = \frac{1}{1 + \dfrac{\kappa^2}{\pi^2 \mu_{11}^2}} = 1 - \alpha_{11} \qquad (53)$$

$$\mu_{11}^2 = \frac{1}{a^2} + \frac{1}{b^2} = \frac{1}{a^2}\left(1 + \frac{a^2}{b^2}\right) \qquad (54)$$

The critical load, $(p_{ve})_{cr}$, of partial composite plates can be obtained

$$\frac{1}{(p_{ve})_{cr}} = \frac{1}{(p_v)_{cr}} + \frac{1}{(p_e)_{cr}} = \frac{1}{(p_v)_{cr}}\left[1 + \frac{(p_v)_{cr}}{(p_e)_{cr}}\right] \qquad (55)$$

Substituting γ_{11} in Eq.(52) into $(p_{ve})_{cr}$ in Eq.(55) becomes

$$\frac{1}{(p_{ve})_{cr}} = \frac{1}{(p_v)_{cr}}(1 + \gamma_{11}) = \frac{1}{(p_v)_{cr}}\left(1 + \beta_{11} \cdot \frac{I_{csv}}{I_{cs}}\right)$$

$$= \frac{1}{(p_v)_{cr}} \frac{(I_{cs} + \beta_{11} \cdot I_{csv})}{I_{cs}} \qquad (56)$$

in which $I_v = I_{cs} + I_{csv}$; $I_{cs} = I_s + \dfrac{I_c}{\bar{n}}$; and $I_{csv} = \dfrac{A_c s_c s}{\bar{n}}$. (57)

The subscripts "v" and "ve" of p_v and p_{ve} in Eq.(55) denote complete and partial composite plates, respectively.

VII. THE ESSENTIAL FEATURES OF TWO-LAYERED COMPOSITE PLATES

It should be kept in mind that the composite plates will be subdivided into the following three major categories, based on their structural action and depending on the value of β_{11} and α_{11} defined by Eq.(44) or (53). $\beta_{11}=0$, i.e., $\alpha_{11}=1$, $\gamma_{11}=0$, signifies complete composite plates. $0<\beta_{11}<1$, i.e., $\alpha_{11}=1-\beta_{11}$, $\gamma_{11}=A_c s_c s/(\bar{n} I_s + I_c)\cdot\beta_{11}$, signifies partial composite plates. $\beta_{11}=1$, i.e., $\alpha_{11}=0$, $\gamma_{11}=A_c s_c s/(\bar{n} I_s + I_c)$, signifies individual plates which do not interact. Composite degree α_{11} is defined by the following expression using Eq.(44) or Eq.(53)

$$\alpha_{11} = \frac{1}{1 + \dfrac{\text{CONST.}}{\kappa^2 a^2}} \tag{58}$$

in which $\text{CONST.} = \pi^2 \cdot (1 + a^2/b^2)$ (59)

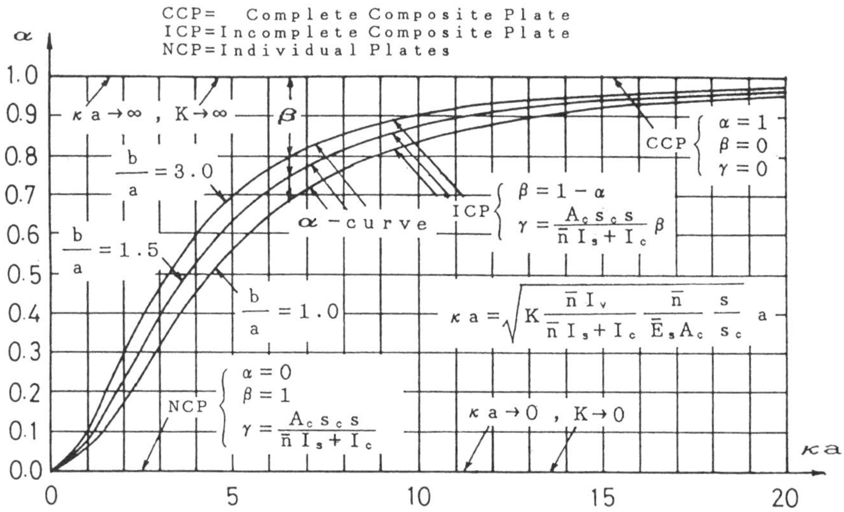

Fig.8. α − curve of composite plates.

It is found that Eqs.(43),(44) and (45) are identical with Eqs.(52),(53) and (54). Fig.8 shows the relationship between α_{11} and κa. In Fig. 8, α_{11} is given by the ordinate and κa is given by the abscissa. α_{11} is defined by Eq.(44) or (53), while κ, defined by Eq.(27), is an important constant and multiplying "a" by κ becomes a dimensionless constant. The following is an explanation of the use of diagram in Fig.8. Firstly, we can determine the value of κ defined by Eq.(27) and can multiply "a" by κ. Secondly, we can find α_{11} graphically by plotting the value of κa. Therefore, $\beta_{11} = 1 - \alpha_{11}$ can also be obtained. Thirdly, γ_{11} can be calculated by $A_c s_c s/(\bar{n} I_s + I_c) \cdot \beta_{11}$. Finally, if we substitute γ_{11} into its equivalent in Eq.(46) or Eq.(56), we can obtain the lateral deflection or the critical load of an partial composite plate by using w_{ve} in Eq.(46) or $(p_{ve})_{cr}$ in Eq.(56). The substript "11" of α_{11}, β_{11}, and γ_{11} in Fig.8 is omitted here.

Furthermore, let us illustrate the practical use of this operation by numerical calculations. The general dimensions of the partial composite plates are as follows (Figs.2 and 3): Rectangular plate of size $a \times b = 3\ \text{m} \times 3\ \text{m}$, $E_s = 2.1 \times 10^7$ ton/m^2, $I_{cs} = 1.42859 \times 10^{-6}$ m^4/m, $I_{csv} = 2.49396 \times 10^{-6}$ m^4/m, $K = 8660$ ton/m/m (assumed to be extremely small), $t = 0.006$ m, $h = 0.05$ m, $\bar{n} = 7.385$, $\nu = 0.3$(assumed), $\nu_s = 0.3$, $\nu_c = 0.2$, $s = 0.028$ m, $s_c = 0.013155$ m, $s_s = 0.014845$ m, $A_c = 0.05$ m^2/m, $A_s = 0.006$ m^2/m, $A_v = 0.0127708$ m^2/m, $I_v = 3.92255 \times 10^{-6}$ m^4/m, $D_v = 90.5205$ ton·m^2/m, $p_z = 10$ ton/m/m.

By using these general dimensions, we can find $\kappa = 5.69$/m, $\kappa a = 17.07$, $\beta_{11} = 0.0634$, $\alpha_{11} = 0.9366$, and $\gamma_{11} = 0.1107$. The lateral deflection at (0.5a,0.5b),

$w_v = 3.6330$ cm, in the case of a complete composite plate can again also be found. The lateral deflection at(0.5a,0.5b), $w_{ve} = 4.0352$ cm, in the case of a partial composite plate can therefore be obtained from mutiplying $w_v = 3.6330$ cm by $(1+\gamma_{11})$;i.e.,1.1107. The critical load, $(p_v)_{cr} = 198.53$ ton/m, in the case of a complete composite plate can again also be found. The critical load, $(p_{ve})_{cr} = 178.74$ ton/m, in the case of a partial composite plate can therefore be obtained from dividing $(p_v)_{cr} = 198.53$ ton/m by $(1+\gamma_{11})$; i.e., 1.1107. Since the lateral deflection w_{ve} and the critical load $(p_{ve})_{cr}$ of partial composite plates can be calculated by the introduction of nondimensionless parameters α_{11}, β_{11}, and γ_{11} without much difficulty, the method presented herein is convenient for practical purposes when determining the lateral deflection and the critical buckling loads.

The computations were carried out with double precision by using a personal computer (NEC PC-9801VM2, made in Japan).

Ⅷ. CONCLUSIONS

The major conclusions of this investigation can be summarized as follows:

(1) This paper offers a set of practical evaluations on composite degree designed to enable static and buckling systematic analyses of partial composite plates consisting of concrete slabs and steel plates.

(2) This paper offers the relationship between lateral deflections and critical loads among complete composite plates, partial composite plates, and individual plates which do not interact by practical evaluations on composite degree, as applied to simply supported rectangular composite plates under fully uniformly distributed loads or compressive axial forces. $\beta_{11} = 0$, i.e., $\alpha_{11} = 1$, $\gamma_{11} = 0$, signifies complete composite plates. $0 < \beta_{11} < 1$, i.e., $\alpha_{11} = 1 - \beta_{11}$, $\gamma_{11} = A_c s_c s/(\bar{n} I_s + I_c) \cdot \beta_{11}$, signifies partial composite plates. $\beta_{11} = 1$, i.e., $\alpha_{11} = 0$, $\gamma_{11} = A_c s_c s/(\bar{n} I_s + I_c)$, signifies individual plates which do not interact.

(3) This paper submits a diagram that represents the essential features of the lateral deflection and the critical buckling load of simply supported rectangular partial composite plates suitable for design purpose. Lateral deflection, w_{ve}, and critical buckling load, $(p_{ve})_{cr}$, can be obtained by the use of the diagram without much difficulty.

Acknowledgments

The author wishes to acknowledge his deep gratitude to emeritus Professor Doctor Noboru Watanabe, Hokkaido University, Sapporo, Japan, for his suggestions and encouragement during the preparation of this manuscript.

REFERENCES

1) Clarke,J.L., and Morley,C.T., Steel-Concrete Composite Plates with Flexible Shear Connectors, Proc. Instn Civ. Engrs, 53, Part2,(1972),pp.557-568.
2) Szilard,R., Theory and Analysis of Plates, Classical and Numerical Methods. (Prentice-Hall,INC.,Englewood Cliffs,1974), New Jersey.
3) Timoshenko,S.P., and Woinowsky-Krieger,S., Theory of Plates and Shells, 2nd ed. (McGraw-Hill Book Company Inc.,1959), New York.
4) Newmark,N.M.,Siess,C.P., and Viest,I.M., Tests and Analysis of Composite Beams with Incomplete Interaction, Proc. of the Society for Experimental

Analysis, Vol.9, No.1(1951),pp.75-93.
5) Shimada,S., and Takagi,R., Theory and Design of Composite Girders.(Sankaido, 1986) (in Japanese).
6) Pister,K.S., and Dong,S.B., Elastic Bending of Layered Plates, J. Engrg. Mech., ASCE, Vol.85, No.4(1959),pp.1-10.
7) Timoshenko,S.P., and Gere,J.M., Theory of Elastic Stability, 2nd ed.(McGraw-Hill Book Company Inc.,1961), New York.

APPENDIX I. NOTATIONS

The following symbols are used in this paper:

A_c, A_s = sectional areas of concrete slab and steel plate per unit length, respectively;
A_v = $A_s + A_c/\bar{n}$: sectional area of composite plate per unit length defined by transforming a concrete part of the section into an equivalent steel plate;
a, b = linear dimensions of plate in X and Y directions, respectively;
D_v = $\bar{E}_s I_v$: flexural rigidity of composite plate;
D_e = defined by Eq.(30c);
E_c, E_s = Young's moduli of concrete and steel, respectively;
$\bar{E}_c$, $\bar{E}_s$ = $E_c/(1-\nu_c^2)$ and $E_s/(1-\nu_s^2)$, respectively;
H = $D_e \cdot \kappa^2 = K \cdot (\bar{n} I_v / A_c s_c)^2$ defined by Eq.(31);
$H_{vex}(x,y)$ = horizontal shearing force in X direction per unit length acting on headed stud connector of partial composite plate;
h = thickness of concrete slab;
I_c, I_s = moments of inertia per unit length about the middle surface of concrete slab and steel plate, respectively;
I_v = $I_s + I_c/\bar{n} + A_v s_c s_s$: moment of inertia per unit length about the middle surface of composite plate by transforming a concrete part of the section into an equivalent steel plate;
K = spring constant of headed stud connector per unit length ($K=0$ signifies individual plates, and $K=\infty$ signifies a complete composite plate.);
M(x,y) = a moment-sum defined by Eq.(6);
$M_{vex}(x,y)$, $M_{vey}(x,y)$ = bending moments in X and Y directions per unit length on the middle surface of partial composite plate, respectively;
$M_{vx}(x,y)$, $M_{vy}(x,y)$ = bending moments in X and Y directions per unit length on the middle surface of complete composite plate, respectively;
$N_{vex}(x,y)$, $N_{vey}(x,y)$ = in-plane forces in X and Y directions per unit length on the middle surface of partial composite plate, respectively;
$N_{vx}(x,y)$, $N_{vy}(x,y)$ = in-plane forces in X and Y directions per unit length on the middle surface of complete composite plate, respectively;
$\bar{n}$ = $\bar{E}_s/\bar{E}_c$

$(p_v)_{cr}$ = critical load per unit length of complete composite plate defined by Eq.(49);

$(p_{ve})_{cr}$ = critical load per unit length of partial composite plate defined by Eq.(56);

$p_z(x,y)$ = lateral live load intensity;

s = distance between the middle surface of concrete slab and steel plate;

s_c = $A_s \cdot s/A_v$: distance between the middle surface of composite plate and concrete slab;

s_s = $A_c \cdot s/(\bar{n} A_v)$: distance between the middle surface of composite plate and steel plate;

t = thickness of steel plate;

$w_e(x,y)$ = $w_{ve}(x,y) - w_v(x,y)$

$w_v(x,y)$ = lateral deflection on the middle surface of complete composite plate;

$w_{ve}(x,y)$ = lateral deflection on the middle surface of partial composite plate;

α_{11}, β_{11}, γ_{11} = dimensionless constants defined by Eq.(44) and Eq.(43) or by Eq.(53) and Eq.(52);

κ = defined by Eq.(27);

$\delta_x(x,y)$ = slip in X direction of headed stud connector;

μ_{11} = defined by Eq.(45) or by Eq.(54);

ν = Poisson's ratio of composite plate;

ν_c, ν_s = Poisson's ratios of concrete and steel,respectively;

∇^2 = two-dimensional Laplacian operator defined by Eq.(3).

Practical Stress-Strain Curve of RC Column and Its Threshold Steel Ratio of Axial Reinforcement

Kiyoshi KATO*, Naoki KATO**, and Norio IWASAKA***

* *Department of Civil Engineering, The National Defense Academy, Yokosuka, Kanagawa*
** *Department of Architectural Engineering, Asano Institute of Technology, Yokohama, Kanagawa*
*** *Neturen Co., Ltd., Tokyo*

Increasing of the earthquake-proof ability of reinforced concrete structures is one of the world-wide problems. Generally, the column of structure may be subjected to both bending moment and torsional force in addition to large compressive load. In this paper, the fundamental case of the short column subject to the concentric loading is dealt with. The previous papers have shown experimentally the effective arrangement of diagonal shear reinforcements for increasing the bearing capacity and toughness, the effect of the increase of the yield strength of all reinforcements and also that the proposed method could unify both the tie bar and the spiral re-bar equations.

This paper describes the mechanism of the strain-hardening of the RC column considering the mechanical properties of the confined concrete by means of lateral confinement, the existence and the significance of the threshold steel ratio for the axial reinforcement and the importance of the lateral reinforcement. Especially, this paper deals with the practical stress-strain curve considering the complex shear strength and its application to gain the threshold steel ratio of the longitudinal reinforcement of the RC column.

I. INTRODUCTION

Increasing of the earthquake-proof ability of reinforced concrete structures is one of the world-wide problems. Generally, the column of structure may be subjected to both bending moment and torsional force in addition to large compressive load. In the prsent paper, the fundamental case of the short column subject to the concentric loading is dealt with as shown in Fig.1.

The previous papers[1),2)] have shown experimentally the arrangement of diagonal reinforcements for increasing the bearing capacity and toughness, the effect of the increase of the yield strength of spiral, tie and axial reinforcements and also that the proposed method could unify both the tie bar and the spiral re-bar equations defined by the JSCE Concrete Standard Specification.

Those papers also described that it was necessary to design so that the trinity of "strengthening of the shear strength of structural concrete-type of lateral reinforcements-strengthening of reinforcements including axial ones" might be satisfied.

This paper, furthermore, describes the mechanism of the strain-hardening of the RC column considering the mechanical properties of the confined concrete by means of lateral confinement, the existence and the significance of the threshold steel ratio for the axial reinforcement discovered by the authors, and the importance of the spacing of the lateral reinforcement.

II. SUMMARY OF PRIMARY ENGINEERING FACTORS FOR DEVELOPMENT OF BEARING CAPACITY OF SHORT COLUMN SUBJECT TO CONCENTRIC LOADING[1),2)]

2.1 Strength Properties of Concrete

(1)Relationship between Compressive Strength and Simple Shear Strength

$$\tau_{uo}/f'_c = -0.200\times10^{-3} f'_c + 0.240 \quad (\gamma=0.980) \simeq 1/6 \qquad (1)$$

where, τ_{uo}:ordinary simple shear strength
f'_c:compressive strength
γ:correlation coefficient

(2)Relationship between Complex Shear Strength Ratio and Confinement Stress Ratio

$$\eta = 1.00 + 17.47\xi - 23.274\xi^2 \quad (\gamma = 0.965) \tag{2}$$

where, τ_{uh}:complex shear strength, that is, shear strength in compression stress field

η:complex shear strength ratio ($\equiv \tau_{uh}/\tau_{uo}$)

σ'_{nd}:confinement stress, that is, normal compressive stress on shear plane

ξ:confinement stress ratio ($\equiv \sigma'_{nd}/\tau_{uo}$)

2.2 Reinforcement Effect

(1)Diagonal Shear Reinforcement

1)Arrangement Type

Figure 1 compares the usual tie bar method with the diagonal shear re-bar method. Figure 2 shows these arrangements of laterally confining reiforcements. Here, the spiral angle of the diagonal shear reinforcement is 20° from the horizontal plane, being at a right angle to diagonal shear cracking planes as shown in Fig.1.

2)Amount

The confinement stress ratio is given by Eq.(3).

$$\xi = 0.375\{1-(1-0.3050\,\mu p f_{yd}\sin^2\theta/\tau_{uo})^{1/2}\} \tag{3}$$

where, p:steel ratio on shear plane

$$\mu = 1.1(f'_{cd})^{1/2}/(pf_{yd}\sin^2\theta + \sigma'_{nd})^{2/3} \leq 0.12(f'_{cd})^{1/2} \tag{4}$$

f'_{cd}:design compressive strength of concrete

f_{yd}:design yield strength of tension reinforcement

σ'_{nd}:confinement stress

θ:intersection angle between shear plane and reinforcement

(2)Axial Reinforcement

Theoretically, the effect of the SBPD130 longitudinal reinforcement of column may be 4.3 times of that of SD30 from the viewpoint of the bearing capacity calculated by using their yield strengths.

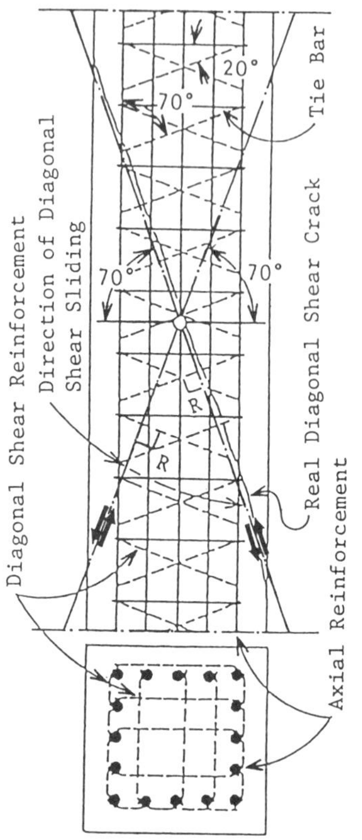

Fig.1. General arrangement of diagonal shear reinforcement and tie bar of column.

III. PRACTICAL STRESS-STRAIN CURVE OF CONFINED CONCRETE

3.1 Mander's Equation

Mander, et al.[3] obtained theoretically the stress-strain curve of the confined concrete which took account of the confining effect, that is to say, the pitch spacing of spiral or tie re-bars quantitatively, as shown by Eq.(5).

$$\sigma(\varepsilon) = \gamma E_{sec}\varepsilon/\{\gamma - 1 + (\varepsilon/\varepsilon_c)^{\gamma}\} \tag{5}$$

$$\text{where, } \gamma = E_c/(E_c - E_{sec}), (\geq 1) \tag{6}$$

E_c:initial tangent modulus of unconfined concrete

E_{sec}:ultimate secant modulus of confined concrete, that is,

$$E_{sec} = f'_{cc}/\varepsilon_{cc} \tag{7}$$

ε_{cc}:ultimate strain of confined concrete

f'_{cc}:ultimate strength of confined concrete

ε:any compressive strain under compressive stress of confined concrete

Equation(5) is derived on the assumption that the lateral confining reinforcement can be regarded as the circular steel cylinder, being too idealized.

3.2 Practical Stress-Strain Curve of Confined Concrete

(1)Study of γ

Fundamentally, the authors will convert the circular cylinder method into the diagonal shear reinforcement method of the column design.

From Eq.(6), Eq.(8) can be obtained as follows.

$$\gamma = 1/(1 - E_{sec}/E_c) = 1/(1-\lambda) \tag{8}$$

(a)Diagonal shear (b)Tie

Fig.2. Experimental setup of reinforcement.

where, $E_{sec}/E_c \equiv \lambda \leq 1$ (9)

E_c in Eq.(6) is converted into the initial tangent modulus of elasticity of unconfined concrete, that is,

$$E_c = \alpha \cdot f'_{co}/\varepsilon_{co} \quad (\text{e.g. } \alpha = 2^{5)}) \qquad (10)$$

where, ε_{co}: ultimate strain ($\fallingdotseq 2000 \times 10^{-6}$, generally) at ultimate strength of unconfined concrete

From Eqs.(1) and (2), the confined concrete strength f'_{cc} can be given by Eq.(11).

$$f'_{cc} = 6\tau_{uh} = \eta f'_{co} = f'_{co}(1.00 + 17.47\xi - 23.274\xi^2) \qquad (11)$$

Figure 3 shows the relationship between the confined concrete strength and the confinement stress ratio in putting the strength of unconfined concrete as a parameter.

On the other hand, the ultimate strain of confined concrete ε_{cc}[3)] is given by Eq.(12) from Eq.(11)

$$\varepsilon_{cc} = \varepsilon_{co}\{1 + 5(f'_{cc}/f'_{co} - 1)\} = \varepsilon_{co}\{1 + 5(\eta - 1)\} \qquad (12)$$

Figure 4 shows the relationship between the ultimate strain at the maximum confined concrete stress and the confinement stress.

From Eqs.(11) and (12), Equation(7) can be transformed into Eq.(13).

$$E_{sec} = \eta f'_{co}/[\varepsilon_{co}\{1 + 5(\eta - 1)\}] = E_c\{1 + 4/(1.00 + 87.35\xi - 116.37\xi^2)\}/5 \qquad (13)$$

$$= \lambda E_c \qquad (14)$$

$$\text{where, } \lambda \equiv \{1 + 4/(1.00 + 87.35\xi - 116.37\xi^2)\}/5 \qquad (15)$$

Figure 5 shows the relationship between the modular ratio ($\lambda \equiv E_{sec}/E_c$) and the confinement stress ratio.

It is characteristic that the modular ratio is insensitive against the confinement stress ratio above 0.1.

Figure 6 shows the enlarged scale of a part of Fig.5.

These Figs.5 and 6 emphasize that even a very small confinement stress by a little lateral reinforcement leads to larger toughness.

From the above-metioned Eqs.(8) and (15), the relationship between γ and the confinement strecs ratio can be

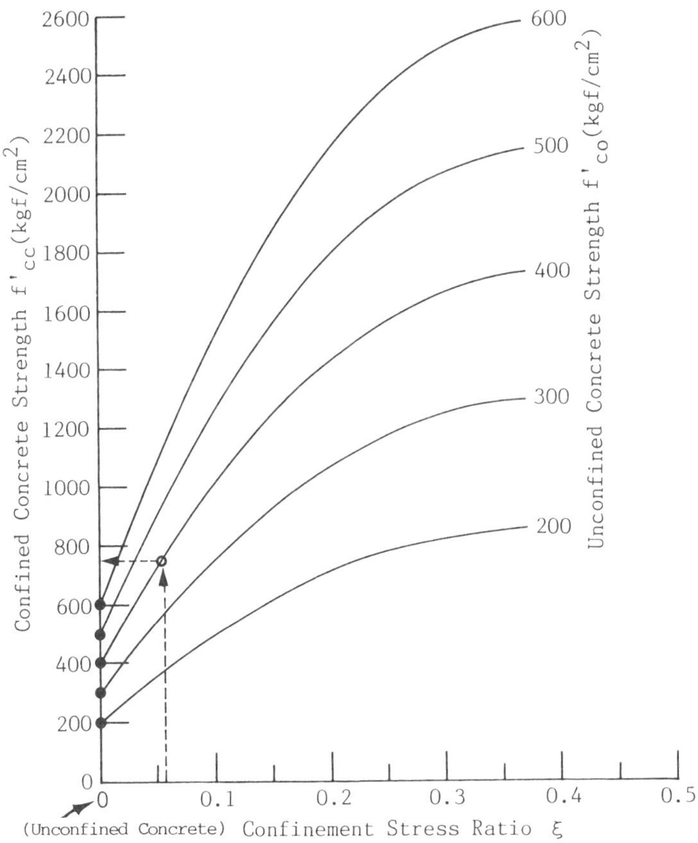

Fig.3. Relationship between confined concrete strength and confinement stress ratio.

shown as in Fig.7.

The enlarged scale of a part of Fig.7 is shown in Fig.6, together.

(2)Relationship between Confinement Stress Ratio and Laterally Confining Steel Ratio

The relationship between the diagonal shear reinforcement steel ratio and the confinement stress ratio is given by Eq.(16).

$$\xi = 0.375\{1-(1-0.305\,\mu f_{yd}\bar{p}/\tau_{uo})^{1/2}\} \qquad (16)$$

where, $\bar{p} \equiv p_t \sin^2\theta$, and p_t: shear steel ratio

Now, if the apparent frictional coefficient μ is the lower limiting value as shown in Eq.(4), μ is expressed by Eq.(17).

$$\mu = 0.12(f'_{cd})^{1/2} \qquad (17)$$

From Eqs.(16) and (17), the idealized laterally confining steel ratio $\bar{p}$ can be given by Eq.(18).

$$\bar{p} = \{145.68(1-1.333\xi)\xi\tau_{uo}\}/\{f_{yd}(f'_{cd})^{1/2}\} \qquad (18)$$

Considering $\tau_{uo} \fallingdotseq f'_{co}/6$ and $f'_{cd} = 0.77f'_{ck} \approx 0.77f'_{co}$, the relationship between the idealized laterally confining steel ratio and the confinement stress ratio can be shown in Fig.8 (here, (a) for $f'_{yd} = 3{,}000$kgf/cm^2 and (b) for $f'_{yd} = 13{,}000$kgf/cm^2).

(3)Practical Modification of Mander's Equation

The practical stress-strain curve of the confined concrete can be expressed by Eq.(19) from Eqs.(5),(8) and (15).

$$\sigma_c(\varepsilon) = \lambda/(1-\lambda)E_c\varepsilon/\{\lambda/(1-\lambda)+(\varepsilon/\varepsilon_{cc})^\gamma\} = \phi E_c\varepsilon/\{\phi+(\varepsilon/\varepsilon_{cc})^\gamma\} \qquad (19)$$

$$\text{where, } \phi \equiv \lambda/(1-\lambda) \qquad (20)$$

The relationship between ϕ given by Eq.(20) and the confinement stress ratio is shown in Figs.5 and 6.

3.3 Case Study of RC Column Model

(1)Setup of Reinforcements and Concrete

A prismatic column (150X150X530mm) was consisted of spiral type diagonal shear reinforcement (SD30; $a_s = 1\phi 6 = 0.283$cm^2, spiral angle = 20°, pitch = 40mm), as shown in Fig.2(a) and Fig.9. The thickness of cover concrete is 15mm. Then, the effective sheared cross-section is $A_{es} = (12\sec 70°)\text{X}12 = 421.0$cm^2. The unconfined concrete strength is

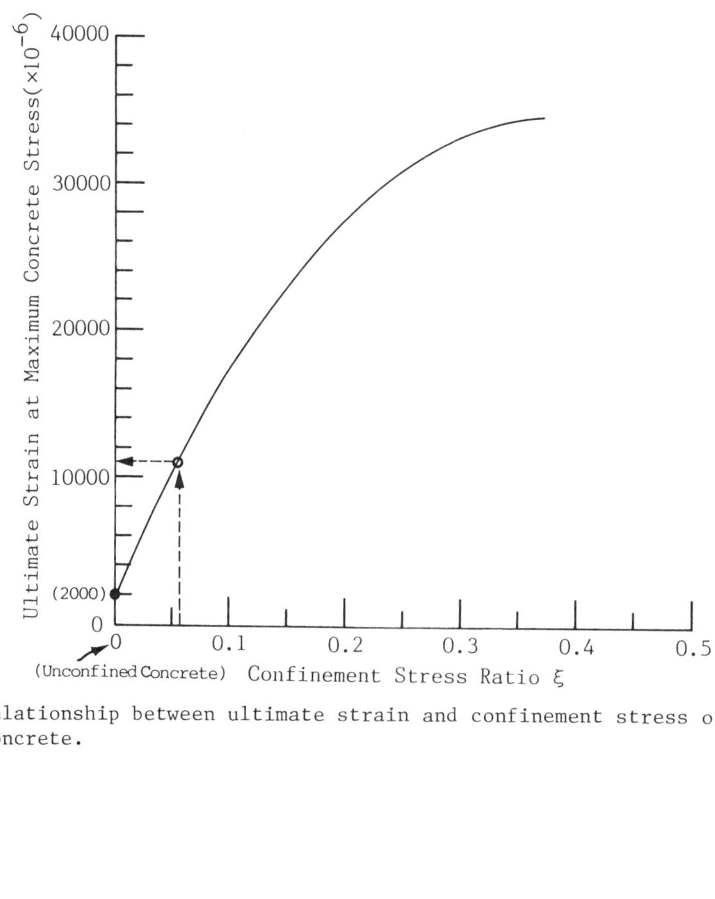

Fig.4. Relationship between ultimate strain and confinement stress of confined concrete.

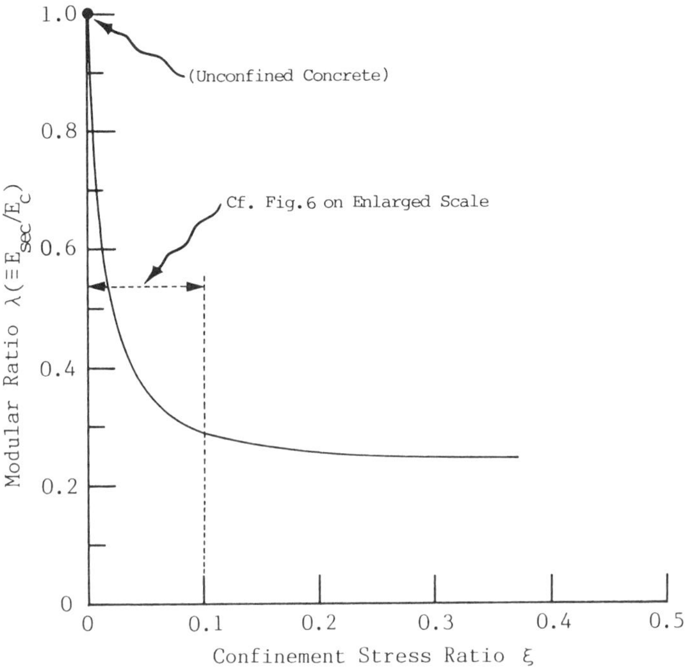

Fig.5. Relationship between modular ratio and confinement stress ratio.

$f'_{co}=400\text{kgf/cm}^2$.

(2)Calculation Process

①Idealized steel ratio of diagonal shear reinforcement

From Fig.9,

$$\bar{p}=p_1\sin^2\theta_1+p_2\sin^2\theta_2$$
$$=a_s n_1/A_{cs}\cdot(1+n_2/n_1\cdot\sin^2 130°)$$
$$=0.283\text{x}11/421\cdot(1+8/11\text{x}0.587)$$
$$=1.06\%$$

②Confinement stress ratio

$\xi=0.060$ for $\bar{p}=1.06\%$ and $f'_{co}=400\text{kgf/cm}^2$ in Fig.8

③Maximum stress of confined concrete

$f'_{cc}=740\text{kgf/cm}^2$ for $\xi=0.060$ and $f'_{co}=400\text{kgf/cm}^2$ in Fig.3

④γ and ϕ

$\gamma=1.50$ and $\phi=0.50$ for $\xi=0.060$ in Figs.(6) and (7)

⑤Ultimate strain at maximum stress

$\varepsilon_{cc}=11{,}000\text{x}10^{-6}$ for $\xi=0.060$ in Fig.4

⑥Elastic modulus of unconfined concrete

$E_c\approx200{,}000\text{kgf/cm}^2$ from Eq.(10)

⑦Stress-strain curve of confined concrete

From Eq.(19),

$$\sigma_c(\varepsilon)=0.50\text{x}200{,}000\text{x}\varepsilon/\{0.50+(\varepsilon/11{,}000\text{x}10^{-6})^{1.50}\}$$
$$=100{,}000\varepsilon/\{0.50+(\varepsilon/11{,}000\text{x}10^{-6})^{1.50}\} \quad (21)$$

⑧Stress-strain diagram

Equation (21) for the confined concrete is shown in Fig.10, comparing with the stress-strain curve of the unconfined concrete.

⑨Summary

According to the above-mentioned calculation procedures, the stress-strain equation for the confined concrete can be easily obtained and its drawing is useful to understand the characteristics of the deformation behavior.

IV. MECHANISM OF STRAIN-HARDENING CONSIDERING STRESS-STRAIN RELATION OF LATERALLY CONFINED CONCRETE

The present paragraph deals with the more realistic characteristics of the laterally confined concrete concerning the deformation behavior of concrete.

Figure 11 shows the compressive stress-strain curves of the RC columns confined by the diagonal shear reinforcement method and the tie bar method, in which the pitches of lateral confinment are 40mm and 80mm. The stress-strain curves by "e-function"[4] and "Desayi's equation"[5], which are often used now, are shown together in Fig.11.

It is well understood that the toughness in case of the pitch, 80mm, by both confined methods decreases rapidly, and also the deformation behavior of the laterally confined concrete can be well expressed by neither "e-function" or "Desayi's equation". Mander[3] obtained theoretically the stress-strain curve of the confined concrete as shown by Eq.(5).

Now, as a case study in use of the lateral reinforcements of SBPD130, Figure 12 shows the compressive stress-strain curves when SD30, SBPD130 and f'_c=300kgf/cm^2. Here, the values in the parentheses in Fig.12(a) show the co-ordinates of the kinks of the stress-strain curves of the reinforcements. And also, Fig.12(b) shows the load-deformation curves, when the effective cross-sectional area A_c=120X120mm, and the steel ratios of axial reinforcement p=3.14% and 6.28%.

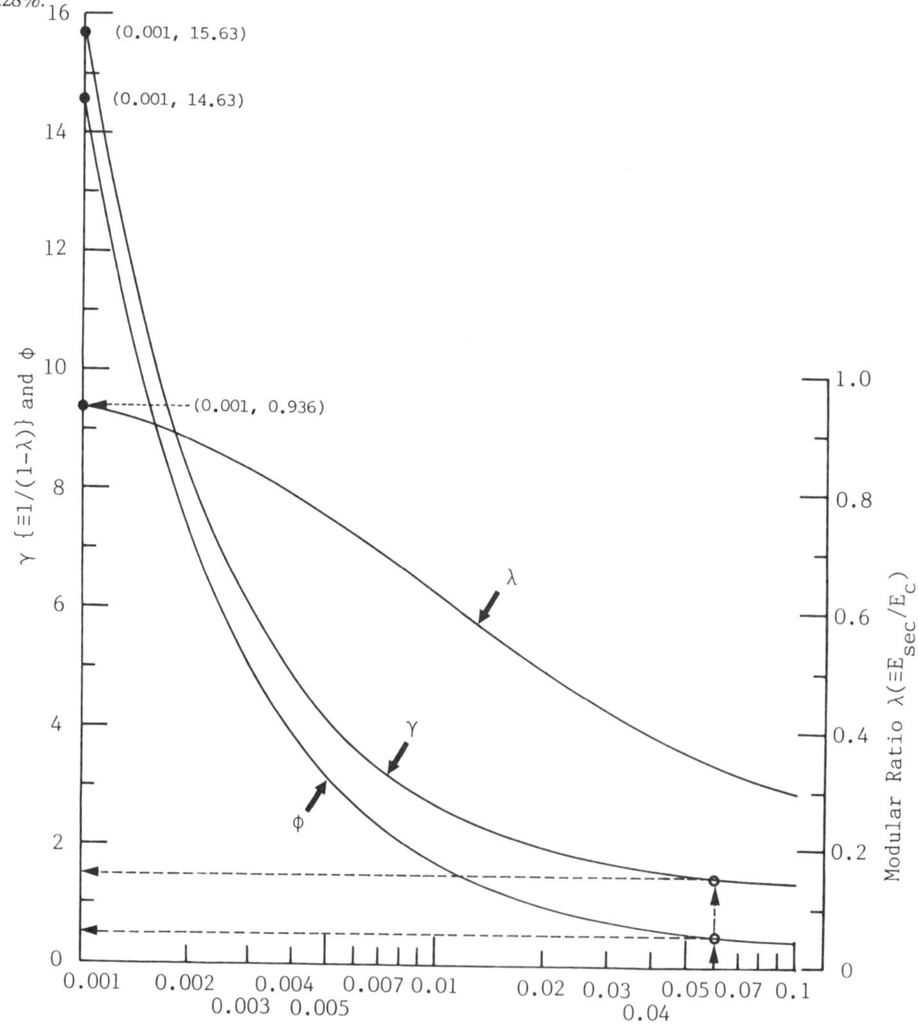

Fig.6. Relationship between λ, γ and ϕ and confinement stress ratio shown on enlarged scale.

The loading capacity can be given by Eq.(22).

$$P_c = \sigma_c(\varepsilon)\cdot A_e + \sigma_s(\varepsilon)\cdot(pA_e) \quad (22)$$

As well understood from Fig.12(b), the SD type axial reinforcement causes the strain-softening to occur at the low compressive strain, and especially the wider pitch spacing causes the earlier strain-softening.

On the other hand, it increases both bearing capacity and toughness of the column to use the SBPD type re-bar in order to strengthen the axial reinforcement, and especially the narrower pitch spacing leads to the more remarkable effects.

V. APPLICATION TO GAIN THRESHOLD STEEL RATIO OF AXIAL REINFORCEMENT

From the conditional equation for the threshold steel ratio given by Eq.(23) and the stress-strain curve of the confined concrete given by Eq.(19), the threshold steel ratio considering the confining effect is given by Eqs.(23) and (24), when putting $\sigma_s(\varepsilon)$ as the stress-strain curve of the longitudinal reinforcement.

$$P_c/\varepsilon = A_e\{\partial\sigma_c(\varepsilon)/\partial\varepsilon + p\,\partial\sigma_s(\varepsilon)/\partial\varepsilon\} \equiv 0 \quad (23)$$

$$p \equiv p_{th} = -\partial\sigma_c(\varepsilon)/\partial\varepsilon / \partial\sigma_s(\varepsilon)/\partial\varepsilon \quad (24)$$

It is very interesting that the threshold steel ratio is given by the ratio of the tangent modulus of confined concrete to that of longitudinal reinforcement of the column.

Figure 13 shows the stress-strain diagram of SBPD130/145 re-bar, given by two Eqs.(25) and (26).

$$\sigma_s(\varepsilon) = 1.351 f'_{pu}\cdot\varepsilon \quad \text{[for elastic range:} 0<\varepsilon<0.61\%] \quad (25)$$

$$\sigma_s(\varepsilon) = (0.749{,}8 + 0.131{,}3\varepsilon - 0.017{,}08\varepsilon^2) f'_{pu} \quad (\gamma = 0.82) \text{ [for plastic range: } 0.61<\varepsilon<4.0\%] \quad (26)$$

where, f'_{pu}:ultimate compressive strength of longtudinal reinforcement.

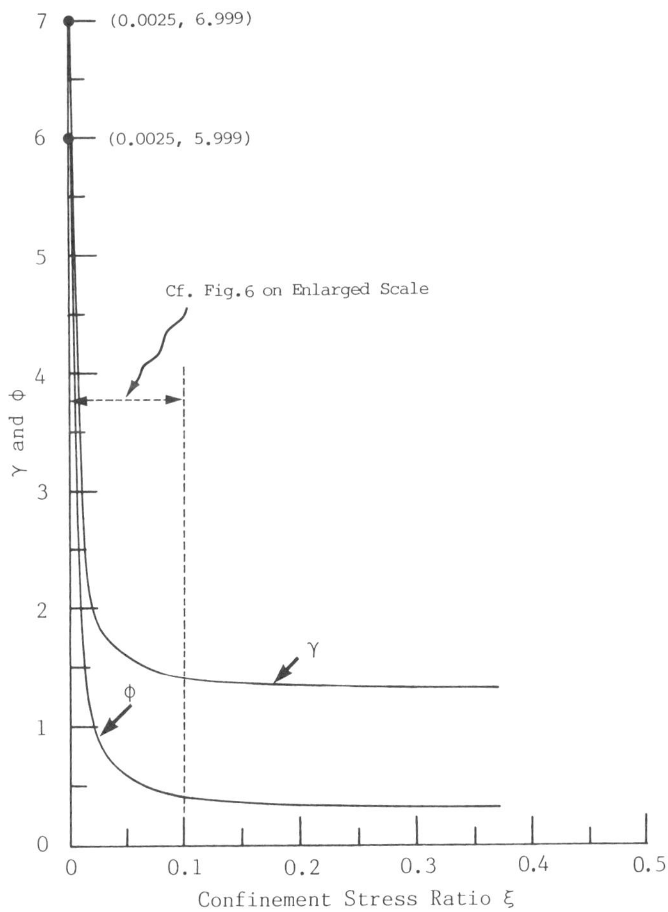

Fig.7. Relationship between γ and φ and confinement stress ratio.

And also, the relationship[8] between the initial tangent modulus of elasticity(E_c) and the characteristic compressive strength of concrete(f'_{ck}) is shown in Fig.14.

Therefore, the threshold steel ratio for the plastic range can be given as Eq.(27) by Eqs.(19),(24) and (26).

$$p_{th}=[\phi E_c/f'_{pu}\{(\gamma-1)(\varepsilon/\varepsilon_{cc})^{\gamma}-\phi\}]/[\{(\varepsilon/\varepsilon_{cc})^{\gamma}+\phi\}^2(0.131,3-0.034,16\varepsilon)] \qquad (27)$$

Figure 15 shows the relationship between the threshold steel ratio and the compressive strain of the column in putting the confinement stress ratio(ξ) and $\kappa(\equiv E_c/f'_{pu})$ as parameters.

It is understood that the threshold steel ratio is very insensitive to κ, in other words, the concrete strength (f'_c) [because E_c is the function of f'_c].

Figure 14 is made on the basis of the condition under 0.8%< the steel ratio of the longitudinal re-bar (p)<6%, when the axial force is dominant to the column.

Ex. when $\xi=0.06$, we wish to occur the strain-hardening as far as $\varepsilon=1.4\%$, p_{th}=? when $f'_{co}=200\text{kgf/cm}^2$.

$E_c=230{,}000\text{kgf/cm}^2$ from Fig.14, and $\kappa=E_c/f'_{pu}=230{,}000/14{,}500=15.862$, then $p_{th}=4.5\%$ from Fig.15.

If $f'_{co}=400\text{kgf/cm}^2$ ($\kappa=310{,}000/14{,}500=21.379$), then $\varepsilon=1.33\%$ from Fig.15. Namely, this fact means that the larger stiffness of concrete constrains the axial deformation of the column.

The threshold steel ratio of the longitudinal reinforcement should be adopted as the minimum reinforcement from the viewpoint of the toughness of the RC column.

VI. RELATIONSHIP BETWEEN EFFECTIVE CROSS-SECTIONAL AREA AND PITCH SPACING

Figure 16 shows quantitatively the effect of the pitch of laterally confining reinforcements on the failure modes along the loading direction of the column axis, as two typical cases out of 8 cases at intervals of 10mm of the hoop spacing.

It is well understood that the core concrete in case of the spacing narrower than 40mm can effectively function and that the laterally confining re-bar contributes to the larger development of the toughness of the RC column. On the other hand, the wider spacing beyond 40mm, for example, as shown in case of 80mm results in spalling off both the cover concrete and the mechanically important core concrete of the column.

"The cross-sectional area surrounded with the center line of the spiral reinforcement" that is to say, "the effec-

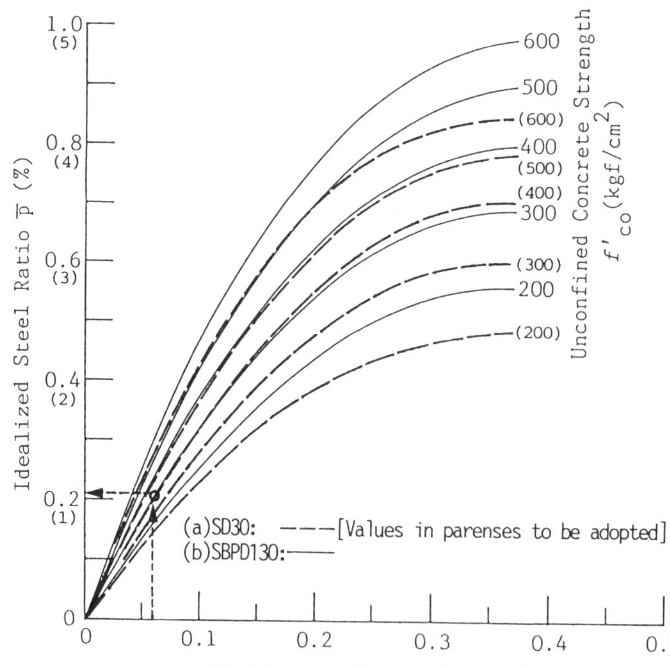

Fig.8. Relationship between idearized laterally confining steel ratio and confinement stress ratio.

tive cross-sectional area" in case of the pitch spacing, 40mm exists nearly in the safety side on computation, but the one in case of the spacing beyond 40mm does in the critical condition.

According to the present "Concrete Standard Specification"[7], the maximum pitches in cases of the tie type and the spiral type are 150mm and 40mm, respectively.

Considering the constructional execution, also, the spacing of the lateral reinforcement of the columns should be uniformly "40mm to the maxmum size of coarse aggregate(mm)X4/3" as the standard pitch.

VII. CONCLUSION

The development of the compressive loading capacity and the toughness of the short RC column subjected to the concentric loading is important to secure the safety. In order to achieve this object, "increasing the complex shear strength of concrete by means of lateral reinforcements and high strengthening from SD to SBPD of all reinforcements" must be taken into the design.

The practical stress-strain equation of the confined concrete including the confinement effect according to both the diagonal shear reinforcement and the tie bar, that is to say, the modified Mander's equation was obtained by using the complex shear strength-confinement stress relation developed already. This equation can be calculated easily and speedily by the monographs developed here.

The fundamental mechanism of strain-hardening contributing to development of the toughness of the confined RC column was analyzed for the first time. And also the threshold steel ratio concerning the realistic confined concrete column was obtained, and so it can be usefully adopted as the mimimum steel ratio of the longitudinal re-bar in case of the design of RC column.

It was made clear that there was the threshold steel ratio for the longitudinal reinforcement and furthermore the pitch spacing must be shortened enough, especially in case of tie bar column, in order to function as the shear transfer reinforcement and that to increase the restraint effect.

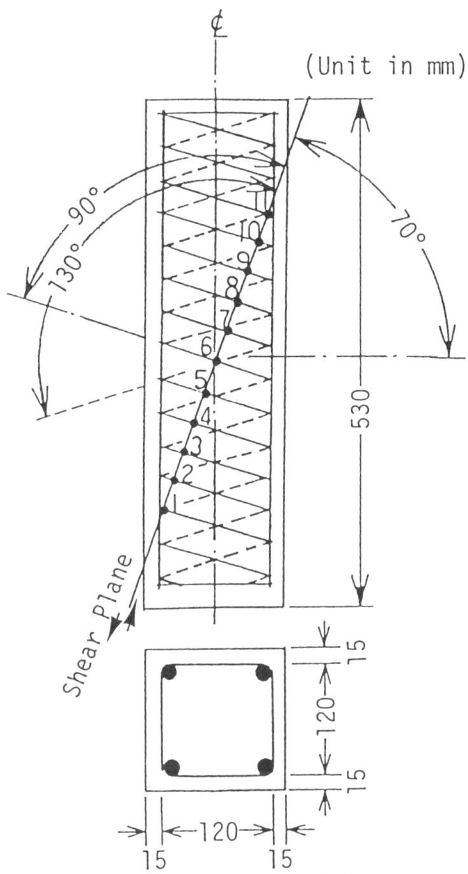

Fig.9. Main dimension of diagonal shear reinforcement square column.

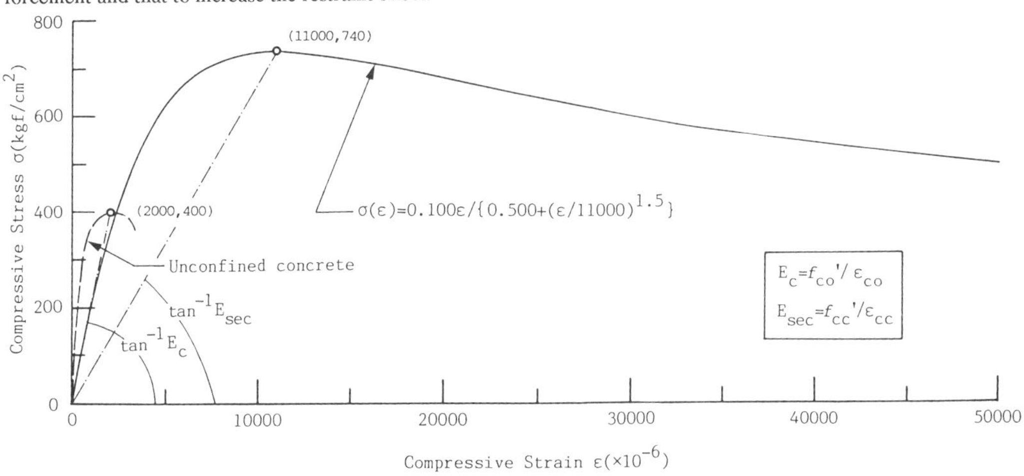

Fig.10. Comparison in stress-strain curve of confined concrete with that of unconfined concrete.

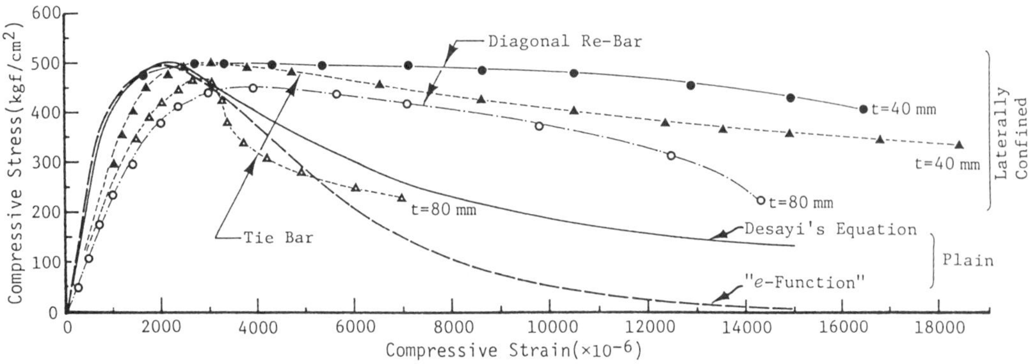

Fig.11. Comparison of stress-strain curve considering pitch effect.

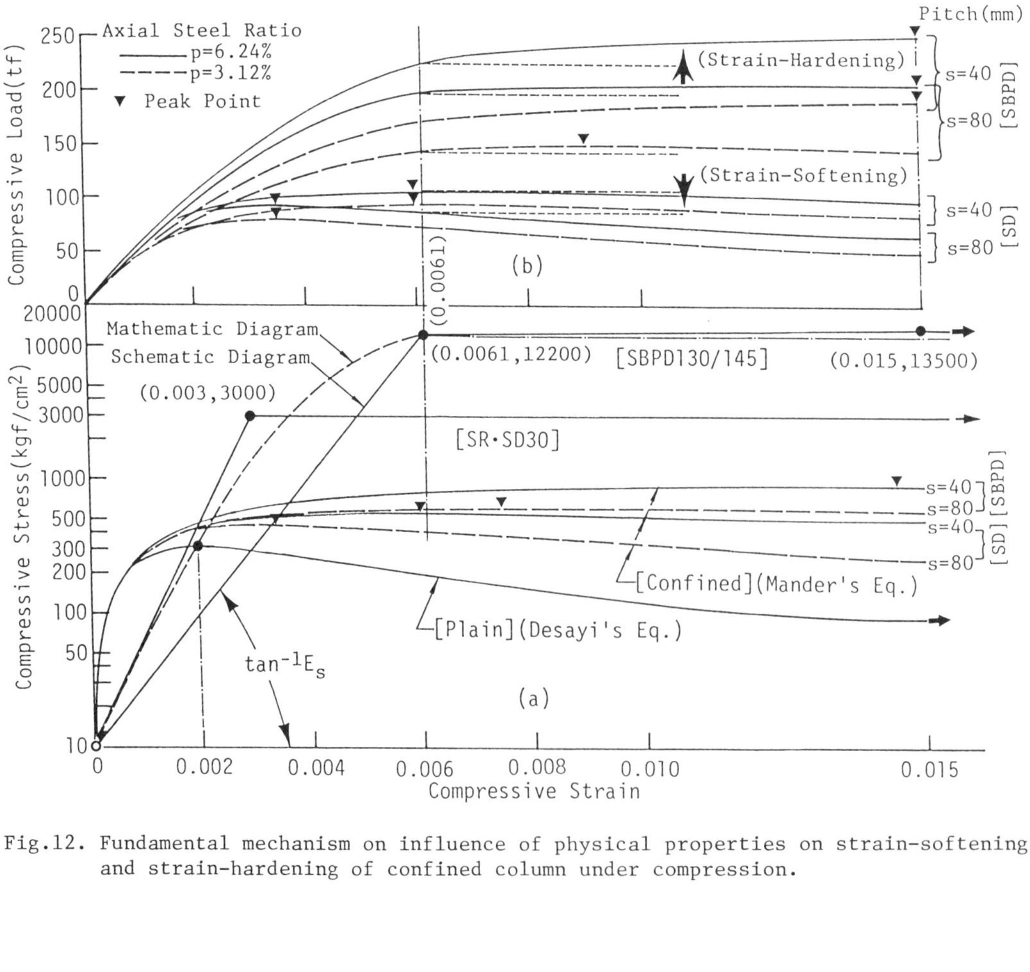

Fig.12. Fundamental mechanism on influence of physical properties on strain-softening and strain-hardening of confined column under compression.

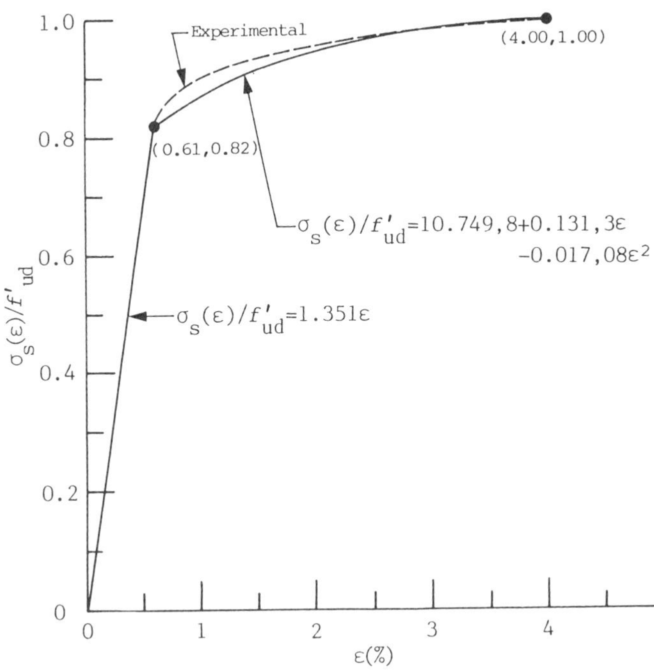

Fig.13. Compressive stress-strain diagram of SBPD re-bar.

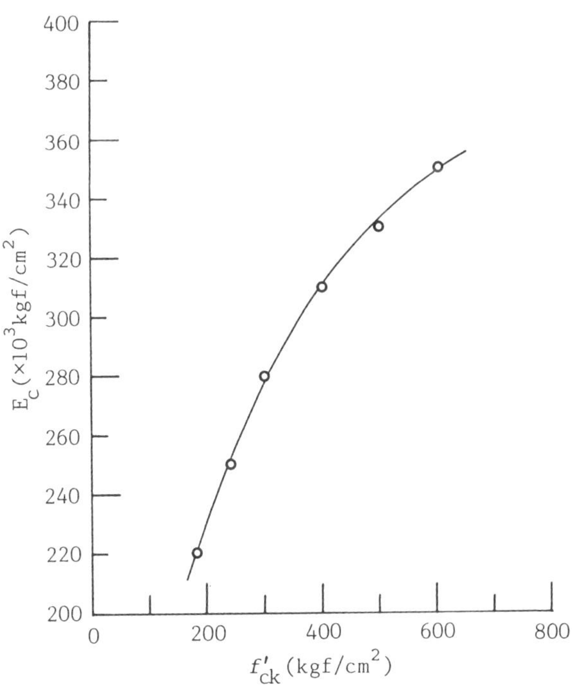

Fig.14. Relationship between characteristic compressive strength and initial tangent modulus of elasticity of concrete(JSCE Specification).

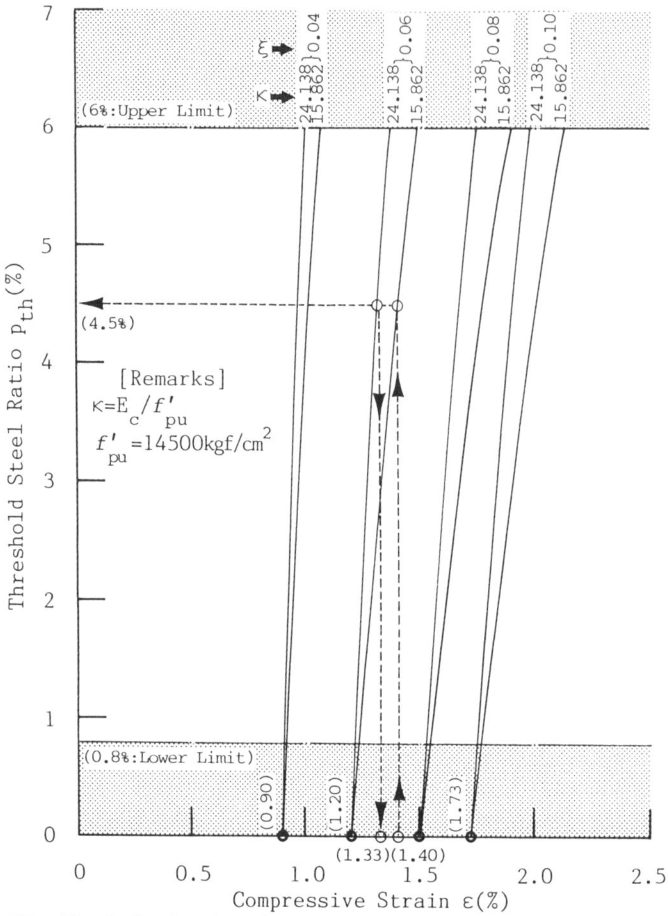

Fig.15. Relationship between threshold steel ratio and compressive strain of column.

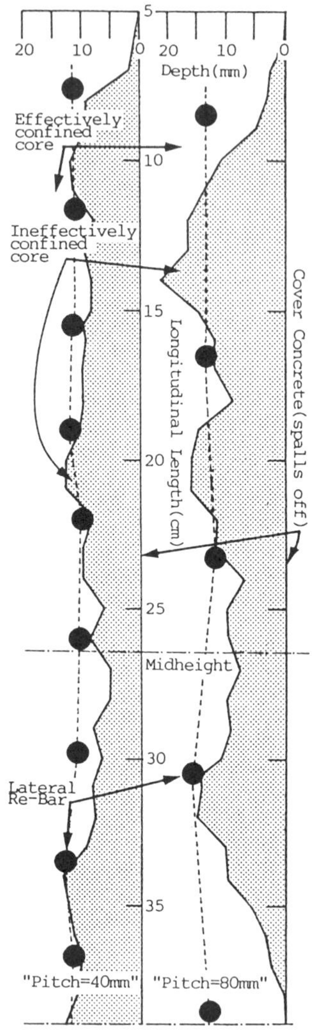

Fig.16. Confined core concrete of RC column.

ACKNOWLEDGMENT

Grateful acknowledgment is made to the assistance of Mr.Kazutaka MINAMI and Mr.Junichi SATO, The National Defense Academy.

REFERENCES

1)Kato,K., Kato,N. and Iwasaka,N.: Development of Loading Capacity of RC Column Considering Complex shear Strength of Concrete and Strength of Shear Reinforcement, Theoretical and Applied Mechanics, Vol.40 (Univ. of Tokyo Press, 1991), pp.233-248.

2)Kato,K., Kato,N. and Iwasaka,N.: Strain-Hardening of RC Column and Development of Its Bearing Capacity, ditto. Vol.41(Univ. of Tokyo Press, 1992), pp.159-168.

3)Mander,J.B., Priestley,M.J.N. and Park,R.:Theoretical Stress-Strain Model for Confined Concrete, Jour. Struct. Engrg., Vol.114, No.8, 1988, pp.1804-1826.

4)Kondo,Y. and Ban,S.: Concrete Engineering Handbook, (Asakura Publishing, 1965), p.343,(in Japanese).

5)Desayi, P. and Krishnan, S.: Equation for the Stress-Strain Curve of Concrete, J. Amer. Conc. Inst., Vol.61, No.3, 1964, pp.345-350.

6)JSCE Concrete Committee: Concrete Standard Specification [Design], 1991, p.54.

7)ibid. 6), pp.165-168.

8)ibid. 6), pp.21.

Impact Response of Cracked Glassfiber Reinforced Plastics at Low Temperatures

Sei UEDA* and Yasuhide SHINDO**

* *Department of Technology, Faculty of Education, Shizuoka University, Shizuoka*
** *Department of Materials Processing, Faculty of Engineering, Tohoku University, Sendai, Miyagi*

This paper considers the impact response of G-10CR glass-epoxy laminates with a crack and temperature dependent properties at low temperatures. The crack is oriented normally to the interfaces. Laplace and Fourier transform techniques are used to reduce the elastodynamic problem to a system of dual integral equations which is solved in the Laplace transform plane for the transform of the dynamic stress intensity factor. The time inversion is accomplished by a numerical scheme. The plastic energy dissipation rate is determined from the simple Dugdale model, and the temperature rise near the crack tip is calculated. Numerical results on the dynamic stress intensity factor, the temperature rise and the dissipated plastic energy at low temperatures are obtained and are presented in a graphic form.

I. INTRODUCTION

G-10CR and G-11CR glass-epoxy laminates(1) are used in magnetic fusion energy structures as thermal insulation, electrical insulation and structural support at low temperatures. Understanding the mechanical behavior of these composites will provide useful information to the magnetic fusion energy designers(2). Taking the temperature dependent properties of the composites into consideration in the thermal stress analysis, the authors have considered the thermal mechanical response of G-10CR glass-epoxy laminates with a crack perpendicular to the interfaces at low temperatures(3). Moreover, the singular stresses of G-10CR glass-epoxy laminates with a T-shaped crack under tension have been treated(4).

High performance superconducting magnets used in applications such as magnetic fusion energy structures are extremely susceptible to premature quenches triggered by small heat inputs generated within the magnet winding, because the heat capacity of the winding materials at low temperatures is very low(5). When a material with a crack is loaded suddenly, the plastic energy dissipation rate near the crack tip is high. According to the theory of irreversible thermodynamics, the dissipated plastic energy will be almost completely converted into heat, which may lead to high temperature rise at the crack tip. Because this high temperature rise can affect the stability of superconducting magnets, quantification of such dissipated energy is important for superconducting magnet technology.

This paper considers the dynamic singular stresses, the temperature rise and the dissipated plastic energy near the crack tip of G-10CR glass-epoxy laminates with a central crack and temperature dependent properties at low temperatures. The analytical method developed herein applies Fourier transform to the space variable and Laplace transform to the time variable. The boundary value crack problem is formulated with stress and displacement conditions. These conditions reduce the elastodynamic problem to the solution of a standard pair of dual integral equations in the Laplace transform plane. The dual integral equations are solved by the integral transform technique and the solution is expressed in terms of a Fredholm integral equation of the second kind having the kernel with finite integrals. A numerical Laplace inversion technique(6) is then also used to recover the time dependence of the solution. The plastic zone at the crack tip is regarded as a distributed source of heat proportional to the plastic energy dissipation rate. The plastic energy dissipation rate is determined from the simple Dugdale model, and the temperature rise near the crack tip is calculated(7). Numerical results on the dynamic stress intensity factor, the temperature rise and the dissipated plastic energy near the crack tip at low temperatures are obtained and are presented in a graphic form.

II. DYNAMIC SINGULAR STRESSES NEAR THE CRACK TIP

Let the layered composite be made of a cracked layer I of width $2h_1$ bonded between two layers II of width $h_2 - h_1$ (3). The ends of the crack of length $2a$ ($a < h_1$) are situated at equal distances away from the interfaces which are directed normal to the crack plane. The composite is assumed to be cooled down to the temperature Φ. The system of rectangular Cartesian coordinates (x, y, z) is introduced in the composite in such way that the crack is centered along the x-axis, and the y-axis is parallel to the interfaces. The layer I is reinforced along the z-axis having the Young's modulus $E_{xI}(\Phi)$, $E_{yI}(\Phi)$, $E_{zI}(\Phi)$, shear modulus $G_{xyI}(\Phi)$ and Poisson's ratio $\nu_{xyI}(\Phi)$, $\nu_{yzI}(\Phi)$, $\nu_{zxI}(\Phi)$. $\nu_{xyI}(\Phi)$ denotes Poisson's ratio for the transverse strain in the y-direction when stressed in the x-direction. The layer II is reinforced along the y-axis possessing the elastic properties $E_{xII}(\Phi)$, $E_{yII}(\Phi)$, $E_{zII}(\Phi)$, $G_{xyII}(\Phi)$, $\nu_{xyII}(\Phi)$, $\nu_{yzII}(\Phi)$, $\nu_{zxII}(\Phi)$. These elastic properties are functions of the temperature. In the following, the subscripts x, y, z will be used to refer to the direction of coordinates and the subscripts I and II indicate the layer I and II, respectively. The elastic properties $E_{x\delta}(\Phi)$, $E_{y\delta}(\Phi)$, $E_{z\delta}(\Phi)$, $\nu_{xy\delta}(\Phi)$, $\nu_{yz\delta}(\Phi)$, $\nu_{zx\delta}(\Phi)$ $(\delta = I, II)$ satisfy the following Maxwell's relations:

$$\frac{\nu_{xy\delta}(\Phi)}{E_{x\delta}(\Phi)} = \frac{\nu_{yx\delta}(\Phi)}{E_{y\delta}(\Phi)}, \quad \frac{\nu_{yz\delta}(\Phi)}{E_{y\delta}(\Phi)} = \frac{\nu_{zy\delta}(\Phi)}{E_{z\delta}(\Phi)}, \quad \frac{\nu_{zx\delta}(\Phi)}{E_{z\delta}(\Phi)} = \frac{\nu_{xz\delta}(\Phi)}{E_{x\delta}(\Phi)} \qquad (\delta = I, II) \tag{1}$$

Let the components of the displacement in the x and y directions be labeled by $u_{x\delta}$ and $u_{y\delta}$ $(\delta = I, II)$, respectively. Under plane strain, the nonzero stress components are found as

$$\left.\begin{aligned} \sigma_{x\delta}/G_{xy\delta}(\Phi) &= c_{11\delta}(\Phi)u_{x\delta,x} + c_{12\delta}(\Phi)u_{y\delta,y} \\ \sigma_{y\delta}/G_{xy\delta}(\Phi) &= c_{12\delta}(\Phi)u_{x\delta,x} + c_{22\delta}(\Phi)u_{y\delta,y} \\ \tau_{xy\delta}/G_{xy\delta}(\Phi) &= u_{x\delta,y} + u_{y\delta,x} \end{aligned}\right\} (\delta = I, II) \tag{2}$$

in which a comma denotes partial differentiation with respect to the coordinate or the time and

$$\left.\begin{aligned} c_{11\delta}(\Phi) &= \{1/E_{z\delta}(\Phi) - \nu_{yz\delta}(\Phi)^2/E_{y\delta}(\Phi)\}/E_{y\delta}(\Phi)G_{xy\delta}(\Phi)c_{0\delta}(\Phi) \\ c_{22\delta}(\Phi) &= \{1/E_{x\delta}(\Phi) - \nu_{zx\delta}(\Phi)^2/E_{z\delta}(\Phi)\}/E_{z\delta}(\Phi)G_{xy\delta}(\Phi)c_{0\delta}(\Phi) \\ c_{12\delta}(\Phi) &= \{\nu_{zx\delta}(\Phi)\nu_{yz\delta}(\Phi)/E_{y\delta}(\Phi) + \nu_{xy\delta}(\Phi)/E_{x\delta}(\Phi)\}/E_{z\delta}(\Phi)G_{xy\delta}(\Phi)c_{0\delta}(\Phi) \\ c_{0\delta}(\Phi) &= \{1 - 2\nu_{xy\delta}(\Phi)\nu_{yz\delta}(\Phi)\nu_{zx\delta}(\Phi) - E_{x\delta}(\Phi)\nu_{zx\delta}(\Phi)^2/E_{z\delta}(\Phi) \\ &\quad - E_{y\delta}(\Phi)\nu_{xy\delta}(\Phi)^2/E_{x\delta}(\Phi) - E_{z\delta}(\Phi)\nu_{yz\delta}(\Phi)^2/E_{y\delta}(\Phi)\}/E_{x\delta}(\Phi)E_{y\delta}(\Phi)E_{z\delta}(\Phi) \end{aligned}\right\} (\delta = I, II) \tag{3}$$

The displacement equations of motion are

$$\left.\begin{aligned} c_{11\delta}(\Phi)u_{x\delta,xx} + u_{x\delta,yy} + \{1 + c_{12\delta}(\Phi)\}u_{y\delta,xy} &= \{1/c_{s\delta}^2(\Phi)\}u_{x\delta,tt} \\ c_{22\delta}(\Phi)u_{y\delta,yy} + u_{y\delta,xx} + \{1 + c_{12\delta}(\Phi)\}u_{x\delta,xy} &= \{1/c_{s\delta}^2(\Phi)\}u_{y\delta,tt} \end{aligned}\right\} \qquad (\delta = I, II) \tag{4}$$

where t denotes the time and $c_{s\delta}(\Phi) = \sqrt{G_{xy\delta}(\Phi)/\rho}$ $(\delta = I, II)$ with ρ being the mass density of the material. In the isotropic solid, $c_{s\delta}(\Phi)$ $(\delta = I, II)$ represent the velocity of the shear wave.

Suppose that the crack is now loaded suddenly by a pair of normal stresses such that the upper and lower crack surfaces move in the opposite directions at the temperature Φ. The composite is assumed to be initially at rest. The symmetry and boundary conditions can be written as

$$\left.\begin{aligned} \sigma_{yI}(x,0,t) &= -\sigma_0 H(t) \qquad (0 \le x < a) \\ u_{yI}(x,0,t) &= 0 \qquad (a \le x \le h_1) \end{aligned}\right\} \tag{5}$$

$$\tau_{xyI}(x,0,t) = 0 \qquad (0 \le x \le h_1) \tag{6}$$

$$\tau_{xyII}(x,0,t) = 0 \qquad (h_1 \le x \le h_2) \tag{7}$$

$$u_{yII}(x,0,t) = 0 \qquad (h_1 \le y \le h_2) \tag{8}$$

$$\tau_{xyI}(0,y,t) = 0 \qquad (0 \le y < \infty) \tag{9}$$

$$u_{xI}(0,y,t) = 0 \qquad (0 \le y < \infty) \tag{10}$$

$$\sigma_{xI}(h_1,y,t) = \sigma_{xII}(h_1,y,t) \qquad (0 \le y < \infty) \tag{11}$$

$$\tau_{xyI}(h_1,y,t) = \tau_{xyII}(h_1,y,t) \qquad (0 \le y < \infty) \tag{12}$$

$$u_{xI}(h_1,y,t) = u_{xII}(h_1,y,t) \qquad (0 \le y < \infty) \tag{13}$$

$$u_{yI}(h_1,y,t) = u_{yII}(h_1,y,t) \qquad (0 \le y < \infty) \tag{14}$$

$$\sigma_{xII}(h_2,y,t) = 0 \qquad (0 \le y < \infty) \tag{15}$$

$$\tau_{xyII}(h_2,y,t) = 0 \qquad (0 \le y < \infty) \tag{16}$$

where $H(t)$ is the Heaviside unit step function in the time and σ_0 is a constant with the dimension of stress.

Define a Laplace transform pair by the equations:

$$f^*(p) = \int_0^\infty f(t)\exp(-pt)dt,\ f(t) = \frac{1}{2\pi i}\int_{Br} f^*(p)\exp(pt)dp \tag{17}$$

in which the second integral is integrated over the Bromwich path. In $x-y$ plane, the layer I is isotropic and the layer II is orthotropic. Applying Laplace and Fourier transforms to equations(4), the displacement field for the layer I may be expressed as[8]

$$\begin{aligned} u^*_{xI}(x,y,p) &= -\frac{2}{\pi}\int_0^\infty \Big[\{sA_1(s,p)\exp(-\gamma_{1I}y) + \gamma_{2I}A_0(s,p)\exp(-\gamma_{2I}y)\}\sin(sx) \\ &- \{\gamma_{1I}A_2(s,p)\sinh(\gamma_{1I}x) + sA_3(s,p)\sinh(\gamma_{2I}y)\}\cos(sy)\Big]ds \end{aligned} \tag{18}$$

$$\begin{aligned} u^*_{yI}(x,y,p) &= -\frac{2}{\pi}\int_0^\infty \Big[\{\gamma_{1I}A_1(s,p)\exp(-\gamma_{1I}y) + sA_0(s,p)\exp(-\gamma_{2I}y)\}\cos(sx) \\ &+ \{sA_2(s,p)\cosh(\gamma_{1I}x) + \gamma_{2I}A_3(s,p)\cosh(\gamma_{2I}y)\}\sin(sy)\Big]ds \end{aligned} \tag{19}$$

where $\gamma_{kI} = \sqrt{s^2 + \{p/c_k(\Phi)\}^2}$ $(k = 1,2)$, $c_1(\Phi) = \sqrt{2G_{xyI}(\Phi)\{1-\nu_I(\Phi)\}/\rho\{1-2\nu_I(\Phi)\}}$ is the velocity of the dilatational wave, $c_2(\Phi) = c_{sI}(\Phi)$ and $\nu_I(\Phi) = 2G_{xyI}(\Phi)\{\nu_{xyI}(\Phi) + \nu_{zxI}(\Phi)\nu_{yzI}(\Phi)\}/E_{yI}(\Phi)$, respectively. Similarly, the displacement field for the layer II can be expresses as[9]

$$\begin{aligned} u^*_{xII}(x,y,p) &= \frac{2}{\pi}\int_0^\infty s\{A_4(s,p)\exp(\gamma_{1II}x) + A_5(s,p)\exp(-\gamma_{1II}x) \\ &+ A_6(s,p)\exp(\gamma_{2II}x) + A_7(s,p)\exp(-\gamma_{2II}x)\}\cos(sy)ds \end{aligned} \tag{20}$$

$$\begin{aligned} u^*_{yII}(x,y,p) &= \frac{2}{\pi}\int_0^\infty \Big[\beta_1\{A_4(s,p)\exp(\gamma_{1II}x) - A_5(s,p)\exp(-\gamma_{1II}x)\} \\ &+ \beta_2\{A_6(s,p)\exp(\gamma_{2II}x) - A_7(s,p)\exp(-\gamma_{2II}x)\}\Big]\sin(sy)ds \end{aligned} \tag{21}$$

where γ^2_{kII} $(k = 1,2)$ are the roots of the characteristic equation:

$$\begin{aligned} &c_{11II}(\Phi)\gamma^4 + [\{c_{12II}(\Phi)^2 + 2c_{12II}(\Phi) - c_{11II}(\Phi)c_{22II}(\Phi)\}s - \{1 + c_{11II}(\Phi)\}\{p/c_{sII}(\Phi)\}^2]\gamma^2 \\ &+[c_{22II}(\Phi)s^2 + \{p/c_{sII}(\Phi)\}^2][s^2 + \{p/c_{sII}(\Phi)\}^2] = 0 \end{aligned} \tag{22}$$

β_k $(k = 1,2)$ are given by

$$\beta_k = \Big[s^2 + \{p/c_{sII}(\Phi)\}^2 - c_{11II}(\Phi)\gamma^2_{kII}\Big] / [\{1 + c_{12II}(\Phi)\}\gamma_{kII}] \quad (k = 1,2) \tag{23}$$

In equations(18)~(21), $A_n(s,p)$ $(n = 0 \sim 7)$ are the unknowns to be solved. The stress expressions are obtained by substituting equations(18)~(21) into equation(2). Note that these stress components and displacement components satisfy the symmetry conditions(7)~(10), automatically.

Making use of the symmetry and boundary conditions(5), (6) and (11)~(16), we have the dual integral equations for the determination of the unknown function $A_0(s,p)$. The dual integral equations are solved by the integral transform technique and the solution is expressed in terms of a Fredholm integral equation of the second kind having the kernel with finite integrals[8].

The dynamic stress intensity factor may be determined by obtaining the asymptotic stress near the crack border in the Laplace transform plane and performing a Laplace inversion. Then the dynamic stress intensity factor $K_I(t)$ is

$$K_I(t) = \frac{1}{2\pi i}\int_{Br} \lim_{x\to a^+} \sqrt{2\pi(x-a)}\sigma^*_{yI}(x,0,p)\exp(pt)dp \tag{24}$$

A numerical Laplace inversion scheme[6] may be used to evaluate the integral in equation(24).

III. TEMPERATURE RISE AND DISSIPATED PLASTIC ENERGY NEAR THE CRACK TIP

We study the temperature rise and the dissipated plastic energy near the crack tip. We use the plastic energy dissipation rate as the heat generation rate. The variations of the thermal properties such as the specific heat and the thermal conductivity of the composite at low temperatures are much larger than the variations of the mechanical properties. Then the influence of the temperature rise and the temperature Φ on the thermal properties also must be taken into consideration in the temperature rise calculation. We consider the temperature rise for the case of $a/h_1 \to 0$. The equation governing the temperature $\phi_E(x,y,t)$ due to heating by the dissipated plastic energy is[10]

$$\phi_{E,t} = \{\kappa_I(\phi_E)\phi_{E,x}\}_{,x} + \{\kappa_I(\phi_E)\phi_{E,y}\}_{,y} + g(x,y,t,\phi_E)/\rho c_{pI}(\phi_E) \tag{25}$$

where $g(x,y,t,\phi_E)$ is the dissipated plastic energy rate, $c_{pI}(\phi_E)$, $\lambda_I(\phi_E)$ and $\kappa_I(\phi_E) = \lambda_I(\phi_E)\ /\rho\ c_{pI}(\phi_E)$ are the specific heat, the thermal conductivity and the thermal diffusivity of the layer I at the temperature ϕ_E, respectively. The initial and symmetry conditions are given by

$$\phi_E(x,y,0) = \Phi, \qquad [\phi_E(x,y,t)_{,x}]_{x=0} = 0 \tag{26}$$

Employing the Dugdale model, the plastic energy dissipation rate $g(x,y,t,\phi_E)$ is[7]

$$g(x,y,t,\phi_E) = \frac{4\{1-\nu_I(\phi_E)\}\sigma_{0Y}(\phi_E)^2}{\pi G_{xyI}(\phi_E)}\sqrt{1-\frac{x-a}{\omega(t,\phi_E)}}\,\delta_D(y)\frac{d\omega(t,\phi_E)}{dt} \quad (a < x < a+\omega(t,\phi_E)) \tag{27}$$

where $\delta_D(\ )$ is the Dirac delta function and $\sigma_{0Y}(\phi_E)$ is an effective yield stress. $\omega(t,\phi_E)$ denotes the small scale yielding plastic zone size:

$$\omega(t,\phi_E) = \frac{\pi}{8}\left\{\frac{K_I(t)}{\sigma_{0Y}(\phi_E)}\right\}^2 \tag{28}$$

To solve equation(25), we divide the time t into subintervals $t_{k-1} < t \le t_k \quad (k = 1,2,..,n,n+1)$, where $0 = t_0 < t_1 < ... < t_n < t_{n+1} = t_{max}$ and t_{max} denotes the time at which the peak value of the dynamic stress intensity factor occurs. In the subinterval $t_{k-1} < t \le t_k \quad (k = 1,2,..,n,n+1)$, we calculate the temperature rise using the mechanical and thermal properties at the crack tip temperature $\phi_D^{(k-1)}$ at the time $t = t_{k-1}$. The equation governing the temperature $\phi_{k-1}(x,y,t)$ due to heating by the dissipated plastic energy in the subinterval $t_{k-1} < t \le t_k \quad (k = 1,2,..,n,n+1)$ is

$$\begin{aligned}\phi_{k-1,t} &= \kappa_I\left(\phi_D^{(k-1)}\right)\{\phi_{k-1,xx} + \phi_{k-1,yy}\} \\ &+ \frac{g\left(x,y,t,\phi_D^{(k-1)}\right)}{\rho c_{pI}\left(\phi_D^{(k-1)}\right)}\{H(t-t_{k-1}) - H(t-t_k)\} \quad (k = 1,2,..,n,n+1)\end{aligned} \tag{29}$$

Superposing the solutions[10] to equation(29), we obtain the temperature $\phi(x,y,t_m) \quad (m = 0,1,2,...,n, n+1)$ as follows:

$$\begin{aligned}\phi(x,y,t_m) &= (1-\delta_{m0})\sum_{k=1}^{m}\phi_{k-1}(x,y,t_m) + \Phi \\ &= (1-\delta_{m0})\sum_{k=1}^{m}\int_{t_{k-1}}^{t_k}\frac{d\tau}{4\pi\kappa_I\left(\phi_D^{(k-1)}\right)(t_m-\tau)}\iint_{A_p(\tau)}\frac{g\left(\xi,\eta,\tau,\phi_D^{(k-1)}\right)}{\rho c_{pI}\left(\phi_D^{(k-1)}\right)} \\ &\times \left[\exp\left\{\frac{-(x-\xi)^2-(y-\eta)^2}{4\kappa_I\left(\phi_D^{(k-1)}\right)(t_m-\tau)}\right\} + \exp\left\{\frac{-(x+\xi)^2-(y-\eta)^2}{4\kappa_I\left(\phi_D^{(k-1)}\right)(t_m-\tau)}\right\}\right]d\xi d\eta \\ &+ \Phi \qquad (m = 0,1,2,..,n,n+1)\end{aligned} \tag{30}$$

where $A_p(\tau)$ is the plastically deforming region at the time τ, δ_{m0} denotes Kronecker's delta and $\phi_D^{(k-1)}$ is

$$\phi_D^{(k-1)} = \left\{\begin{array}{ll}\Phi & (k=1) \\ \phi(a,0,t_{k-1}) & (k = 2,3,...,m)\end{array}\right\} \tag{31}$$

For the sake of convenience, we perform the following non-dimensionalization:

$$\left.\begin{array}{l}\alpha = \tau/t_{max}, \quad \alpha_k = t_k/t_{max}, \quad \varphi_k = 2\sqrt{\kappa_I\left(\phi_D^{(k-1)}\right) t_k/\omega\left(t_k, \phi_D^{(k-1)}\right)} \\ \Omega_k(\alpha) = \omega\left(\tau, \phi_D^{(k-1)}\right)/\omega\left(t_k, \phi_D^{(k-1)}\right) = \{K_I(\tau)/K_I(t_k)\}^2\end{array}\right\} \quad (k = 1, 2, .., n, n+1) \qquad (32)$$

In general, t_{max} is a very short time so that φ_k $(k = 1, 2, .., n, n+1)$ is small compared to unity. Substituting equation(27) to equation(30), we obtain the temperature $\phi_D = \phi(a, 0, t_{max})$ at the crack tip as follows:

$$\phi_D = \sum_{k=1}^{n+1} \phi^{(k-1)} \int_{\alpha_{k-1}}^{\alpha_k} \frac{1}{\sqrt{1-\alpha}} \left\{\frac{d\Omega_k(\alpha)}{d\alpha}\right\} d\alpha + \Phi \qquad (33)$$

where

$$\phi^{(k-1)} = \frac{\left\{1 - \nu_I\left(\phi_D^{(k-1)}\right)\right\} K_I(t_k)^2}{4G_{xyI}\left(\phi_D^{(k-1)}\right)\sqrt{\pi\rho c_{pI}\left(\phi_D^{(k-1)}\right)\lambda_I\left(\phi_D^{(k-1)}\right) t_{max}}} \quad (k = 1, 2, .., n, n+1) \qquad (34)$$

Let r_0 denote the distance from the crack tip, the temperature $\phi_{DR}(r_0)$ on a circle of radius r_0 at t_{max} is

$$\begin{aligned}\phi_{DR}(r_0) &= \sum_{k=1}^{n+1} \frac{3\phi^{(k-1)}}{4r_0\bar{\Omega}} \sqrt{\frac{\kappa_I\left(\phi_D^{(k-1)}\right) t_{max}}{\pi}} \int_{\alpha_{k-1}}^{\alpha_k} \frac{1}{\sqrt{1-\alpha}} \left\{\frac{d\Omega_k(\alpha)}{d\alpha}\right\} d\alpha \\ &+ \Phi \quad \left(r_0 \gg 2\sqrt{\kappa_I\left(\phi_D^{(k-1)}\right) t_{max}}\right)\end{aligned} \qquad (35)$$

where

$$\bar{\Omega} = \sum_{k=1}^{n+1} \int_{\alpha_{k-1}}^{\alpha_k} \frac{1}{\sqrt{1-\alpha}} \left\{\frac{d\Omega_k(\alpha)}{d\alpha}\right\} d\alpha \qquad (36)$$

The dissipated plastic energy G_{Dp} in the plastic zone is

$$\begin{aligned}G_{Dp} &= \sum_{k=1}^{n+1} \int_{t_{k-1}}^{t_k} d\tau \iint_{A_p(\tau)} g\left(\xi, \eta, \tau, \phi_D^{(k-1)}\right) d\xi d\eta \\ &= \sum_{k=1}^{n+1} \frac{\pi\left\{1 - \nu_I\left(\phi_D^{(k-1)}\right)\right\} K_I(t_k)^4}{48G_{xyI}\left(\phi_D^{(k-1)}\right)\sigma_{0Y}\left(\phi_D^{(k-1)}\right)^2} \left[\Omega_k(\alpha_k)^2 - \Omega_k(\alpha_{k-1})^2\right]\end{aligned} \qquad (37)$$

For the case of taking the influence of the temperature Φ only on the mechanical and thermal properties into consideration, the temperatures at the crack tip and on the circle of radius r_0 are denoted by ϕ_I and $\phi_{IR}(r_0)$, respectively. Substituting $n = 0$ to equations(33) and (35), ϕ_I and $\phi_{IR}(r_0)$ are given by

$$\phi_I = \phi^{(0)} \int_0^1 \frac{1}{\sqrt{1-\alpha}} \frac{d\Omega_1(\alpha)}{d\alpha} d\alpha + \Phi \qquad (38)$$

$$\phi_{IR}(r_0) = \frac{3\phi^{(0)}}{4r_0}\sqrt{\frac{\kappa_I(\Phi)t_{max}}{\pi}} + \Phi \quad \left(r_0 \gg 2\sqrt{\kappa_I(\Phi)t_{max}}\right) \qquad (39)$$

Similarly the dissipated plastic energy for this case is denoted by G_{Ip}. Substituting $n = 0$ to equation(37), G_{Ip} is given by

$$G_{Ip} = \frac{\pi\{1 - \nu_I(\Phi)\} K_I(t_{max})^4}{48G_{xyI}(\Phi)\sigma_{0Y}(\Phi)^2} \qquad (40)$$

Note that the values of G_{Ip} are independent of the temperature rise and are valid for any value of a/h_1.

IV. NUMERICAL RESULTS AND DISCUSSION

First, we examine the effect of the temperature Φ and the crack length $2a$ on the dynamic stress intensity factors $K_I(t)$. The elastic properties of the glassfiber and the matrix are approximated by the exponential functions[3]. The composite elastic properties $E_l(\Phi)$, $E_t(\Phi)$, $G_{lt}(\Phi)$, $G_{tt}(\Phi)$, $\nu_{lt}(\Phi)$, $\nu_{tl}(\Phi)$ and $\nu_{tt}(\Phi)$ are calculated using equations of Hashin[11] and are related to the elastic properties $E_{x\delta}(\Phi)$, $E_{y\delta}(\Phi)$, $E_{z\delta}(\Phi)$, $\nu_{xy\delta}(\Phi)$, $\nu_{yz\delta}(\Phi)$, $\nu_{zx\delta}(\Phi)$ $(\delta = I, II)$ by the relations:

$$\left.\begin{array}{l} E_{zI}(\Phi) = E_{yII}(\Phi) = E_l(\Phi),\ E_{xI}(\Phi) = E_{yI}(\Phi) = E_{xII}(\Phi) = E_{zII}(\Phi) = E_t(\Phi), \\ G_{xyI}(\Phi) = G_{tt}(\Phi),\ G_{xyII}(\Phi) = G_{lt}(\Phi), \\ \nu_{zxI}(\Phi) = \nu_{yzII}(\Phi) = \nu_{lt}(\Phi),\ \nu_{yzI}(\Phi) = \nu_{xyII}(\Phi) = \nu_{tl}(\Phi),\ \nu_{xyI}(\Phi) = \nu_{zxII}(\Phi) = \nu_{tt}(\Phi) \end{array}\right\} \quad (41)$$

where the subscripts l and t indicate the longitudinal and transverse directions to the fiber axis, respectively. The volume fraction of the fiber is $V_f = 0.5$. Temperatures of $\Phi_s = 395[K]$(cure temperature of the thermal strain free state)[2], $\Phi = 77[K]$(liquid nitrogen temperature) and $\Phi = 4[K]$(liquid helium temperature) are chosen. Fig.1 shows the variation of the normalized dynamic stress intensity factor $K_I(t)/\sigma_0(\pi a)^{1/2}$ with normalized time $c_2(\Phi_s)t/a$ for $a/h_1 = 0.1, 0.9$, $h_2/h_1 = 2.0$ and $\Phi = 4, 77[K]$. The peak values of $K_I(t)/\sigma_0(\pi a)^{1/2}$ are seen to increase with decreasing the values of a/h_1 and the effect of a/h_1 is more pronounced for the case of $\Phi = 77[K]$ than the case of $\Phi = 4[K]$. It is observed that the peak values of $K_I(t)/\sigma_0(\pi a)^{1/2}$ for the case of $\Phi = 4[K]$ occur at earlier time than the case of $\Phi = 77[K]$.

Next, we consider the dissipated plastic energy and the temperature rise near the crack tip. The effective yield stress and the thermal properties are approximated by the following elementary functions from experimental values[12]~[14].

$$\sigma_{Y0}(\phi) = 466.9 + 0.91\phi - 5.002 \times 10^{-3}\phi^2 \quad [MPa] \quad (42)$$

$$\lambda_I(\phi) = 0.02891\phi^{0.4151} + 0.03360 \quad [W/mK] \quad (43)$$

$$c_{pI}(\phi) = \left\{\begin{array}{ll} 2.757\phi + 3.217\sqrt{\phi} & (30[K] \leq \phi) \\ 2.635 \times 10^{-3} + 0.1776\phi^{3/2} & (10[K] \leq \phi < 30[K]) \\ 8.250 \times 10^{-3}\phi^3 & (\phi < 10[K]) \end{array}\right\} \quad [J/kgK] \quad (44)$$

where ϕ denotes temperature. Fig.2 exhibits the influence of a/h_1 on the dissipated plastic energy G_{Ip} for $\Phi = 4,\ 77[K]$. As the a/h_1 ratio increases, the values of normalized plastic energy $G_{Ip}G_{xyI}(\Phi_s)\sigma_{0Y}(\Phi_s)^2/\sigma_0^4 a$ and the influence of the temperature on the normalized plastic energy decrease. Fig.3 shows the distribution of the temperature rise $\phi_{IR}(r_0)$ with r_0/a for $a/h_1 \to 0$ and $\Phi = 4,\ 77[K]$. The temperature decreases rapidly with increasing r_0. Temperature rise for the case of $\Phi = 4[K]$ is much larger than the case of $\Phi = 77[K]$. Fig.4 indicates the influence of temperature dependent properties on the temperature rise at the crack tip with $\sigma_0^2 a^{1/2}$ $[Pa^2m^{1/2}]$ for $a/h_1 \to 0$ and $\Phi = 4,\ 77[K]$. As $\sigma_0^2 a^{1/2}$ increases, the influence of temperature dependent properties on the temperature rise increases. The effect of $\sigma_0^2 a^{1/2}$ for the case of $\Phi = 4[K]$ is more pronounced than the case of $\Phi = 77[K]$. The normalized plastic energy $G_{Dp}G_{xy1}(\Phi_s)\sigma_{0Y}(\Phi_s)^2/\sigma_0^4 a^2$ is plotted in Fig.5 as a function of $\sigma_0^2 a^{1/2}$ for $a/h_1 \to 0.0$ and $\Phi = 4,\ 77[K]$. As $\sigma_0^2 a^{1/2}$ increases, the normalized plastic energy increases. The normalized plastic energy for the case of $\Phi = 77[K]$ is 10 ~ 14% larger than the case of $\Phi = 4[K]$.

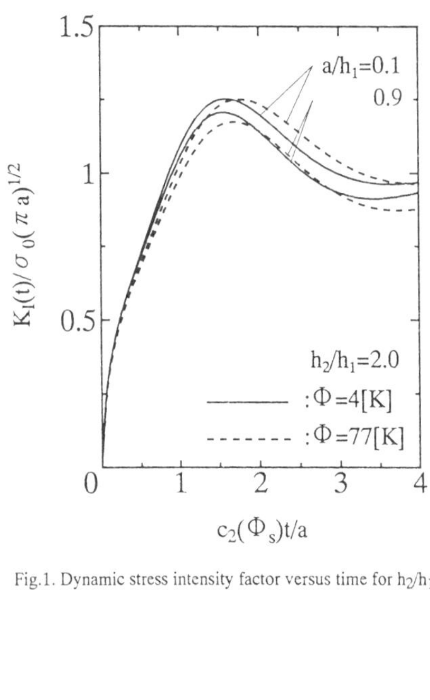

Fig.1. Dynamic stress intensity factor versus time for h_2/h_1=2.0.

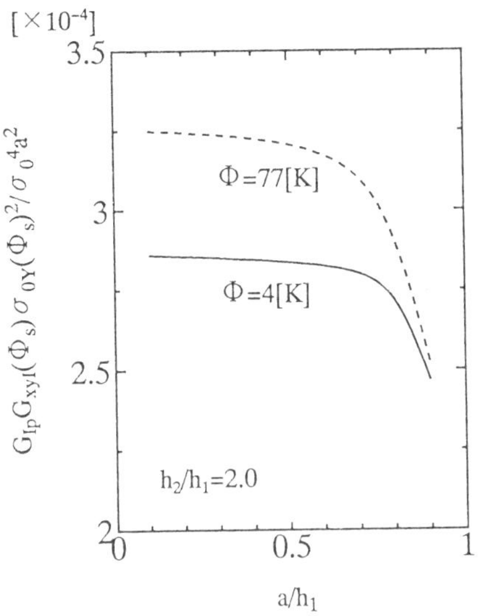

Fig.2. Dissipated plastic energy G_{Ip} versus a/h_1 for $h_2/h_1=2.0$.

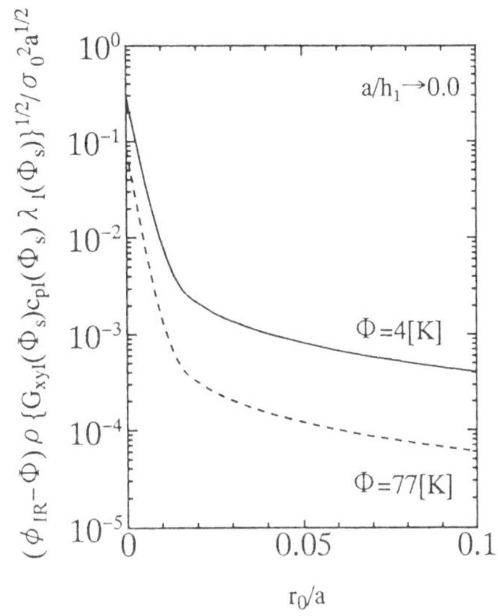

Fig.3. Temperature rise near the crack tip versus r_0/a for $a/h_1 \to 0$.

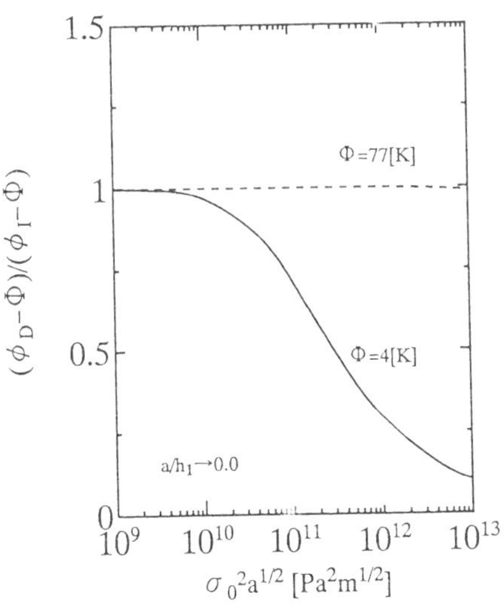

Fig.4. Influence of temperature dependent properties on the temperature rise at the crack tip versus $\sigma_0{}^2 a^{1/2}$ for $a/h_1 \to 0$.

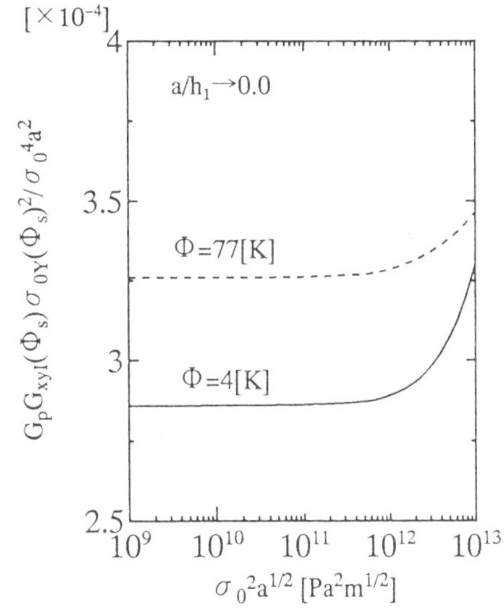

Fig.5. Dissipated plastic energy G_{Dp} versus $\sigma_0{}^2 a^{1/2}$ for $a/h_1 \to 0$.

REFERENCES

(1) Kasen, M.B., MacDonald, G.R., Beekman, D.H. and Schramm, R.E., "Mechanical, electrical and thermal characterization of G-10CR and G-11CR glass/epoxy laminates between room temperature and 4K," *Adv. Cryog. Eng.* **26**,235-244(1980).
(2) Kriz, R.D., "Influence of damage on mechanical properties of woven composites at low temperatures", *J. Compos. Tech.&Res.* **7**, 55-58(1985).
(3) Ueda, S. and Shindo, Y., "Thermal singular stresses and mechanical properties in cracked glass-fiber reinforced plastics at low temperatures," *Theor. Appl. Mech.* **40**,249-257(1991).
(4) Ueda, S. and Shindo, Y., "Glassfiber reinforced plastics with a T-shaped crack under tension at low temperatures," *Theor. Appl. Mech.* **41**, 169-179(1992).
(5) Iwasa, Y., "Experimental and theoretical investigation of mechanical disturbances in epoxy-impregnated superconducting coils. 1. General introduction," *Cryogenics* **25**,305-306(1985).
(6) Papoulis, A., "A new method of inversion of the Laplace transform," *Q. Appl. Math.* **14**, 405-414(1957).
(7) Rice, J.R. and Levy, N., "Local heating by plastic deformation at a crack tip," *Physics and Strength and Plasticity(ed.by A. S. Argon),* M.I.T. Press, Cambridge, Massachusetts,277-293(1969).
(8) Nozaki, H., Shindo, Y. and Astumi, A., "Impact response of a strip composite with a finite crack," *Eng. Fract. Mech.* **23**,943-952(1986).
(9) Chen, E.P. and Sih, G.C., *Mechanics of Fracture 6,*Noordhoff, Legden, 304-328(1981).
(10) Carslaw, H.S. and Jaeger, J.C., *Conduction of Heat in Solids,* Oxford, Clarendon Press(1959).
(11) Hashin, Z., "Theory of fiber reinforced materials," *NASA-CR-1974* (1972).
(12) Hartwing, G., "Low-temperature properties of epoxy resins and composites," *Adv. Cryog. Eng.,* **24**,17-36(1978).
(13) Kasen, M.B., "Mechanical and thermal properties of filamentary-reinforced structural composites at cryogenic temperatures," *Cryogenics,* **15**,327-349(1975).
(14) Dahlerup-Pertersen, K. and Perrott, A., "Properties of organic composite materials at cryogenic temperatures," *CERN ISR-BON/79-39* (1975).

IV
DYNAMICAL PROBLEMS

Coefficient of Restitution between a Ball and a Tennis Racket

Yoshihiko KAWAZOE

Department of Mechanical Engineering, Saitama Institute of Technology, Okabe, Saitama

The performance of tennis racket in terms of the coefficient of restitution (CO R) is closely related to the impact phenomena. In the present paper, the distribution of the coefficient of restitution of a tennis racket when the racket vibrates is predicted using an approximate non-linear impact model relating to an experimentally identified vibration model of a tennis racket on the basis of the idea that the COR can be derived from the energy loss during ball-racket impact. The calculated COR, considering the main sources of energy loss such things as the impact between the ball and strings, rotation of the racket and vibration of the racket-frame, can explain the experimental results very well. It was found that each racket having different physical property has its own characteristics with respect to the coefficient of restitution.

I. INTRODUCTION

With sport equipment, engineering technology has advanced to enable manufacturers to discover and synthesize new materials and new design. There are rackets of all compositions, sizes, weights, shapes and string tension. However, little is ever said about what equipment type is best suited for what individual.

The performance of the tennis racket in terms of the coefficient of restitution (COR) is closely related to the impact phenomena. Nevertheless, there are a number of unclarified points regarding the impact phenomenon between a ball and a racket as well as regarding optimum design of rackets[1),2)], because the impact is a complicated strong nonlinear phenomenon with an extremely short contact time and a very large contact force and deformation. It is the purpose of this study to investigate the impact phenomena for the optimum design of tennis rackets.

The previous paper[3)] has investigated the mechanism of a ball/racket impact, such as the impact force, the contact duration and the racket vibrations, on the basis of the idea that the contact duration, which has a strong influence on the racket-frame vibration and is not much affected by the frame stiffness, is determined by the natural period of a whole system composed of the mass of a ball, the nonlinear stiffness of a ball and strings, and the reduced mass of a racket at the impact location on the string face. The present paper proposes a simple nonlinear impact model for the prediction of the distribution of the coefficient of restitution of a tennis racket during impact using an experimentally identified vibration model of tennis rackets, considering the main sources of energy loss such as the impact between the ball and strings, rotation of the racket and vibration of the racket-frame.

Although the spin of ball seems to be very important for ball control, it is not dealt with here for the simplification of impact phenomena concerning the coefficient of restitution.

II. MAIN FACTORS ASSOCIATED WITH THE ENERGY LOSS AND COEFFICIENT OF RESTITUTION DURING IMPACT

2.1 Energy Loss in a Collision between a Ball and Strings

The measured coefficient of restitution versus the incident velocity when a ball strikes the rigid wall is shown in Fig.1, while the measured coefficient of restitution (COR) e_{BG} when a ball strikes the strings with a racket head clamped is shown in Fig.2. Although the COR in Fig.1 decreases with increasing incident velocity, the coefficient e_{BG} with a racket head clamped is almost independent of ball velocity and string tension. Thus, this constant value of COR can be regarded as COR inherent to the materials of ball and strings. This is due to the nonlinear characteristics of a composed ball/strings system[3)4)], and the experimental value agrees well with that of the literature[5)] being independent of the materials such as glassfibers, aluminum and wood.

Accordingly, the energy loss due to impact between a ball and strings with a racket head firmly clamped can be related to the coefficient e_{BG}.

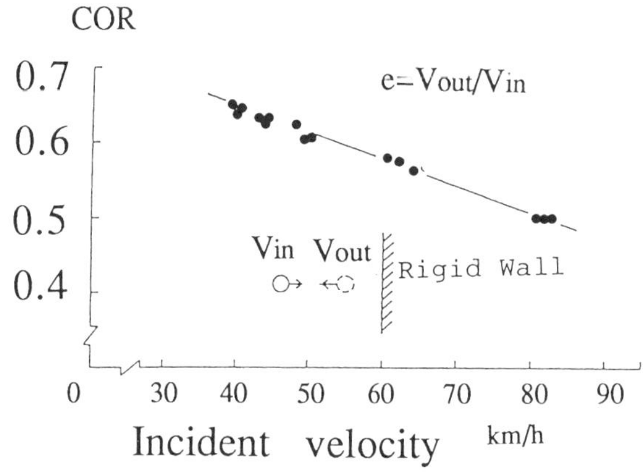

Fig.1. Measured COR between a ball and a rigid wall.

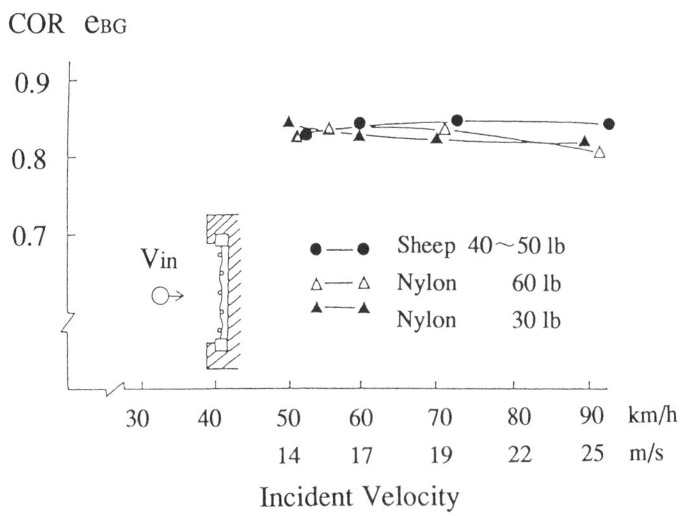

Fig.2. Measured COR between a ball and strings with racket head clamped.

2.2 Nonlinear Restoring Force Characteristics of a Ball and Strings and a Composed Ball/String System

Figure 3 shows a racket being struck by a ball where the racket handle can be rotated. It seems that the ball deforms only at the side which contact to the strings. Figure 4 shows the curves of restoring force F_B vs. ball deformation, restoring force F_G vs. string deformation, and the restoring force F_{GB} vs. deformation of the composed ball/string system, where the curves are determined so as to satisfy a number of experimental data using the least square method assuming that a ball with concentrated mass deforms only at the side in contact with the strings[2]. Figure 5 shows the curves of the corresponding stiffness K_B, K_G, K_{GB} derived by differentiation of the restoring force with respect to deformation. The nonlinear springs K_B of a ball, K_G of strings and K_{GB} of a composed ball/strings system become stiffer for larger deformation.

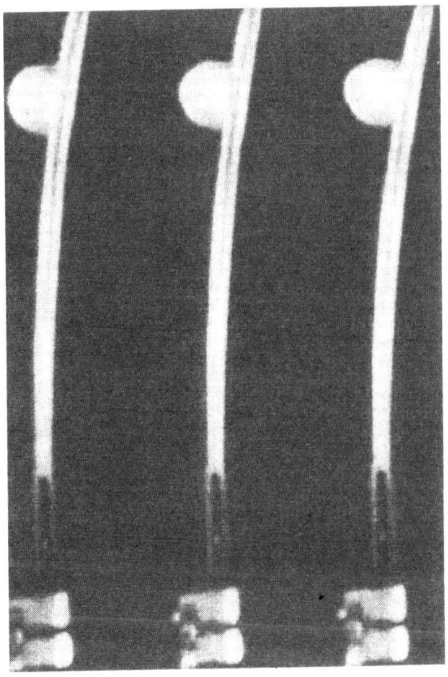

Fig.3. A racket being struck by a ball where the racket handle can be rotated.

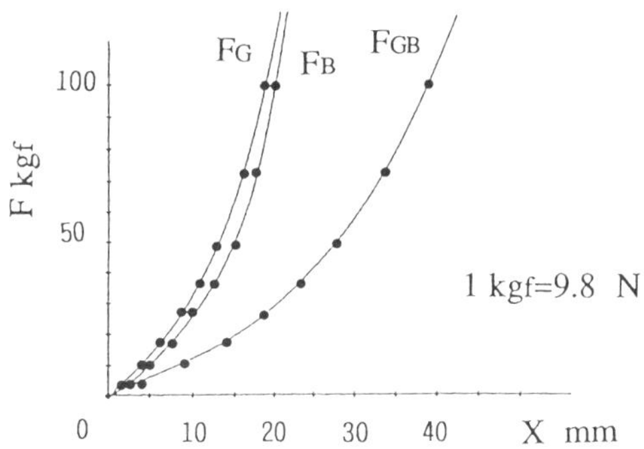

Fig.4. Restoring forces vs. deformation of a ball, strings, and a composed system of a ball and strings.

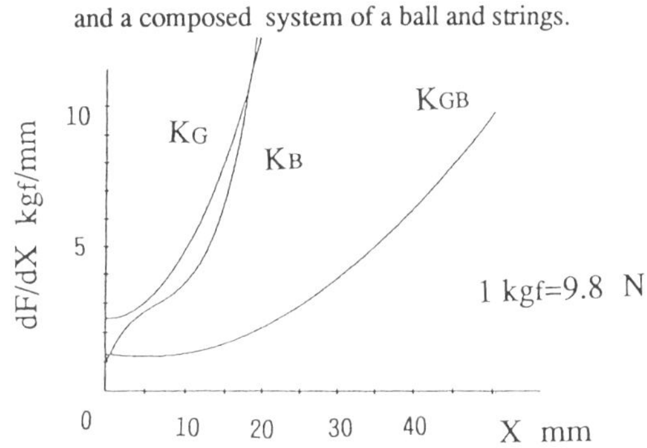

Fig.5. Stiffnesses vs. deformation of a ball, strings, and a composed system of ball and strings.

2.3 Contact Duration between a Racket and a Ball during Impact

The result of measured contact duration, which means how long the ball stays on the strings, with a normal racket and with a wide-body racket (stiffer) shows that the stiffness of the racket frame does not affect the contact duration much[2]. Accordingly, the masses of a ball and a racket as well as the nonlinear stiffness of a ball and strings are the main factors in the deciding of a contact time.

Therefore, the contact duratin(contact time) can be calculated using a model assuming that a ball with a concentrated mass m_B and nonlinear stiffness K_B collides with the nonlinear spring K_G of strings supported by a frame without vibration, where the measured coefficient of restitution inherent to the materials of ball-strings impact is employed as one of the sources of energy loss.

2.4 Energy Loss due to the Rotation of the Racket and an Introduction of the Reduced Mass of a Racket

In case a racket rotates around the grip portion at the racket handle, the reduced mass Mr at the impact location on the string face can be derived as Eq.(1.a) from the principle of the conservation of angular momentum, where I_A is the moment of inertia about the grip portion and L_A is the distance between a impact location and a grip portion[2].

$$M_r = I_A/L_A^2 \qquad (1a)$$

Also in case a freely-supported racket (not constrained at the grip) rotates around the center of gravity, the reduced mass can be calculated in the same way as Eq.(1.b), where $I_G(= M_R k_G^2)$ is the moment of inertia about the center of grvity, L_{GA} is the distance between the impact location and the center of gravity, and M_R is the total mass of a racket[2].

$$M_r = I_G/(L_{GA}^2 + k_G^2) \qquad (1b)$$

Fig.6 shows an example of reduced mass of a normal (standard) racket when the racket rotates about the grip portion and about the center of gravity. There is no big difference between them.

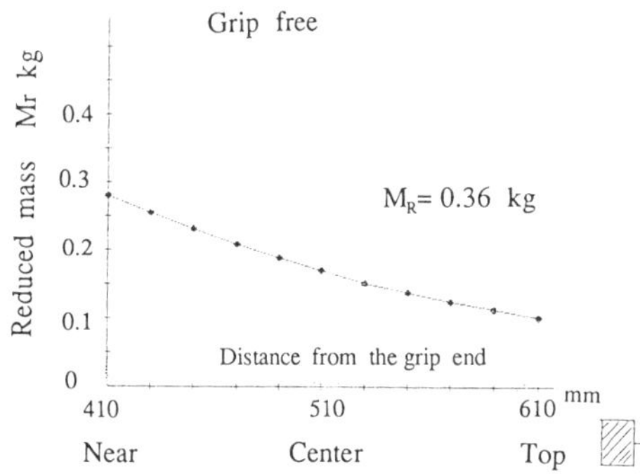

(a) Freely-supported racket (Grip free)

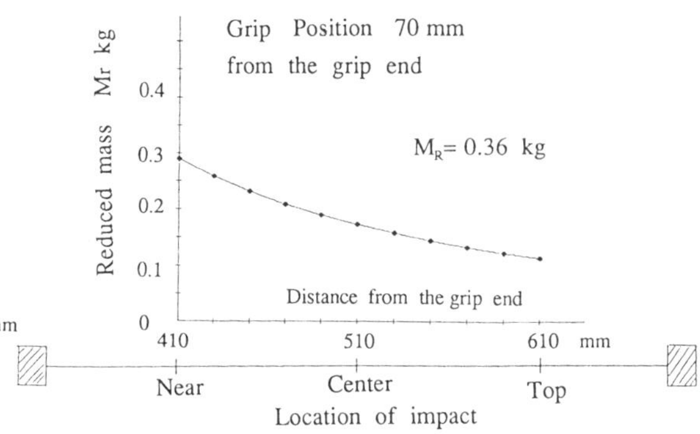

(b) Grip is supported at 70 mm from the grip end

Fig.6. Reduced mass M_r of a normal racket (M_R=0.36 kg) vs. location of impact.

2.5 Support Condition of a Racket Handle

The result of the experimental modal analysis[6)7)] showed that the fundamental vibration mode of a racket supported by a hand has two nodes being similar to the mode of a freely-supported racket, although the frequency of hand-held racket is a little lower than that of freely supported racket as shown in Fig.7 and Fig.8 with two different type of rackets[8]. Furthermore,

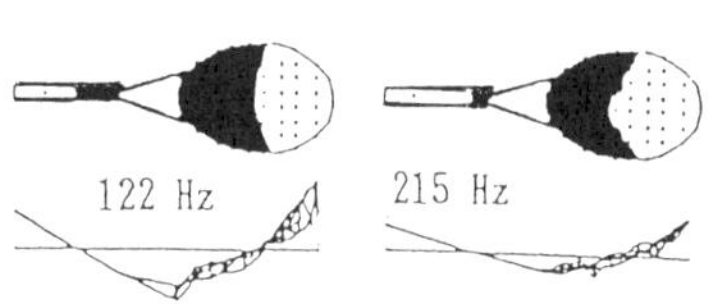

(a) Normal (b) Wide-body

Fig.7. Fundamental Vibration mode of freely supported rackets. (a) Normal racket , (b) Wide body racket

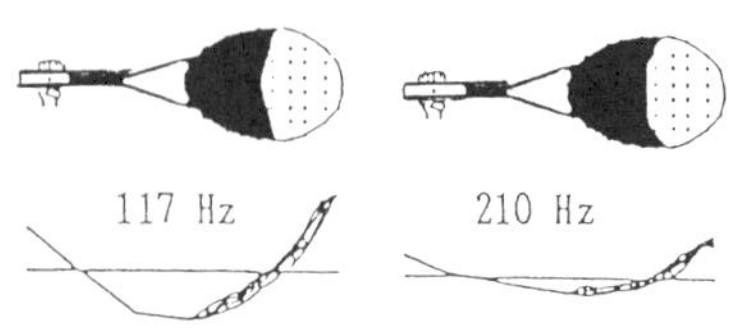

(a) Normal (b) Wide-body

Fig.8 Fundamental Vibration mode of hand-held rackets. (a) Normal racket, (b) Wide body racket there is no significant difference between the reduced mass of a freely-supported racket and that of a hand-held racket.

Accordingly, the racket is assumed to be supported freely or pin-jointed at the grip portion.

III. APPROXIMATE NONLINEAR ANALYSIS FOR DERIVATION OF THE IMPACT FORCE AND CONTACT DURATION

3.1 Outline of Approximate Nonlinear Impact Model

In order to predict the coefficient of restitution, the vibrational response of a racket has to be calculated. The impact force-time curve at the hitting portion on the string face has a strong influence on the amplitude of the vibrations of a racket[2)4)6)]. Therefore, in this chapter, the impact force and contact time will be calculted using an approximate nonlinear impact model assuming that a ball with a concentrated mass m_B and nonlinear stiffness k_B collides with the nonlinear spring k_G of strings supported by a rigid frame without vibrations, where the measured coefficient of restitution inherent to the materials of ball-strings is employed as one of the sources of energy loss.

3.2 Nonlinear Analysis of Impact Force and Contact Duration

The coefficient of restitution e_{BG} in the collision between a ball and strings with the racket head clamped could be regarded as the case in which the mass of racket frame is infinitely large. In case the vibration of racket frame is neglected, the momentum equation and the coefficient e_{BG} give the post-impact velocity V_B of a ball and V_R of a racket at the impact location. When a ball collides with a racket, post-impact velocities V_B and V_R are given as

$$V_B = -V_{Bo}(e_{BG} - m_B/M_r)/(1+ m_B/M_r) + V_{Ro}(1+ e_{BG})/(1+ m_B/M_r) \quad (2)$$

$$V_R = V_{Ro}(M_r/m_B - e_{BG})/(1+ M_r/m_B) + V_{Bo}(1+ e_{BG})/(1+ M_r/m_B) \quad (3)$$

where m_B is the mass of a ball, M_r is the reduced mass of a racket at the hitting location considering the rotation of the racket, and V_{Bo} is the ball velocity before impact.

The impulse could be described as

$$\int F(t)dt = m_B V_{Bo} - m_B V_B$$
$$= (V_{Bo} - V_{Ro})(1+e_{BG}) m_B /(1+m_B/M_r) \quad (4)$$

Assuming the contact duration during impact to be half the natural period of a whole system composed of the mass m_B of a ball, the stiffness K_{GB} of ball/strings, and the reduced mass of the racket, it could be obtained as

$$T_c = \pi\sqrt{m_B}/\sqrt{K_{GB}(1+ m_B/M_r)} \quad (5)$$

Actually, the stiffness K_{GB} has a strong nonlinearity and its value changes during impact also depending on the impact velocity. In order to make the analysis simpler, the equivalent force F_{mean} can be introduced during impact time T_C, which is described as

$$\int F(t)dt = F_{mean} T_c . \quad (6)$$

Thus, from Eq.(4), Eq.(5) and Eq.(6), the relationship between F_{mean} and corresponding K_{GB} against the velocity before impact is given by

$$F_{mean} = (V_{Bo}-V_{Ro})(1+ e_{BG})\sqrt{m_B}\sqrt{K_{GB}} /(\pi\sqrt{1+m_B/M_r}) \quad (7)$$

On the other hand, from Fig.4 and Fig.5 using the least square method, the relationship between F_{GB} and K_{GB} can be expressed in the form

$$F_{mean} = \text{func.}(K_{GB}). \quad (8)$$

From Eq.(7) and Eq.(8), the parameters K_{GB} and F_{mean} against impact velocity can be obtained as shown graphically in Fig.9, accordingly the contact duration T_C can also be determined against the velocity before impact from Eq.(5).

Figure 10 is a comparison between the measured contact times during actual forehand strokes[9] and the calculated ones when a ball hits the center of the strings face, showing a good agreement between them.

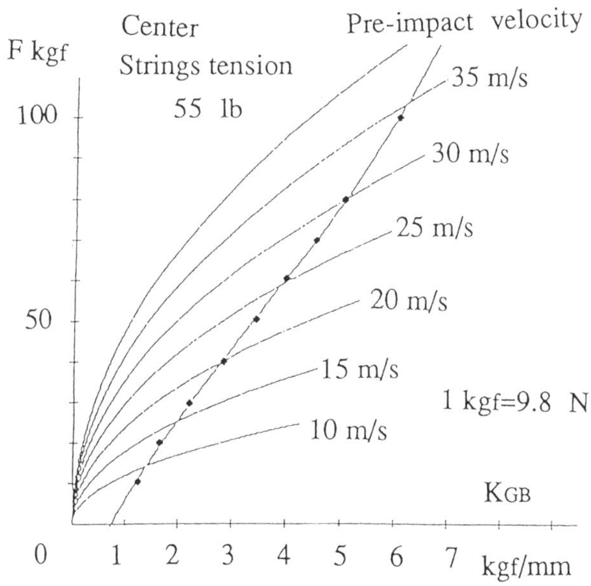

Fig.9. Graphical description of the derivation of impact force and the stiffness of the composed ball/strings system against the impact velocity

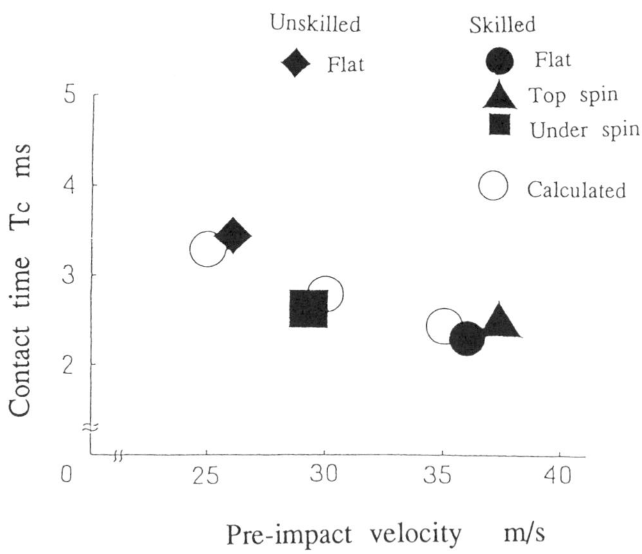

Fig.10. Comparison between the measured contact times during strokes and the calculated results.

3.3 Approximation of Impact Force-time Curve

Since the force-time curve of impact has an influence on the magnitude of racket frame vibrations, it is approximated as a half-sine pulse shown on the left in Fig.11, which is similar in shape to the actual impact force. The mathematical expession is

$$F(t)= F_{max}\sin(\pi t/T_c) \qquad (0\leqq t\leqq Tc) \qquad (9)$$

where $F_{max}= \pi F_{mean}/2$.

The fourier spectrum of Eq.(9) is shown graphically on the right in Fig.11 and represented as

$$S(f)= 2F_{max}T_c|\cos(\pi f T_c)|/[\pi|1-(2fT_c)^2|]. \qquad (10)$$

where f is the frequency.

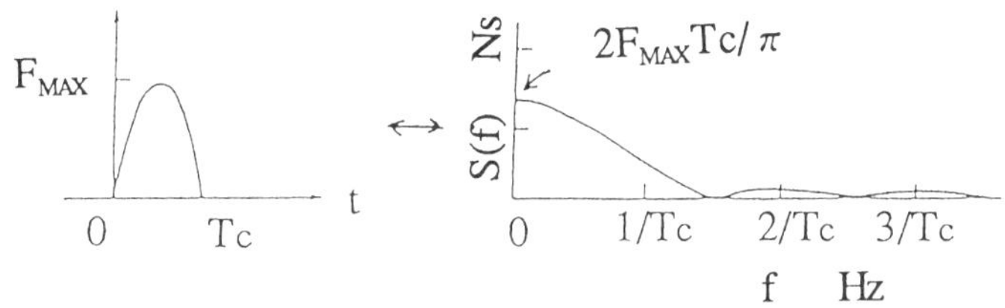

Fig.11. Shock shape and its spectrum during ball/racket impact.

IV. PREDICTION OF THE RACKET VIBRATION AND ENERGY LOSS INDUCED BY IMPACT

4.1 Prediction of the Racket Vibration

The vibration characteristics of a racket can be identified using the experimental modal analysis[6,7] and the racket vibrations can be simulated by applying the impact force-time curve to the hitting portion on the string face of the identified vibration model of the racket.

When the impact force $S_j(\omega_k)$ applies to the point j on the racket face, the summation of amplitudes X_{ij} at point i is expressed as

$$X_{ij}= \sum_{k=1}^{n} r_{ijk}S_j(\omega_k) \qquad (11)$$

where $r_{ij,k}$ denotes the residue of k-th mode between arbitrary point i and j derived on the basis of the experimental modal analysis, and $S_j(\omega_k)$ is the impact force component of k-th frequency ω_k in the frequency region[2,6,7]. When the hitting location and the pre-impact velocity are given, the amplitude of racket vibrations can be simulated by using Eq.(10) and Eq.(11).

Figure 12 is the predicted vibration amplitude of a racket struck by a ball at a velocity of 30 m/s, showing that the amplitude of racket vibration is very small when a ball hits the center, whereas it is large when a ball hits the off-center (the top or the place close to the throat) on the string face.

Figure 13 shows the predicted amplitude of vibration of rackets with different string tension when a ball hits the top (off-center) on the strings at a velocity of 30 m/s. It is seen that the amplitude of a racket strung with higher tension is larger than that of a racket strung with lower tension[10].

4.2 Energy Loss due to Racket Vibration Induced by Impact

The energy loss due to the ràcket vibration induced by impact can be derived from the amplitude distribution of the vibration velocity and the mass distribution along a racket frame when a ball collides with a racket at a certain location on the string face and at a certain impact velocity.

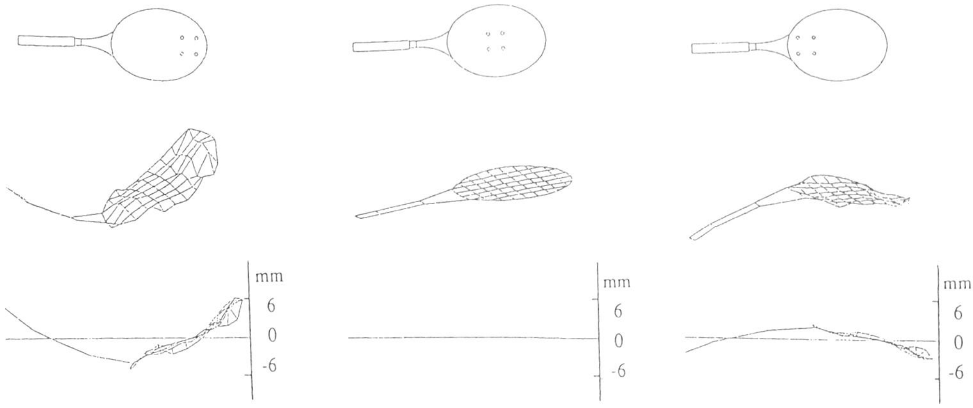

Fig.12. Prediction of amplitude of racket vibration when a ball hits the racket at a velocity of 30 m/s.

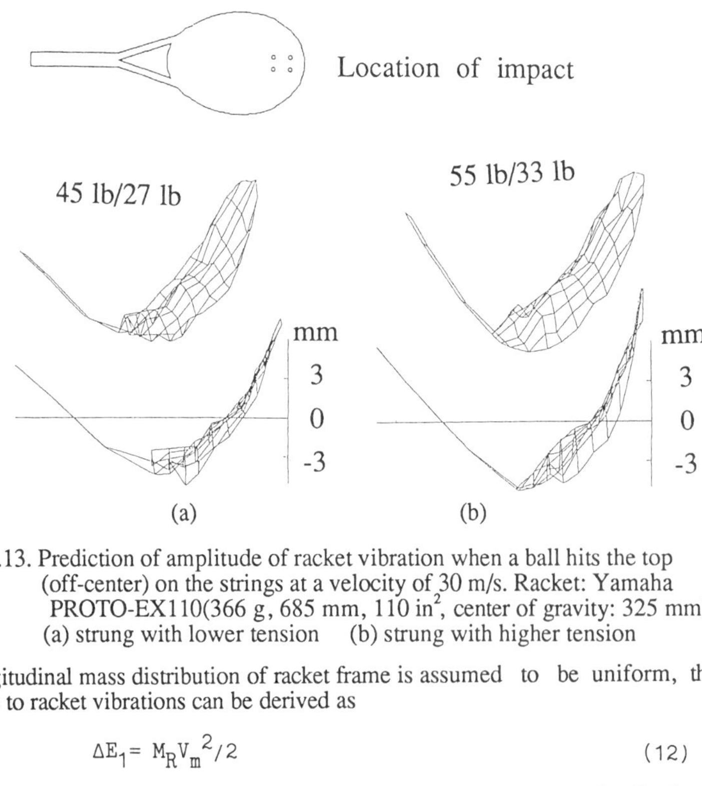

Fig.13. Prediction of amplitude of racket vibration when a ball hits the top (off-center) on the strings at a velocity of 30 m/s. Racket: Yamaha PROTO-EX110(366 g, 685 mm, 110 in^2, center of gravity: 325 mm)
(a) strung with lower tension (b) strung with higher tension

If the longitudinal mass distribution of racket frame is assumed to be uniform, the energy loss ΔE_1 due to racket vibrations can be derived as

$$\Delta E_1 = M_R V_m^2/2 \tag{12}$$

where V_m is the mean velocity amplitude calculated from the amplitude distribution of the post-impact vibrations of a racket frame, and M_R is the total mass of a racket.

V. DERIVATION OF THE COEFFICIENT OF RESTITUTION CONSIDERING THE ENERGY LOSS DURING IMPACT

5.1 Energy Loss during Impact

The coefficient of restitution (COR) might be related to the energy loss during impact.

The COR can be derived considering the main sources of energy loss such as the impact between the ball and strings, rotation of the racket and vibration of racket frame.

Since the measured coefficient of restitution e_{BG} inherent to the materials of ball-string system is considered as one of the sources of energy loss, another energy loss besides ΔE_1 is the energy loss ΔE_2 due to the rotational motion of the racket as a rigid body. If a ball collides with a racket at rest ($V_{Ro} \doteq 0$), the energy loss ΔE_2 could be obtained as

$$\Delta E_2 = [m_B V_{Bo}^2 - (m_B V_B^2 + M_r V_R^2)]/2$$
$$= [m_B M_r/(m_B + M_r)](1 - e_{BG}^2) V_{Bo}^2/2 \qquad (13)$$

5.2 Derivation of the Coefficient of Restitution

When a ball strikes a racket at rest ($V_{Ro} = 0$), the coefficient of restitution $e_r = (V_R - V_B)/V_{Bo}$ with respect to the relative velocities corresponds to the total energy loss ΔE ($= \Delta E_1 + \Delta E_2$) being expressed as

$$\Delta E = [m_B V_{Bo}^2 - (m_B V_B^2 + M_r V_R^2)]/2$$
$$= [m_B M_r/(m_B + M_r)](1 - e_r^2) V_{Bo}^2/2 \qquad (14)$$

From the above equation, the coefficient of restitution e_r is obtained as

$$e_r = (V_R - V_B)/V_{Bo}$$
$$= \sqrt{1 - 2\Delta E(m_B + M_r)/(m_B M_r V_{Bo}^2)} \qquad (15)$$

If the rebound velocity relative to the incident velocity of a ball is defined as the coefficient of restitution $e = -V_B/V_{Bo}$, which is often used for the comparison to the experiments, it is written as Eq.(16) using Eq.(3) and Eq.(15).

$$e = -V_B/V_{Bo}$$
$$= \sqrt{1 - 2\Delta E(m_B + M_r)/(m_B M_r V_{Bo}^2)} - (1 + e_{BG})/(1 + M_r/m_B) \qquad (16)$$

VI. ANALYTICAL RESULTS AND THE EXPERIMENTAL RESULTS

The energy loss due to the vibrations is very small when a ball hits the center, whereas it is large when a ball hits the off-center (the top or the near of the racket face). Furthermore, the energy loss due to vibrations of a racket with higher string tension is larger than that of a racket with lower tension.

Figure 14 is a comparison between the measured distribution of the COR (impact velocity: 26.4 m/s) and the calculated one(30.0 m/s) when a ball hits the string face of a freely-supported racket, showing a good agreement between them. The COR tends to be maximized along the longitudinal axis of the racket and peaks close to the throat due to the mass distribution of a racket.

Figure 15 is the predicted distribution of the COR of a racket (mass:366 g, hitting area:110 in^2) made of carbon graphite, showing that the COR tends to become lower at the tip and at the near on the string face with an increase of impact velocity due to the vibration modes of racket frame.

Figure 16 is the predicted coefficient of res titution when a ball strikes rackets with various physical properties at the top on the string face (off-center impact) at different velocities. The rackets in Fig.14 are a Yamaha racket EX-II (standard, 360 g, length 680 mm, area 100 in^2, center of gravity: 308 mm from grip end, 1st vibration mode: 122 Hz), a Yamaha PROTO-02 (very stiff frame called wide body, 370 g, 680 mm, 100 in^2, center of gravity:317 mm, 215 Hz), a Yamaha PROTO-EX110 (366 g, 685 mm, 110 in2, center of gravity:325 mm, 132 Hz), and a Yamaha EOS100 (super light weight, 290 g, 680 mm, 100 in^2, center of gravity:350 mm, 171 Hz). It is seen that each racket has its own characteristics with respect to the coefficient of restitution. The coefficients of restitution with PROTO-02 (wide body racket) and EOS100 (super light and top-heavy) are not affected by an increase of

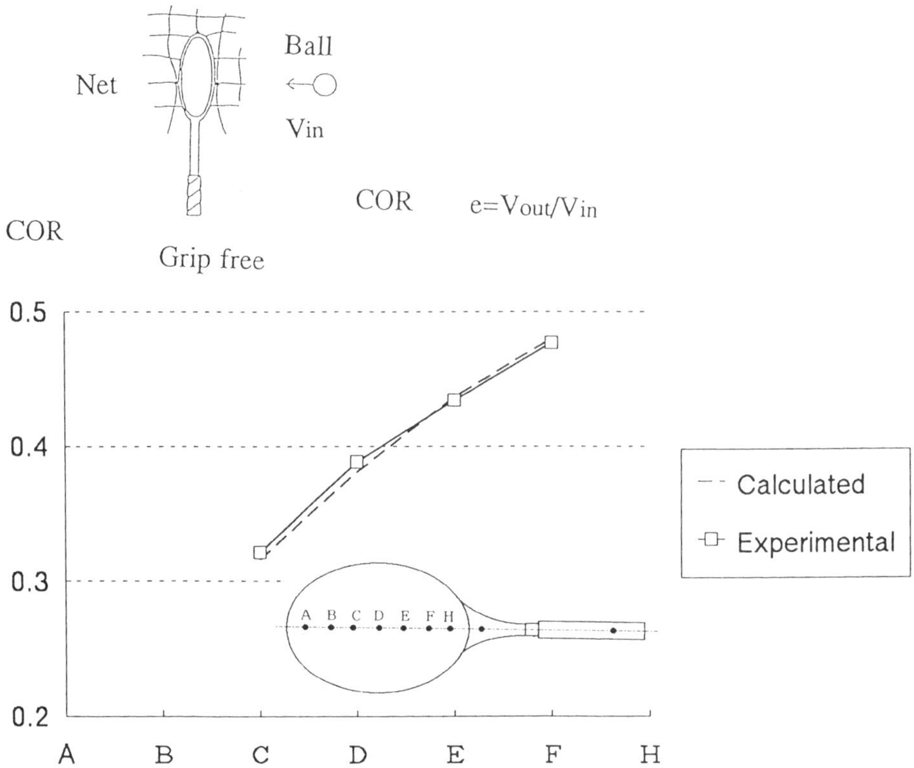

Fig.14. Comparison between the measured distribution of the COR (impact velocity 26.4 m/s) and the calculated results (30.0 m/s).

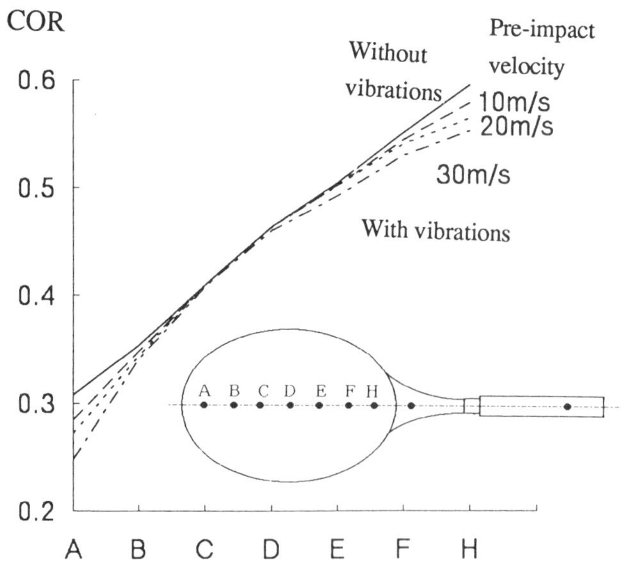

Fig.15. Predicted distribution of the COR vs. impact velocity

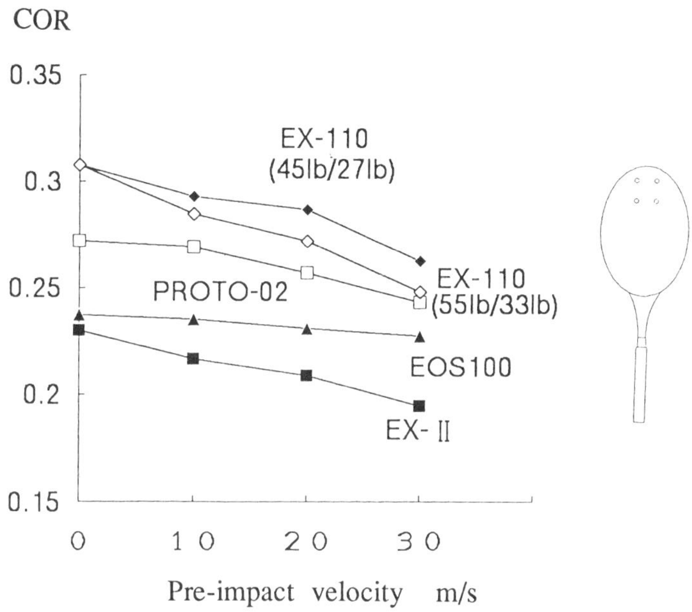

Fig.16. Predicted coefficient of restitution when a ball strikes rackets with various physical properties at the top on the string face (off-center impact) at different velocities.

impact velocity at the top and higher than the standard EX-II. Although the coefficient with PROTO-EX110 strung loosely is the highest, it decreases with increasing impact velocity. Particularly with PROTO-EX110 strung tightly it remarkably decreases in the off-center impact.

VII. CONCLUSIONS

The distribution of the coefficient of restitution (COR) of a tennis racket is predicted using a simple nonlinear impact model and an experimentally identified vibration model of tennis racket, considering the main sources of energy loss such as the impact between the ball and strings, rotation of the racket and vibration of racket-frame. The calculated COR can explain the experimental results very well. It was found that each racket having different physical property has its own characteristics with respect to the coefficient of restitution. The main results are summarized as follows:

(1) The calculated contact time between a ball and a racket agrees well with the measured one during actual forehand stroke by a player.

(2) The calculated COR on the racket face agrees well with the measured one.

(3) The COR tends to be maximized along the longitudinal axis of the racket and peaks close to the throat (very near on the string face) due to the mass distribution of a racket.

(4) The COR tends to become lower at the tip and at the near of the racket face with increasing impact velocity due to the vibration modes of racket frame.

(5) When the center of the racket face is struck, the value of COR is almost independent of the impact velocity, accordingly it is very close to that of racket without frame vibrations.

Acknowledgments

The author is grateful to Yamaha Company for their help in this study and also to Messers T.Maekawa and T.Midorikawa for their help in carrying out the study as senior students during the 1991 academic year.

REFERENCES

1) Groppel,J.L.,Shin,I.S.,Thomas,J.A. and Welk,G.J., The effects of string type and tension on impact in midsized and oversized tennis racquets, Int.J.Sport Biomech.,Vol.3(1987),pp.40-46.
2) Kawazoe,Y., CAE of tennis rackets with impact phenomena (Prediction of racket response and a view of restitution in racket-ball impact), Trans.JSME, 58-552 (1992),pp.143-150.(in Japanese)
3) Kawazoe,Y., Impact phenomena between racket and ball during tennis stroke, Theoretical and Applied Mechanics,Vol.41(1992),pp.3-13.
4) Kawazoe,Y., Analysis of coefficient of restitution during a nonlinear impact between a ball and strings considering vibration modes of racket frame, JSME D&D'92 symposium, No.920-55(1992),pp.77-81.(in Japanese)
5) Asami,T.,Miyasita,M. and Watanabe,T., as editors,Handbook of physical education & sports in the future,Vol.7, (Kohdansha,1984),p.277.(in Japanese)
6) Kawazoe,Y., Dynamics and computer aided design of tennis racket, Proc.Int.Sympo. on Advanced Computer for Dynamics and Design'89(1989),pp.243-248.
7) Kawazoe,Y.,Dynamics and computer aided design of tennis racket (The influence of some factors on the mode characteristics and a view of optimum design),Trans. JSME, Vol.56,No.526(1990),pp.1511-1517.(in Japanese)
8) Kawazoe,Y. and Tomosue,R.,Impact vibrations of a racket handle and the wrist joint in tennis, Trans. JSME, Vol.59,No.560(1993),pp.1001-1008.(in Japanese)
9) Nagata,A., Analysis of tennis movement, J.J.Sports Sci.,2-4(1983),pp.245-259. (in Japanese)
10) Kawazoe,Y., CAE of tennis rackets with impact phenomena (The effect of strings tension on a ball-racket impact), Trans. JSME, Vol.59,No.558(1993),pp.521-528. (in Japanese)

Active Fin Control System for Wind-Induced Structural Vibrations

Yoichi MUKAI, Eizaburo TACHIBANA, and Yutaka INOUE

Department of Architectural Engineering, Osaka University, Osaka

In this paper, an active fin system is proposed as a new type of active control system for reducing wind-induced vibrations of structures. The active fin system uses wind forces as control forces by opening or closing the active fin depending on the responses of structure. The active fin system has been investigated from experimental approach. In this experiment, CTAC which developed as a compact testing system in Osaka university is used and a three-story structural model equipped the active fin device is tested in a wind tunnel. The active fin system is controlled by very simple algorithm. Through this investigation, it is found that the active fin system is considerably effective for reducing structural vibrations induced by wind forces. In the present, these experimental results are described in detail.

I. INTRODUCTION

Recently, various structural vibration control systems are discussed widely in order to reduce wind-induced structural vibrations of high-rise buildings or towers. Mainly those systems are introduced in order to accomplish more comfortable residence. Passive control systems have been popularized, for example, a tuned pendulum mass damper (TMD) [1)], a tuned liquid damper by sloshing (TLD) [2)] and so on. Although, active control systems aim to get more effective to reduce structural vibrations than that of passive control systems. For example, an active mass damper (AMD) [3,4)], an active tendon [5,6)], a semi-active damper [7)] and so on. However, the active control systems are still having many problems concerning about cost performance or about robustness. In this paper, an active fin system is proposed as a new type of active control system for reducing wind-induced structural vibrations. In order to investigate the active fin system, in a wind tunnel, CTAC which developed as a compact testing system for active control systems [8)-11)] is used, and a three-story structural model equipped the active fin device is adopted as a tested structure. Through this experimental study, it is appeared that the active fin system can be controlled by comparatively a simple algorithm and can make available control effects. In this present, the results which observed on a series of those experiments are described in detail.

II. CONTROL SYSTEM OF ACTIVE FIN

The active fin control device for equipping on an experimental model is shown in Fig.1. The active fin is controlled by using the difference of resisting forces caused from two configurations as follows. One is a ‘close mode’ which receives wind forces as seen in Fig.1 (a), and the other is an ‘open mode’ which doesn’t receive wind forces as seen in Fig.1 (b). Those configurations are given by rotating the active fin. The active fin system is a new type of active control technique, in this present, it is conveniently classified into a type of the structural vibration control system by ASC (Active Selection of Configuration). The flow diagram of the active fin control system is shown in Fig.2. As the testing apparatus, CTAC is used for this aim. This system is composed from four parts which are sensor, controller, driver and monitor. By replacing target structural models, control devices or control algorithms according to the active control system to investigate, various types of active control systems can be tested easily with CTAC. In this experiment, a three-story structural model, which equipped the active fin control device on the top floor, is adopted as a tested structure. Relative displacements between each story of the structural model are measured by three lazer sensors (in order to avoid electric noise, low-pass filter (10 Hz) are used). Those data are transferred to micro computers via A/D converters. The absolute displacements, velocities or accelerations, which are needed in order to determine control forces, are computed by arithmetic sum or finite difference from those data, approximately. According to those digital data and according to assigned control algorithms (which are mentioned in following articles), pulse rates are calculated. According to the pulse rates, control pulse signals are generated at PC modules. According to the control pulse signals, the driving sequences of stepping motor are generated at motor driver, and then the stepping motor is rotated and the control movements for the active fin are given.

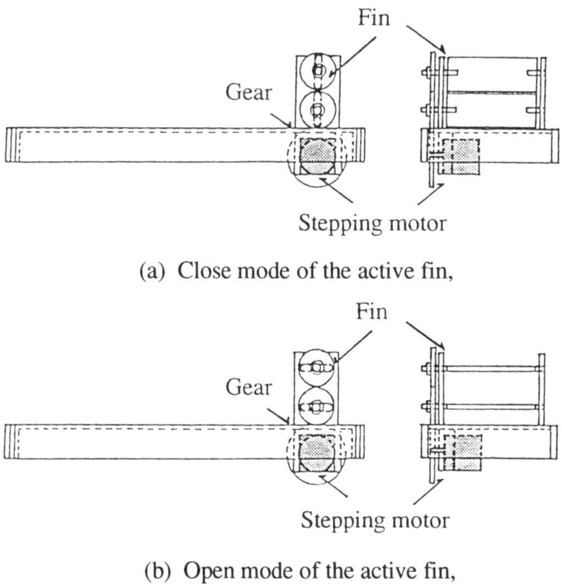

Fig. 1. Control device of the active fin.

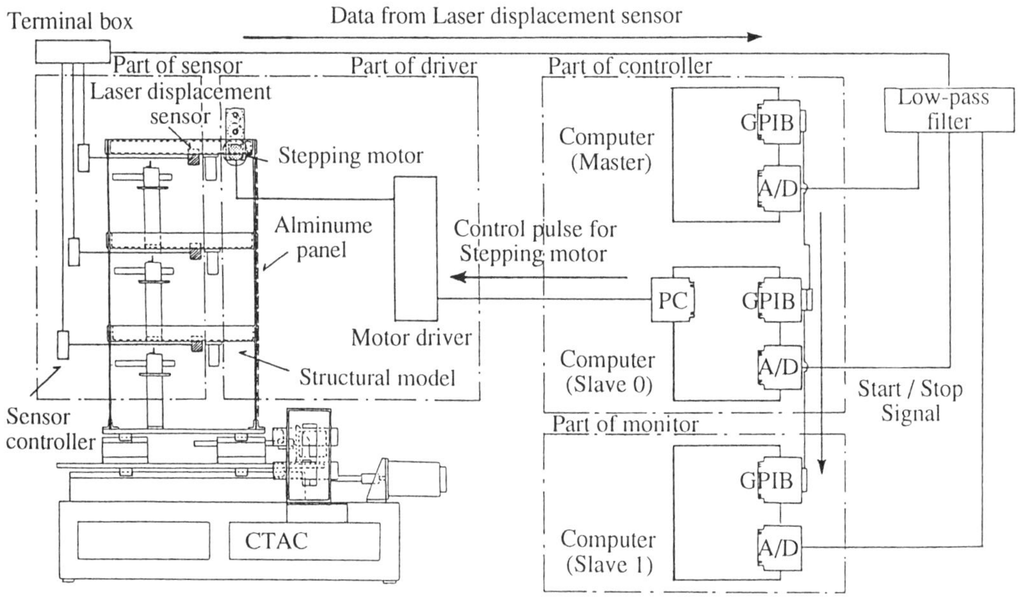

Fig. 2. Flow diagram of the active fin control system.

III. CONTROL ALGORITHM

The control algorithm of the active fin introduced here is very simple. Control forces are generated by the active fin which is attached on the top floor of the structural model. The configuration of the active fin is selected from only two modes that is the close mode or the open mode as seen in Fig.1 (a) or (b), respectively. The basic idea of this control algorithm as follows. If the structural model will move to windward, the active fin is closed and becomes to catch wind forces, or if the structural model will move to leeward, the active fin is opened and become not to catch wind forces. By such a method, the wind forces work as if those are control forces to stop the movement of the structural model. Whether the direction of the movement is windward or leeward is judged from the absolute velocity of the floor equipped the active fin device at each discrete time step. The time interval for each step is assigned as about 0.1 second. In this value, the time interval required for the control action of the active fin to change its mode is included as about 0.05 second. Let denote the absolute displacement of the *k*-th floor at the *i*-th time step be $X_{k,i}$ (cm). The approximate absolute velocity $\Delta X_{k,i}$ (cm/sec) can be expressed as

$$\Delta X_{k,i} = (X_{k,i} - X_{k,i-1}) / \Delta t ,$$

in which the Δt means the time interval for each step. In this case, the positive sign is the direction to windward. It may be reasonable to stop the active control, when the vibrations of the structural model become small. Then, let denote the cut-off coefficient be α, which defines the minimum velocity ΔX_{min} (cm/sec) as the trigger to start or stop control.

$$\Delta X_{min} = \alpha \times \delta_{min} / \Delta t ,$$

where the δ_{min} means the minimum resolution of the A/D converters, in here, this value is 0.00244 cm. The control algorithm adopted here is as follows.

Control Algorithm (A)

If	$\Delta X_{3,i} > \Delta X_{min}$	***then,***	close mode.
If	$\Delta X_{3,i} < -\Delta X_{min}$	***then,***	open mode.
If	$\lvert\Delta X_{3,i}\rvert \leq \Delta X_{min}$	***then,***	keeping before mode.

In this aim, the approximate absolute velocity of the top floor is used in order to select the mode of the active fin. As shown in Fig.3, when the active fin is rotated, two patterns can be considered. One is that the upper side of two fins is turned to right when the configuration of the active fin is changing from the open mode to the close mode, in which the parts of [b] and [c] shown in Fig.3 (a) are closed to leeward (as seen in Fig. 3 (b)). And the other is the opposite for this pattern (as seen in Fig. 3 (c)). Those control patterns are named here as an upward pattern and as a downward pattern, respectively.

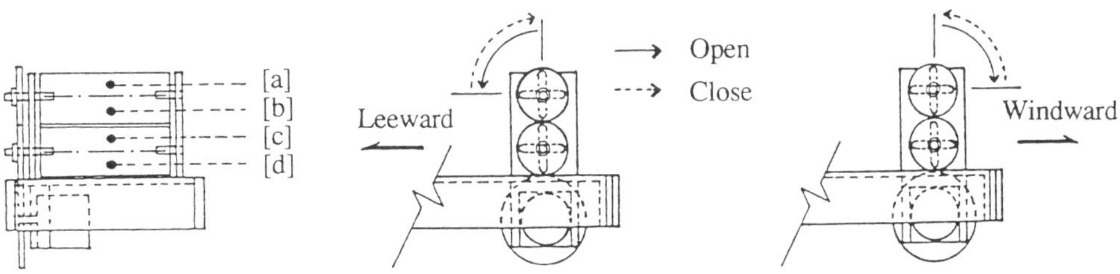

(a) Turning parts, (b) Upward pattern, (c) Downward pattern,

Fig. 3. Turning patterns of the active fin.

IV. DYNAMIC PROPERTIES OF STRUCTURAL MODEL

Free vibration tests are examined in order to know the dynamic characteristics of the structural model. Natural periods and damping ratio are shown in Table 1. Those values are estimated by Fast Fourier Transform analysis and by logarithmic decrement, respectively.

Table 1. Natural period and damping ratio of the structural model.

Natural period* (sec)		Damping ratio (%)
First	1.138	0.56
Second	0.379	
Third	0.253	

* Natural periods over the third order are omitted.

In this experiment, the average of the velocity of the wind is as 10.2 m/sec. By inserting a grid filter in the wind tunnel, laminar flow is changed to semi-turbulent flow in order to imitate practical wind flow. It may be consider that some aerodynamic effects are arisen by setting the control device on the structural model, whatever if it is controlled or not [12]. So, to make sure, two cases are examined. One is for the case of the close mode shown as Fig.1 (a), and the other is for the case of the open mode shown as Fig.1 (b). The displacements of the top floor corresponding to the close mode and the open mode are shown in Fig.4 (a) and (b), respectively. But significant difference can not be recognized in those figures. In the following discussions, both are referred as non controlled response.

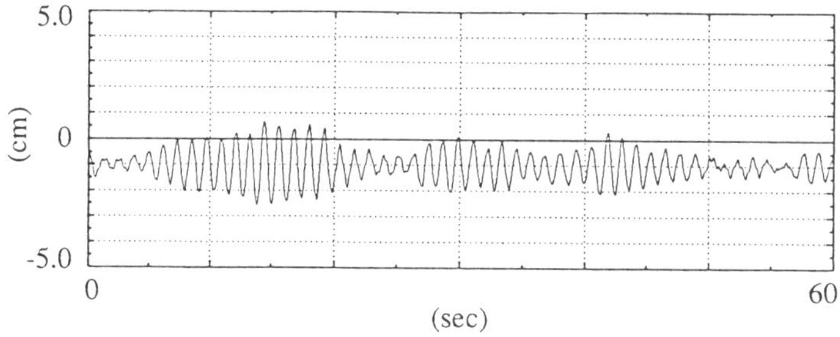

(a) Case of the close mode,

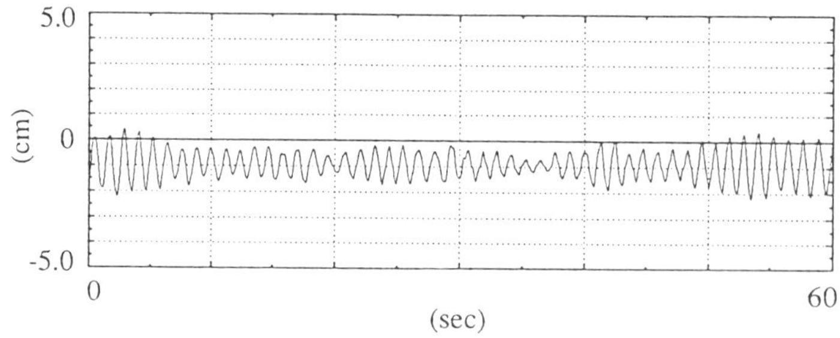

(b) Case of the open mode,

Fig. 4. Displacements of the top floor for the cases of non control.

V. EXPERIMENTAL RESULTS AND CONSIDERATIONS

At first, introducing to the control algorithm (A), the control effects of the active fin system are investigated. In order to test the influence of the turning direction of the active fin, the upward pattern and the downward pattern are examined. In those cases, the cut-off coefficient α is fixed on 3. The displacements of the top floor for the upward pattern and the downward pattern are shown in Fig.5 (a) and (b), respectively. But significant difference can not be recognized in those figures. So, in the following discussions, both are referred as controlled responses. By comparing controlled responses (as seen in Fig.5) with non controlled responses (as seen in Fig.4), good reductions of vibrations can be observed.

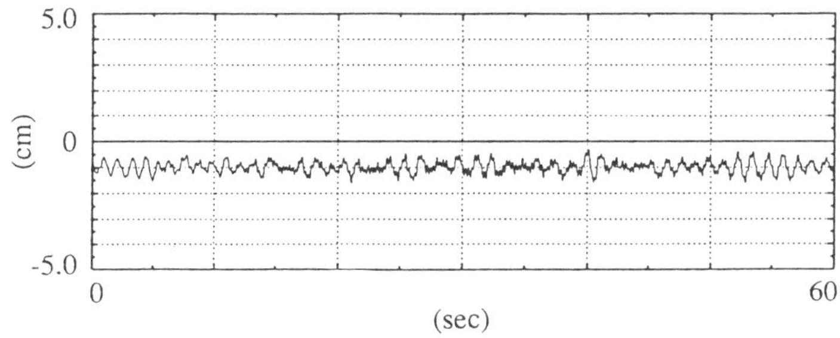

(a) Case of the upward pattern,

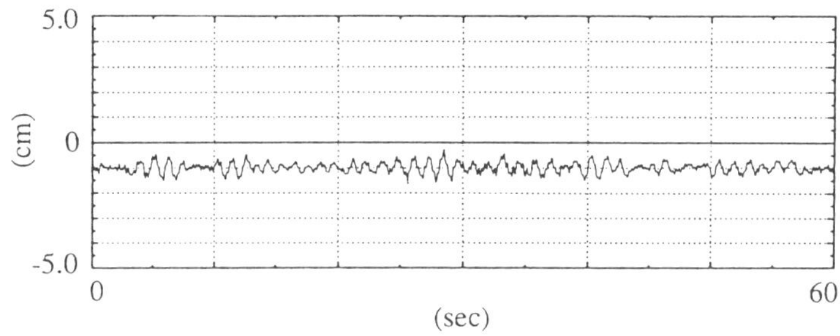

(b) Case of the downward pattern,

Fig. 5. Displacements of the top floor for the cases of control.

The spectrum diagram of the top floor for the case of non control (the open mode) and the case of control (the upward pattern) is shown in Fig.6. And the running spectrum diagrams for each case are shown in Fig.7 (a) and (b), respectively. In those figures, the fundamental frequency (near 0.8 Hz) is extremely dominant at the case of non control. By using the control algorithm (A), the frequency responses near 0.8 Hz are reduced to about 20 %. But, the frequency responses near 4.0 Hz are increased by control. As the reason of this, it is considered that the control actions of the active fin affect to high frequency components. However, as considering the controlled responses of the displacements of the top floor as time-history and running spectrum (as seen in Fig.5 and Fig.7 (b), respectively), it is found that stable reductions of responses are observed. In those results, it is appeared that the effective control forces are given by the control movements of the active fin by using the control algorithm (A).

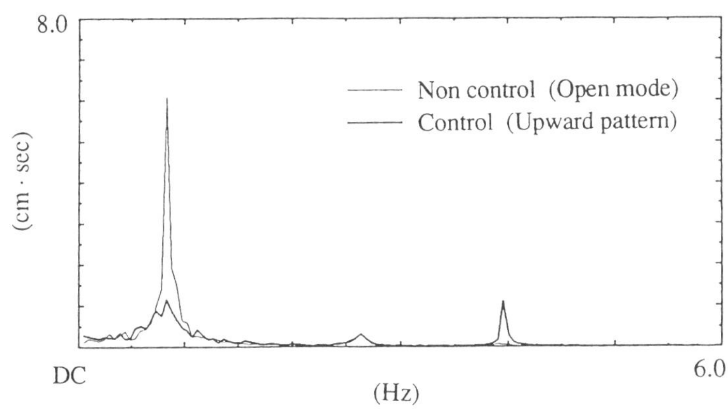

Fig. 6. Spectrum of the displacements of the top floor (Time duration : 90 sec)

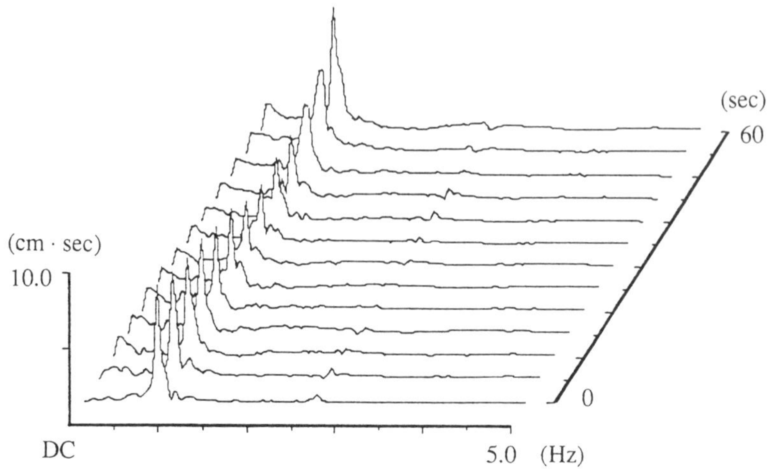

(a) Case of non control (the open mode),

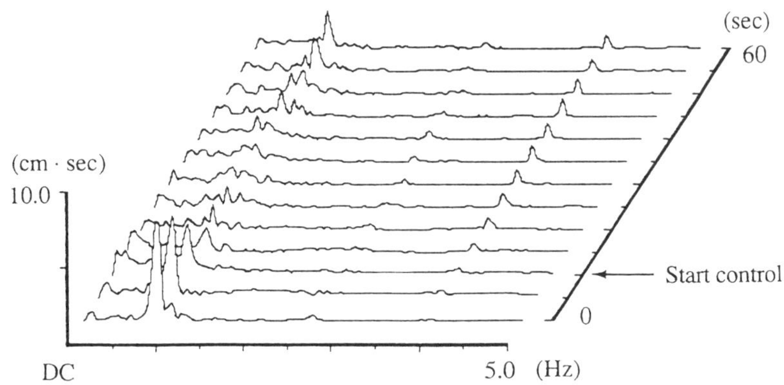

(b) Case of control (the upward pattern),

Fig. 7. Running spectrum of the displacements of the top floor.

In order to examine the influences of the electric noise, it is tested for the case of removing the low-pass filter. The displacement of the top floor for this case is shown in Fig.8. By comparing with Fig.5, it is clear that the control effects decrease on the case of removing the low-pass filter. It is found that this system is very sensitive for the electric noise.

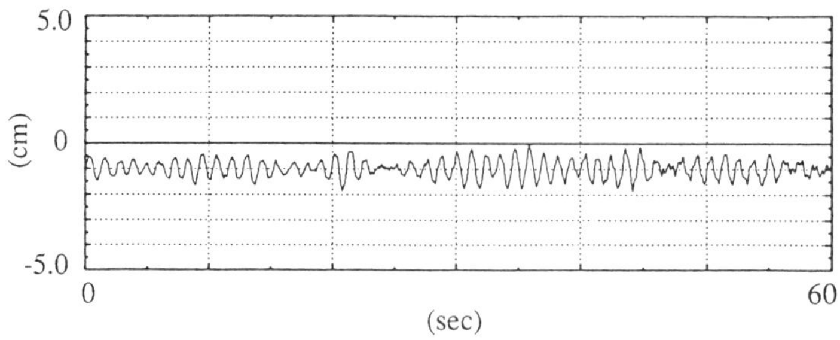

Fig. 8. Displacement of the top floor for the case of removing the low-pass filter.

Another test is examined to ensure the implicit ability of the active fin. That is 'It may be reasonable that the control algorithm can be used for amplifying vibrations'. So, following algorithm is used for this aim.

Control Algorithm (B)

If	$\Delta X_{3,i} > \Delta X_{min}$	***then,***	open mode.
If	$\Delta X_{3,i} < -\Delta X_{min}$	***then,***	close mode.
If	$\lvert \Delta X_{3,i} \rvert \le \Delta X_{min}$	***then,***	keeping before mode.

This control algorithm can be easily gained by replacing the word 'open' with 'close' in the control algorithm (A).

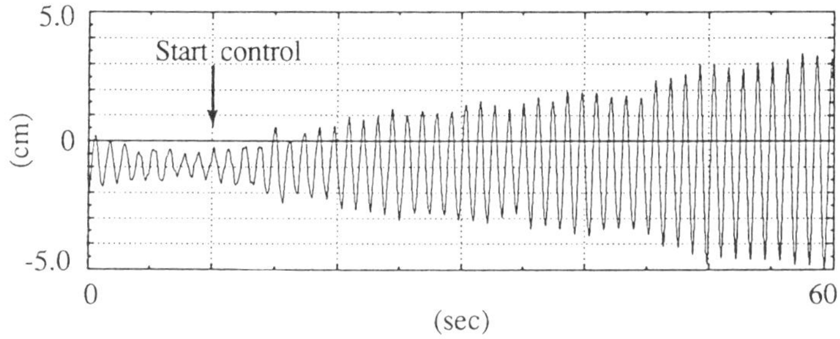

Fig. 9. Displacement of the top floor for the case of using the control algorithm (B).

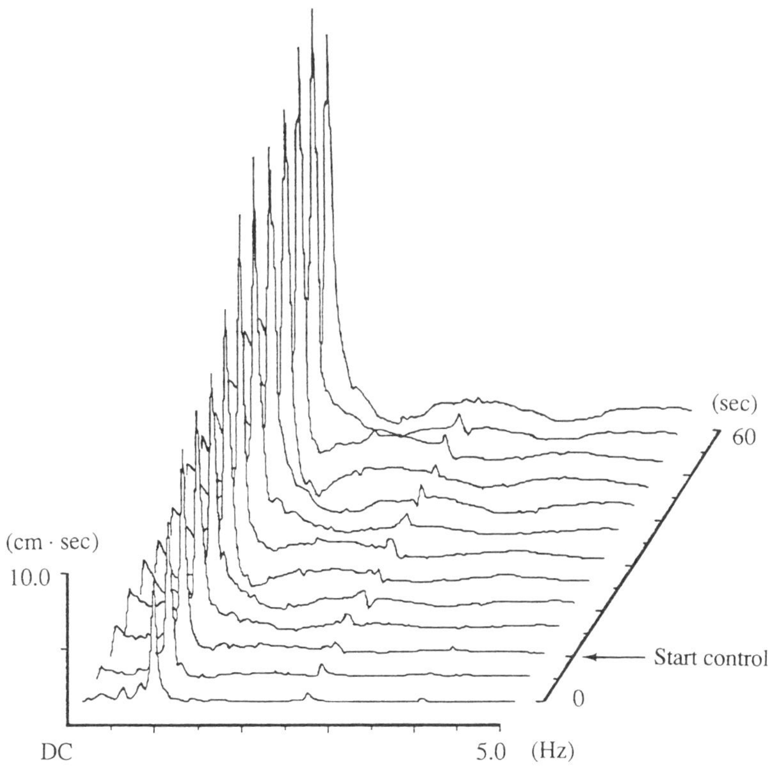

Fig. 10. Running spectrum of the displacement of the top floor for the case of using the control algorithm (B).

The displacement of the top floor, and the running spectrum diagram for the case of using the control algorithm (B) are shown in Fig.9 and Fig.10, respectively. In those figures, expected amplifying effects can be gained. As seen in Fig.10, fundamental frequency responses (near 0.8 Hz) are amplified radically, but high frequency components are under little influence of control.

In order to examine the influence of the cut-off coefficient α, four cases are tested (α=0, 1, 3 or 5). The displacements of the top floor for those cases are shown in Fig.11. By comparing those figures with the case of α=3 (as seen in Fig.5), it is found that the control effects are decreasing whenever if the coefficient α is smaller or bigger than this value. The reasons of this are explained as follows. At the cases of α=0 and 1, the control procedures are often affected by the electric noise. And at the case of α=5, the timing of changing the mode of the active fin are delayed from the real turning points of the sign of velocity of the structural model. So that, it is necessary to select an adequate cut-off coefficient α, in order to operate good performances of control.

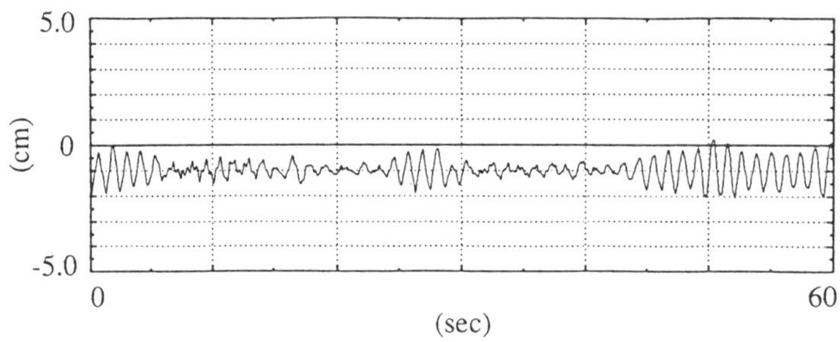

(a) Case of $\alpha = 0$,

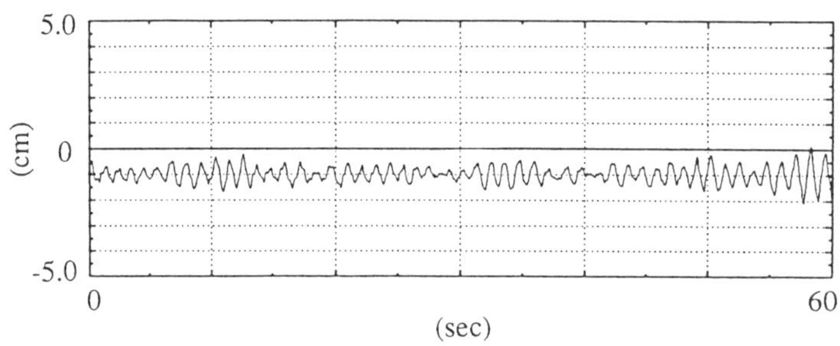

(b) Case of $\alpha = 1$,

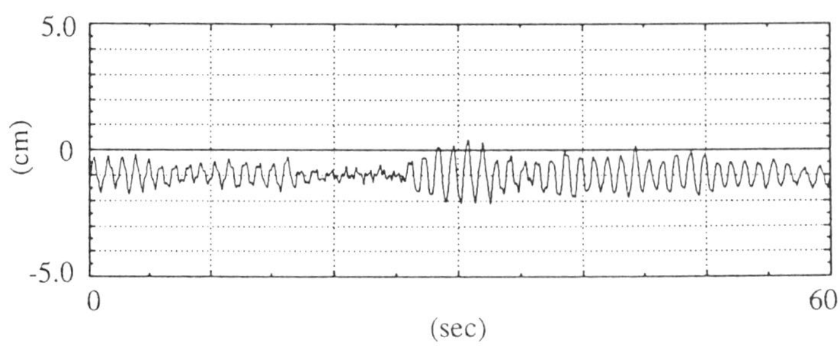

(c) Case of $\alpha = 5$,

Fig. 11. Displacements of the top floor for the cases of changing the cut-off coefficient.

Those experimental results are indicated in Table 2 as the values of the average (the center around vibrations), the root-mean-square and the maximum amplitude of the displacements of the top floor. On this table, Case-1 [1] and [2] mean the cases of the close mode and the open mode of non control, respectively. Case-2 [1] and [2] mean the cases of the upward pattern and the downward pattern of control, respectively. Case-3 means the case of removing the low-pass filter, and Case-4 means the case of using the control algorithm (B). Case-5 [1], [2] and [3] mean the cases of α=0, α=1 and α=5, respectively.

Table 2. Experimental results in wind tunnel.

Absolute Displacements of the top floor (cm)									
	Non control		Control						
	Case-1 [1]	Case-1 [2]	Case-2 [1]	Case-2 [2]	Case-3	Case-4	Case-5 [1]	Case-5 [2]	Case-5 [3]
Average	-1.025	-0.919	-0.989	-0.986	-0.993	-0.889	-0.995	-1.007	-0.983
RMS	0.463	0.616	0.254	0.239	0.343	2.106	0.403	0.314	0.453
Max. amplitude	3.224	3.392	1.857	1.495	1.783	8.448	2.273	2.239	2.539

VI. CONCLUDING REMARKS

In this paper, the active fin system is proposed as a new type of active control system. As concluding remarks, it is appeared that the active fin system is very effective for reduction of the wind-induced structural vibrations. Moreover, the algorithm used in this system is very simple. Since the required mechanical power for the active fin may be considered as small, it can be said that the active fin system is a very practical system.

ACKNOWLEDGEMENTS

The authors gratefully thank to Messrs. Kazutaka Igarashi and Jun Kawakami for having cooperated in this experimental research, and also thank to Messrs. Hiroaki Yokoyama and Yushi Yamada for their helpful comments.

REFERENCES

1) McNamara, R. J. : Tuned Mass Dampers for Buildings, Journal of the Structural Division, ASCE, Vol. 103, No. ST9, Feb., 1977, pp. 1785-1798.
2) Fujino, Y., Pacheco, B., Piyawat, C. and Fujii, K. : An Experimental Study on Tuned Liquid Damper Using Circular Containers, Journal of Structural Engineering, Vol. 34A, Mar., 1988, pp. 603-616. (in Japanese).
3) Chang, J. C. H. and Soong, T. T. : Structural Control Using Active Tuned Mass Dampers, Journal of the Engineering Mechanics Division, ASCE, Vol. 106, No. EM6, Dec., 1980, pp. 1091-1098.
4) Samali, B., Yang, J. N. and Yeh, C. T. : Control of Lateral-Torsional Motion of Wind-Excited Buildings, Journal of Engineering Mechanics, ASCE, Vol. 111, No. 6, Jun., 1985, pp. 777-796.
5) Roorda, J. : Tendon Control in Tall Structures, Journal of the Structural Division, ASCE, Vol. 101, No. ST3, Mar., 1975, pp. 505-521.
6) Yang, J. N. and Giannopoulos, F. : Active Tendon Control of Structures, Journal of the Engineering Mechanics Division, ASCE, Vol. 104, No. EM3, Jun., 1978, pp. 551-568.
7) Sugimoto, H., Abe, I., Hayakawa, K. and Yasui, K. : A study on A Semi-Active Vibration Control System for Structural Control, Summaries of Technical Papers of Annual Meeting Architectural Institute of Japan 1989 - B Structures I, Oct., 1989, pp. 533-534. (in Japanese).
8) Mukai, Y., Tanaka, R., Yamada, Y., Baba, K., Tachibana, E. and Inoue, Y. : A study on developments of experimental apparatus for the active vibration control systems (Part 1. Developments

of experimental apparatus), Summaries of Technical Papers of Annual Meeting Architectural Institute of Japan 1991 - B Structures I, Sep., 1991, pp. 1137-1138. (in Japanese).

9) Yamada, Y., Mukai, Y., Tanaka, R., Baba, K., Tachibana, E. and Inoue, Y. : A study on developments of experimental apparatus for the active vibration control systems (Part 2. Active control tests with the apparatus by the control algorithms of external damping type), Summaries of Technical Papers of Annual Meeting Architectural Institute of Japan 1991 - B Structures I, Sep., 1991, pp. 1139-1140. (in Japanese).

10) Tachibana, E., Inoue, Y. and Yamada, Y. : Development of compact, experimentally testing system for active control algorithms of building vibration, Proceeding of 10th World Conference on Earthquake Engineering, Jul., 1992, pp. 2637-2642.

11) Tachibana, E., Mukai, Y., Yamada, Y. and Inoue, Y. : Development of Compact Experimental System for evaluation of active control algorithm, Transactions of the Japan National Symposium on Active Structural Response Control, Mar., 1992, pp. 159-166.

12) Kwok, K. C. S. and Bailey, P. A. : Aerodynamic Devices for Tall Buildings and Structures, Journal of Engineering Mechanics, ASCE, Vol. 113, No. 3, Mar., 1987, pp. 349-365.

V
FLUID DYNAMICS

Slip Flow of a Binary Gas Mixture over a Catalytic Wall

Kyoji YAMAMOTO*, Toshiyuki NISHITANI*, Takashi NAKAO*, and
Michio NISHIDA**

* *Department of Mechanical Engineering, Faculty of Engineering, Okayama University, Okayama*
** *Department of Aeronantics and Astronautics, Faculty of Engineering, Kyushu University, Fukuoka*

Flow of a slightly rarefied binary gas mixture (a molecule and its atom) over an infinite plane catalytic wall is investigated on the basis of a Krook type Boltzmann equation when atoms are recombined into molecules at the wall surface. The near-continuum equations together with their slip boundary conditions and the behaviors of the flow in the Knudsen layer are obtained up to the first order in the Knudsen number. The slip flow over the catalytic wall is analysed by using these near-continuum equations and boundary conditions.

I. INTRODUCTION

Behind the strong shock wave caused by the flight of the hypervelocity vehicle such as a space shuttle, AOTV, a space plane etc. through the atomosphere, the temperature is elevated so highly that molecules are dissociated to atoms. These atoms are then recombined to form molecules at the vehicle surface where the temperature is low. In the process of recombination, the atoms release some of dissociation energy and hence the reaction may affect the heat transfer to the surface. Especially, if the wall is catalytic for recombination, the reaction proceeds at a much faster rate and will raise the heat transfer very much. This process may appear even at high altitude and then we must take the effect of rarefaction of the gas into consideration.

Scott et al.[1∿3] studied the flow of multicomponent gas mixtures over the catalytic wall and derived the boundary conditions at the wall for the Navier-Stokes equation when the Knudsen number is small. They used the first order Chapmann-Enskog distribution function and assumed that the fluxes of mass, momentum and energy across the Knudsen layer are constant. Consequently,they did not analyse the Knudsen layer correctly.

In the present study, we consider a semi-infinite expanse of a binary gas mixture composed of atoms and diatomic molecules over a plane catalytic wall. We analyse both the Knudsen layer and the near-continuum region correctly on the basis of the Krook type Boltzmann equation for gas mixtures proposed by Hamel[4] and obtain the slip boundary conditions at the catalytic wall. Using the near-continuum equations accompanied with the slip boundary conditions obtained, we analyse the slip flow over the catalytic wall.

II. FUNDAMENTAL EQUATIONS

We assume that there is neither dissociation of the molecule nor recombination of the atom in the gas phase. The recombination takes place only at the catalytic wall. We also assume that the molecules are regarded as possessing no internal energy. Taking the y axis normal to the wall, we have the fundamental equations[4] as

$$K_i v \frac{\partial \phi_i}{\partial y} = -(\sigma_i + k_i\sigma_j)\phi_i + \sigma_i{}^2\theta_i{}^{-3/2}\exp\{\mu_i \mathbf{v}^2 - \mu_i \bar{\mathbf{v}}_i{}^2/\theta_i\}$$

$$+ k_i\sigma_i\sigma_j\theta_{ij}{}^{-3/2}\exp\{\mu_i \mathbf{v}^2 - \mu_i \bar{\mathbf{v}}_{ij}{}^2/\theta_i\} \qquad (1)$$

$$\begin{pmatrix} \sigma_i \\ \sigma_i q_i \\ \sigma_i r_i \\ 3\sigma_i\theta_i/2 \end{pmatrix} = \left(\frac{\mu_i}{\pi}\right)^{3/2} \int \begin{pmatrix} 1 \\ u \\ v \\ \mu_i \bar{\mathbf{v}}_i{}^2 \end{pmatrix} \exp(-\mu_i \bar{\mathbf{v}}_i{}^2)\phi_i d\mathbf{v} \qquad (2)$$

$$p_i = \sigma_i\theta_i, \quad \mathbf{v}^2 = u^2 + v^2 + w^2, \quad \bar{\mathbf{v}}_i{}^2 = (u - q_i)^2 + (v - r_i)^2 + w^2 \qquad (3)$$

$$\bar{\mathbf{v}}_{ij}{}^2 = (u - q_{ij})^2 + (v - r_{ij})^2 + w^2$$

$$q_{ij} = q_i + \mu_j(q_j - q_i), \quad r_{ij} = r_i + \mu_j(r_j - r_i) \qquad (4)$$

$$\theta_{ij} = \theta_i + 2\mu_i\mu_j(\theta_j - \theta_i) + \frac{2}{3}\mu_i\mu_j{}^2\{(q_i - q_j)^2 + (r_i - r_j)^2\}$$

where Ly is the coordinate, $a_0 q_i$ and $a_0 r_i$ the flow velocities parallel and normal to the wall, respectively, $T_0\theta_i$ the temperature, $n_{0,i}\sigma_i$ the number density, $p_0 p_i$ the pressure, $f_{0,i}\phi_i$ the velocity distribution function of the molecule, $a_0\mathbf{v}$ = $a_0(u,v,w)$ the molecular velocity. Here, we take the reference length L, the reference temperature T_0, the reference number density $n_{0,i}$, the reference pressure $p_0 = n_{0,i}kT_0$, and the reference speed a_0 defined by

$$a_0 = (2kT_0/m_{12})^{1/2}, \quad m_{12} = m_1 + m_2 \qquad (5a)$$

where k is the Boltzmann constant, m_1 the mass of a molecule, m_2 $(=m_1/2)$ the mass of an atom and

$$\mu_i = m_i/m_{12} \qquad (5b)$$

The reference distribution function is defined by

$$f_{0,i} = n_{0,i}\left(\frac{m_i}{2\pi kT_0}\right)^{3/2}\exp(-\mu_i \mathbf{v}^2) \qquad (5c)$$

The subscripts (i,j) in Eqs.(1)∿(4) are taken as

i = 1 and j = 2 for molecules

i = 2 and j = 1 for atoms.

Further, we put

$$K_1 = \frac{a_0}{Ln_{0,1}\kappa_{11}}, \quad K_2 = \frac{a_0}{Ln_{0,2}\kappa_{22}} = \left(\frac{k_2}{k_1}\right)n_0 K_1$$

$$n_0 = n_{0,2}/n_{0,1}, \quad k_1 = (n_{0,2}\kappa_{12})/(n_{0,1}\kappa_{11}), \quad k_2 = (n_{0,1}\kappa_{12})/(n_{0,2}\kappa_{22}) \qquad (6)$$

Here, the constants K_i (i=1,2) correspond to the Knudsen number of the molecule and the atom, respectively, and κ_{ij} represent collisional parameters and are given for Maxwell molecules as

$$\kappa_{ii} = 2.906(p_{ii}m_i)^{1/2}, \quad \kappa_{ij} = 2.66(p_{ij}m_{12})^{1/2} \qquad (7)$$

where p_{ij} are constants related to the intermolecular force.[4]

On the catalytic wall, we assume that the molecule is reflected diffusely. A part of atoms coming to the wall is reflected diffusely, while the rest part is recombined to form molecules, which also leave the wall diffusely. Then, we have

$$\phi_i = \phi_{i,w} = \sigma_{i,w}\theta_w^{-3/2}\exp[\ \mu_i(\ 1 - 1/\theta_w\ )\mathbf{v}^2\ ]\ (\ v > 0,\ i=1,2\ ) \tag{8}$$

where $T_0\theta_w$ is the wall temperature, and $\sigma_{i,w}$ is the number density to be determined by the conservation equation of number density at the wall surface. We define the atom recombination coefficient γ as

$$\gamma = \{ \int_{v<0} v\ \exp(-\mu_2\mathbf{v}^2)\phi_2 d\mathbf{v} + \int_{v>0} v\ \exp(-\mu_2\mathbf{v}^2)\phi_{2,w} d\mathbf{v}\ \} \Big/ \int_{v<0} v\ \exp(-\mu_2\mathbf{v}^2)\phi_2 d\mathbf{v} \tag{9}$$

We get the following equation from the conservation of the number flux of the molecule at the surface:

$$\int_{v<0} v\ \exp(-\mu_1\mathbf{v}^2)\phi_1 d\mathbf{v} + \int_{v>0} v\ \exp(-\mu_1\mathbf{v}^2)\phi_{1,w} d\mathbf{v} + \frac{\gamma}{2} n_0 (\ \frac{\mu_2}{\mu_1}\ )^{3/2} \int_{v<0} v\ \exp(-\mu_2\mathbf{v}^2)\phi_2 d\mathbf{v} = 0 \tag{10}$$

In the present analysis, we consider the flow of small Knudsen number ($K_1 \ll 1$) and also the case of small recombination coefficient of the order of K_1. Then, we may put

$$\gamma = K_1\gamma_0 \tag{11}$$

where γ_0 is an arbitrary constant.

III. NEAR CONTINUUM REGION

The method of analysis is quite similar to that developed for a single gas.[5] In the near continuum region far from the wall compared with the mean free path, we apply the Hilbert expansion to solve equation (1), that is, we expand ϕ_i and the mean quantities $\sigma_i, \cdots$ in a power series of $K \equiv K_1$

$$\phi_i = \phi_{i,H} = \phi_{i,H}^{(0)} + K\phi_{i,H}^{(1)} + K^2\phi_{i,H}^{(2)} + \cdots \tag{12}$$

$$\sigma_i = \sigma_{i,H} = \sigma_{i,H}^{(0)} + K\sigma_{i,H}^{(1)} + K^2\sigma_{i,H}^{(2)} + \cdots \tag{12a}$$
$$\cdots\ \cdot$$

Substituting these expansions into Eq.(1) and equating like powers of K, we have $\phi_{i,H}^{(n)}$ in terms of the mean quantities $\sigma_{i,H}^{(n)}$, $\cdots$ and their derivatives. Putting the distribution function obtained thus together with the expansion (12a) into the Maxwell transport equation, we obtain the near-continuum equations satisfied by $\sigma_{i,H}^{(n)}$, $q_{i,H}^{(n)}$, $\cdots$. The equations are given by (up to the first order)

$$r_{i,H}^{(0)} = 0 \tag{13}$$

$$q_{1,H}^{(n)} = q_{2,H}^{(n)} \qquad (\ n = 0,\ 1\ ) \tag{14}$$

$$\theta_{1,H}^{(n)} = \theta_{2,H}^{(n)} \qquad (n = 0, 1)$$

$$\sigma_{i,H}^{(0)} r_{i,H}^{(1)} = C_i^{(1)} \tag{15}$$

$$\frac{\partial}{\partial y}(\sigma_{1,H}^{(0)}\theta_{1,H}^{(0)}) = 2k_1\mu_1\mu_2(\sigma_{1,H}^{(0)}\sigma_{2,H}^{(0)})(r_{2,H}^{(1)} - r_{1,H}^{(1)}) \tag{16}$$

$$n_0\frac{\partial}{\partial y}(\sigma_{2,H}^{(0)}\theta_{2,H}^{(0)}) = 2k_1\mu_1\mu_2(\sigma_{1,H}^{(0)}\sigma_{2,H}^{(0)})(r_{1,H}^{(1)} - r_{2,H}^{(1)}) \tag{17}$$

$$\left[\frac{\sigma_{1,H}^{(0)}\theta_{1,H}^{(0)}}{\sigma_{1,H}^{(0)} + k_1\sigma_{2,H}^{(0)}} + \frac{\sigma_{2,H}^{(0)}\theta_{1,H}^{(0)}}{\sigma_{2,H}^{(0)} + k_2\sigma_{1,H}^{(0)}}\left(\frac{k_2}{k_1}n_0^2\right)\right]\frac{\partial q_{1,H}^{(0)}}{\partial y} = B^{(0)} \tag{18}$$

$$\begin{aligned}\frac{5}{2}(\sigma_{1,H}^{(0)}r_{1,H}^{(1)} + n_0\sigma_{2,H}^{(0)}r_{2,H}^{(1)})\theta_{1,H}^{(0)} \\ -\left[\frac{\sigma_{1,H}^{(0)}\theta_{1,H}^{(0)}}{\mu_1(\sigma_{1,H}^{(0)} + k_1\sigma_{2,H}^{(0)})} + \frac{\sigma_{2,H}^{(0)}\theta_{1,H}^{(0)}}{\mu_2(\sigma_{2,H}^{(0)} + k_2\sigma_{1,H}^{(0)})}\left(\frac{k_2}{k_1}\right)n_0^2\right] \\ \times \frac{\partial}{\partial y}\left[\frac{5}{4}\theta_{1,H}^{(0)} \right] - B^{(0)}q_{1,H}^{(0)} = D^{(0)}\end{aligned} \tag{19}$$

$$\sigma_{i,H}^{(0)}r_{i,H}^{(2)} + \sigma_{i,H}^{(1)}r_{i,H}^{(1)} = C_i^{(2)} \tag{20}$$

$$\begin{aligned}\frac{\partial}{\partial y}\left\{\sigma_{1,H}^{(0)}\theta_{1,H}^{(0)}\left(\frac{\sigma_{1,H}^{(1)}}{\sigma_{1,H}^{(0)}} + \frac{\theta_{1,H}^{(1)}}{\theta_{1,H}^{(0)}}\right)\right\} =2k_1\mu_1\mu_2\times \\ \times\{\sigma_{1,H}^{(0)}\sigma_{2,H}^{(0)}(r_{2,H}^{(2)}-r_{1,H}^{(2)}) \\ +(\sigma_{1,H}^{(0)}\sigma_{2,H}^{(1)}+\sigma_{1,H}^{(1)}\sigma_{2,H}^{(0)})(r_{2,H}^{(1)}- r_{1,H}^{(1)})\}\end{aligned} \tag{21}$$

$$\begin{aligned}n_0\frac{\partial}{\partial y}\left\{\sigma_{2,H}^{(0)}\theta_{2,H}^{(0)}\left(\frac{\sigma_{2,H}^{(1)}}{\sigma_{2,H}^{(0)}} + \frac{\theta_{2,H}^{(1)}}{\theta_{2,H}^{(0)}}\right)\right\} =2k_1\mu_1\mu_2\times \\ \times\{\sigma_{1,H}^{(0)}\sigma_{2,H}^{(0)}(r_{1,H}^{(2)}-r_{2,H}^{(2)}) \\ +(\sigma_{1,H}^{(0)}\sigma_{2,H}^{(1)}+\sigma_{1,H}^{(1)}\sigma_{2,H}^{(0)})(r_{1,H}^{(1)}- r_{2,H}^{(1)})\}\end{aligned} \tag{22}$$

$$\begin{aligned}\left[\frac{\sigma_{1,H}^{(0)}\theta_{1,H}^{(0)}}{\sigma_{1,H}^{(0)}+k_1\sigma_{2,H}^{(0)}} + \frac{k_2}{k_1} n_0^2 \frac{\sigma_{2,H}^{(0)}\theta_{1,H}^{(0)}}{\sigma_{2,H}^{(0)}+k_2\sigma_{1,H}^{(0)}}\right] \frac{\partial q_{1,H}^{(1)}}{\partial y} \\ + \left[\frac{\sigma_{1,H}^{(0)}\theta_{1,H}^{(0)}}{\sigma_{1,H}^{(0)}+k_1\sigma_{2,H}^{(0)}}\left(\frac{\sigma_{1,H}^{(1)}}{\sigma_{1,H}^{(0)}} + \frac{\theta_{1,H}^{(1)}}{\theta_{1,H}^{(0)}} - \frac{\sigma_{1,H}^{(1)}+k_1\sigma_{2,H}^{(1)}}{\sigma_{1,H}^{(0)}+k_1\sigma_{2,H}^{(0)}} \right)\right.\end{aligned}$$

$$+\frac{k_2}{k_1}n_0^2\frac{\sigma_{2,H}^{(0)}\theta_{1,H}^{(0)}}{\sigma_{2,H}^{(0)}+k_2\sigma_{1,H}^{(0)}}\left(\frac{\sigma_{2,H}^{(1)}}{\sigma_{2,H}^{(0)}}+\frac{\theta_{1,H}^{(1)}}{\theta_{1,H}^{(0)}}-\frac{\sigma_{2,H}^{(1)}+k_2\sigma_{1,H}^{(1)}}{\sigma_{2,H}^{(0)}+k_2\sigma_{1,H}^{(0)}}\right)]$$

$$\times\left(\frac{\partial q_{1,H}^{(0)}}{\partial y}\right)=B^{(1)} \qquad (23)$$

$$\frac{5}{2}\{(\sigma_{1,H}^{(0)}r_{1,H}^{(1)}+n_0\sigma_{2,H}^{(0)}r_{2,H}^{(1)})\theta_{1,H}^{(1)}+[\sigma_{1,H}^{(0)}r_{1,H}^{(2)}$$

$$+\sigma_{1,H}^{(1)}r_{1,H}^{(1)}+n_0(\sigma_{2,H}^{(1)}r_{2,H}^{(2)}+\sigma_{2,H}^{(1)}r_{2,H}^{(1)})]\theta_{1,H}^{(0)}\}$$

$$-[\frac{\sigma_{1,H}^{(0)}\theta_{1,H}^{(0)}}{\mu_1(\sigma_{1,H}^{(0)}+k_1\sigma_{2,H}^{(0)})}+\frac{\sigma_{2,H}^{(0)}\theta_{1,H}^{(0)}}{\mu_2(\sigma_{2,H}^{(0)}+k_2\sigma_{1,H}^{(0)})}\left(\frac{k_2}{k_1}\right)n_0^2]\times\frac{\partial}{\partial y}[\frac{5}{4}\theta_{1,H}^{(1)}]$$

$$-[\frac{\sigma_{1,H}^{(0)}\theta_{1,H}^{(0)}}{\sigma_{1,H}^{(0)}+k_1\sigma_{2,H}^{(0)}}\left(\frac{\sigma_{1,H}^{(1)}}{\sigma_{1,H}^{(0)}}+\frac{\theta_{1,H}^{(1)}}{\theta_{1,H}^{(0)}}-\frac{\sigma_{1,H}^{(1)}+k_1\sigma_{2,H}^{(1)}}{\sigma_{1,H}^{(0)}+k_1\sigma_{2,H}^{(0)}}\right)\frac{1}{\mu_1}$$

$$+\frac{k_2}{k_{11}}n_0^2\frac{\sigma_{2,H}^{(0)}\theta_{1,H}^{(0)}}{\sigma_{2,H}^{(0)}+k_2\sigma_{1,H}^{(0)}}\left(\frac{\sigma_{2,H}^{(1)}}{\sigma_{2,H}^{(0)}}+\frac{\theta_{1,H}^{(1)}}{\theta_{1,H}^{(0)}}-\frac{\sigma_{2,H}^{(1)}+k_2\sigma_{1,H}^{(1)}}{\sigma_{2,H}^{(0)}+k_2\sigma_{1,H}^{(0)}}\right)\frac{1}{\mu_2}]$$

$$\times\left(\frac{\partial\theta_{1,H}^{(0)}}{\partial y}\right)-B^{(0)}q_{1,H}^{(1)}-B^{(1)}q_{1,H}^{(0)}=D^{(1)} \qquad (24)$$

Here $B^{(k)}$, $C_i^{(k)}$, and $D^{(k)}$ (i=1 or 2; k=0, 1 or 2) are unknown constants. Equation (13) is verified by the analysis of the Knudsen layer to follow.

IV. KNUDSEN LAYER

We next analyse the Knudsen layer adjacent to the wall, whose thickness is of the order of the mean-free path. We introduce a stretched coordinate η defined by

$$y = K\eta \qquad (25)$$

and look for a solution of Eq.(1) in the form:

$$\phi_i = \phi_{i,H}(\mathbf{v},y) + K\phi_{i,K}(\mathbf{v},\eta)$$
$$\sigma_i = \sigma_{i,H}(\mathbf{v},y) + K\sigma_{i,K}(\mathbf{v},\eta) \qquad (26)$$
$$\cdots\cdots$$

The correction terms $\sigma_{i,K}(\eta)$, ••• in the Knudsen layer should vanish as $\eta \to \infty$. The functions $\phi_{i,K}$ and the mean quantities $\sigma_{i,K}$ ••• are also expanded as

$$K\phi_{i,K} = K\phi_{i,K}^{(1)} + K^2\phi_{i,K}^{(2)} + \cdots$$
$$K\sigma_{i,K} = K\sigma_{i,K}^{(1)} + K^2\sigma_{i,K}^{(2)} + \cdots \qquad (27)$$
$$\cdots\cdots$$

We put these expansions (26) and (27) into Eqs.(1) ∿ (4),(9) and (10) and rewrite them in new variable η. We expand the Hilbert solution in power series of K using the Taylor series. Then, we get the system of equation for $\phi_{i,K}^{(n)}$, $\sigma_{i,K}^{(n)}$, $q_{i,K}^{(n)}$, $r_{i,K}^{(n)}$ and $\theta_{i,K}^{(n)}$. From the zeroth order terms in K, we get

$$\theta_{i,H}^{(0)}(0) = \theta_{2,H}^{(0)}(0) = \theta_w$$

$$\sigma_{i,H}^{(0)}(0) = \sigma_{i,w}^{(0)}, \quad q_{i,H}^{(0)}(0) = 0 \quad (i = 1, 2) \tag{28}$$

These constitute the no-slip boundary conditions for the continuum equations.

The analysis of the first order solution is much more complicated, but we here give only the result for brevity. The normal velocity to the wall is given by

$$r_{i,K}^{(1)}(\eta) = 0 \tag{29}$$

and also we obtain the following relations from the integral equations for $\Sigma_{iA}^{(1)}$ and $\Sigma_{iB}^{(1)}$ shown below :

$$r_{1,H}^{(1)}(0) = \frac{1}{4\sqrt{\pi}}\left(\frac{\theta_w}{\mu_2}\right)^{1/2}\frac{\sigma_{2,H}^{(0)}(0)}{\sigma_{1,H}^{(0)}(0)}\gamma_0 n_0$$

$$r_{2,H}^{(1)}(0) = -\frac{1}{2\sqrt{\pi}}\left(\frac{\theta_w}{\mu_2}\right)^{1/2}\gamma_0 \tag{30}$$

Hence, we have

$$\mu_1\sigma_{1,H}^{(0)}r_{1,H}^{(1)} + n_0\mu_2\sigma_{2,H}^{(0)}r_{2,H}^{(1)} = 0 \tag{31}$$

We put

$$\omega_{1,H}^{(0)}(0) = \sigma_{1,H}^{(0)}(0) + k_1\sigma_{2,H}^{(0)}(0)$$

$$\omega_{2,H}^{(0)}(0) = \frac{k_1}{k_2 n_0}\left(\sigma_{2,H}^{(0)}(0) + k_2\sigma_{1,H}^{(0)}(0) \right) \tag{32}$$

$$\omega_{1,H}^{(0)}(0)\left(\frac{\mu_1}{\theta_w}\right)^{1/2}\eta = z, \qquad \omega_{2,H}^{(0)}(0)\left(\frac{\mu_2}{\theta_w}\right)^{1/2}\eta = \alpha z$$

$$\alpha = \left(\frac{\mu_2}{\mu_1}\right)^{1/2}\frac{\omega_{2,H}^{(0)}(0)}{\omega_{1,H}^{(0)}(0)} \tag{33}$$

$$\begin{pmatrix} q_{1,H}^{(1)}(0) \\ \\ q_{i,K}^{(1)}(z) \end{pmatrix} = \left(\frac{\theta_w}{\mu_1}\right)^{1/2} \frac{1}{\omega_{1,H}^{(0)}(0)}\left(\frac{\partial q_{1,H}^{(0)}}{\partial y}\right)_0 \begin{pmatrix} \kappa^{(1)} \\ \\ Y_i^{(1)}(z) \end{pmatrix} \tag{34}$$

$$\begin{pmatrix} \theta_{1,H}^{(1)}(0)/\theta_w \\ \sigma_{i,K}^{(1)}(z)/\sigma_{i,H}^{(0)}(0) \\ \theta_{i,K}^{(1)}(z)/\theta_w \end{pmatrix} = \gamma_0 \begin{pmatrix} \alpha_A^{(1)} \\ \Sigma_{iA}^{(1)}(z) \\ \Theta_{iA}^{(1)}(z) \end{pmatrix} + \frac{1}{\omega_{1,H}^{(0)}(0)\theta_w}\left(\frac{\theta_w}{\mu_1}\right)^{1/2}\left(\frac{\partial \theta_{1,H}^{(0)}}{\partial y}\right)_0 \begin{pmatrix} \alpha_B^{(1)} \\ \\ \Sigma_{iB}^{(1)}(z) \\ \\ \Theta_{iB}^{(1)}(z) \end{pmatrix} \tag{35}$$

Then, we find that ($\alpha_A^{(1)}$, $\Sigma_{iA}^{(1)}$, $\Theta_{iA}^{(1)}$) satisfy the following integral equations:

$$\sqrt{\pi}\Sigma_{1A}^{(1)}(z) = \frac{n_0}{2\sqrt{\pi}}\left(\frac{\mu_1}{\mu_2}\right)^{1/2}\frac{\sigma_{2,H}^{(0)}(0)}{\sigma_{1,H}^{(0)}(0)}\left[\frac{\sqrt{\pi}}{2}J_0(z) - J_1(z)\right]$$
$$- \alpha_A^{(1)}\left[J_2(z) - \frac{1}{2}J_0(z)\right]$$
$$+ 2J_0(z)I[\Sigma_{1A}^{(1)}, \Theta_{1A}^{(1)}, \Theta_{2A}^{(1)}, \sigma_{2,H}^{(0)}(0)/\omega_{1,H}^{(0)}(0), 1]$$
$$+ L_1[\Sigma_{1A}^{(1)}, \Theta_{1A}^{(1)}, \Theta_{2A}^{(1)}, \sigma_{2,H}^{(0)}(0)/\omega_{1,H}^{(0)}(0), 1] \qquad (36)$$

$$\frac{3\sqrt{\pi}}{2}\Theta_{1A}^{(1)}(z) = \frac{n_0}{2\sqrt{\pi}}\left(\frac{\mu_1}{\mu_2}\right)^{1/2}\frac{\sigma_{2,H}^{(0)}(0)}{\sigma_{1,H}^{(0)}(0)}\left[\frac{\sqrt{\pi}}{2}\left(J_2(z) - \frac{1}{2}J_0(z)\right)\right.$$
$$\left.- \frac{1}{2}\left(2J_3(z) - J_1(z)\right)\right] - \alpha_A^{(1)}\left[J_4(z) - \frac{3}{2}J_2(z) + \frac{3}{2}J_0(z)\right]$$
$$+ 2\left[J_2(z) - \frac{1}{2}J_0(z)\right]I\left[\Sigma_{1A}^{(1)}, \Theta_{1A}^{(1)}, \Theta_{2A}^{(1)}, \frac{\sigma_{2,H}^{(0)}(0)}{\omega_{1,H}^{(0)}(0)}, 1\right]$$
$$+ L_2[\Sigma_{1A}^{(1)}, \Theta_{1A}^{(1)}, \Theta_{2A}^{(1)}, \sigma_{2,H}^{(0)}(0)/\omega_{1,H}^{(0)}(0), 1] \qquad (37)$$

$$\sqrt{\pi}\Sigma_{2A}^{(1)}(z) = -\frac{1}{\sqrt{\pi}}\left[\frac{\sqrt{\pi}}{2}J_0(\alpha z) - J_1(\alpha z)\right] - \alpha_A^{(1)}\left[J_2(\alpha z) - J_0(\alpha z)\right]$$
$$+ 2\alpha J_0(\alpha z)I[\Sigma_{2A}^{(1)}, \Theta_{2A}^{(1)}, \Theta_{1A}^{(1)}, \sigma_{1,H}^{(0)}(0)/(n_0\omega_{2,H}^{(0)}(0)), \alpha]$$
$$+ \alpha L_1[\Sigma_{2A}^{(1)}, \Theta_{2A}^{(1)}, \Theta_{1A}^{(1)}, \sigma_{1,H}^{(0)}(0)/(n_0\omega_{2,H}^{(0)}(0)), \alpha] \qquad (38)$$

$$\frac{3\sqrt{\pi}}{2}\Theta_{2A}^{(1)}(z) = \frac{1}{\sqrt{\pi}}\left[-\frac{\sqrt{\pi}}{2}\left(J_2(\alpha z) - \frac{1}{2}J_0(\alpha z)\right) + J_3(\alpha z) - \frac{1}{2}J_1(\alpha z)\right]$$
$$- \alpha_A^{(1)}\left[J_4(\alpha z) - \frac{3}{2}J_2(\alpha z) + \frac{3}{2}J_0(\alpha z)\right]$$
$$+ 2\alpha\left[J_2(\alpha z) - \frac{1}{2}J_0(\alpha z)\right]I[\Sigma_{2A}^{(1)}, \Theta_{2A}^{(1)}, \Theta_{1A}^{(1)},$$
$$\sigma_{1,H}^{(0)}(0)/(n_0\omega_{2,H}^{(0)}(0)), \alpha]$$
$$+ \alpha L_2[\Sigma_{2A}^{(1)}, \Theta_{2A}^{(1)}, \Theta_{1A}^{(1)}, \sigma_{1,H}^{(0)}(0)/(n_0\omega_{2,H}^{(0)}(0)), \alpha] \qquad (39)$$

where

$$I[\Sigma_1, \Theta_1, \Theta_2, \sigma, \alpha] = \int_0^\infty \{ \Sigma_1(z_0)J_0(\alpha z_0) + \Theta_1(z)[J_2(\alpha z_0) - \frac{1}{2}J_0(\alpha z_0)]$$

$$+ 2k_1\mu_1\mu_2\sigma[\Theta_2(z_0) - \Theta_1(z_0)][J_2(\alpha z_0) - \frac{1}{2}J_0(\alpha z_0)]\}dz_0 \quad (40)$$

$$L_1[\Sigma_1, \Theta_1, \Theta_2, \sigma, \alpha]$$
$$= \int_0^\infty \{ \Sigma_1(z_0)J_{-1}(\alpha|z-z_0|) + \Theta_1(z_0)[J_1(\alpha|z-z_0|) - \frac{1}{2}J_{-1}(\alpha|z-z_0|)]$$

$$+ 2k_1\mu_1\mu_2\sigma[\Theta_2(z_0) - \Theta_1(z_0)][J_1(\alpha|z-z_0|) - \frac{1}{2}J_{-1}(\alpha|z-z_0|)]\}dz_0 \quad (41)$$

$$L_2[\Sigma_1, \Theta_1, \Theta_2, \sigma, \alpha]$$
$$= \int_0^\infty \{ \Sigma_1(z_0)[J_1(\alpha|z-z_0|) - \frac{1}{2}J_{-1}(\alpha|z-z_0|)]$$
$$+ \Theta_1(z_0)[J_3(\alpha|z-z_0|) - J_1(\alpha|z-z_0|) + \frac{5}{4}J_{-1}(\alpha|z-z_0|)]$$
$$+ 2k_1\mu_1\mu_2\sigma[\Theta_2(z_0) - \Theta_1(z_0)]$$
$$\times [J_3(\alpha|z-z_0|) - J_1(\alpha|z-z_0|) + \frac{5}{4}J_{-1}(\alpha|z-z_0|)]\}dz_0 \quad (42)$$

$$J_n(z) = \int_0^\infty \xi^n \exp(- \xi^2 - z/\xi)d\xi \quad (43)$$

The equations for ($\kappa^{(1)}$, $Y_i^{(1)}$) and ($\alpha_B^{(1)}$, $\Sigma_{iB}^{(1)}$, $\Theta_{iB}^{(1)}$) are essentially the same as those studied by Onishi[6] for the flow of slightly rarefied gas mixture over plane boundaries.

The analysis of the second order in K is very complicated, but we can get the following relations at the surface from the integral equations for the number density :

$$\sigma_{1,H}^{(0)}(0)r_{1,H}^{(2)}(0) + \sigma_{1,H}^{(1)}(0)r_{1,H}^{(1)}(0)$$
$$= \frac{\gamma_0 n_0}{\sqrt{\pi}}\left(\frac{\theta_w}{\mu_2}\right)^{1/2} \sigma_{2,H}^{(0)}(0)\left\{\frac{\gamma_0}{4} + \alpha S + \frac{\theta_{1,H}^{(1)}(0)}{4\theta_w} + \frac{\sigma_{2,H}^{(1)}(0)}{2\sigma_{2,H}^{(0)}(0)}\right\} \quad (44)$$

$$\sigma_{2,H}^{(0)}(0)r_{2,H}^{(2)}(0) + \sigma_{2,H}^{(1)}(0)r_{2,H}^{(1)}(0)$$
$$= - \frac{\gamma_0}{\sqrt{\pi}}\left(\frac{\theta_w}{\mu_2}\right)^{1/2} \sigma_{2,H}^{(0)}(0)\left\{\frac{\gamma_0}{2} + 2\alpha S + \frac{\theta_{1,H}^{(1)}(0)}{2\theta_w} + \frac{\sigma_{2,H}^{(1)}(0)}{\sigma_{2,H}^{(0)}(0)}\right\} \quad (45)$$

$$S=\int_0^\infty \frac{\sigma_{2,K}^{(1)}(z)}{\sigma_{2,H}^{(0)}(0)}J_0(\alpha z)dz+\int_0^\infty \frac{\theta_{2,K}^{(1)}(z)}{\theta_w}[J_2(\alpha z)-\frac{1}{2}J_0(\alpha z)]dz$$

$$+\frac{\mu_1\mu_2 k_1}{\omega_{1,H}^{(0)}(0)}\sigma_{1,H}^{(0)}(0)\int_0^\infty [\frac{\theta_{2,K}^{(1)}(z)}{\theta_w}-\frac{\theta_{1,K}^{(1)}(z)}{\theta_w}]\times$$

$$\times[2J_2(\alpha z)-J_0(\alpha z)]dz \qquad (46)$$

Hence, we have

$$\mu_1[\sigma_{1,H}^{(0)}r_{1,H}^{(2)}+\sigma_{1,H}^{(1)}r_{1,H}^{(1)}]$$

$$+\mu_2 n_0[\sigma_{2,H}^{(0)}r_{2,H}^{(2)}+\sigma_{2,H}^{(1)}r_{2,H}^{(1)}]=0 \qquad (47)$$

V. NUMERICAL CALCULATION OF THE INTEGRAL EQUATIONS

The simultaneous integral equations (36) $\sim$ (39) were solved numerically by applying the Galerkin method.[7] We assumed $p_{ii}=p_{ij}$ in the collisional parameter. Then, only one parameter, i.e., the number density ratio at the wall surface

$$n_r=\frac{n_{0,2}\sigma_{2,H}^{(0)}(0)}{n_{0,1}\sigma_{1,H}^{(0)}(0)} \qquad (48)$$

can be given for the numerical calculation. It may be noted that the convergence of the Galerkin method is very good. Figure 1 shows the variations of the slip coefficients $\kappa^{(1)}$, $\alpha_A^{(1)}$ and $\alpha_B^{(1)}$ with n_r. These coefficients are also given in Table I. The distributions of the correction of the velocity, number density and temperature in the Knudsen layer for several values of n_r from 0.1 to 10 are shown in Fig.2. For reference, we also show in Fig.1 and Table I the slip coefficients in which $\kappa_{ii}=\kappa_{ij}$ is taken .

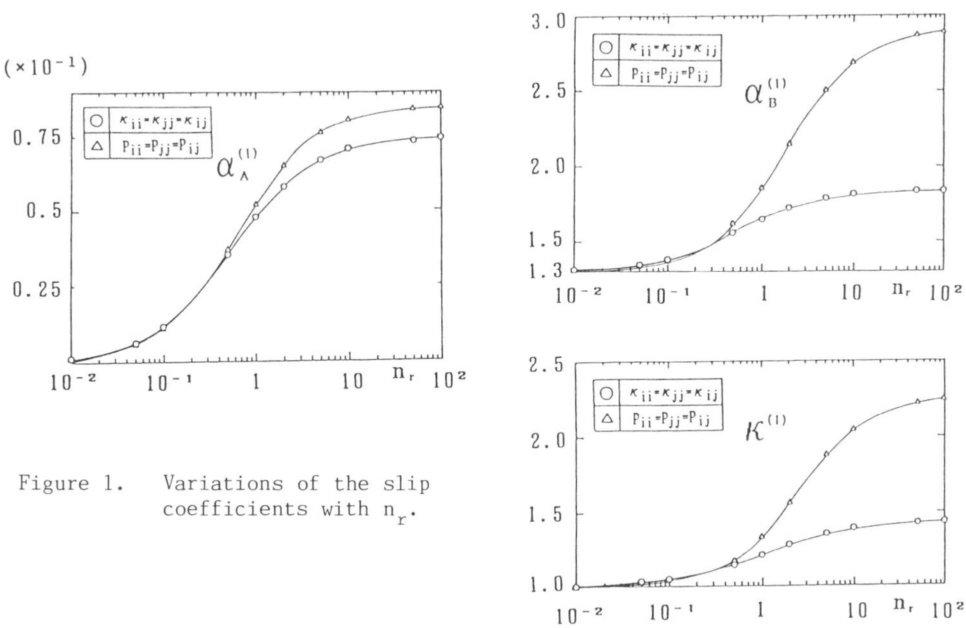

Figure 1. Variations of the slip coefficients with n_r.

Table I. Values of slip coefficients $\alpha_A^{(1)}$, $\alpha_B^{(1)}$, $\kappa^{(1)}$.

	case (1) $p_{ii}=p_{jj}=p_{ij}$			case (2) $\kappa_{ii}=\kappa_{jj}=\kappa_{ij}$		
n_r	$\alpha_A^{(1)}$	$\alpha_B^{(1)}$	$\kappa^{(1)}$	$\alpha_A^{(1)}$	$\alpha_B^{(1)}$	$\kappa^{(1)}$
0.01	1.166×10^{-3}	1.308	1.018	1.321×10^{-3}	1.313	1.019
0.05	5.582×10^{-3}	1.332	1.028	6.165×10^{-3}	1.347	1.034
0.1	1.058×10^{-2}	1.363	1.041	1.137×10^{-2}	1.384	1.050
0.5	3.662×10^{-2}	1.613	1.175	3.523×10^{-2}	1.557	1.145
1.0	5.200×10^{-2}	1.854	1.337	4.718×10^{-2}	1.648	1.213
2.0	6.513×10^{-2}	2.149	1.569	5.823×10^{-2}	1.723	1.285
5.0	7.612×10^{-2}	2.500	1.879	6.705×10^{-2}	1.787	1.359
10.0	8.048×10^{-2}	2.681	2.048	7.062×10^{-2}	1.813	1.394
50.0	8.424×10^{-2}	2.865	2.225	7.382×10^{-2}	1.836	1.427
100.0	8.473×10^{-2}	2.890	2.249	7.429×10^{-2}	1.839	1.432

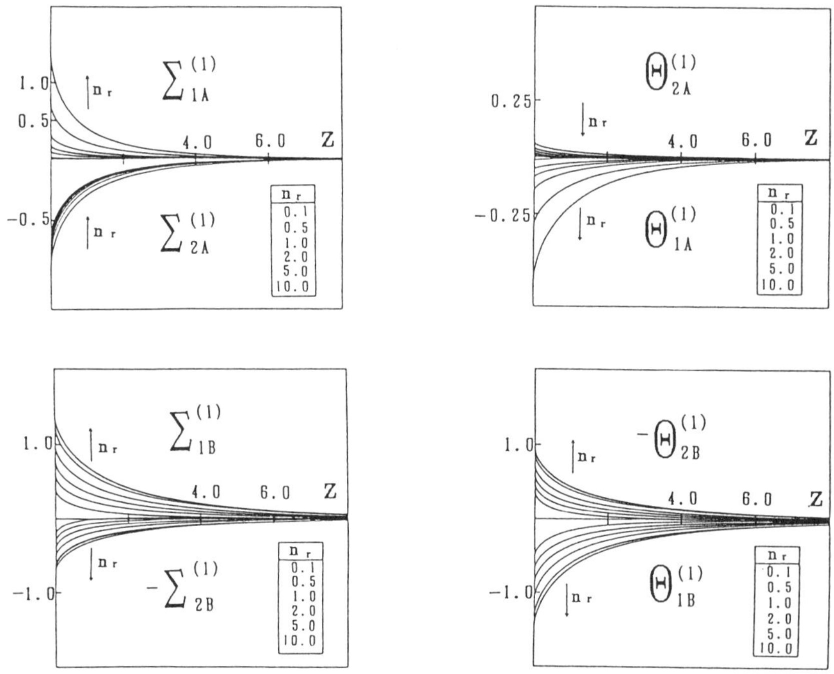

Figure 2. Knudsen layer corrections.

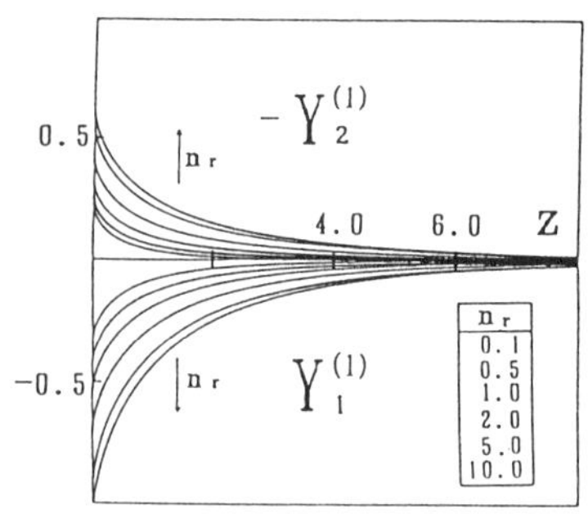

VI. SLIP FLOW OVER THE CATALYTIC WALL

Applying the results obtained in the preceeding sections, we shall investigate the flow over the catalytic plane wall when the Knudsen number is small. We take the reference point at y=1 and fix the flow quantities there irrespective of the Knudsen number. Further, we shall take $n_0 = 1$ and $\gamma_0 = 1$. We assume that there is no temperature difference between the wall and the reference point, and that the flow speed is a_0 at the reference point. Namely, the boundary conditions at y=1 are given by

$$\theta_{i,H}^{(0)} = 1,\ \sigma_{i,H}^{(0)} = 1 \text{ and } q_{i,H}^{(0)} = 1 \qquad (i=1,2) \tag{49}$$

$$\theta_{i,H}^{(1)} = 0,\ \sigma_{i,H}^{(1)} = 0 \text{ and } q_{i,H}^{(1)} = 0 \tag{50}$$

The near continuum equations of the zeroth order are given by Eqs.(15)∿(19) and (31). The boundary conditions at the wall surface for these equations are given by Eqs.(28) and (30) with θ_w=1. After getting the zeroth order continuum solution, we can solve the integral equations for the correction terms in the Knudsen layer, i.e. the equations for $Y_i^{(1)}$, $(\Sigma_{iA}^{(1)}, \Theta_{iA}^{(1)})$, $(\Sigma_{iB}^{(1)}, \Theta_{iB}^{(1)})$. Then, we obtain the boundary conditions for the first order continuum equations, which are given by

$$q_{1,H}^{(1)}(0) = \kappa^{(1)}\left(\frac{\theta_w}{\mu_1}\right)^{1/2} \frac{1}{\omega_{1,H}^{(0)}(0)} \left(\frac{\partial q_{1,H}^{(0)}}{\partial y}\right)_0 \tag{51}$$

$$\theta_{1,H}^{(1)}(0)/\theta_w = \gamma_0 \alpha_A^{(1)} + \frac{\alpha_B^{(1)}}{\omega_{1,H}^{(0)}(0)\theta_w} \left(\frac{\theta_w}{\mu_1}\right)^{1/2} \left(\frac{\partial \theta_{1,H}^{(0)}}{\partial y}\right)_0 \tag{52}$$

together with Eqs.(44)∿(46). The first order continuum equations are given by Eqs.(20)∿(24) and (47). The near-continuum equations were solved numerically using a finite difference method. The numerical results are shown in Figs.3 and 4. The dotted lines mean the near-continuum solution without correction of the Knudsen layer. The rarefaction effect will be seen clearly in these figures.

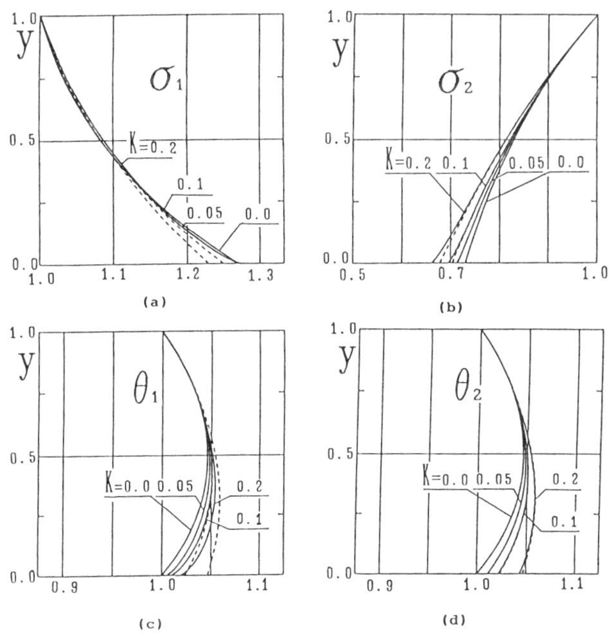

Figure 3. Flow behaviors.

(a) Molecular density.

(b) Atom density.

(c) Molecular temperature.

(d) Atom temperature.

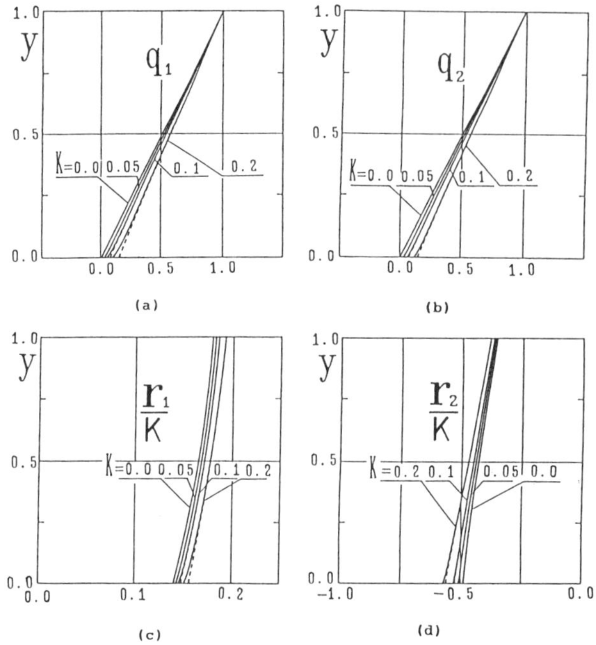

Figure 4. Flow behaviors.

(a) Molecular flow velocity along the wall.

(b) Atom flow velocity along the wall.

(c) Molecular flow velocity normal to the wall.

(d) Atom flow velocity normal to the wall.

VII. RESULTS AND DISCUSSION

We have analyzed the slip flow of a reactive binary gas mixture over a catalytic plane wall and obtained the slip boundary conditions. The slip boundary condition for the tangential velocity $q_{1,H}^{(1)}$(Eq.(51)) is essentially the same as the ordinary slip condition for the single gas. The temperature jump condition (Eq.(52)) also has the same form as that of the single gas except for the term of γ_0. The slip coefficients included in these conditions are depicted in Fig.1. The slip coefficients which are calculated under the condition of equal collisional parameters ($\kappa_{ii}=\kappa_{jj}=\kappa_{ij}$) are different from those which are calculated under equal coefficients of the intermolecular force ($p_{ii}=p_{jj}=p_{ij}$). The difference is considerably large for large n_r. The temperature jump coefficient $\alpha_A^{(1)}$ is directly related with the recombination of the atom and this value is seen to be small.

In the present analysis, we assumed that the recombination coefficient γ is as small as the order of K. This coefficient is difficult to measure, but it is reported that this value is actually small.[3)]

REFERENCES

1) Scott, C. D., Wall boundary equations with slip and catalysis for multi-component, nonequilibrium gas flows, NASA TM X-58111 (1973).
2) Gupta, R. N., Scott, C. D. and Moss, J. N., Slip-boundary equations for multi-component nonequilibrium airflow, NASA TP- 2452 (1985).
3) Scott,C.D.,Effects of thermochemistry, nonequilibrium, and surface catalysis on the design of hypersonic vehicles, Hypersonics, Vol.1,Birkhauser, Boston,(1989).
4) Hamel, B. B., Kinetic model for binary gas mixtures, Phys, of Fluids, **8** (1965), pp.418-425.
5) Sone, Y. and Yamamoto, K., Flow of rarefied gas over plane wall, J. Phys. Soc. Jpn., **29** (1970), pp.495-508.
6) Onish, Y., On the behavior of a slightly rarefied gas mixture over plane boundaries,J.Appl.Maths. and Phys.(ZAMP),**37** (1986), pp.573-596.
7) Yamamoto, K. Galerkin method to an integral equation in the kinetic theory, Mem. School of Engng., Okayama Univ., **19-2** (1985),pp.49-61.

Numerical Computations to Interfaces in Porous Medium Flow

Tatsuyuki NAKAKI* and Kenji TOMOEDA**

* *Fukuoka University of Education, Munakata, Fukuoka*
** *Osaka Institute of Technology, Asahi-ku, Osaka*

In this paper, we consider some nonlinear reaction-diffusion equations describing finite extinction phenomena. For a solution v, there appear *interfaces* between $v > 0$ and $v = 0$. To trace the interfaces, we introduce a finite difference scheme and show the stability and convergence of the numerical interfaces as well as numerical solutions. We also demonstrate a numerical simulation which suggests splitting of the support of v.

I. INTRODUCTION

In this paper, we are concerned with numerical computations for an isentropic flow of an ideal gas through the one-dimensional homogeneous porous medium with volumetric absorption. Such a flow is described by the following equation:

$$v_t = (v^m)_{xx} - cv^p, \quad x \in \mathbf{R}^1, \quad t > 0 \tag{1}$$

with an initial condition

$$v(0, x) = v^0(x), \qquad x \in \mathbf{R}^1, \tag{2}$$

where $m(> 1)$, $p(> 0)$ and $c(\geq 0)$ are constants, and $v^0(x)$ is a given nonnegative function. Physically, v, v^{m-1} and $(v^{m-1})_x$ represent the density of the gas, the pressure, and the velocity, respectively.

The most interesting property of the solution of (1)-(2) is a finite propagation of the initial support. Let supp v^0 be compact in $\mathbf{R}^1$. Then supp $v(t, \cdot)$ also becomes compact in $\mathbf{R}^1$ for all $t > 0$, and there are three cases in the behavior of supp $v(t, \cdot)$:

Case 1. Positivity. For $c = 0$ and $m > 1$, or $c > 0$ and $p \geq m > 1$ supp $v(t, \cdot)$ expands as t increases and $\lim_{t\to\infty}$ supp $v(t, \cdot) = \mathbf{R}^1$ [1),2),3),7),8),14)].

Case 2. Localization. For $c > 0$ and $m > p \geq 1$ supp $v(t, \cdot)$ expands as t increases and is uniformly bounded with respect to t [6),8),10),11)].

Case 3. Total Extinction. For $c > 0$, $m > 1$ and $0 < p < 1$ supp $v(\cdot, \cdot)$ is compact in $[0, \infty) \times \mathbf{R}^1$ and $v(t, x)$ extincts in finite time [8),9),10)].

Some particular solutions, which demonstrate each property stated above, are known and described explicitly [2),3),6),9),14)]. But for an arbitrary initial value, the explicit solution is not obtained in general. This motivates us to develop numerical methods which realize the behavior of supp $v(t, \cdot)$, that is, *interfaces* appearing between $v > 0$ and $v = 0$.

Along such a motivation, several difference schemes to (1)-(2) in Cases 1-2 have been constructed [4),5),12)]. In Case 3 Rosenau and Kamin[15)] tried numerical computations. Their numerical evidence suggests the pulse splitting into several sub-pulses. But they do not theoretically discuss this property. In Case 3 with $m+p = 2$ Nakaki[13)] constructed the difference scheme, and proved that not only numerical solutions but also numerical interfaces converge to exact ones under

Condition A. $(v^0(x))^{m-1}$ is concave downward on its support.

From numerical points of view, it is necessary to improve his result so that the convergence of numerical interfaces also holds for the functions $(v^0(x))^{m-1}$ which are not concave. In the following, our aim is to establish the convergence of numerical interfaces without Condition A for Case 3 with $m+p=2$. However, instead of Condition A we have to impose the following condition on numerical results.

Condition B. There exist positive numbers M, $\tilde{T}$ and $\tilde{h}$ such that

$$\dot{\ell}_h(t) < M \quad \text{and} \quad \dot{r}_h(t) > -M \quad \left(\dot{} = \frac{d}{dt}\right) \tag{3}$$

for almost all $t \in [0,\tilde{T}]$ and for all $h \in (0,\tilde{h})$. Here $\ell_h(t)$ and $r_h(t)$ denote the left and right numerical interfaces, respectively, and h is a space mesh width.

Finally, in Theorem 3 we state a sufficient condition under which Condition B is satisfied.

II. DIFFERENCE SCHEMES

We set $u = v^{m-1}$ and rewrite (1)–(2) as

$$u_t = muu_{xx} + a(u_x)^2 - c', \tag{4}$$

$$u(0,x) = u^0(x) \equiv (v^0(x))^{m-1}, \tag{5}$$

where $a = m/(m-1)$ and $c' = (m-1)c$. Our difference scheme approximates the problem (4)–(5) instead of (1)–(2), and is constructed based on splitting the equation (4) into three parts:

$$u_t = Du \equiv -c', \tag{6}$$

$$u_t = Hu \equiv a(u_x)^2, \tag{7}$$

$$u_t = Pu \equiv muu_{xx}. \tag{8}$$

Then our difference scheme is described as follows:

Find the sequence $\{u_h^n\}_{n=0,1,2,\cdots} \subset V_h$ such that

$$u_h^{n+1} = S_{h,k}u_h^n \equiv (P_{h,k/\mu})^\mu \cdot (\prod_{j=1}^{\nu} H_{h,\tau_j}) \cdot (D_{h,k})u_h^n \quad \text{for } n = 0,1,2,\cdots. \tag{9}$$

Here $k \equiv k_{n+1} \equiv t_{n+1} - t_n$ is a variable time step, $\mu \equiv \mu_{n+1}$ and $\nu \equiv \nu_{n+1}$ are integers depending on k_{n+1}, $\tau_j (j = 1,2,\cdots,\nu)$ are positive numbers satisfying $\sum_{j=1}^{\nu} \tau_j = k$ and V_h is the set of the nonnegative continuous functions $u_h = u_h(x)$ with the following properties:

(i) u_h has compact support $[\ell(u_h), r(u_h)]$;

(ii) u_h is linear on each interval $[x_i, x_{i+1}]$ $(i \in \mathbf{Z})$,

where

$$x_i = \begin{cases} ih & \text{for } i \in \mathbf{Z} \setminus \{L-1, R+1\}, \\ \ell(u_h) & \text{for } i = L-1, \\ r(u_h) & \text{for } i = R+1, \end{cases} \tag{10}$$

$$L = \min\{i \in \mathbf{Z} | ih > \ell(u_h)\}, \qquad R = \max\{i \in \mathbf{Z} | ih < r(u_h)\}. \tag{11}$$

Now we construct difference operators $D_{h,k}$, $H_{h,\tau}$ and $P_{h,k/\mu}$ mapping from V_h to V_h, which are approximating (6), (7) and (8), respectively. Let $h_i = x_{i+1} - x_i$ and $u_i = u_h(x_i)$.

Difference operator $D_{h,k}$

$$\ell(D_{h,k}u_h) = \sup\{\xi | u_h - c'k \le 0 \text{ on } (-\infty,\xi]\}, \tag{12}$$

$$r(D_{h,k}u_h) = \inf\{\xi | u_h - c'k \le 0 \text{ on } [\xi,\infty)\}, \tag{13}$$

$$(D_{h,k}u_h)(x_i) = \{[u_h - c'k]^+\}(x_i) \quad \text{for } i \in \mathbf{Z}, \tag{14}$$

where $[f]^+ = \max\{f, 0\}$.

Difference operator $H_{h,\tau}$

$$\ell(H_{h,\tau}u_h) = \ell(u_h) - a\delta u_{L-1}\tau, \tag{15}$$
$$r(H_{h,\tau}u_h) = r(u_h) - a\delta u_R\tau, \tag{16}$$
$$(H_{h,\tau}u_h)(x_i) = \text{solution } u(\tau, x_i) \text{ of } u_t = Hu \text{ with } u(0,x) = u_h(x) \tag{17}$$

where L and R are defined by (11).
Stability Condition: τ is a number satisfying the following inequality:

$$a\|(u_h)_x\|_\infty\tau \leq \min\{h/4,\ Lh - \ell(u_h),\ r(u_h) - Rh\}. \tag{18}$$

Difference operator $P_{h,k/\mu}$

$$\ell(P_{h,k/\mu}u_h) = \ell(u_h), \qquad r(P_{h,k/\mu}u_h) = r(u_h), \tag{19}$$
$$(P_{h,k/\mu}u_h)(x_i) = u_i + \frac{k}{\mu}mu_i\delta^2 u_i \quad \text{for all } i \in \mathbf{Z}, \tag{20}$$
$$\delta^2 u_i = 2(\delta u_i - \delta u_{i-1})/(h_i + h_{i-1}),$$

Stability Condition: μ satisfies the following inequalities:

$$m\|u_h\|_\infty \frac{k}{\mu}\left[\frac{1}{h^2} + \frac{2}{h(h+h_i)}\right] \leq 1 \quad (i = L-1, R), \tag{21}$$
$$4m\|(u_h)_x\|_\infty \frac{k}{\mu}\frac{1}{h+h_i} \leq 1 \quad (i = L-1, R). \tag{22}$$

To start the scheme (9), we take $t_0 = 0$ and

$$\ell_0 = \ell(u_h^0), \quad r_0 = r(u_h^0), \quad u_h^0(x_i) = u^0(x_i), \tag{23}$$

$$k_{n+1} = \min\Big\{\theta h, \max\{u_L^n, u_{L+1}^n\}, \ \max\{u_R^n, u_{R-1}^n\}\Big\}/c', \tag{24}$$

where θ is an arbitrary positive constant. When $D_{h,k}u_h^n(x) = 0$ for $x \in \mathbf{R}^1$, we denote the numerical extinction time by $T_h^* = t_n + k_{n+1}$, and stop the numerical computation.

III. STABILITY AND CONVERGENCE

We introduce the following condition.
Condition C. $u^0 \in C^0(\mathbf{R}^1) \cap BV(\mathbf{R}^1)$ is a nonnegative function with compact support $\mathbf{I} = [\ell(u^0), r(u^0)]$ and $u_x^0 \in BV(\mathbf{R}^1)$ is absolutely continuous on $\mathbf{I}$. Moreover, for some constant $C_1 > 0$

$$\text{ess.inf}_{\mathbf{I}} u_{xx}^0(x) \geq -C_1. \tag{25}$$

Theorem 1 (Stability) *Assume Condition C. Then*

$$0 \leq u_h^n \leq \|u^0\|_\infty \quad \textit{for all} \quad n \geq 0, \tag{26}$$

where $\|\cdot\|_\infty = \|\cdot\|_{L^\infty(\mathbf{R}^1)}$. *Moreover, the following estimates hold for all* $n \geq 0$:

$$\|(u_h^n)_x\|_\infty \leq \|(u_h^0)_x\|_\infty, \qquad \|(u_h^n)_x\|_{L^1(\mathbf{R}^1)} \leq \|(u_h^0)_x\|_{L^1(\mathbf{R}^1)}, \tag{27}$$
$$V((u_h^n)_x) \leq V((u_h^0)_x), \tag{28}$$
$$\ell_0 - a\|u_x^0\|_\infty t_n \leq \ell_n < r_n \leq r_0 + a\|u_x^0\|_\infty t_n, \tag{29}$$
$$\left\|\frac{u_h^{n+1} - u_h^n}{k_{n+1}}\right\|_{L^1(\mathbf{R}^1)} \leq m\|u^0\|_\infty V(u_x^0) + a\|u_x^0\|_\infty\|u_x^0\|_{L^1(\mathbf{R}^1)} + c'(r_0 - \ell_0 + 2a\|u_x^0\|_\infty t_n), \tag{30}$$
$$\inf_i (\delta u_i^n - \delta u_{i-1}^n) \geq -C_1 h, \tag{31}$$

where $\ell_n = \ell(u_h^n)$, $r_n = r(u_h^n)$ *and* $V(f)$ *denotes the total variation of* f *on* $\mathbf{R}^1$.

We define a function $u_h(t,x)$ by

$$u_h(t,x) = u_h^n(x) \quad \text{on} \quad [t_n, t_{n+1}) \times \mathbf{R}^1, \tag{32}$$

and define the left (resp. right) numerical interface $\ell_h(t)$ (resp. $r_h(t)$) by piecewise-linearly interpolating (t_n, ℓ_n) (resp.(t_n, r_n)) $(n \geq 0)$.

Theorem 2 (Convergence) *Assume Conditions B and C. Then*

$$\|v_h - v\|_{L^\infty(H)} \to 0 \quad \text{as} \quad h \to 0, \tag{33}$$
$$|T_h^* - T^*| \to 0 \quad \text{as} \quad h \to 0, \tag{34}$$
$$\|\ell_h - \ell^*\|_{L^\infty([0,\tilde{T}))} \to 0 \quad \text{as} \quad h \to 0, \tag{35}$$
$$\|r_h - r^*\|_{L^\infty([0,\tilde{T}))} \to 0 \quad \text{as} \quad h \to 0, \tag{36}$$

where $H = [0,\infty) \times \mathbf{R}^1$, $v_h \equiv (u_h)^{1/(m-1)}$, $v \equiv v(t,x)$ *is the unique weak solution of (1)-(2),* T^* *is the extinction time of* v, *and* $\ell^*(t)$ *and* $r^*(t)$ *are the left and right interfaces of* v, *respectively.*

We state a sufficient condition under which Condition B holds.

Theorem 3 *Under Condition C, let* K *and* ε *be positive constants such that*

$$|u_x^0(x)| \geq K \tag{37}$$

holds for $x \in [\ell_0, \ell_0 + \varepsilon) \cup (r_0 - \varepsilon, r_0]$, *where* $\ell_0 = \ell(u^0)$ *and* $r_0 = r(u^0)$. *Then there exist some positive constants* $M, \tilde{T}$ *and* $\tilde{h}$ *satisfying Condition B.*

IV. NUMERICAL EXPERIMENTS

We carry on numerical computations with $m = 1.5$, $p = 0.5$, $h = 0.01$ and $\theta = 0.2$ for the following initial function.

$$v^0(x) = \begin{cases} 0.1, & \text{for} \quad |x| < 0.6, \\ 4.5(x - 0.6) + 0.1, & \text{for} \quad 0.6 \leq |x| < 0.8, \\ 1, & \text{for} \quad 0.8 \leq |x| < 0.9, \\ -10(x-1), & \text{for} \quad 0.9 \leq |x| < 1, \\ 0, & \text{for} \quad |x| \geq 1. \end{cases} \tag{38}$$

Figure 1 shows numerical solutions and interfaces with $c = 1$. We find that supp $v(t,\cdot)$ is connected for $0 \leq t < T_h^*$.

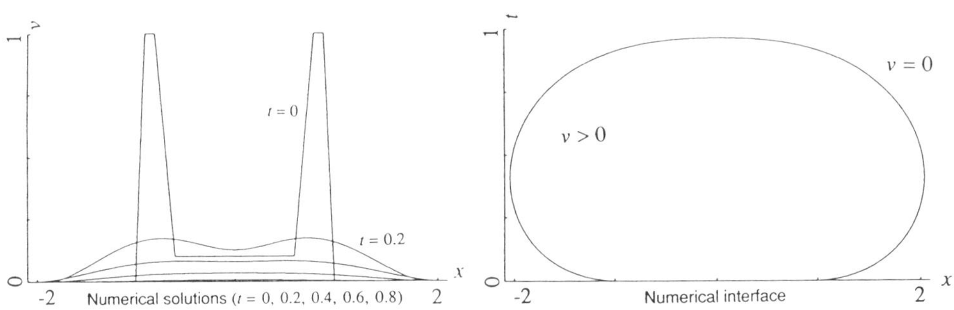

Fig.1. Numerical simulation with $c = 1$.

On the other hand, when $c = 10$, Figure 2 shows that a number of peaks of $v(t,\cdot)$ is invariant and supp $v(t,\cdot)$ is splitted.

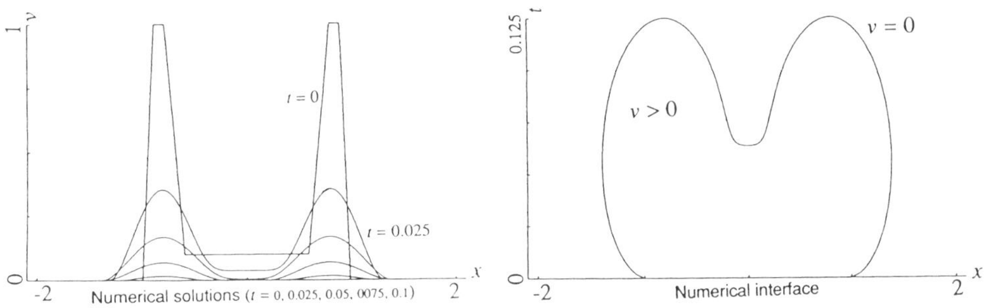

Fig.2. Numerical simulation with $c = 10$.

Next we compute the numerical extinction time $T_h^*$and

$$\tilde{T}_M = \sup\{T \mid \dot{\ell}_h(t) < M, \dot{r}_h(t) > -M \text{ on } [0,T)\}. \tag{39}$$

when $m = 1.5$, $p = 0.5$, $c = 1$, $\theta = 0.2$ and

$$v^0(x) = \left[-0.25(x^2 - 1)(13x^2 + 1)\right]^+ . \tag{40}$$

Table 1. The numerical extinction time T_h^* and $\tilde{T}_M$.

h	T_h^*	$\tilde{T}_{50}$	$\tilde{T}_{25}$
2^{-3}	1.2962	1.2826	1.2720
2^{-4}	1.2950	1.2893	1.2893
2^{-5}	1.2908	1.2886	1.2853
2^{-6}	1.2874	1.2861	1.2820
2^{-7}	1.2851	1.2838	1.2804
2^{-8}	1.2837	1.2825	1.2790
2^{-9}	1.2830	1.2818	1.2783
2^{-10}	1.2825	1.2813	1.2779

We may observe that $\tilde{T}_{50}$ gives a good approximation of T_h^* and that T_h^* converges to the exact extinction time as $h \to 0$ by Theorem 2.

REFERENCES

1) Aronson, D.G., Regularity properties of flows through porous media: The interfaces. Arch. Rat. Mech. Anal., **37** (1970), pp. 1–10.

2) Barenblatt, G.I., On some unsteady motions of a liquid and a gas in a porous medium. Prikl. Math. Meh., **16** (1952), pp. 67–78.

3) Bertsch, M., Kersner, R., and Peletier, L.A., Positivity versus localization in degenerate diffusion equations. Nonlinear Anal., **9** (1985), pp. 987–1008.

4) DiBenedetto, E., and Hoff, D., An interface tracking algorithm for the porous medium equation. Trans. Amer. Math. Soc., **284** (1984), pp. 463–500.

5) Graveleau, J.L., and Jamet, P., A finite difference approach to some degenerate nonlinear parabolic equations. SIAM J. Appl. Math., **20** (1971), pp. 199–223.

6) Gurtin, M.E., and MacCamy, R.C., On the diffusion of biological populations. Math. Biosci., **33** (1977), pp. 35–49.

7) Herrero, M.A., and Vázquez, J.L., The one-dimensional nonlinear heat equation with absorption: Regularity of solutions and interfaces. SIAM J. Math. Anal., **18** (1987), pp. 149–167.

8) Kalashnikov, A.S., The propagation of disturbances in problems of non-linear heat conduction with absorption. Zh. Vychisl. Mat.i Mat. Fiz., **14** (1974), pp. 891–905.

9) Kersner, R., The behavior of temperature fronts in media with nonlinear thermal conductivity under absorption. Vestnik. Mosk. Univers. Mat., **33** (1978), pp. 44–51.

10) Kersner, R., Nonlinear heat conduction with absorption: Space localization and extinction in finite time. SIAM J. Appl. Math., **43** (1983), pp. 1274–1285.

11) Knerr, B.F., The behaviour of the support of solutions of the equation of nonlinear heat conduction with absorption in one dimension. Trans. Amer. Math. Soc., **249** (1979), pp. 409–424.

12) Mimura, M., Nakaki, T., and Tomoeda, K., A numerical approach to interface curves for some nonlinear diffusion equations. Japan J. Appl. Math., **1** (1984), pp. 93–139.

13) Nakaki, T., Numerical interfaces in nonlinear diffusion equations with finite extinction phenomena. Hiroshima Math. J., **18** (1988), pp. 373–397.

14) Pattle, R.E., Diffusion from an instantaneous point source with a concentration-dependent coefficient. Quart, J. Mech. Appl. Math., **12** (1959), pp. 407–409.

15) Rosenau, P., and Kamin, S., Thermal waves in an absorbing and convecting medium. Physica, **8D** (1983), pp. 273–283.

16) Tomoeda, K., and Mimura, M., Numerical approximations to interface curves for a porous media equation. Hiroshima Math. J., **13** (1983), pp. 273–294.

17) Tomoeda, K., and Nakaki, T., Difference approximations to interface curves for nonlinear diffusion equations with absorption. Appl. Math. Lett., **5** (1992), pp. 41–47.

Oblique Shock Reflection over Porous Media with Large Porosity

Susumu KOBAYASHI*, Takashi ADACHI*, and Tateyuki SUZUKI**

* *Department of Mechanical Engineering, Saitama Institute of Technology, Okabe-machi, Saitama*
** *Department of Mechanical Systems Engineering, Toyama Prefectural University, Kosugi-machi, Toyama*

The regular reflection of a shock wave over porous media was investigated both theoretically and experimentally. Two types of models were considered. One may be called a simple sink theory, in which a deflection angle is the only parameter taking the sink effect of the porous medium into consideration and the detail of the flow in the porous medium is not taken into account. The other model is a more realistic one, in which the pressure wave and the gas flow in the porous phase are considered as well as in the gas phase. Using the former, we assumed the deflection angle and compared the theoretical results with the experimental. With the latter, however, since the coupled problem for both phases was too complicated to analyze theoretically, so we analyzed two limiting cases and showed on a physical plane the domain where the solution of the coupled problem was possible.

I. INTRODUCTION

When a plane shock wave is incident on a slope, a reflected wave develops. The overall wave configuration is roughly classified into two categories[1], i.e., regular reflection and Mach reflection. In regular reflection, ends of both incident and reflected waves coincide on the surface of slope. In Mach reflection, however, an intersection of the incident and reflected waves (which is called a *triple point*) is above the slope surface, and a third shock wave connecting the triple point and the slope surface appears. This shock characteristic of Mach reflection is called a *Mach stem*. The well-known von Neumann's theory is a simple application of oblique shock relation to such wave configuration by approximating curved shocks with plane ones. For regular reflection, von Neumann's theory is often termed *two-shock theory* because two shocks (i.e., incident and reflected shocks) are involved[2, 3]. Due to similar reason, his theory for Mach reflection is called *three-shock theory*, where the Mach stem is also involved. If the slope is smooth and flat, the wave configuration is determined by the incident shock Mach number and the inclination angle of slope, as well as a thermodynamic property of fluid (say, the ratio of specific heats). In ordinary reflection over smooth plane surface, the behavior of the triple point can be regarded pseudosteady, and not only an overall wave configuration but also a flow field is self-similar. Such a problem as an oblique shock reflection over smooth plane surface has long been studied by many investigators[3, 4], and generally well understood in both theory and experiment.

On the other hand, shock reflection problem over surfaces other than smooth one has attracted, considering its practical importance, less attention so far. When the incident shock is reflected on the inclined porous medium, the velocity component perpendicular to the solid boundary is possible and as a result the boundary acts as a sink (i.e., *sink effect*). Since von Neumann's theory considers uniform flow fields bounded by plane discontinuities such as shocks and slipstream, the effect of the boundary on which the incident shock is reflected is not considered. Therefore it is not surprising that the theory and the experimental result are not in good agreement when the effect of boundary is prominent. In the early stage of investigation, the authors had reported[5-7] that the structure of a group of waves reflected from a two-dimensional rough surface ("multi-guttered wedge") and pointed out the importance of a role played by a compression wave formed by accumulation of reflected waves. Moreover, comparison of reflected wave structures for multi-guttered wedge and porous layer had revealed that they were quite similar in spite of the irregularity of particle disposition[8]. In the present paper, the experiment has been performed with deep porous medium in order to neglect the influence of waves reflected from the bed. Results are compared with numerical calculation for two limiting cases.

II. EXPERIMENT

The experimental setup is the same as that shown in our previous paper[7] and its schematic diagram is omitted here. The principal part of the shock tube used is composed of a 79-mm-diameter driver section 1200 mm long and a 65 × 30 mm rectangular driven section 3900 mm long. At the end of the driven section is a test section with a pair of 62 × 94 mm optical viewing windows on either side. A dump tank, 155 mm in inner diameter and 450 mm in length, is connected to the test section to moderate shock pressure.

The incident shock Mach number M is estimated by measuring the time interval of passage of an incident shock between two pressure gauges ahead of the test section. Waves are visualized through shadowgraphy using a xenon flash lamp with a pulse width of 180 ns. The flash lamp is triggered by the pressure gauge nearer to the test section via a delay circuit. Through regulation of the delay circuit, the wave structure at any desired instant can be either recorded with a TV camera via a monochromatic image analyzer (SVA-1, Sugawara Laboratories, Inc.) or photographed.

Two kinds of porous media have been used. One is a dusty layer composed of spherical glass particles whose mean diameter is 60 μm and its porosity ϕ is 0.44. The other is made of foam rubber whose porosity is 0.98.

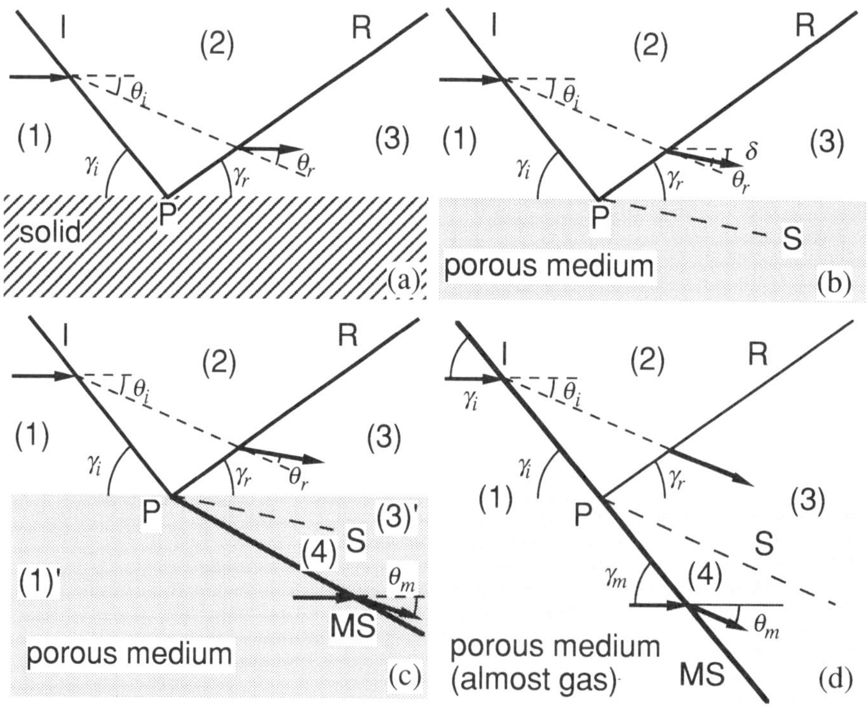

Fig. 1. Flow field relative to the reflection point, (a) ordinary regular reflection over solid surface, (b) regular reflection over porous medium; simple sink model, (c) realistic model, (d) limiting model when the porosity is almost unity and the reflected wave is acoustically weak.

The working gas is air (its ratio of specific heats $\kappa = 1.4$). The experimental parameters are the reflecting wedge angle θ_w and the incident shock Mach number M, i.e., θ_w ranges from 30° to 50° at intervals of 5° for dusty layer; from 25° to 47.5° at intervals of 2.5° for foam rubber, M = 1.20 and 1.41. The initial conditions of the driven section are room temperature and atmospheric pressure.

III. ANALYSIS

3.1 Models for Analysis

Figure 1 (a) shows the flow field relative to the reflection point P in the case of regular reflection over solid smooth boundary. The sign of the deflection angle of the flow across a discontinuity is consistently defined; i.e., counterclockwise deflection is positive, clockwise negative. The incoming flow parallel to the solid boundary is deflected clockwise by θ_i across the incident shock I and it is again deflected by θ_r across the reflected shock R. The boundary condition in region (3) requires $\theta_i + \theta_r = 0$. On the other hand, in the case of regular reflection over porous layer, the velocity component perpendicular to its surface is possible because of the permeability of the porous layer. As a result, the boundary condition is modified as $\theta_i + \theta_r = -\delta$, where $-\delta$ is a deflection angle, induced by the sink effect, of the flow behind the reflected wave (see Fig. 1 (b)). The value of δ can be a function of the incident shock Mach number M and the angle of incidence γ_i as well as the porosity ϕ of the porous medium. It is easily inferred that the sink effect is marked for regular reflection rather than Mach reflection whose triple point is above the surface of porous layer. We may refer to this model as *simple sink model*, since only the flow into the porous layer is considered while the propagation of shock wave is disregarded. In reality, however, the air-borne porous layer as well as the upstream gas phase (1) goes on in the reference frame fixed to the reflection point, and the incident shock propagates into the porous layer. Such flow field illustrated in Fig. 1 (c) may be called *realistic model*. Realistic as it is, it is not easy to solve the whole flow field since we must deal with the coupled problem between the pure gas phase and the porous phase.

3.2 Numerical Results Based on the Simple Sink Model

By assuming the value of δ in the two-shock theory (Fig. 1 (b)), we can easily obtain the relation between angles of incidence and reflection. The results are shown by dotted curves in Fig. 2 for M = 1.20 and 1.41 when the fluid is air. The dashed curve represents the two-shock theory, i.e., $\delta = 0$, while the solid curve stands for the three-shock theory.

3.3 Two Limiting Models

In order to analyze the phenomenon, two limiting modes of regular reflection are introduced. One is an ordinary regular reflection over a solid surface (i.e., porosity ϕ is zero, see Fig. 1 (a)). The other is a regular reflection over a "porous" medium in the limit of $\phi \to 1$. The latter is illustrated in Fig. 1 (d) and requires explanation. As the porosity is almost unity (i.e., the layer is substantially an imaginary one), the incident shock propagates into the "porous" medium under least interference and so the angle of incidence in the porous medium γ_m is identical with that in the gas phase γ_i. The reflected wave from the interface is no more than acoustic. Therefore, the flow in region (2) is not deflected by the reflected wave. The flow in porous medium is deflected by θ_m, which is also identical with θ_i, across the incident shock (which is a degenerated Mach stem in the three-shock theory) and the boundary condition on the slipstream S is satisfied. This is the trivial case for three-shock theory and the corresponding solution may be termed *trivial solution*. It is inferred that the real situation lies between these both limiting cases according as its porosity (*trivial solution hypothesis*).

3.4 Trivial Solution of Three-Shock Theory

The trivial solution for Fig. 1 (d) is easily obtained by setting $\theta_r = 0$, $\gamma_m = \gamma_i$, $\theta_m = \theta_i$, $p_2 = p_3 = p_4$ in the three-shock theory:

$$\tan\theta_i = \frac{2\cot\gamma_i(M_1^2\sin^2\gamma_i - 1)}{M_1^2(\kappa + \cos 2\gamma_i) + 2}, \tag{1}$$

$$M_2^2 = \frac{(\kappa - 1)M_1^2\sin^2\gamma_i + 2}{\sin^2(\gamma_i - \theta_i)\left\{2\kappa M_1^2\sin^2\gamma_i - (\kappa - 1)\right\}}, \tag{2}$$

$$\frac{2\kappa M_2^2 \sin^2(\gamma_r + \theta_i) - (\kappa - 1)}{\kappa + 1} = 1. \tag{3}$$

The last equation reduces to

$$\gamma_r = \sin^{-1}(1/M_2) - \theta_i. \tag{4}$$

When the Mach number M_1 of the flow in region (1) and the ratio of specific heats κ are specified, the deflection angle θ_i across the incident shock and thus the Mach number M_2 of the flow in region (2) are functions of γ_i, so Eq. (4) gives the analytical relation between angles of incidence and reflection. In Fig. 2, the dash-and-dotted curve represents the trivial solution thus obtained.

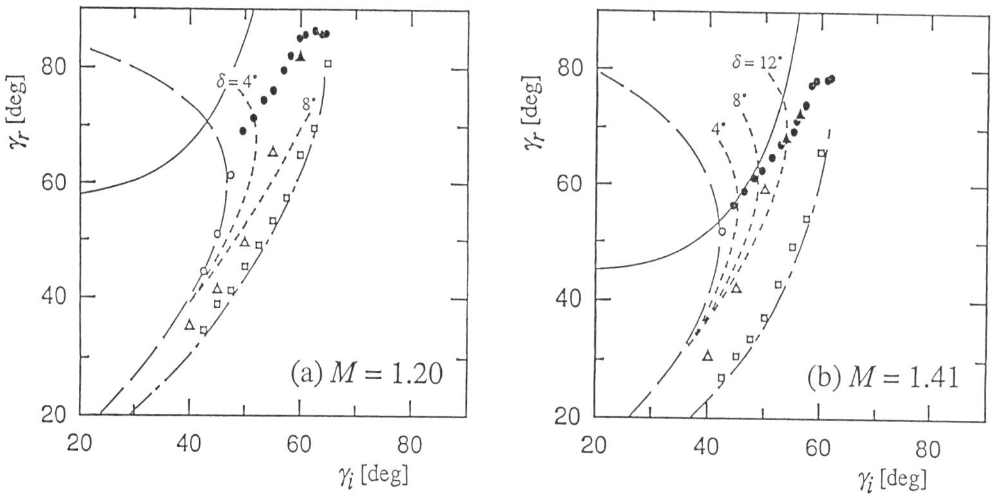

Fig. 2. Relation between angles of incidence and reflection, (a) $M = 1.20$, (b) $M = 1.41$. Experiment: smooth plane surface ○ ●, dusty layer △▲, foam rubber □ (open symbols denote regular reflection, solid ones Mach reflection). Theory: —— —— the two-shock theory, ———— the three-shock theory, – – – – – simple sink model, —— – —— trivial solution.

IV. RESULTS AND DISCUSSION

The experimental data are shown in Fig. 2 together with theoretical and numerical results. Circles, triangles and squares correspond to the results for smooth plane surface, dusty layer and foam rubber, respectively. Open symbols stand for regular reflection, while solid symbols Mach reflection. Note that for regular reflection the relationship $\gamma_i = 90° - \theta_w$ holds. Concerning porous medium data, the distance from the wedge tip to the root of the incident shock (or the Mach stem) is around 30 mm. For $M = 1.20$ (Fig. 2 (a)), the results of dusty layer (porosity $\phi = 0.44$) do not agree with the two-shock theory. This is more evident for foam rubber ($\phi = 0.98$).

Although the discrepancy between theory and experiment is explained to some extent by considering the sink effect, it still remains especially for $M = 1.41$ since however the value of δ is large there exists a region where the dotted curves cannot cover.

It is remarkable that the foam rubber results are in good agreement with the trivial solution. In either case, the experimental results lie in the limited region between the solid (two-shock theory) and the dash-and-dotted (trivial solution) curves as mentioned in the previous section. It should be noted that the curve of the trivial solution comes to an end where $M_2 = 1$. This is because the derivative of Eq. (4):

$$\frac{d\gamma_r}{d\gamma_i} = \frac{1}{2\sqrt{M_2^2 - 1}} \frac{dM_2^2}{d\gamma_i} - \frac{d\theta_i}{d\gamma_i}, \tag{5}$$

increases indefinitely as M_2 approaches unity.

It is interesting that the trivial solution hypothesis on the basis of the three-shock theory explains the phenomenon of regular reflection over porous medium better than the simple sink model which is in itself the two-shock theory. Physically this means that the effect of porous material is not restricted to the sink effect but it is essential to take the porous phase into consideration. In this respect, the present result is in accordance with Clarke[9] who insisted the effect other than the air flow within porosities (i.e., sink effect) should be considered.

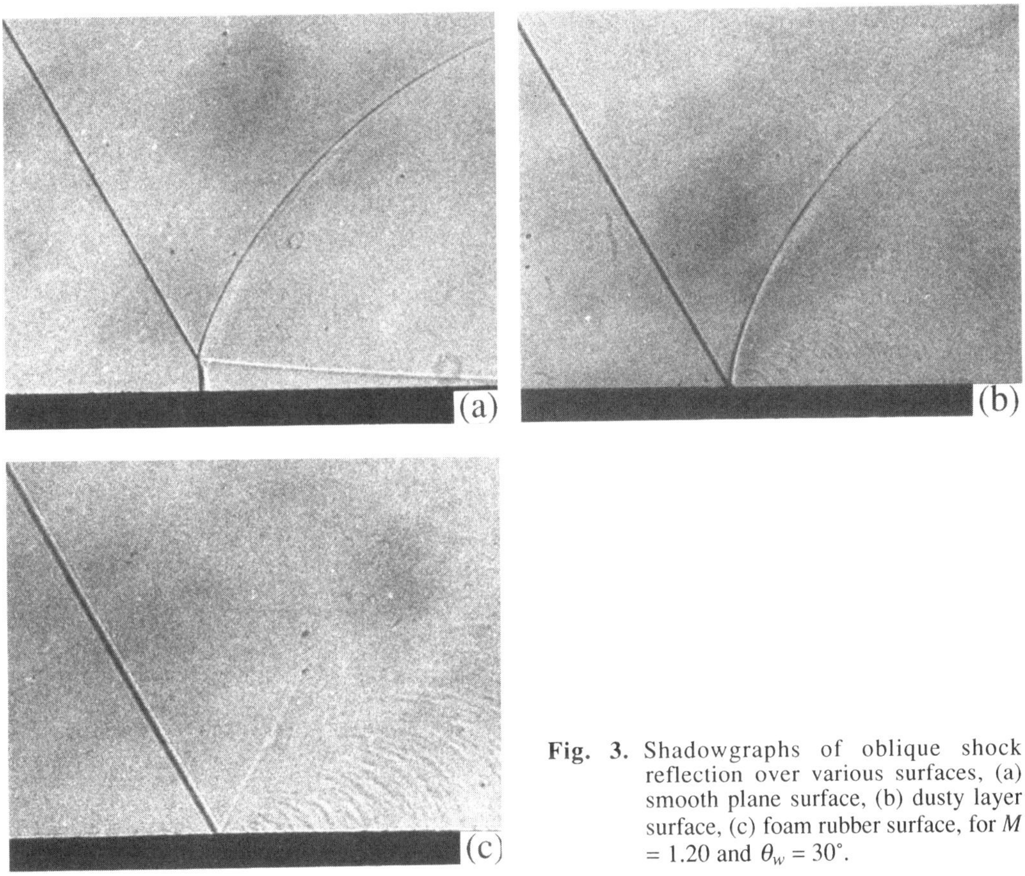

Fig. 3. Shadowgraphs of oblique shock reflection over various surfaces, (a) smooth plane surface, (b) dusty layer surface, (c) foam rubber surface, for $M = 1.20$ and $\theta_w = 30°$.

Figures 3 (a) to (c) show shadowgraphs of oblique shock reflection over three different wedges, i.e., plane smooth wedge, dusty layer and foam rubber, for $M = 1.20$ and $\theta_w = 30°$. The Mach reflection takes place for plane smooth wedge (Fig. 3 (a)). Although the Mach stem is quite short (Fig. 3 (b)), it still persists in the case of dusty layer. For larger porosity, the reflection type is regular and it is characteristic that the reflected wave is quite weak (Fig. 3 (c)). This fact supports the trivial solution hypothesis in which the reflected wave was assumed acoustic.

V. CONCLUSION

The regular reflection of a shock wave over porous layer was investigated both theoretically and experimentally. The following conclusions are made:

(1) The simple sink theory explains the experiment to some extent.
(2) The trivial solution hypothesis proposed here is quite effective. It is essential to take the flow in porous layer into consideration.

REFERENCES

1) von Neumann, J., Oblique reflection of shocks. *Explosive Res. Rep.*, No. 12 (1943), also reprinted in *John von Neumann Collected Works*, Vol. 6 (ed. by A. H. Taub, Pergamon, 1963), pp. 238-299.
2) Ikui, T. and Matsuo, K., *Mechanics of Shock Waves* (Corona-sha, 1983), pp. 129-132 (in Japanese).
3) Ben-Dor, G., *Shock Wave Reflection Phenomena* (Springer-Verlag, 1992).
4) Hornung, H., Regular and Mach reflection of shock waves. *Ann. Rev. Fluid Mech.*, **18** (1986), pp. 33-58.
5) Adachi, T., Kobayashi, S. and Suzuki, T., Unsteady behavior of Mach reflection of a planar shock wave over a wedge with surface roughness. *Theoret. Appl. Mech.*, **39** (1990), pp. 285-291.
6) Suzuki, T., Adachi, T. and Kobayashi, S., Experimental analysis of reflected shock behavior over a wedge with surface roughness: local behavior. *JSME Int. J.*, Ser. B, **36** (1993), pp. 130-134.
7) Adachi, T., Kobayashi, S. and Suzuki, T., An experimental analysis of oblique shock reflection over a two-dimensional multi-guttered wedge, *Fluid Dynamics Res.*, **9** (1992), pp. 119-132.
8) Adachi, T., Kobayashi,S. and Suzuki,T., An experimental analysis on the reflected wave structure over a rough surface. *Theoret. Appl. Mech.* **40** (1991), pp. 173-180.
9) Clarke, J. F., The reflection of weak shock waves from absorbent surfaces. *Proc. Roy. Soc. Lond.*, **A396** (1984), pp. 365-382.

Numerical Analysis of Three-Dimensional Natural Convective Heat Transfer from an Inclined, Isothermally Heated, Square Plate

Katsuhisa NOTO and Tsuyoshi NAKAJIMA

Department of Mechanical Engineering, Kobe University, Nada, Kobe, Hyogo

The three-dimensional(3-D), laminar, natural convective heat transfer from an inclined, isothermally heated, square plate with zero-thickness placed in an infinite space is analysed numerically by finite difference discretizations of the 3-D continuity equation, the 3-D Navier-Stokes equations, and the 3-D energy equation applying the vector potential satisfying the 3-D continuity equation. A 3-D coordinate transformation is introduced in order to preserve a large number of grid points needed for the 3-D infinite open space. The minimum value of the local Nusselt number is always at 0.13L above the plate center at any angle of the inclination. The local Nusselt number in the central region of the plate is smaller than that predicted by the 2-D boundary layer theory. The mean Nusselt number is constant at any angle of the inclination at the given Grashof number.

I. INTRODUCTION

The 2-D, laminar, natural convective heat transfer from a plate with an infinite width and a finite height has been investigated experimentally by Gilmore et al.[1] and Saunders[2] and numerically by Suriano and Yang[3] and Miyamoto and Kurima[4]. However, if the plate width, which is generally finite, becomes narrow, it is not possible to disregard the effect of the side edges. Furthermore, a 3-D effect may occur, which may have considerable influence on the thermo-fluid fields.

The 3-D, natural convective heat transfer has been investigated as follows: Doston[5] measured the local Nusselt Number on the center line in a vertical plate with the aspect ratio 3.75, the inclination angle 0°, and uniform surface heat flux. Matsumura[7] measured the local Nusselt number on the center line in a vertical plate with the small aspect ratio, the inclination angle 0°, and uniform surface heat flux. Hassani and Hollands[8] measured the mean Nusselt number of the vertical plate with the aspect ratio 1.0, uniform surface heat flux, and the inclination angle 0° and 45°. Noto[9] computed the 3-D, laminar, natural convection from a vertical plate with the inclination angle 0°, assuming the flow symmetry with respect to the center line of the plate.

As described above, the distribution of the local Nusselt number on the surface of vertical plate have not been made clear in the previous studies.

The local Nusselt number on the vertical, isothermal plate with the aspect ratio 1.0 and various angle of inclination is analyzed by 3-D computation in the present investigation.

II. 3-D PHYSICAL MODEL

2.1 Heated Plate and 3-D Coordinate

A 3-D physical model and a 3-D coordinate system are shown in Fig.1. A vertical *isothermal flat plate with height 2L, width 2γL, and surface temperature θ_w is placed in an infinite space. The plate inclines with the angle α with respect to the gravitational direction, as shown in Fig.1. Here, γ represents the aspect ratio of the plate, width/height. Nomenclatures $\vec{x}(x, y, z)$ and $\vec{v}(u, v, w)$ denote, respectively, the 3-D coordinate and the 3-D velocity vector as shown in Fig. 1.*

2.2 Plate Thickness

The plate thickness used is generally quite small. For example, the thickness 2δ of the plate used by Saunders[2] has a relation that the order of δ/L is $0.33\times10^{-4}\sim 8.00\times10^{-4}$ and this fact means that the value of δ/L is nearly zero. Matsumura[7] found that the thickness effect on the local Nusselt number can be disregarded when δ/γL is between 4.16×10^{-5} and 5.00×10^{-2}. It is, therefore, assumed herein that the plate thickness in the present investigation is equal to zero.

2.3 Steady State

The swaying motion[10] of a thermal plume above a heated plate may occur. However, Hardwick and Levy[11] showed that the feature of the swaying motion can be negligible in plate plume. In the present investigation, the author, therefore, assumes that a plume does not sway, i.e. the present natural convective flow is assumed to be steady.

2.4 Laminar State

Since the plate is short, the Grashof number governing a thermal flow in the present case becomes small, and the thermal flow in the present study is laminar.

2.5 Symmetrical Condition

Since the thermal flow in the present investigation is assumed to be steady, the distributions of velocity and temperature in Fig.1 are, therefore, symmetrical with respect to the x-z plane. Accordingly, the presentinvestigation deals only with Region I ($-\infty<x<\infty$, $0\le y$, $-\infty<z<\infty$) in Fig.1.

III. MATHEMATICAL FORMULATION

3.1 3-D Governing Equations

Since the temperature variation is small in the convective field, Boussinesq's approximation can be employed for physical properties. The 3-D governing equations, therefore, can be written as follows:

$$\nabla\cdot\vec{V} = 0, \tag{1}$$

$$(\vec{V}\cdot\nabla)\vec{V} = -\nabla P + 2\nabla^2\vec{V} + (\vec{G}r/2)\Theta, \tag{2}$$

$$\vec{V}\cdot\nabla\Theta = (2/Pr)\nabla^2\Theta, \tag{3}$$

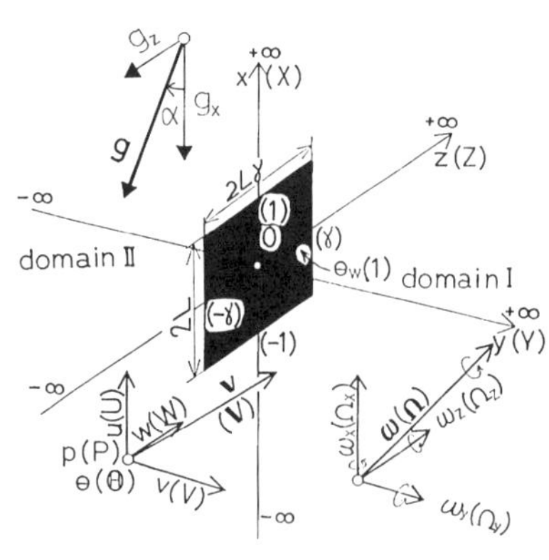

Figure 1. 3-D physical model and 3-D coordinate system.

where, $\vec{Gr}$ is the Grashof number vector defined as follows:

$$\vec{Gr} = (2L)^3\vec{g}\beta(\theta_w-\theta_\infty)/\nu^2,$$

where, $\vec{g}$ is the gravitational acceleration vector, β the thermal coefficient of expansion, ν the kinematic viscosity, θ_∞ the temperature at infinity, and Pr the Prandtl number of the fluid. In Eqs. (1)~(3), $\vec{X}$ is the non-dimensional 3-D coordinate vector, $\vec{V}$ the 3-D non-dimensional velocity vector, P the non-dimensional pressure, and Θ the non-dimensional temperature. They are defined as follows:

$$\vec{X} = \vec{x}/L,\ \vec{V} = 2L\vec{v}/\nu,\ P = (2L)^2p/\rho\nu^2,\ \Theta = (\theta-\theta_\infty)/(\theta_w-\theta_\infty), \tag{4}$$

where, ρ is the fluid density, and p the static pressure.

Eliminating the pressure p from Eq.(2), a 3-D vorticity transport equation can be obtained, such that:

$$\vec{V}\cdot\nabla\vec{\Omega} - \vec{\Omega}\cdot\nabla\vec{V} = 2\nabla^2\vec{\Omega} + (Gr/2)\begin{bmatrix} \partial\Theta/\partial Y\cdot\sin\alpha \\ \partial\Theta/\partial Z\cdot\cos\alpha-\partial\Theta/\partial X\cdot\sin\alpha \\ -\partial\Theta/\partial Y\cdot\cos\alpha \end{bmatrix}, \tag{5}$$

where, Gr is the Grashof number defined as follows:

$$Gr = (2L)^3g\beta(\theta_w-\theta_\infty)/\nu^2.$$

The 3-D non-dimensional vorticity vector $\vec{\Omega}$ is defined as follows:

$$\vec{\Omega} = \nabla\times\vec{V}. \tag{6}$$

To obtain the 3-D velocity from the 3-D vorticity, the following 3-D vector potential $\vec{\Phi}$ is introduced:

$$\vec{V} = \nabla\times\vec{\Phi}. \tag{7}$$

According to Aziz and Hellums[12], the vector potential should satisfy the following solenoidal condition:

$$\nabla\cdot\vec{\Phi} = 0. \tag{8}$$

Substituting Eq.(7) into Eq.(6) and with the aid of Eq.(8), the following relation can be obtained:

$$\vec{\Omega} = -\nabla^2\vec{\Phi}. \tag{9}$$

Equations (3),(5),(7), and (9) constitute the system of 3-D governing equations in the present subject.

3.2 3-D Boundary Condition

For an external natural convective flow, two kinds of methods have been proposed to express the boundary conditions at infinity, i.e. far from the plate[13]. One is the Dirichlet-type condition and the other the Neumann-type one. Although Kuehn and Goldstein[14] applied the Neumann-type condition to the 2-D natural convective flow around a heated, circular cylinder, it was found to be impossible to obtain a circulating vortex flow occurring far from a heat source, if the Neumann-type condition is employed. In the present problem, the Dirichlet-type condition is, therefore, applied at the infinite boundary locations.

The 3-D boundary conditions in Region I are, thus, defined as follows:

(i) on the wall ($|X|\leq 1,\ Y=0,\ |Z|\leq \gamma$):

$$U_i=\Phi_X=\partial\Phi_Y/\partial Y=\Phi_Z=0,\ \Omega_X=\partial W/\partial Y,\ \Omega_Y=0,\ \Omega_Z=-\partial U/\partial Y,\ \Theta=1, \tag{10-1}$$

(ii) on Plane $Y=0$ excluding the wall ($1<|X|$, $Y=0$; $|X|<1$, $Y=0$, $\gamma<|Z|$):

$$\partial U/\partial Y=V=\partial W/\partial Y=\Phi_X=\partial\Phi_Y/\partial Y=\Phi_Z=0,\ \Omega_X=\partial\Omega_Y/\partial Y=\Omega_Z=\partial\Theta/\partial Y=0, \qquad (10\text{-}2)$$

(iv) at infinity ($X=\pm\infty$; $Y=\infty$; $Z=\pm\infty$): $U_i=\Phi_i=\Omega_i=\Theta=0$, (10-3)

where, the subscript $i(=1, 2, 3)$ denotes the number of the 3-D coordinate axis O_{Xi} (e.g., $U_3=W$).

IV. 3-D COORDINATE TRANSFORMATION

The natural convective flow treated in this investigation is an elliptic-type in a 3-D open infinite space. To avoid the use of a large number of grid points, the infinite open space X-Y-Z is transformed to a 3-D comparatively small finite space $\bar{X}$-$\bar{Y}$-$\bar{Z}$ by the following 3-D coordinate transformation:

$$\vec{\bar{X}} = (2/\pi)\tan^{-1}\vec{X}. \qquad (11)$$

In other words, the physically infinite position $(X, Y, Z)=(\pm\infty, \pm\infty, \pm\infty)$ can be replaced to the mathematically finite position $(\bar{X}, \bar{Y}, \bar{Z})=(\pm 1, \pm 1, \pm 1)$, using Eq. (11).

Since the 3-D governing equations of Eqs.(3), (5), (7), and (9) and the boundary conditions of Eq.(10) are transformed by Eq.(11), the corresponding equations and the boundary conditions can be obtained, respectively.

V. COMPUTATIONAL ANALYSIS

5.1 3-D Grid Structure

The 3-D grid structure in the 3-D open space X-Y-Z of Region I ($-\infty<X<\infty$, $0\leq Y$, $-\infty<Z<\infty$) is not explicitly employed in the present computation. A 3-D computation was carried out using a 3-D grid structure in the comparatively small finite enclosed space $\bar{X}$-$\bar{Y}$-$\bar{Z}$ of the transformed Region I ($-1\leq\bar{X}\leq 1$, $0\leq\bar{Y}\leq 1$, $-1\leq\bar{Z}\leq 1$). Computational values in the region except for grid points were obtained from values at surrounding grid points.

5.2 3-D Discretization

The transformed governing equations and the boundary conditions were discretized by the central finite difference scheme with the second-order accuracy. The Dirichlet type boundary conditions were discretized by the three-point approximation, which has the second-order accuracy.

5.3 3-D Computational Procedures

The computational solutions of the transformed equations of Eqs. (3), (5), (7), and (9) were obtained by 3-D SOR (Successive Over Relaxation) method with an under relaxation parameter. The computational solutions were obtained in the order of $\vec{\Phi}$, $\vec{V}$, $\vec{\Omega}$, and Θ from algebraic equations with discretization of the transformed equations of Eqs. (3), (5), (7), and (9).

The 3-D computation of the present investigation is a 3-D boundary value problem. The boundary conditions should be satisfied not only the Dirichlet-type condition, Eq.(10-3), but also the Neumann-type condition at the infinite region, such that:

$$\partial(\vec{\Phi}, \vec{V}, \vec{\Omega}, \Theta)/\partial X_j=0, \qquad (12)$$

where, the subscript j means the direction perpendicular to a surface of the outer boundary at infinity.

When the error of the Neumann-type condition was less than 1×10^{-4}, the value obtained in the present 3-D computation was regarded as a convergent solution. The convergent constraint of the solenoidal condition of the 3-D vector potential $\vec{\Phi}$ was selected to be

$$|\nabla(\nabla\cdot\vec{\Phi})/\nabla^2\vec{\Phi}| \le 4\times10^{-2}. \qquad (13)$$

The distributions in the 3-D finite enclosed space $\bar{X}$-$\bar{Y}$-$\bar{Z}$ were transformed in the 3-D open infinite space X-Y-Z by the following inversion of Eq.(11):

$$\vec{X} = \tan(\pi\vec{\bar{X}}/2). \qquad (14)$$

As a result, 3-D computational solutions of the 3-D governing equations (1) to (3) with the 3-D boundary conditions of Eq.(10) was obtained by 3-D computation.

VI. THREE DIMENSIONAL HEAT TRANSFER CHARACTERISTICS

6.1 Local Nusselt Number Obtained by 3-D Analysis

The local Nusselt number on the plate surface is defined as follows:

$$Nu = 2Lh/\lambda = -2(\partial\Theta/\partial Y)_{Y=0} = -4/\pi\cdot(\partial\Theta/\partial\bar{Y})_{\bar{Y}=0}, \qquad (15)$$

where, h is the local heat transfer coefficient.

Figure 2 shows the equi-value lines of the local Nusselt number at Gr=800 and α= 0°, 15°, 30°, and 45°. The minimum value of the local Nusselt number is at 0.13L above the plate center. The local Nusselt number increases near the side edges.

6.2 Comparison of 3-D Local Nusselt Number with That by 2-D Boundary Layer Theory

The local Nusselt number predicted by the 2-D, laminar, natural convective boundary layer theory[15] becomes as follows:

$$Nu = f(Pr)\cdot\{2Gr/(X+1)\}^{1/4}, \qquad (16)$$

where, f(Pr) is the value determined with the Prandtl number; e.g., f(Pr)=0.3568 for Pr=0.72.

Figure 3 shows a comparison with the 3-D local Nusselt number on the center line of the plate and that by the 2-D boundary layer theory. The former is found to be larger than the latter near the upper edge of the plate and to be smaller than the latter in the central region of the plate.

6.3 Rear Edge Effect on the Local Nusselt Number

The velocity vector combined the velocity components in the X- and Y-directions in the X-Y plane by the present 3-D analysis is shown in Fig.4. The flow is free from the wall shear stress above the rear edge. This causes the acceleration and the velocity component with the horizontal direction toward to the plane y=0 near the upper edge of the plate. These rear edge effects lead to the increase of the local Nusselt number near the upper edge, as shown in Fig.3.

6.4 Reason for the Local Nusselt Number less than That by the 2-D Boundary Layer Theory in the Central Region

The distributions of the velocity vector in the 3-D natural convection are quite different from those obtained by the 2-D boundary layer theory in the following manner: The 3-D flow has entrainment below the bottom edge, and it has not a flow with the horizontal direction far away the plate. On the other hand, the flow determined by the 2-D boundary layer theory is not entrained to the boundary layer from below the plate and has flows with the horizontal direction outside the boundary layer. For example, the boundary layer thickness on the upper edge is almost equal to 1.0L at Gr=800.

Thus, the 3-D flow differs from the 2-D boundary layer flow: the 3-D flow region with the x-component of the velocity, i.e. higher temperature than that of the ambient air, is larger than that by the boundary layer theory. This leads to the 3-D lo-

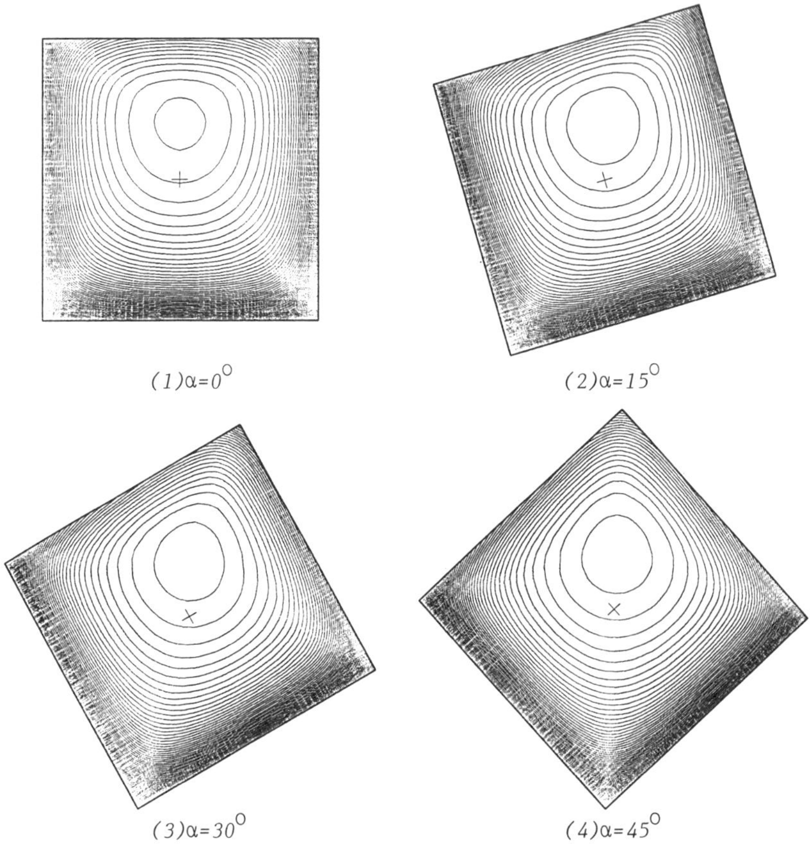

Figure 2. Local Nusselt number (Gr=800).

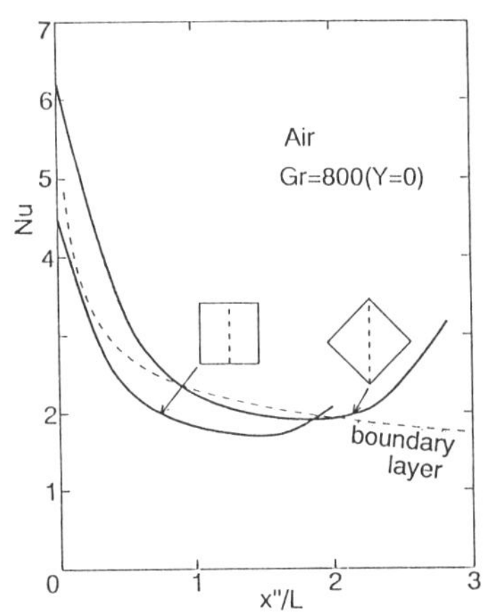

Figure 3. Comparison with the 3-D local Nusselt number and the 2-D boundary layer theory (Gr=800, X" is the vertical distance from the bottom edge of the plate.).

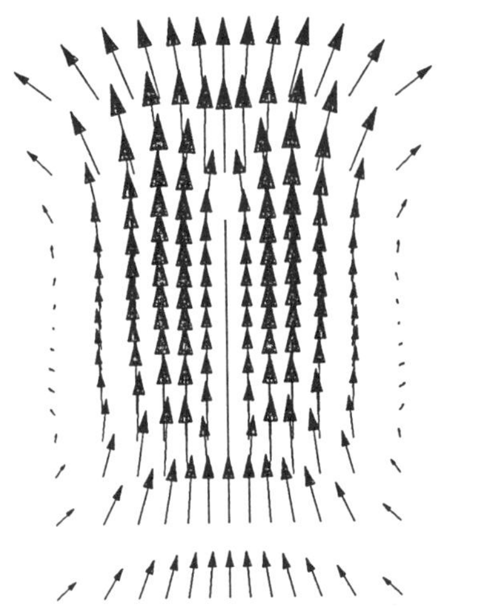

Figure 4. Velocity vector in the X-Y plane (Gr=800, Z=0, γ=0°).

cal Nusselt number less than that by the 2-D boundary layer theory in the central region on the center line of the plate, as shown in Fig.3.

6.5 Mean Nusselt Number Obtained by 3-D Analysis

The mean Nusselt number on one side of the plate is obtained as follows:

$$Nu_m = 2Lh_m/\lambda = \iint Nu dX dZ/4, \qquad (17)$$

where, h_m is the mean heat transfer coefficient and λ the heat conductivity.

The mean Nusselt number at Gr=800 was 2.585, 2.584, 2.584, and 2.584 at $\alpha=0^\circ$, 15°, 30°, and 45° respectively, and it is, therefore, independent of the angle of inclination. The mean Nusselt number at Gr=80 is 2.559 at $\alpha=0^\circ$.

6.6 Comparison with the Mean Nusselt Number in the Existing Experiment

The result by the present analysis agrees well with that measured by Hassani and Hollands[8] who showed that the mean Nusselt number at $\alpha=45^\circ$ is equal to that at $\alpha=0^\circ$ and is 2.49~2.69 at Gr=800 and $\alpha=0^\circ$ and 45°.

6.7 Comparison of 3-D Mean Nusselt Number with That by 2-D Boundary Layer Theory

The mean Nusselt number at $\alpha=0^\circ$ by the 2-D boundary layer theory is as follows:

$$Nu_m = 4f(Pr)Gr^{1/4}/3. \qquad (18)$$

The mean Nusselt numbers 1.423 and 2.530 at Gr=80 and 800, respectively, determined by Eq.(18) are smaller than those by 3-D computation, as described in Sec.6.4. That is, the 3-D mean Nusselt number is larger by 79.8% and 2.18% at Gr=80 and 800, respectively, than that by the 2-D boundary layer theory. The mean Nusselt number, therefore, approaches the value by the 2-D boundary layer theory, with increasing the Grashof number.

VII. CONCLUSIONS

(1) The local Nusselt number has the minimum value at 0.13L above the plate center and increases near the side edges of the plate, regardless of the angle of inclination.

(2) The local Nusselt number is smaller than that determined by the 2-D boundary layer theory in the central region of the plate. The reason is that the 3-D flow has entrainment below the bottom edge and has not flow with the horizontal direction far away the plate.

(3) The mean Nusselt number, which agrees well with the previously measured result, approaches a value determined by the 2-D boundary layer theory with increasing the Grashof number. It is also found to be independent of the angle of inclination.

REFERENCES

1) Gilmore,R.J., Yelmgren,K.E., Szewczyk,A.A. and Yang,K.T., Experimental investigation of laminar free convection about short vertical flat plates and a horizontal disc at small Grashof numbers. Proc. 5th Int. Heat Transfer Conf., Vol.3 (1974), pp.25-29.
2) Saunders,O.A., The effect of pressure upon natural convection in air. Proc. of the Royal Society, Ser.A, No.157(1936), pp. 278-291.
3) Suriano,F.J. and Yang,K.T., Laminar free convection about vertical and horizontal plates at small and moderate Grashof numbers. Int. J. Heat Mass Transfer, Vol.11(1968), pp.473-490.
4) Miyamoto,M. and Kurima,J., Laminar free convection heat transfer from a short vertical flat plate. Proc. 18th National Heat Transfer Symposium of Japan, Vol. 18(1981), pp.298-300 (in Japanese).
5) Doston,J.P., see page 439 in Ref.(7).

6) Sparrow,E.M. and Gregg,J.L., Laminar free convection from a vertical plate with uniform surface heat flux, Trans. ASME, Vol.78(1965), pp.435-440.
7) Matsumura,H., The effect of plate width on natural convection heat transfer from vertical plates. Trans. JSME, Ser.B, Vol.47(1981), pp.911-914(in Japanese).
8) Hassani,A.V. and Hollands,K.G.T., On natural convection heat transfer from three-dimensional bodies of arbitrary shape. Trans. ASME, J. Heat Transfer, Vol. 111(1989), pp.363-371.
9) Noto,K. and Matsumoto,R., Three-dimensional analysis of the Navier-Stokes equations on laminar natural convection around a vertical flat plate. Numerical Methods in Laminar and Turbulent Flow, Vol.4(1985), pp.865-877.
10) Noto,K., Swaying motion in thermal plume above a horizontal line heat source. AIAA J. Themophysics Heat Transfer, Vol.3(1989), pp.428-434.
11) Hardwick,N.E. and Levy,E.K., Study of the laminar free convection wake above an isothermal vertical plate. Trans. ASME, J. Heat Transfer, Vol.95(1973), pp.289-294.
12) Aziz,K. and Hellums,J.D., Numerical solution of the three-dimensional equations of motion for laminar natural convection. Physics of Fluid, Vol.10(1967), pp. 314-324.
13) e.g., Roache,P.J., Computational Fluid Dynamics, revised edition, (Hermosa Pub., U.S.A., 1982).
14) Kuehn,T.H. and Goldstein,R.J., Numerical solution to the Navier-Stokes equations for laminar natural convection about a horizontal isothermal circular cylinder. Int. J. Heat Mass Transfer, Vol.23(1980), pp.971-979.
15) Kakac.S. Aung,W. and Viskanta,R.(ed.), Natural convection, (Springer-Verlag, 1985), pp.11.

Petrov-Galerkin Finite Element Analysis Using Exponential Functions for Natural Convection at High Rayleigh Numbers

Kazuhiko KAKUDA*, Yuko KAWAHARA**, and Nobuyoshi TOSAKA*

* *Department of Mathematical Engineering, College of Industrial Technology, Nihon University, Narashino, Chiba*
** *Taiyo Kogyo Corporation, Yodogawa-ku, Osaka*

A Petrov-Galerkin finite element method using exponential functions for the computation of natural convection problems at high Rayleigh numbers is presented. The unsteady incompressible Navier-Stokes equations and energy equation are discretized by means of a semi-explicit scheme with respect to time variable. As the time-marching scheme, the fractional step method is used effectively. Numerical results demonstrate that the present method is capable of solving the complex transient phenomena accurately and in a stable manner for Rayleigh numbers up to 10^{10}.

I. INTRODUCTION

Buoyancy-driven flow within differentially heated cavity in enclosures is a suitable typical problem for testing and verifying computational codes developed in a variety of practical thermal problems. Numerical simulations of the natural convection in square enclosures have been performed by many researchers using finite difference method[1], finite element method[1] and boundary element method[2]. For the computation of two-dimensional flow, De Vahl Davis[3] produced the bench mark solutions up to Rayleigh numbers of 10^6 using finite-difference schemes of second-order accuracy on meshes of different size. De Vahl Davis and Jones[1] also provided a basis for the assessment of the various numerical methods for the solutions of the problem.

Recently, the natural convection fields for Rayleigh numbers up to 10^8 have been increasingly simulated[4)-8)]. Le Quere and De Roquefort[4] demonstrated that the spectral method used generates very accurate solutions through comparison with the bench mark solutions[3] up to Rayleigh numbers of 10^6. Moreover, they shown some interesting phenomena for Rayleigh numbers of 10^7 and 4×10^7. Chenoweth and Paolucci[5] numerically solved the unsteady Navier-Stokes equations under the small-Mach-number approximation by means of an explicit predictor-corrector finite difference method. They compared in detail their numerical solutions with those of De Vahl Davis[3] and also shown numerical performances for Rayleigh numbers ranging from 10^7 to 8×10^7. Mizukami[6] shown the solutions for wide ranges of Rayleigh numbers from 10^3 to 10^8, which were computed effectively by the penalty/multiplier finite element method. Finite element solutions up to Rayleigh numbers of 10^8 were produced by Itakura et al.[7] using a third-order accurate upwind technique based on the choice of upwind and downwind points. However, details of the temporal evolution of the flow pattern can not be fully presented in the above-mentioned papers. In order to make clear the transient phenomena at high Rayleigh numbers, Okanaga and Tanahashi[8] investigated numerically the flow fields for Rayleigh numbers ranging from 10^6 to 10^8 using the GSMAC finite element method. They also discussed the validity of BTD (balancing tensor diffusivity) and the multi-pass explicit algorithm. However, there are few attempts to obtain numerical solutions for Rayleigh numbers larger than 10^8. Markatos and Pericleous[9] produced the steady-state solutions for values of the Rayleigh numbers ranging from 10^3 to 10^{16} by using the finite-domain method, and then they used conventionally a two-equation turbulence model for Rayleigh numbers larger than 10^6.

Recently, we have developed a novel finite element scheme based on the Petrov-Galerkin formulation using exponential weighting functions for solving numerically the unsteady incompressible Navier-Stokes equations for high Reynolds numbers[10),11)]. The numerical results demonstrated that the method is capable of solving the incompressible Navier-Stokes equations accurately and in a stable manner for a wide range of Reynolds numbers. More recently, the Petrov-Galerkin finite element method using exponential functions was also applied to natural convection problems in two-dimensional enclosed cavity[12)]. The unsteady incompressible Navier-Stokes equations and energy equation were discretized by means of a semi-explicit scheme in which velocity and temperature were treated explicitly and pressure was treated implicitly with respect to time variable. Moreover, the fractional step method was adopted effectively in our study[12)]. The validity of the present method was demonstrated for numerical solutions at Rayleigh numbers ranging from 10^6 to 10^8 through comparison with other existing solutions.

The purpose of this paper is to evaluate the workability and the numerical accuracy of the present method for heigher values of the Rayleigh number (Ra). For the computation at $Ra = 10^8$, two cases on meshes of different minimum size are performed, and our computational results are compared with other ones. In addition the transient phenomena at higher Rayleigh number of 10^{10} are successfully observed. As a result, the complex transient flows can be obtained and a steady state of the flow can not be recognized for $Ra = 10^{10}$. It is also found through our numerical performances that the Nusselt number is nearly proportional to Rayleigh numbers up to 10^8, whereas the value obtained for $Ra = 10^{10}$ is not on its straight line.

Throughout this paper, the summation convention on repeated indices is employed. A comma following a variable is used to denote partial differentiation with respect to space variable, and a dot over a variable denotes differentiation with respect to time variable.

II. GOVERNING EQUATIONS

Let Ω be a bounded domain in Euclidean space $\boldsymbol{R}^2$ with a piecewise smooth boundary Γ. The unit outward normal vector to Γ is denoted by $\boldsymbol{n}$. And also, T denotes a closed time interval.

The motion of the natural convection with the Boussinesq approximations is given by the following governing differential equations in dimensionless form[8)] :

momentum equation

$$\dot{u}_i + u_j u_{i,j} = -p_{,i} - \frac{1}{Gr}(\theta e_i - u_{i,jj}) \qquad in\ T \times \Omega \tag{1}$$

continuity equation

$$u_{i,i} = 0 \qquad in\ T \times \Omega \tag{2}$$

energy equation

$$\dot{\theta} + u_i \theta_{,i} = \frac{1}{Ra}\theta_{,ii} \qquad in\ T \times \Omega \tag{3}$$

where u_i is the velocity vector component, p is the hydrostatic pressure, θ is the temperature, $e_1 = 0$ and $e_2 = -1$. In the above equations Gr is the Grashof number and Ra is the Rayleigh number of the fluid.

By adopting a semi-explicit formulation with respect to time variable of equations (1) to (3) and applying the fractional step scheme[13)] to those equations, we obtain the following set of semi-discretized equations :

$$\frac{\bar{u}_i - u_i^n}{\Delta t} + u_j^n u_{i,j}^n = -\frac{1}{Gr}(\theta^n e_i - u_{i,jj}^n) \tag{4}$$

$$\phi_{,ii} = -\bar{u}_{i,i} \tag{5}$$

$$u_i^{n+1} = \bar{u}_i + \phi_{,i} \tag{6}$$

$$p^{n+1} = -\frac{1}{\Delta t}\phi \tag{7}$$

$$\frac{\theta^{n+1} - \theta^n}{\Delta t} + u_i^{n+1}\theta_{,i}^n = \frac{1}{Ra}\theta_{,ii}^n \tag{8}$$

where Δt is a time increment, $\bar{u}_i$ is an auxiliary velocity vector, ϕ is the modified velocity potential, and u_i^n denotes the value of u_i at time $n\Delta t$.

III. PETROV-GALERKIN FORMULATION USING EXPONENTIAL FUNCTIONS

In order to solve equations (4) and (8) in a stable manner up to high Rayleigh number, we shall adopt the Petrov-Galerkin finite element method using exponential functions[10]. On the other hand, the conventional Galerkin finite element formulation can be applied to solve numerically equations (5) to (7).

First of all, let us consider the Petrov-Galerkin weak formulation using exponential functions for equation (4). By applying the divergence theorem to the weighted residual statement of equation (4), we obtain the following weak form :

$$\int_{\Omega_i}\left\{\frac{\bar{u}_i-u_i^n}{\Delta t}+u_j^n u_{i,j}^n\right\}M_\alpha d\Omega+\int_{\Omega_i}\frac{1}{Gr}\theta^n e_i M_\alpha d\Omega+\int_{\Omega_i}\frac{1}{Gr}u_{i,j}^n M_{\alpha,j}d\Omega-\int_{\Gamma_i}\tau_i^n M_\alpha d\Gamma=0 \quad (9)$$

in which Ω_i denotes a subdomain in the whole domain Ω and $\tau_i^n \equiv u_{i,j}^n n_j/Gr$.

It is well known that the conventional Galerkin finite element approximation leads to spurious oscillatory solutions for natural convection at high Rayleigh number. In order to overcome such oscillations, we adopt the following exponential-type weighting functions[12] as M_α :

$$\left.\begin{aligned} M_\alpha(x_1,x_2) &= \sum_\gamma N_\alpha(x_1,x_2)e^{-\{a_1(N_\gamma x_1^\gamma - x_1^\alpha)+a_2(N_\gamma x_2^\gamma - x_2^\alpha)\}} \\ a_1 &= \frac{\alpha_1}{|L_1|}sgn(v_1^n) \quad , \quad a_2=\frac{\alpha_2}{|L_2|}sgn(v_2^n)\end{aligned}\right\} \quad (10)$$

where N_α is the shape function in two dimension, v_i^n is the velocity vector averaged in Ω_i, L_1 and L_2 are the reference lengths for x_1- and x_2-directions, respectively, as shown in Fig.1, α_1 and α_2 are the scaling parameters which control an effect of the upwinding, and $sgn(\ )$ denotes the signum function.

According to the same formulation as equation (4), the weak form of equation (8) can be derived as follows :

$$\int_{\Omega_i}\left\{\frac{\theta^{n+1}-\theta^n}{\Delta t}+u_i^{n+1}\theta_{,i}^n\right\}M_\alpha^* d\Omega+\int_{\Omega_i}\frac{1}{Ra}\theta_{,i}^n M_{\alpha,i}^* d\Omega-\int_{\Gamma_i}q^n M_\alpha^* d\Gamma=0 \quad (11)$$

where $q^n \equiv \theta_{,i}^n n_i/Ra$. In the above equation, M_α^* is the exponential-type weighting functions which can be replaced v_i^n, α_1 and α_2 in equation (10) with v_i^{n+1}, α_3 and α_4, respectively.

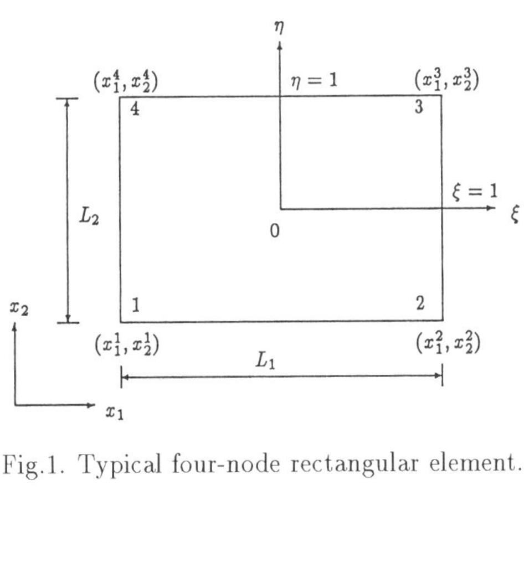

Fig.1. Typical four-node rectangular element.

On the other hand, by applying the Galerkin-type finite element formulation to equation (5), we obtain the following weak form :

$$\int_{\Omega_i} \phi_{,i} N_{\alpha,i} d\Omega - \int_{\Omega_i} \bar{u}_{i,i} N_\alpha d\Omega = \int_{\Gamma_i} \phi_{,n} N_\alpha d\Gamma \tag{12}$$

By applying the finite element discretization to equations (9), (11) and (12), we obtain the following finite element equations :

$$M_{\alpha\beta} \frac{\{\bar{u}_i\}_\beta - \{u_i^n\}_\beta}{\Delta t} + K_{\alpha\beta}(u_j^n)\{u_i^n\}_\beta + \tilde{M}_{\alpha\beta} e_i \theta_\beta^n = F_{\alpha\beta}\{\tau_i^n\}_\beta \tag{13}$$

$$M_{\alpha\beta}^* \frac{\theta_\beta^{n+1} - \theta_\beta^n}{\Delta t} + K_{\alpha\beta}^*(u_i^{n+1})\theta_\beta^n = F_{\alpha\beta}^* q_\beta^n \tag{14}$$

$$H_{\alpha\beta}\phi_\beta - G_{\alpha\beta i}\{\bar{u}_i\}_\beta = f_\alpha \tag{15}$$

In the above equations, each matrix is defined as

$$\left.\begin{aligned}
&M_{\alpha\beta} = Gr\tilde{M}_{\alpha\beta} = \int_{\Omega_i} M_\alpha N_\beta d\Omega \\
&M_{\alpha\beta}^* = \int_{\Omega_i} M_\alpha^* N_\beta d\Omega \\
&K_{\alpha\beta}(u_j^n) = \int_{\Omega_i} M_\alpha u_j^n N_{\beta,j} d\Omega + \int_{\Omega_i} \frac{1}{Gr} M_{\alpha,j} N_{\beta,j} d\Omega \\
&K_{\alpha\beta}^*(u_i^{n+1}) = \int_{\Omega_i} M_\alpha^* u_i^{n+1} N_{\beta,i} d\Omega + \int_{\Omega_i} \frac{1}{Ra} M_{\alpha,i}^* N_{\beta,i} d\Omega \\
&F_{\alpha\beta} = \int_{\Gamma_i} M_\alpha \tilde{N}_\beta d\Gamma \quad , \quad F_{\alpha\beta}^* = \int_{\Gamma_i} M_\alpha^* \tilde{N}_\beta d\Gamma \\
&H_{\alpha\beta} = \int_{\Omega_i} N_{\alpha,i} N_{\beta,i} d\Omega \\
&G_{\alpha\beta i} = \int_{\Omega_i} N_\alpha N_{\beta,i} d\Omega \\
&f_\alpha = \int_{\Gamma_i} N_\alpha \phi_{,n} d\Gamma
\end{aligned}\right\} \tag{16}$$

where $\tilde{N}_\beta$ denotes the shape function on the boundary.

Setting up equations (13), (14) and (15) for all the elements, respectively, we obtain the final set of equations as follows :

$$\boldsymbol{C} \frac{\bar{\boldsymbol{U}} - \boldsymbol{U}^n}{\Delta t} = \boldsymbol{F}^n \tag{17}$$

$$\boldsymbol{C}^* \frac{\boldsymbol{\Theta}^{n+1} - \boldsymbol{\Theta}^n}{\Delta t} = \boldsymbol{F}^* \tag{18}$$

$$\boldsymbol{B\Phi} = \bar{\boldsymbol{F}} \tag{19}$$

where $\boldsymbol{C}$ and $\boldsymbol{C}^*$ are the diagonal coefficient matrices constructed by using the mass lumping, $\boldsymbol{B}$ is the sparse coefficient matrix, and $\boldsymbol{F}^n$, $\boldsymbol{F}^*$ and $\bar{\boldsymbol{F}}$ denote the known vectors, respectively.

The computational cycle of this approach can now be given as follows :

Step 1 : Give the velocity vector and the temperature at time level $n\Delta t$, and get the auxiliary velocity vector $\bar{\boldsymbol{U}}$ from equation (17).
Step 2 : Determine the potential function $\boldsymbol{\Phi}$ by applying SCG method to equation (19).
Step 3 : Get u_i^{n+1} and p^{n+1} at new time level $(n+1)\Delta t$ from the weighted residual statements of equations (6) and (7), respectively.
Step 4 : Get $\boldsymbol{\Theta}^{n+1}$ at time level $(n+1)\Delta t$ by solving explicitly equation (18) for temperature, and go to step 1.

IV. NUMERICAL EXAMPLES

In order to demonstrate the validity of our method, we have computed the unsteady-state solutions of the natural convection for higher values of the Rayleigh number $Ra = 10^8$ and 10^{10}. The Prandtl number Pr is set to 0.71. In our numerical performances, we adopt the lowest interpolation functions in which the velocity, the scaler potential and the temperature are piecewise bilinear, and the pressure is constant over each element. The initial velocities and temperature are assumed to be zero everywhere in the interior domain.

Fig.2 shows the geometrical configuration and a finite element mesh which becomes fine near the boundaries. For the boundary conditions of the modified velocity potential ϕ, we assume the same ones as described in ref.11. The parameters that characterize the finite element approximation are summarized in Table I. In our computation at the Rayleigh number of 10^8, we perform the numerical simulations for different size of the minimum mesh, namely, cases (a) and (b) shown in Table I.

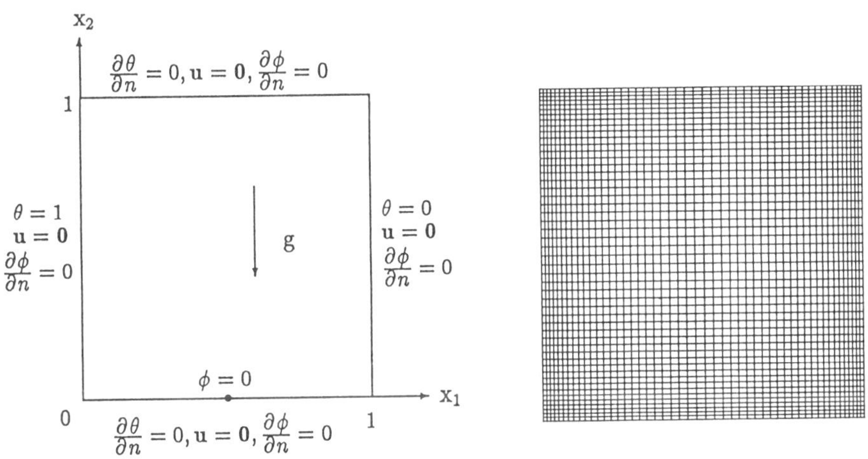

(a) Geometrical configuration (b) Finite element mesh

Fig.2. Natural convection in an enclosed cavity.

Table I. A summary of the parameters.

Ra		number of elements	minimum mesh size	Δt	$\alpha_i(i = 1 \sim 4)$
10^8	case (a)	50×50	0.011569	50.0	2.0
	case (b)	50×50	0.001969	50.0	
10^{10}		50×50	0.001162	1000.0	

In the following, we present some results for $Ra = 10^8$ and 10^{10}. Figs.3 and 4 show the velocity vector and isothermal fields obtained for cases (a) and (b) at $Ra = 10^8$, respectively. An interval of the isovalues in these cases is 0.1. It can be seen from these figures that the transient behaviour for both cases is generally similar and very complex, whereas there exists a stable steady-state solution to the set of equations at $Ra = 10^8$. In addition, the boundary layer of case (b) is slightly thinner than that of case (a). We show in Fig.5 the time histories of the Nusselt numbers at the hot wall and the cold one, respectively. It is clear that the Nusselt numbers at both walls tend progressively to its final value.

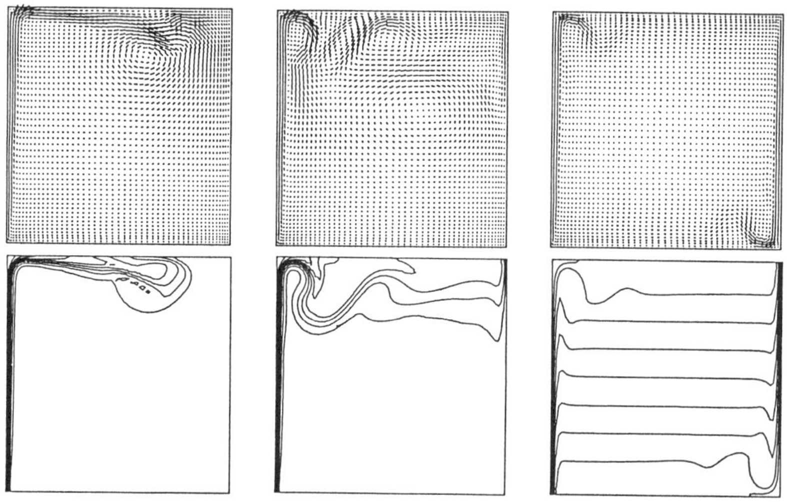

$t = 8.0 \times 10^4$ $t = 20.0 \times 10^4$ $t = 500.0 \times 10^4$

Fig.3. Velocity vector and isothermal fields at $Ra = 10^8$: case (a).

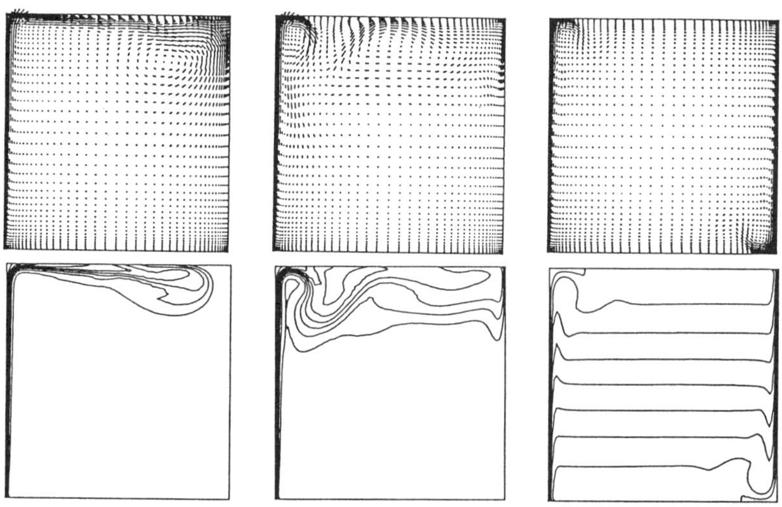

$t = 8.0 \times 10^4$ $t = 20.0 \times 10^4$ $t = 500.0 \times 10^4$

Fig.4. Velocity vector and isothermal fields at $Ra = 10^8$: case (b).

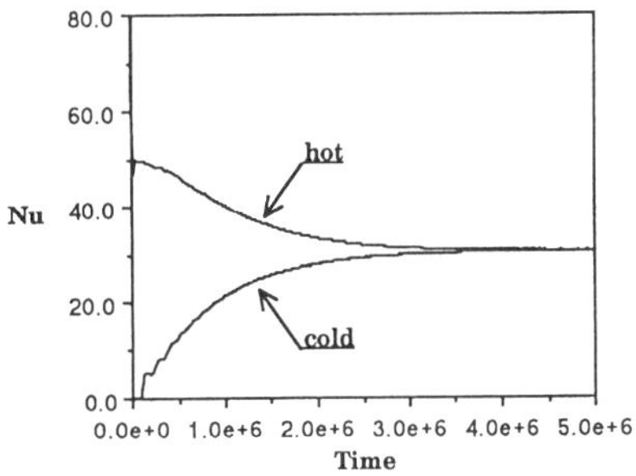

(a) Results of case (a)

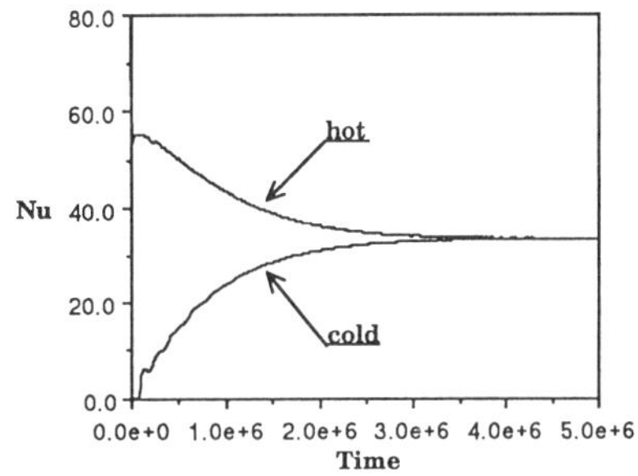

(b) Results of case (b)

Fig.5. Time histories of the Nusselt numbers at $Ra = 10^8$.

Fig.6 shows the velocity vector and isothermal fields for the higher Rayleigh number of 10^{10}. The transient behaviour in this case is more complex than that of the Rayleigh number set to 10^8. The boundary layer for $Ra = 10^{10}$ is also thinner than those for $Ra = 10^8$. We show in Fig.7 the behaviour of the Nusselt numbers at both vertical walls. We expect that both Nusselt numbers will tend in an oscillatory manner to about 115. As can be seen from these results, the behaviour for $Ra = 10^{10}$ does not attain clearly to a steady state.

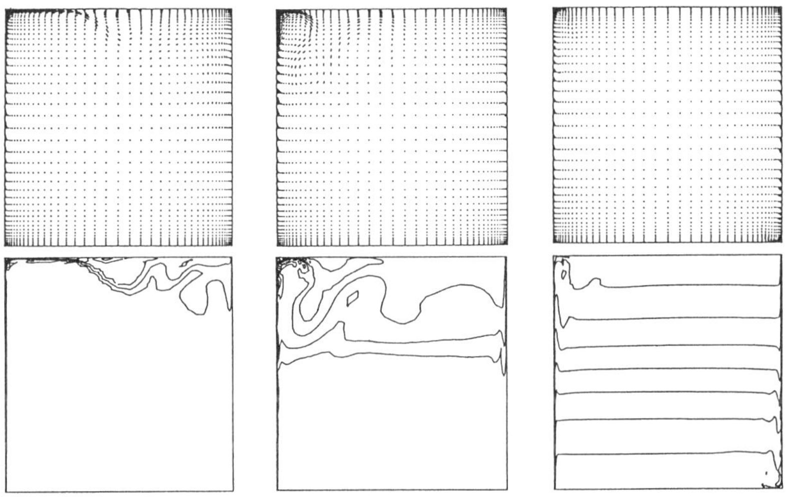

$t = 2.0 \times 10^6$ $t = 8.0 \times 10^6$ $t = 100.0 \times 10^6$

Fig.6. Velocity vector and isothermal fields at $Ra = 10^{10}$.

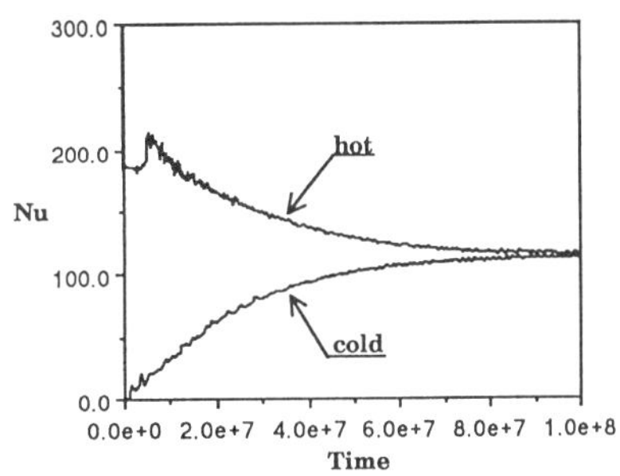

Fig.7. Time histories of the Nusselt numbers at $Ra = 10^{10}$.

In Table II we show two cases of results at $Ra = 10^8$ produced with our approach together with other existing numerical solutions. In the table, Nu_{av} denotes the Nusselt number on the hot wall. In order to obtain the values Nu_{av} for both cases, we average the fields over the last 10000 time steps shown in Fig.5. It is interesting to note that the value Nu_{av} of case (b) is larger than that of case (a). The agreement between our numerical results and other existing ones[6),8)] appears generally satisfactory except for the results in Markatos and Pericleous[9)] resorted to a two-equation model of turbulence.

Numerical results for $Ra = 10^{10}$ are shown in Table III. Similarly, we obtain the value Nu_{av} by averaging the fields over the last 10000 time steps shown in Fig.7. The fact that the characteristic values for $Ra = 10^{10}$ are larger than those for $Ra = 10^8$ is also noteworthy[14)].

Fig.8 shows a comparison of the mean Nusselt number Nu_{av} as a function of Rayleigh number. As can be seen clearly from this figure, the value Nu_{av} is nearly proportional to Rayleigh numbers up to 10^8, whereas there is the value obtained for $Ra = 10^{10}$ above the dotted line.

Table II. Comparison of characteristic values at $Ra = 10^8$.

		U_{max} x_2 ($x_1 = 0.5$)	V_{max} x_1 ($x_2 = 0.5$)	Nu_{av}	Nu_{max} x_2 ($x_1 = 0.0$)	Nu_{min} x_2 ($x_1 = 0.0$)
Present	case (a)	306.53 0.9381	2036.72 0.01157	30.402	66.242 0.02314	1.510 1.0000
	case (b)	287.59 0.9434	2250.82 0.00894	33.312	86.104 0.01874	1.466 1.0000
Okanaga & Tanahashi[8)]		320.72 0.924	2221.16 0.01358	30.063	84.934 0.00998	1.904 1.0000
Mizukami[6)]				28.93		
Markatos & Pericleous[9)]		514.3 0.941	1812 0.0135	32.045	61.06 0.0075	5.2246 0.997

Table III. Summary of characteristic values at $Ra = 10^{10}$.

	U_{max} x_2 ($x_1 = 0.5$)	V_{max} x_1 ($x_2 = 0.5$)	Nu_{av}	Nu_{max} x_2 ($x_1 = 0.0$)	Nu_{min} x_2 ($x_1 = 0.0$)
Present	1055.24 0.9755	22724.08 0.00349	115.351	377.422 0.00349	5.497 1.0000

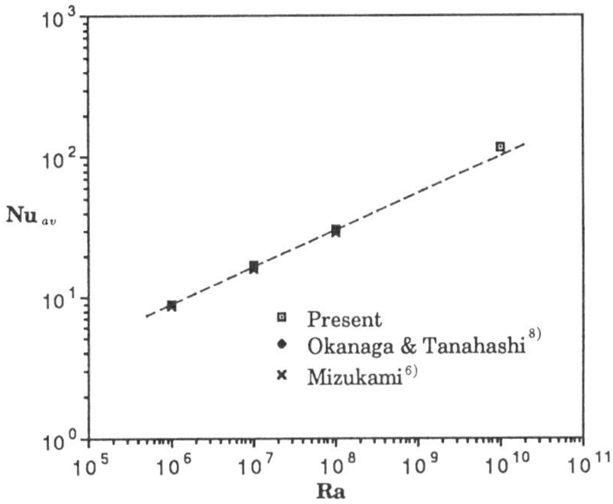

Fig.8. Mean Nusselt number as a function of Ra.

V. CONCLUSIONS

We have presented the Petrov-Galerkin finite elemet approach using exponential weighting functions for solving numerically the unsteady Navier-Stokes equations of a Boussinesq fluid for high Rayleigh numbers. In this approach, the fractional step method based on a semi-explicit scheme as a time-marching one is applied to the incompressible Navier-Stokes equations and energy equation, and then the semidiscrete system of equations obtained is formulated by the Petrov-Galerkin finite element method using exponential functions. In order to reduce the overall computing time, we also adopt a mass-lumping technique in our formulation. On the other hand, the Poisson's equation in terms of a scaler potential is formulated by means of the conventional Galerkin-type finite element method, and a final set of equations is solved effectively using SCG method.

For the computation of natural convection in an enclosed cavity, we have demonstrated some numerical performances for high Rayleigh numbers of 10^8 and 10^{10}. We have also shown the correlation between the Nusselt and Rayleigh numbers. As those results we belive that we may conclude that the behaviour at $Ra = 10^8$ tends progressively to a steady state, whereas the one at $Ra = 10^{10}$ does not attain to a steady state. This study demonstrate that numerical solutions can be obtained accurately and in a stable manner for natural convection problems at high Rayleigh numbers.

This approach will be immediately extended to three-dimensional natural convection problems, and we will present some typical examples of 3D flows in a future paper.

REFERENCES

1) De Vahl Davis, G and Jones, I.P., Natural convection in a square cavity : a comparison exercise. Int. J. Num. Meths. Fluids, 3 (1983), pp.227-248.

2) Wrobel, L.C. and Brebbia, C.A. (Eds.), Boundary Element Methods in Heat Transfer. Computational Mechanics Pub., (1992).

3) De Vahl Davis, Natural convection in a square cavity : a bench mark numerical solution. Int. J. Num. Meths. Fluids, 3 (1983), pp.249-264.

4) Le Quere, P. and De Roquefort, T.A., Computation of natural convection in two-dimensional cavities with Chebyshev polynomials. J. Comput. Phys, 57 (1985), pp.210-228.

5) Chenoweth, D.R. and Paolucci, S., Natural convection in an enclosed vertical air layer with large horizontal temperature differences. J. Fluid Mech., 169 (1986), pp.173-210.

6) Mizukami, A., Finite element analysis of natural convection in a square cavity at high Ra number. (in Japanese), Proc. of Symposium on Computational Methods in Structural Engineering and Related Fields, 16 (1992), pp.433-438.

7) Itakura, K., Fujima, S. and Tabata, M., Numerical computation of natural convection by a finite element method based on the choice of upwind and downwind points. (in Japanese), Proc. of Symposium on Numerical Fluid Dynamics, 6 (1992),pp.427-430.

8) Okanaga, H. and Tanahashi, T., Numerical analysis of natural convection in a square cavity at high Rayleigh numbers using the GSMAC finite-element method. (in Japanese), Trans. of JSME, (B),56 (1990), pp.94-101.

9) Markatos, N.C. and Pericleous, K.A., Laminar and turbulent natural convection in an enclosed cavity. Int. J. Heat Mass Transfer, 27 (1984), pp.755-772.

10) Kakuda, K. and Tosaka, N., Petrov-Galerkin finite element method using exponential functions for unsteady incompressible viscous flow problems. (in Japanese), J. Struct. Constr. Engng, AIJ, No.439 (1992), pp.189-198.

11) Kakuda, K. and Tosaka, N., Finite element approach for high Reynolds number flows. Theoretical and Applied Mechanics, 41 (1992), pp.223-232.

12) Kawahara, Y., Kakuda, K. and Tosaka, N., Unsteady analysis of natural convection by the Petrov-Galerkin finite element method using exponential functions. (in Japanese), Proc. of Symposium on Computational Mechanics, 6 (1992),pp.127-132.

13) Donea, J., Giuliani, S., Laval, H. and Quartapelle, L., Finite element solution of the unsteady Navier-Stokes equations by a fractional step method. Comput. Meths. Appl. Mech. Eng., 30 (1982), pp.53-73.

14) Tsuji,T. and Nagano, Y., Characteristics of a turbulent natural convection boundary layer along a vertical flat plate. Int. J. Heat Mass Transfer, 31 (1988), pp.1723-1734.

Numerical Study of Natural Convection along Wavy Wall

Nabil. Y. ABDEL-SHAFI*, Nobuhide NISHIKAWA**, and Yoshinori KIMURA*

* *Graduate Student, Faculty of Engineering, Chiba University, Yayoi 1-33, Chiba*
** *Professor, Faculty of Engineering, Chiba Univesity, Yayoi 1-33, Chiba*

Two-dimensional natural convection flows in enclosures with partitions and sinusoidal wall are studied numerically. The numerical method is based on MAC scheme. In the case of enclosures with partitions, Cartesian coordinates are used, while in the case of enclosures with sinusoidal wall, the employed method is an application of BFC (Boundary Fitted Coordinate). The finite differencing of the convection terms in Navier-Stokes equation employs a third order upwind difference. These solutions cover the parameter range, Pr=0.72, $1000 \leq Gr \leq 5x10^5$ and the aspect ratio, $1 \leq A \leq 2.8$. A comparison of the present study for partitioned enclosure partitions with the previous experimental work is made.

I. INTRODUCTION

Natural convection in enclosures is a subject that has been extensively treated in the literatures. Geophysical systems, nuclear reactor, solar collectors, and electronic equipment, are just a few applications motivating research in this area. A number of numerical and experimental studies are available on natural convection in rectangular enclosures with no partitions or obstructions are reported by De Vahl Davis [1)], Schinkel [2)], and Bejan [3)], among others. Comprehensive review of both experimental and theoretical studies have been given by Ostrach [4)]. Most investigation deal with the dependence of heat transfer on the relevant parameters. Relatively little attention has been given to the flow behavior in the enclosures with complex geometry. As evident from the above reviews of the literature, natural convection in enclosures, which have sinusoidal wall has not received adequate research attention.

The flow characteristics over wavy wall have been investigated over a long period of time in the literature. Patel et al. [5)] consider two model problems, the first is the flow past a flat plate with a sinusoidal cavity, and the second is the flow in a straight duct with one of the walls becoming wavy after a development length sufficient to establish a fully developed flow. ABDEL-SHAFI et al. [6)] performed a numerical model of flow over wavy wall using the HSMAC scheme.

The aim of the present investigation is to study the flow and heat transfer in an enclosure with partition or sinusoidal wall. In this study, the flow and heat transfer in a square enclosure with partitions is first studied to establish a reference for the second part of the study, which focuses on an enclosure with sinusoidal wall. The present calculation are performed in two different cases: One, enclosure with a two vertical side walls are held at constant but different temperatures, while the connecting horizontal top and bottom walls are considered adiabatic (case I). The other is the side walls are assumed adiabatic, while the top and bottom walls are isothermal (case II).

Results presented for enclosure, which have partitions, are compared with the previous work, which it given by Bajorek et al. [8)]. A finite difference technique based on the HSMAC method is used to obtain solutions of an enclosure with sinusoidal wall. The numerical method for enclosures with sinusoidal wall is the mixed form of Cartesian and contravariant velocity components expression that was described by Akune et al. [7)] .

II. GOVERNING EQUATIONS

In the present study, the flow is assumed to be two-dimensional, and laminar, and the fluid properties are assumed to be constant. The variation of density with temperature is calculated using the Boussinesq approximation. The governing equations are nondimensionalized using the following definitions of the dimensionless variables.

$$X = \frac{x}{L}, Y = \frac{y}{L}, u = \frac{u^*}{u_C}, v = \frac{v^*}{u_C},$$
$$P = \frac{p}{\rho u_C^2}, T = \frac{\theta - \theta_c}{\theta_h - \theta_c} \tag{1}$$

where,

$$u_C = \sqrt{gL\beta\Delta\theta}, \frac{u_C L}{\nu} = \sqrt{\frac{g\beta L^3 \Delta\theta}{\nu^2}} = \sqrt{Gr}$$

and L, u, v, p, θ, ν, $\Delta\theta$, u_C, g, and Gr are the width of enclosure, velocity components, pressure,temperature,kinematic viscosity, temperature difference, characteristic velocity, gravitational acceleration, and Grashof number, respectively. The subscripts c and h are cold and hot walls temperatures, respectively.

The dimensionless expression of the governing equations can be written in the cartesian form as :

Continuity equation ;

$$\frac{\partial u}{\partial X} + \frac{\partial v}{\partial Y} = 0 \tag{2}$$

Momentum equations ;

$$\frac{Du}{Dt} = -\frac{\partial P}{\partial X} + T + \frac{1}{\sqrt{Gr}}\nabla^2 u \tag{3}$$
$$\frac{Dv}{Dt} = -\frac{\partial P}{\partial Y} + \frac{1}{\sqrt{Gr}}\nabla^2 v \tag{4}$$

Energy equation ;

$$\frac{DT}{Dt} = \frac{1}{Pr\sqrt{Gr}}\nabla^2 T \tag{5}$$

where,

$$\frac{D}{Dt} = \frac{\partial}{\partial t} + u\frac{\partial}{\partial X} + v\frac{\partial}{\partial Y}$$

Differential equations in the Cartesian coordinate system are transformed to those in the general curvilinear coordinate system. The physical domain is transformed to a rectangular domain of general coordinate ξ and η. The method applied by Akune et at. [7] is adopted here for the numerical procedure. In this method the mixed form of Cartesian and contravariant velocity components is used. The general transformation from the physical plan (x,y) to the transformed plane (ξ, η) is given by $\xi = \xi(x, y), \eta = \eta(x, y)$. Then, Navier-Stokes equation for BFC method can be written in non-dimensional form as :

$$q_t + E_\xi + F_\eta = Gr^{-\frac{1}{2}}(R_\xi + S_\eta) + G \tag{6}$$

where,

$$q = J^{-1}\begin{bmatrix} 0 \\ u \\ v \\ T \end{bmatrix}, E = J^{-1}\begin{bmatrix} U \\ Uu + \xi_x p \\ Uv + \xi_y p \\ UT \end{bmatrix},$$

$$R = J^{-1}\begin{bmatrix} 0 \\ g^{11}u_\xi + g^{12}u_\eta \\ g^{11}v_\xi + g^{12}v_\eta \\ (g^{11}T_\xi + g^{12}T_\eta)/Pr \end{bmatrix},$$

$$S = J^{-1}\begin{bmatrix} 0 \\ g^{21}u_\xi + g^{22}u_\eta \\ g^{21}v_\xi + g^{22}v_\eta \\ (g^{21}T_\xi + g^{22}T_\eta)/Pr \end{bmatrix}, G = \begin{bmatrix} 0 \\ T \\ 0 \\ 0 \end{bmatrix}$$

and,

$$\begin{aligned} g^{11} &= \xi_x^2 + \xi_y^2, \\ g^{22} &= \eta_x^2 + \eta_y^2, \\ g^{12} &= \xi_x\eta_x + \xi_y\eta_y, \\ J &= \xi_x\eta_x - \xi_y\eta_y \end{aligned}$$

Equations (2-4) are used in the case of enclosures with partitions, while equations (6) are used in the case of enclosures with sinusodial wall.

III. NUMERICAL SCHEME

It has been an important subject how tactically a staggered differencing is applied on non-cartesian grids without oscillatory pressure or velocity. The reason for the oscillation is that the transformed velocity component are disaligned from coordinate directions. In the present study the velocity components are defined at the middle point of η and ξ lines of mesh, respectively. The scalar variables such as pressure, P and temperature, T are located at the geometric center of four vertices defining the mesh.

The time integration for Navier-Stokes equation is made explicitly by Euler method and gives the renewed velocities at a new time level. The pressure and velocities are adjusted to satisfy the continuity equation. The relaxation process is the iteration between the pressure expression, which is given by the diagonal dominant approximation, and the new estimate for the velocities as shown below.

In this method the residual is expressed with $D = div\ V$, and its differential expression is given by:

$$D = J[(J^{-1}\tilde{U})_\xi + (J^{-1}\tilde{V})_\eta] \tag{7}$$

Here, $\tilde{U}$ and $\tilde{V}$ are the contravariant velocity components, which are defined as:

$$\tilde{U} = \xi_x u + \xi_y v \tag{8}$$

$$\tilde{V} = \eta_x u + \eta_y v$$

and, where $\tilde{U}$ and $\tilde{V}$ are the renewed velocities.

The Poisson equation for pressure is given as:

$$[J^{-1}(g^{11}\delta p_\xi + g^{12}\delta p_\eta)]_\xi + [J^{-1}(g^{12}\delta p_\xi + g^{22}\delta p_\eta)]_\eta = \frac{D}{\Delta t} \tag{9}$$

where, $g^{11}, g^{12}, g^{12}, g^{22}$ are the metric coefficients.

The diagonal dominant approximation for the Laplacian becomes:

$$\begin{aligned} \delta p_{i,j} &= \omega D/\Delta t g d g \\ gdg &= 2(g^{11} + g^{22}) \\ &= 0.5(g^{11}_{i,j} + g^{11}_{i+1,j} + g^{11}_{i,j+1} + g^{11}_{i+1,j+1} + g^{22}_{i,j} + g^{22}_{i+1,j} + g^{22}_{i,j+1} + g^{22}_{i+1,j+1} \end{aligned} \tag{10}$$

Then the new estimate for the velocities on the cell sides becomes:

$$\begin{aligned} \delta u &= -\Delta t(\xi_x \delta p_\xi + \eta_x \delta p_\eta) \\ \delta v &= -\Delta t(\xi_y \delta p_\xi + \eta_y \delta p_\eta) \end{aligned} \qquad (11)$$

The new velocities obtained are substituted into continuity equation (equation 2 for case I and equation 7 for case II), and then the pressure value is renewed with $p_{k+1} = p_k + \delta p$. The iteration loop is iterated until convergence criterion $(p_{k+1} - p_k)_{ij}/\rho < 10^{-4}$ is attained. When the velocity field has been converged, the new temperature field is calculated using the energy equation. The velocity, pressure and temperature fileds are then used as the starting values for the next time step.

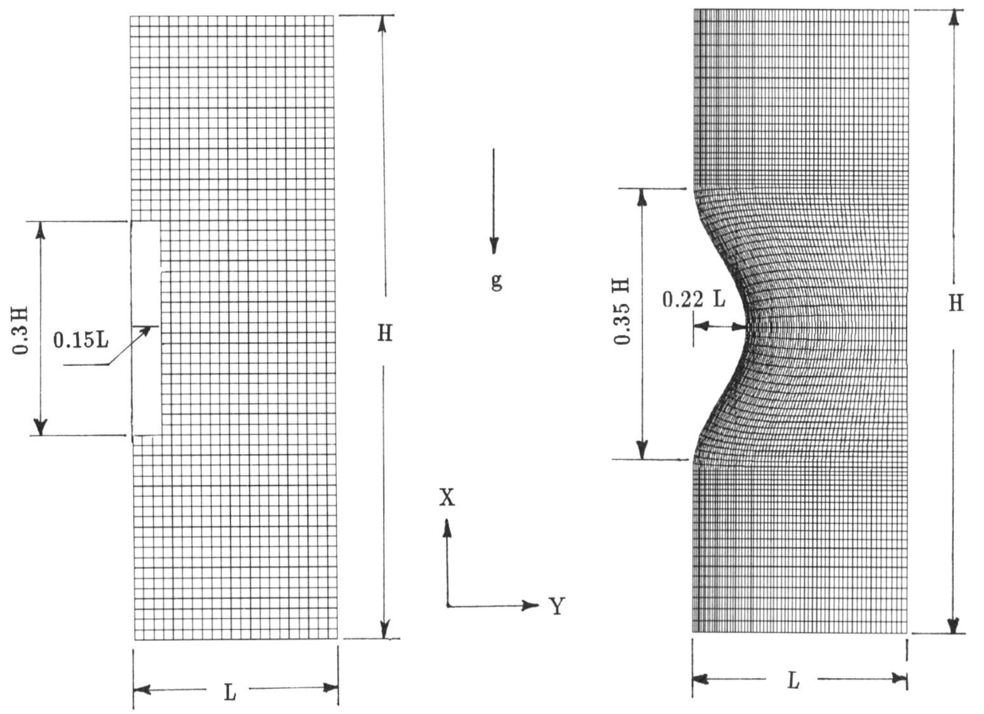

(a) Enclosure with partition (30x80).

(b) Enclosure with sinusoidal wall.

Fig.1. Typical grid schematics of the flow configuration.

IV. RESULTS AND DISCUSSION

Figures 1-a and 1-b show two examples of typical grid schematics of the flow in enclosures with partition (case A) and sinusoidal wall (case B), respectively. All the computations are performed on 30x80 grid in the case A and 100x100 in the case B. Figure 2 shows the schematic diagram of enclosure with two identical partitions located vertically at the top and bottom walls. In this figure the vertical walls are maintained isothermal at different temperatures, while the horizontal walls and the partitions are insulated. Figure 3 shows the comparsion of mid-height dimensionless temperature profiles of this study and experimental results for partitioned enclosure (Fig.2) at Gr=4.93×10^5, which was given by Bajorek et al. [8]. The figure shows good agreement between the results in the present study and this previous experimental work. In this figure, the both experimental and numerical temperature profiles for $0.1 \leq Y \leq 0.45$ is higher than the temperature profiles in the range $0.65 \leq Y \leq 0.9$. This is due to the deflection of the relatively warm fluid downword and the cold fluid upward by the top and bottom partitions, respectively.

Figure 4-a and 4-b show the velocity vectors of flows in enclosures with partition for case I and case II, respectively. In Figure 4-a we can see two vortices, one is near the center of enclosure, while the smaller one on the upstream region of the partition. In Figure 4-b we can see two vortices too, the larger one is near the bottom of enclosure (hot wall), while the smaller one near the top of the enclosure. Also, the velocity of flow in the cold region is small compared with the velocity in the hot region.

Typical local heat transfer data for partitioned enclosure (case I) at Gr= 2.8×10^5 are presented in Fig.5. The local Nusselt number, Nu along hot wall is plotted versus the nondimensional distance X. As can be seen in the figure, Nu attains its highest values up to X=0.1. This is attributable to the cold fluid, which descends from the cold wall, moves along the lower surface, and climbs up the hot wall near its base. Also, the spikes of Nu apear in the edge of obstacles. This figure also present a shear stress along the same wall.

Figure 6-a show the velocity vectors of flows in the enclosures with sinusoidal wall, where the two vertical side walls are isothermal, while the top and bottom walls are considered adiabatic. Figure 6-b shows the isotherm patterns for the same enclosure. On the other hand figure 7 shows the isotherm patterns for the enclosure with sinusoidal wall, where the side walls are assumed adiabatic, while the top and bottom walls are isothermal. In these figures we can see the vortices at both upstream and downstream region of wavy part of wall in small dimensionless time.

V. CONCLUSIONS

Based on the MAC method, the flow in the enclosures with partitions and sinusoidal wall is numercilly studied. These studies show a very significant flow behaviour due to the presence of partition and wavy wall in the convective heat transfer in an enclosure. For enclosures with partitions good agreement was found with the expermental results of the previous study [8]. The flow pattern in the enclosure with sinusoidal wall exhibits two eddies, one on either side of the wavy part. The caluculation for the enclosures, which have wavy wall with few humps, has been started by the present authors.

REFERENCES

1) De Vahl Davis,G., Laminar Natural Convection in an Enclosed Rectangular Cavity. Int.J.Hea Mass Transfer, Vol. 11(1968), pp.1675-1693.

2) Schinkel,W., Linthorst,S., and Hoogendoorn,C., The Stratification in Natural Convection in Vertical Enclosures. J. Heat Transfer, Vol. 105(1983), pp. 267-272.

3) Bejan,A., and Tien,C.L., Laminar Natural Convection Heat Transfer in a Horizontal Cavity with Different End Temperatures. ASME J. Heat Transfer, Vol. 100(1978), pp. 641-647.

4) Ostrach.S., Natural Convection in Enclosures. ASME J. Heat Transfer, Vol. 110(1988), pp. 1175-1190.

5) Patel,V.C.,Tyndall.J., and Yoon,J.Y., Laminar Flow Over Wavy Walls. ASME J. Heat Transfer, Vol. 100(1978), pp. 641-647.

6) Abdel-Shafi,N.Y., Nishikawa,N., and Sudou,H., Numerical Analysis of Laminar Flows Over Walls with Protuberance. Theoretical and Applied Mech., Vol. 41(1992), pp. 213-222.

7) Akune,O.,and Nishikawa,N., Visual Monitoring of CFD Jobs on Graphic Computer and Mainframes. Proc. 2nd Symp. CFD(1988)pp.487-490(in Japanese).

8) Bajorek,S., and Lloyd,J., Experimental Investigation of Natural Convection in Partitioned Enclosures. ASME J. Heat Transfer,Vol.104(1982), pp.527-532.

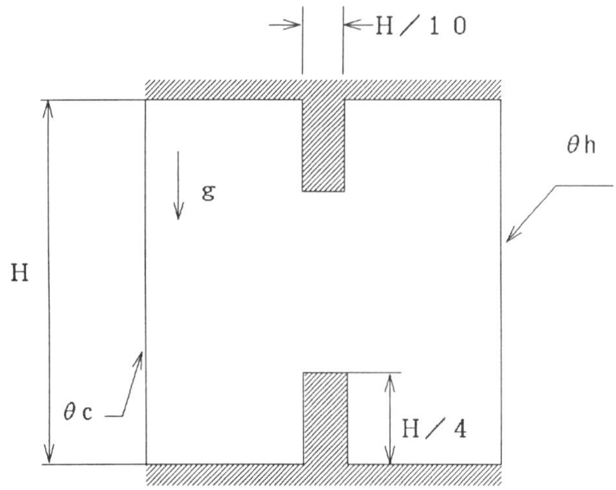

Fig. 2. Schematic diagram of enclosure with two partitions.

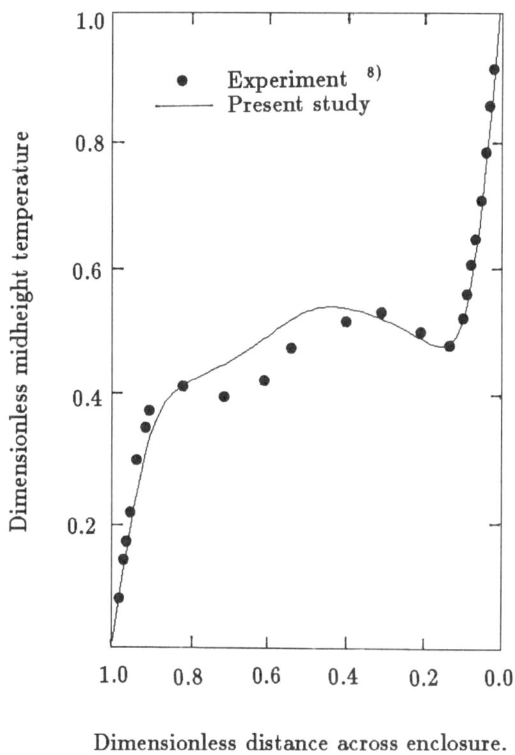

Fig. 3. Comparison of midheight temperature profiles for present study and expermental work [8] ($Gr=4.93x10^5$, $Pr=0.72$, $A=1$).

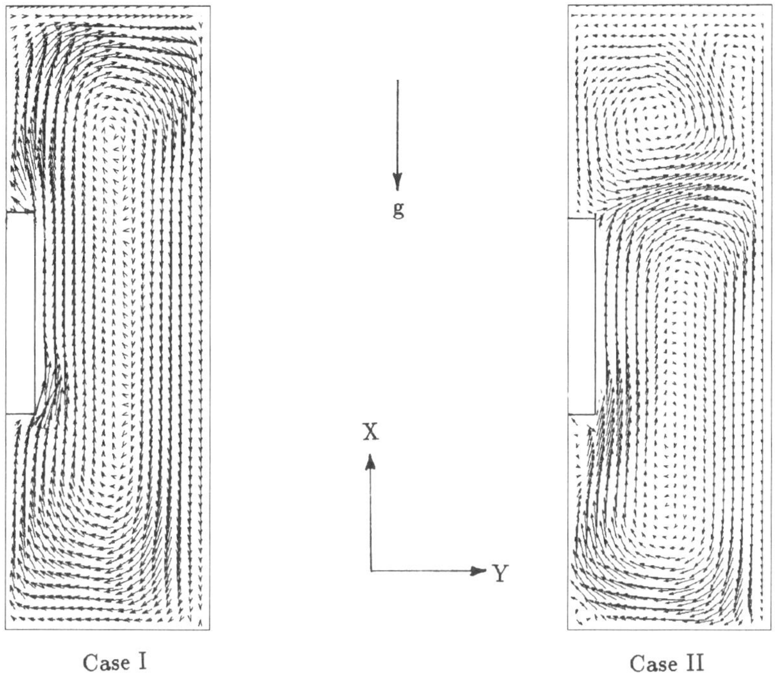

Fig. 4.The velocity vectors of flow in enclosure with partition (Gr=3.5x10^5, A=2.8).

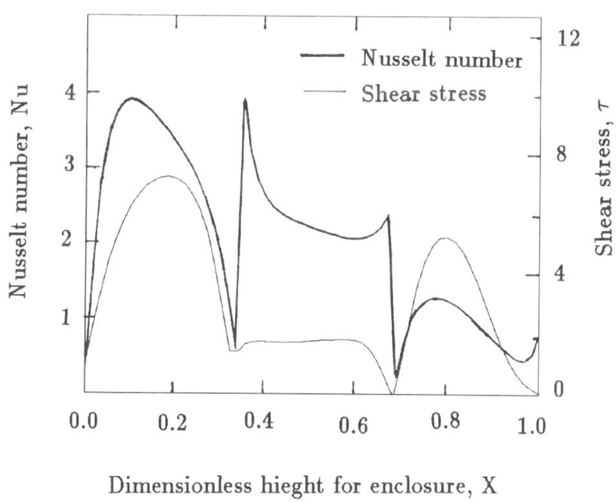

Fig. 5.Variation of local Nusselt number, Nu and shear stress for vertical hot wall for enclosure with partition, case I (Gr=2.8x10^5, A=2.8).

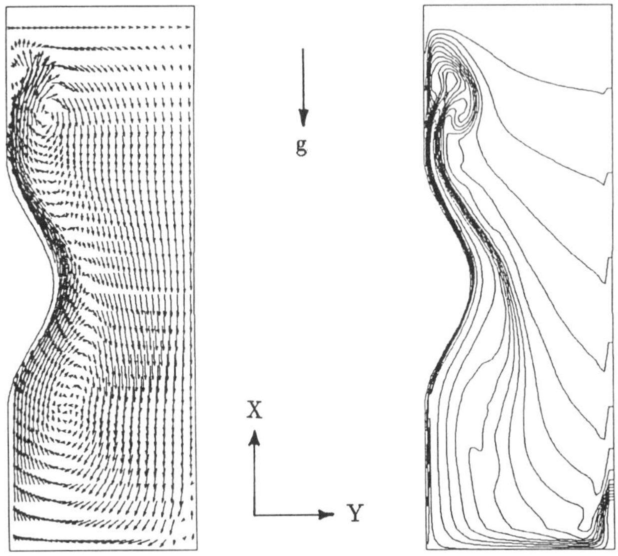

(a) Velocity vectors (b) Isotherms

Fig. 6. The results for enclosure with sinusoidal wall, case I ($Gr=3.5x10^5$, A=2.8).

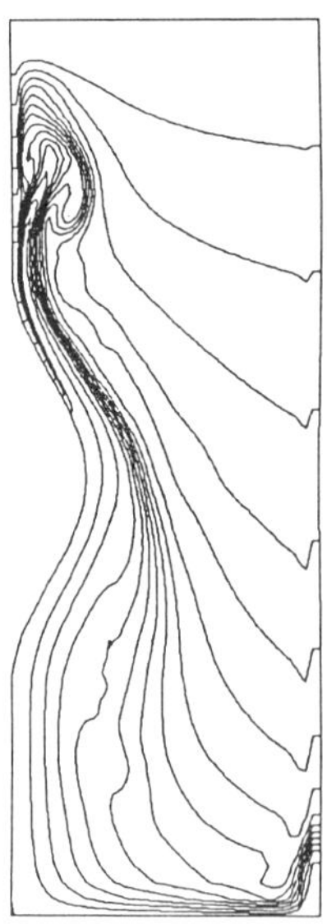

Isotherms

Fig. 7. The results for enclosure with sinusoidal wall, case II ($Gr=3.5x10^5$, A=2.8).

Analysis of Windward Performance of Sailing Crafts

Takeshi SUGIMOTO

Department of Mechanical Engineering, Saitama Institute of Technology, Okabe, Saitama

The conventional wing theory is applied to analyze windward performance of sailing crafts. Sailing crafts gain the available power due to the sail thrust and lose the necessary power due to the hull-keel resistance. Steady-state sailing holds at two kinds of trim velocities when the available and necessary powers are in equilibrium. While the smaller trim velocity of the two is unstable, the larger trim velocity is stable. The convergence of numerical scheme for the velocity predicting program (VPP) is discussed on the same basis for this static stability of basic equations.

I. INTRODUCTION

The most important performance of sailing crafts is their ability to run windward[1]. The equilibrium between aerodynamic and hydrodynamic forces and moments leads to the steady-state sailing to windward. Hence analysis of the performance needs understanding of the complex interactions of forces and moments. Classical analyses are quite empirical and analytical approach to the problem belongs to rather new idea[2-3]. These studies start even from analytical approach but results in numerical, parametrical calculations. The most recent approach is full simulation[4]. Nonlinear basic equations are solved numerically by using computers. Therefore it is still open to analytically understand mechanics of windward sailing. It is also an open question why and how iterations in numerical schemes can be convergent.

Introducing the conventional wing theory into analyses of sail aerodynamics and hull-keel hydrodynamics, the present author discusses the windward performance of sailing crafts: the static stability with respect to crafts' speed. He makes some remarks on the convergence of iterative calculations included in the velocity prediction program (VPP) used in yacht design.

II. BASIC EQUATIONS

We assume that the air flow is steady and inviscid, and that the sea surface is flat. Figure 1 shows the geometry of velocity vectors and the equilibrium of forces. Let V_T, V_R, V, λ, θ, and ϕ denote the true wind velocity, the relative wind velocity, the craft's speed, the ratio of the craft's speed to the true wind velocity, the angle of the sailing course to the true wind, and the angle of the sailing course to the relative wind, respectively, and we have

$$V = \lambda V_T , \tag{1}$$

$$V_R = \sqrt{\lambda^2 + 2\cos\theta\,\lambda + 1}\; V_T , \tag{2}$$

$$\sin\phi = V_T \sin\theta / V_R = \sin\theta / \sqrt{\lambda^2 + 2\cos\theta\lambda + 1}, \tag{3}$$

$$\cos\phi = (V_T \cos\theta + V)/V_R = (\cos\theta + \lambda)/\sqrt{\lambda^2 + 2\cos\theta + 1}. \tag{4}$$

The true wind velocity shall be the value just outside the atmospheric boundary layer on the sea surface.

The conventional wing theory assures that the drag coefficient is in direct proportion to the square of the lift coefficient. This idea is applicable to flow of inviscid and viscous fluids, because the lift of wings without camber is an odd function of the incidence and the drag is an even function of the incidence. As far as the sea surface is approximately flat, we can use the reflection technique. Therefore the use of the conventional wing theory is quite suited for our problem.

Let us consider hydrodynamic forces produced by a hull and a keel. A keel is a wing beneath the sea with its mirror image reflected by the sea surface. The side force corresponds to the lift. For the brevity of the theory, we can include the effect of the wave drag into the form drag coefficient. Hence we have the following expression of the hydrodynamic resistance:

$$R = \frac{1}{2}\rho_{aq}V_T{}^2S_KC_{DK} + \frac{S^2}{\frac{1}{2}\rho_{aq}V_T{}^2S_Ke_K\pi AR_K}, \quad (5)$$

where ρ_{aq}, S_K, C_{DK}, e_K, and AR_K denote the water density, the projected keel area, the form drag coefficient of hull and keel, the wing efficiency of keel, and the aspect ratio of keel, respectively.

Secondly we consider the forces acting on sails. The flexible sail in flow deforms and has a positive camber at the positive incidence, and *vice versa.* Therefore the lift and the drag are odd and even functions of the incidence, respectively, although sails don't have symmetric camber. This fact is confirmed through the comparison between experiments and the theory[5]. Since the same argument is applicable to the sail aerodynamics, we have the following relation between the aerodynamic drag D and the sail lift L:

$$D = \frac{1}{2}\rho_{air}V_R{}^2S_SC_{DS} + \frac{L^2}{\frac{1}{2}\rho_{air}V_R{}^2S_Se_S\pi AR_S}, \quad (6)$$

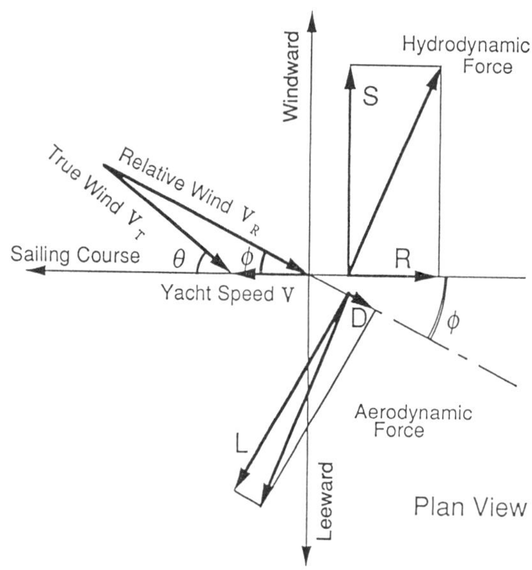

Fig. 1. Definition and nomenclature for the analysis.

where ρ_{air}, S_S, C_{DS}, e_S, and AR_S denote the air density, the projected sail area, the form drag coefficient of sail, the wing efficiency of sail, and the aspect ratio of sail, respectively.

The sail thrust, the sailing course component of the aerodynamic force, is given by

$$L \sin\phi \; - D \cos\phi \; . \quad (7)$$

The aerodynamic side force, the leeward component of the aerodynamic force, is given by

$$L \cos\phi \; + D \sin\phi \; . \quad (8)$$

III. ANALYSIS OF POWERS AND STATIC STABILITY OF CRAFTS

The necessary power P_n due to the hydrodynamic resistance is given by

$$P_n = RV \, . \quad (9)$$

The available power P_a due to the sail thrust is given by

$$P_a = (L \sin\phi - D \cos\phi)V \quad (10)$$

In the steady-state sailing, the aerodynamic and hydrodynamic side forces are in equilibrium:

$$S = L \cos\phi + D \sin\phi \quad (11)$$

Substituting Eqs.(1), (2), (3), (4), (5), (6), and (11) into Eqs.(9) and (10), we can eliminate V, V_R, ϕ, R, S, and D from Eqs.(9) and (10). After some algebra, we have the non-dimensional expression of powers, $\overline{P_n}$ and $\overline{P_a}$, defined as

$$\overline{P_n} \equiv \frac{P_n}{\frac{1}{2}\rho_{air}V_T{}^3 S_S} = \frac{C_{DK}\lambda^3}{\sigma\beta} + \frac{\sigma\beta}{e_K\pi AR_K}\left\{C_L(\cos\theta + \lambda) + \left(C_{DS} + \frac{C_L{}^2}{e_S\pi AR_S}\right)\sin\theta\right\}^2\left(\lambda + 2\cos\theta + \frac{1}{\lambda}\right), \quad (12)$$

and

$$\overline{P_a} \equiv \frac{P_a}{\frac{1}{2}\rho_{air}V_T{}^3 S_S} = \left\{C_L\sin\theta - \left(C_{DS} + \frac{C_L{}^2}{e_S\pi AR_S}\right)(\cos\theta + \lambda)\right\}\lambda\sqrt{\lambda^2 + 2\cos\theta\,\lambda + 1}, \quad (13)$$

where

$$\sigma \equiv \frac{\rho_{air}}{\rho_{aq}}, \quad (14)$$

$$\beta \equiv \frac{S_S}{S_K}, \quad (15)$$

and the lift coefficient:

$$C_L \equiv \frac{L}{\frac{1}{2}\rho_{air}V^2 S_S}. \quad (16)$$

Equations (12) and (13) show how non-dimensional powers, $\overline{P_n}$ and $\overline{P_a}$, depend on the velocity ratio λ., and physical and design parameters. The physical parameter is σ. Design parameters are C_{DK}, e_K, AR_K, C_{DS}, e_S, AR_S and β. The lift coefficient C_L and the angle of the sailing course to the true wind θ are parameters to specify conditions between sails and winds.

The necessary power $\overline{P_n}$ consists of the term in proportion to λ^{-1} and polynomial terms of λ up to the third power. The term of λ^{-1} is monotone decreasing for positive λ. Polynomial terms are monotone increasing for positive λ. As the necessary power is the sum of monotone increasing and decreasing functions, this power is the positive function of λ with the single minimum. The power diverges in proportion to λ^{-1}, as λ tends to zero. The power diverges in proportion to λ^3, as λ tends to infinity.

The available power $\overline{P_a}$ has two zeros at λ=0 and λ_z:

$$\lambda_z = \frac{C_L\sin\theta}{C_{DS} + \frac{C_L{}^2}{e_S\pi AR_S}} - \cos\theta \quad (17)$$

The power above is positive in the segment of $\lambda \in [0, \lambda_z]$, and has the single maximum.

Hence $\overline{P_n}$ and $\overline{P_a}$ cross each other at two points at most. These points correspond to the equilibria.

Figure 2 shows a numerical example of powers against the velocity ratio. Assumptions are as follows:

σ=0.001225, β=10, $e_K = e_S$=1, AR_K=1, AR_S=10, $C_{DK} = C_{DS}$=0.01, C_L=1, and $\theta = \pi/4$.

In this example, the smaller velocity ratio of equilibrium λ_l and the larger one λ_u are 0.06221 and 1.239, respectively.

Let us study the static stability of windward sailing by using Fig. 2. The equilibrium, $\lambda = \lambda_l$, is *unstable* because of the following reason. If this equilibrium is disturbed to speed up the craft, the excess available power accelerates the craft further. If the craft is made slow down, the excess necessary power decelerates the craft.

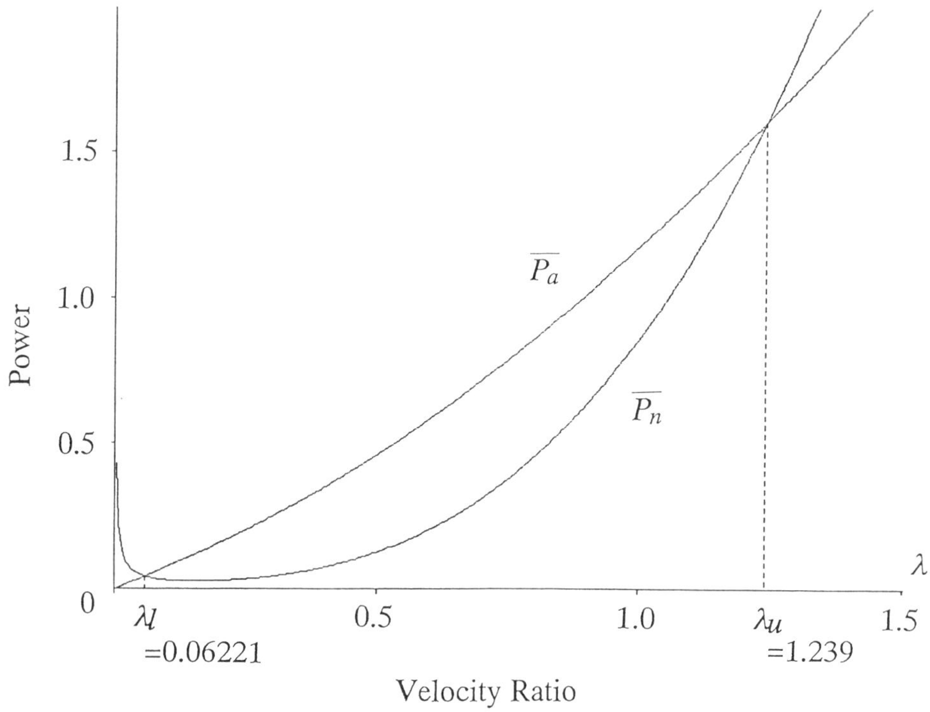

Fig. 2. Powers against the velocity ratio.

The equilibrium, $\lambda = \lambda_u$, is *stable*. If this equilibrium is disturbed to speed up the craft, the excess necessary power slows down the craft. If the craft is made slow down, the excess available power accelerates the craft. Hence this equilibrium is sustained.

If the initial velocity ratio is less than λ_l, the craft will stop. If the initial velocity ratio is more than λ_l, the craft runs at the larger velocity ratio λ_u in the end.

IV. REMARKS ON ITERATIONS IN VELOCITY PREDICTION PROGRAMS

The polar plot summarizes the performance of yachts. The polar plot gives the speed of the yacht against the true wind angle. Yacht designers use velocity prediction programs (in short VPP) to get the polar plot. The procedure of VPP is quite complicated, because programs take account of the velocity dependence of C_{DS} and C_{DK}. They obtain the trim velocity through iterations starting from the initially guessed velocity. The convergence of iterations is closely related to the static stability of crafts against their speed. If $\overline{P_n}$ is greater than $\overline{P_a}$, we shall guess the smaller speed than the temporal speed in the next step. If $\overline{P_n}$ is smaller than $\overline{P_a}$, we shall guess the larger speed than the temporal speed in the next step. Then the stability of numerical calculations becomes similar to the physical nature of crafts. Hence we expect to obtain the stable trim velocity in the end of calculations.

Introducing the Newton-Raphson method, we can construct the more efficient scheme for prediction of the trim velocity. Let $f(\lambda)$ be $\overline{P_a} - \overline{P_n}$, and we have the relation between λ_n and λ_{n-1}, approximate values of the trim velocity at n-th and (n-1)th steps:

$$\lambda_n = \lambda_{n-1} - f(\lambda_{n-1}) / \frac{df(\lambda_{n-1})}{d\lambda}. \tag{18}$$

If we use the initial guess:

$$\lambda_0 = \lambda_z , \tag{19}$$

then we have the result:

$$\lambda_n \to \lambda_t \quad \text{as} \quad n \to \propto . \tag{20}$$

This scheme is better in efficiency of convergence than simple iterations. We have better not take account of the velocity dependence of C_{DS} and C_{DK} to calculate the derivative on the right-hand side of Eq.(18), because their contribution to derivatives is little. It is a matter of course that we should take account of their direct contribution to the function $f(\lambda)$ itself.

REFERENCES

1) Marchaj, C.A., Aero-Hydrodynamics of Sailing, 2nd ed. (Adlard Coles, 1988).
2) Letcher, J. S. Jr., Optimum Windward Performance of Sailing Craft. J. Hydronaut. **10** (1976), pp. 140-144.
3) Curtiss, H. C. Jr., Upright Sailing Craft Performance and Optimum Speed to Windward. J. Hydronaut. **11** (1977), pp. 42-48.
4) Larsson, L., Scientific Methods in Yacht Design. Annu. Rev. Fluid Mech. **22** (1990), pp. 349-385.
5) Sugimoto, T. and Sato, J., Aerodynamic Characteristics of Two-dimensional Membrane Airfoils. Trans. Jpn. Soc. Aeronaut. & Space Sci. **34** (1991), pp. 88-100.

A Thin Airfoil Experiment in a Cryogenic Wind Tunnel

Tadashi SAKAUE and Yutaka YAMAGUCHI

Department of Aerospace Engineering, National Defense Academy, Yokosuka, Kanagawa

Two thin airfoil models were tested at $108K$ of total temperature in NDA cryogenic wind tunnel and the pressure distributions were measured. The airfoil models, which have the chord length of $120mm$, and the aspect ratio of 0.5, were tested in the range of Mach number and that of Reynolds number, 0.3 to 0.7, 5 million to 12 million, respectively. As the test section area of this tunnel is small, the experimental data should be corrected by the sequential correction method which combines the corrections for the sidewall boundary layer and the top and bottom wall interference. The corrected results were compared with the results computed by a panel method and the experimental data obtained at an ambient low speed wind tunnel. These comparisons showed that the present sequential correction method works well for the present tunnel, and that the NDA cryogenic tunnel has a possibility to perform two-dimensional airfoil test with a model of low aspect ratio.

I. INTRODUCTION

The cryogenic wind tunnel concept has been proven as the best way to increase the Reynolds number capability of wind tunnels. This concept can be applicable to a relatively small wind tunnel used for basic aerodynamic research. The Japanese National Defense Academy (NDA) designed and constructed a two dimensional, continuous flow high speed cryogenic wind tunnel with a $0.06 \times 0.30m$ test section in 1985. It uses liquid nitrogen as a coolant, and the cooling system is based on the same concept of the NASA 0.3 m TCT.

As the second stage of the research work with the NDA tunnel, a cryogenic airfoil testing has been undertaken. Generally, the blockage facter FB, which is a ratio of model's front area and the test section area, should be less than 0.01 to minimize the top and bottom wall effects for tests in high subsonic and transonic Mach number ranges[1]. Furthermore, Ganzer, et al.[2] showed that the aspect ratio of the model should be greater than 2.0 to avoid the strong sidewall interferences of the test section when characteristics of an airfoil are obtained by measuring the pressure distribution along the mid span of the model in a two dimension wind tunnel.

The preliminary results of the airfoil testing with a R4 airfoil were in fair agreement with those obtained in the NASA 0.3 m TCT[3]. They indicated fairly stronger test section wall interfernces than those of NASA. The aspect ratio of the NDA's airfoil model was 0.5, and almost one-third of the NASA's model, for a model of $120mm$ chord legth was the smallest one, in which 43 pressure taps could be provided on the upper and lower surfaces with our method[3].

In the previous papers[4,5], the wall interferences in the NDA cryogenic tunnel was preliminarily treated by considering effects of the top and bottom walls and of the sidewalls separately, and it was indicated that the sidewall boundary layer effects might dominate the interferences.

The purpose of this paper is more precisely to assess the wall interferences in the NDA cryogenic tunnel and to evaluate the validity of the airfoil experiment with low aspect ratio models in this cryogenic wind tunnel by comparing with experimental data obtained at a low speed wind tunnel and a numerical computation.

II. WIND TUNNEL SYSTEM

The NDA cryogenic wind tunnel is a closed circuit tunnel with a centrifugal compressor driven by a $75kW$ electric motor. Fig.1 shows a schematic drawing of the present tunnel. The materials of the pressure shell are the stainless cast steel, SCS13, for the contraction section and the 1st diffuser, and the stainless steel, SUS304, for the other sections.

The tunnel has external insulation which mainly consists of $100mm$ of formed urethane foam. The final vapor barrier is the zinc plated steel of $0.3mm$ thickness. No air purging system for insulation material is provided.

This test section is $60mm$ wide, $300mm$ high and $1000mm$ long including the diffuser flap section. Fig.2 shows a schematic drawing of this tunnel's test section. Both of the top and bottom walls have two slots of $3.6mm$ width and 9 static pressure orifices.

Liquid nitrogen is directly injected in front of the compressor for cooling, and the resulted nitrogen gas is the working gas. The major characteristics of the present tunnel are presented in Table 1.

The data acquisition system consists of a 8-bit personal computer (CPU : $MC6809$) having $64KB$ RAM, two 5.25 inch floppy disk drives, a multi-function thermo-volt meter with an IEE-488 interface bass (GPIB). It records seven temperatures and four pressures. The most significant data are the flow total temperature, the total and the free stream pressures, for the flow parameters are calculated with them. As the present personal computer is a rather old 8-bit computer and the acquisition program is written in BASIC, the overall cycle time of data acquisition is now about 15 seconds.

III. AIRFOIL MODELS

In this study, two generic thin airfoil models were made. They have the chord length of $120mm$ and 6 % thickness ratio. Therefore, the aspect ratio of the models is only 0.5 as same as the previous model, R4 airfoil. The airfoil models (Type I and Type II) are made of the stainless steel of SUS304. Type I and Type II airfoils are only different at the trailing edge region. The thickness distribution at trailinge edge region of Type I is thicker than that of Type II. Type I model has 17 pressure orifices on the upper surface, 16 on the lower surface and one at the leading edge. Type II model has 17 pressure orifices on the upper surface, 15 on the lower surface and one at the leading edge. These pressure orifices were distributed on the mid span station. The diameter of a pressure orifice is $0.3mm$.

IV. WALL INTERFERENCE CORRECTIONS

Researches on wall interference effects up to transonic speeds have indicated that a correction to the free stream Mach number is sufficient for a large number of flows, and the section normal force coefficient should be rather emphasized instead of the experimental angle of attack[6] for two dimensional tests. In addition, the sidewall boundary layer effect should be considered for two dimensional airfoil tests.

As described in Refs. 4 and 5, the top and bottom walls effects, namely blockage effect, will be evaluated by the Blackwell's empirical method[7]. Though the Blackwell's method gives only Mach number correction, it requires simply the wall pressure distributions over the region of the airfoil. In the previous papers[4,5], available numbers of wall pressure taps were only 4 on the top wall and 3 on the bottom wall. But, the wall pressure distributions of top and bottom walls could be obtained with measuring pressures at 9 stations on each wall in this study. The blockage correction is given by the following equations shown in Reference 7. The blockage Mach number correction, ΔM_B, is calculated by

$$\Delta M_B = \overline{M}_E - \overline{M}_T \tag{1}$$

where $\overline{M}_E$ and $\overline{M}_T$ are the averaged experimental and theoretical Mach numbers at the wall over the airfoil, respectively.

$$\overline{M}_E = \frac{1}{c}\int_0^c \frac{M_u + M_l}{2}\,dx \tag{2}$$

$$\overline{M}_T = \left\{ \frac{5 + M_\infty^2}{1 - 0.2M_\infty^2\left[\left(\frac{\overline{u}}{U_\infty} + 1\right)^2 - 1\right]} - 5 \right\}^{1/2} \tag{3}$$

and

$$\frac{\overline{u}}{U_\infty} = \frac{A}{2\pi\beta}\left\{\left[\frac{0.75}{(0.75)^2 + \overline{Z}^2}\right] + \left[\frac{0.25}{(0.25)^2 + \overline{Z}^2}\right]\right\} \tag{4}$$

where $\beta = \sqrt{1 - M_\infty^2}$, and A is nondimensional airfoil cross-sectional area.

The second interference correction is for the sidewall boundary layers. Barnwell developed a method accounting for the effect of the attached sidewall boundary layer thickness at subsonic regime[8]. Sewall extended the Barnwell's method to transonic regime[9]. Furthermore, Murthy also extended the Barnwell-Sewall method to a method to account for the airfoil aspect ratio[10]. It is shown in Ref.10 that the Murthy's method gives the best correction when the aspect ratio of a model is greater than unity.

As the aspect ratio of the present airfoil models is 0.5, the Barnwell-Sewall type correction method was adopted for the present study. According to the theory, the relationship between the measured free stream Mach number, M_∞ and an interference-free equivalent Mach number, $\overline{\overline{M}}_\infty$ is given as

$$M_\infty - \overline{\overline{M}}_\infty = -\Delta M_W \simeq \frac{3M_\infty}{2 + M_\infty^2}\left(2 + \frac{1}{H} - M_\infty^2\right)\frac{\delta^*}{b} \tag{5}$$

ΔM_W and b are the sidewall Mach number correction and the tunnel width, respectively. And the relationship between the measured and equivalent pressure coefficients (C_p and $\overline{\overline{C}}_p$) is

$$\overline{\overline{C}}_p = \left(\frac{\overline{\beta}}{\sqrt{1 - \overline{\overline{M}}_\infty^2}}\right) C_p \tag{6}$$

where

$$\overline{\beta} = \left\{1 - M_\infty^2 + \frac{2\delta^*}{b}\left(2 + \frac{1}{H} - M_\infty^2\right)\right\}^{1/2} \tag{7}$$

Applying this theory to the correction of airfoil pressure distributions, the displacement thickness δ^* and the shape factor H of the boundary layer on the tunnel side walls must be known. In the previous paper[4] as the rough and preliminary study, the cryogenic boundary layer characteristics were estimated by using the ambient ones assuming that the boundary layer thickness at a fixed position on the tunnel sidewall is only proportional to the free stream unit Reynolds number, and the shape factor is invariable even in cryogenic conditions. It was needed to measure the wall boundary layer of the present tunnel for the more accurate assessment in cryogenic conditions.

In this study, a boundary layer mouse was designed, which has no moving parts, and is of trouble-free in cryogenic environment. The side wall boundary layer on the present test section was measured with it at the cryogenic conditions.

Four wall interference corrections were also attempted by combining the corrections for the side wall boundary layer and the top and bottom wall interference. The procedure employed was a regular sequential one. In this procedure, the side wall corrections were made first, and the top and bottom wall corrections were applied to the first corrected results. Therefore, the corrected free stream Mach numbers (M_I) and the corrected interference-free pressure coefficient ($C_{p,c}$) are given by the following equations.

$$M_I = \overline{\overline{M}}_\infty - \Delta M_B \tag{8}$$

$$C_{p,c} = \left(\frac{\overline{\beta}}{\sqrt{1 - \overline{\overline{M}}^2_\infty}} \right) C_p \tag{9}$$

V. DISCUSSION OF RESULTS

In the present cryogenic experiment, the total temperature was kept at about $108K$, and the maximum total pressure was $170kPa$. The range of the free stream Mach number and that of the angle of attack were 0.3 to 0.7, and $-2.0°$ to $3.0°$, respectively. The Reynolds number based on the airfoil chord length was in the range of 4.8×10^6 to 12×10^6.

The boundary layer measurement with the mouse revealed that the displacement thickness of the sidewall boundary layer δ^* is about $1mm$ and the shape factor H is about 1.4. They change little with the free stream Mach number. The sidewall Mach number correction ΔM_W was obtained with the equation (5) and plotted in Fig.3. Fig.3 shows that ΔM_W increases from about -0.02 to -0.36 with Mach number M_∞ though the previously estimated value was about $-0.03 \sim -0.02$.

As the chord length and the thickness ratio of the present models are $12cm$ and 0.06, respectively, the resulted FB of Type I and II airfoils is 0.024. This value of FB may be permissible up to high subsonic regime. Their blockage corrections to Mach number were calculated with the measured wall pressure distributions by equations (1) to (5). The corrections at the angle of attack of 0° are also plotted in Fig.3. Fig.3 indicates that the blockage correction to Mach number of Type II is almost negligible, but that of Type I is substantial. This difference between Type I and II may be caused by their different thickness distributions in near trailing edge region. Fig.3 also shows that the tunnel wall interferences on the present thin models is almost due to the sidewall boundary layer effects.

As mentioned in the previous chapter, the sequential correction by equations (8) and (9) gives corrected pressure distribution which is considered to be of no wall effects. Therefore, all experimental data were corrected with the sequential manner. The resulted Mach number correction increases from -0.005 to -0.047 when the nominal free stream Mach number, M_∞ increases from 0.300 to 0.700.

The effectiveness of the present corrections may be evaluated by comparing corrected test results with the corresponding ones of a computational method. In this study, a vortex panel method was employed for such a purpose. Vortex panels are of linearly varying strength, and the number of panels is 40. Computed incompressible results are converted into compressible ones by the Prandtl-Glauert rule.

Figure 4 shows comparison of corrected experimental results with the computed results of Type I and type II at the corrected Mach number of 0.56. The experimental and the computed results are generally in good agreement for both of the airfoils. For other Mach numbers and normal force coefficients, similar agreement between the experiment and the computed data was obtained.

The present airfoils have relatively sharp leading edge, and their distributions show the following feature. The lower surface pressure distributions of Type I and II models are nearly constant except near leading and trailing edges. The upper surface pressure of Type II is a gradual pressure distribution with pressure peak at almost 50 % chord point. Type I's upper surface pressure distribution is also gradual one, but the pressure peak shifts toward the trailing edge. This movement of the pressure peak position may be one of causes why the blockage effect becomes stronger for Type I airfoil than for Type II.

Figure 5 shows comparison of the present corrected experiment results with the ambient low speed wind tunnel results[11]. The test section of the low speed wind tunnel is a closed one. The results were, therefore, corrected with the standard method of Ref.12. The chord length and the aspect ratio of the models tested in the low speed tunnel are $300mm$ and 2, respectively. Mach number difference was also corrected by the Prandtl-Glauert rule as before. In Fig.5, octagonal symbols indicate the present corrected experimental pressure coefficients. The asterisks are of corrected data obtained in the ambient low speed tunnel. Though the Reynolds number of the present cryogenic experimental data of Fig.5-(a) is about ten times higher than that of the ambient data and about five times higher for Fig.5-(b), they are in good agreement. This is true for another values of normal force coefficients.

Thus, good correlations obtained from the above mentioned comparisons indicate that the present measured wall interference parameters are almost correct and the sequential correction works well for predicting two dimensional characteristics of an airfoil having an aspect ratio even as low as 0.5.

VI. CONCLUDING REMARKS

In this study, two thin airfoils were tested in subsonic flow at cryogenic conditions, and the cryogenic experimental data corrected by the empirical methods were compared with data computed by the vortex panel method, and also with data obtained at a low speed wind tunnel. The following remarks are obtained.

(1) Blockage Mach number correction is up to -0.01 for airfoils of 6 % thickness, and the sidewall boundary layer effects dominate the wall interferences of the NDA cryogenic tunnel.

(2) Corrected experimental data and calculated data are in good agreement.

(3) The sequential correction method gave good correlation of the present experiments for models of low aspect ratio with other experiment for those of higher aspect ratio.

These results may indicate that by adopting suitable wall correction, the present wind tunnel has a possibility to perform two-dimensional airfoil test with a model of low aspect ratio. Further experimental works, however, should be performed to validate this expectation. It is also required that the test section should be modified to decrease the sidewall boundary layer thickness as much as possible.

REFERENCES

[1] Tani, I., Kobashi Y. and Sato, H., "Fluid Dynamic Exparimental Method (in Japanese),"

Iwanami book Inc., Dec. 1977.

[2] Ganzer, U., Stanewsky, E. and Ziemann, J., "Sidewall Effects on Airfoil Test," *AIAA Journal,* Vol.22, No2, Feb. 1984.

[3] Yamaguchi, Y., Nakauchi, Y., Yorozu, M. and Saito, T., "Preliminary Airfoil Testing Experience in the NDA Cryogenic Wind Tunnel," *Proceedings of IEEE ICIASF'91,* October 1991, pp.223-230.

[4] Yorozu, M., Nakauchi, Y. and Yamaguchi, Y., "Preliminary study on Wall Interfererence Effect of The Cryogenic Wind Tunnel (in Japanese)," *Proceedings of JSASS Seibu Branch Annual Conference,* 1991, pp.17-20.

[5] Yorozu, M., Nakauchi, Y. and Yamaguchi, Y., "Study on Wall Interfererence Effect of The Cryogenic Wind Tunnel," *Preprint of Japan National Congress of Theoretical and Applied Mechanics,* 1992, pp.463-464.

[6] Nixson, D., "Transonic Aerodynamics," *Progress in Astronautics and Aeronautics,* Vol.81, 1981.

[7] Blackwell, J.A. Jr, "Wind-Tunnel Blockage Correction for Two-Dimensional Transonic Flow," *J. Aircraft,* Vol.16, No.4, 1979, pp.256-263.

[8] Barnwell, R.W., "Similarity Rule for Sidewall Boundary-Layer Effect in Two- Dimensional Wind Tunnels," *AIAA Journal,* Vol.18, September 1980, pp.1149-1151.

[9] Sewall, W. G., "Effects of Sidewall Boundary Layers in Two-Dimensionl Subsonic and Transonic Wind Tunnels," *AIAA Journal,* Vol.20, No.9, 1982, pp.1253-1256.

[10] Murthy, A.V., "Effect of Aspect Ratio on Sidewall Boundary-Layer Influence in Two-Dimensional Airfoil Testing," *NASA CR*-4008, 1986.

[11] Fujimoto, T., Ishikawajima-Harima Heavy Industries Co.,Ltd., private communication.

[12] Rae, W.H.,Jr. and Pope, A., "Low-Speed Wind Tunnel Testing," *John Wiley & Sons, Inc.* Second Edition, 1984.

Table 1. Major characteristics of NDA cryogenic tunnel.

Type	closed circuit, centrifugal compressor
Material of construction	SUS 304, SCS 13 stainless steel
Insulation	external
Cooling	liquid nitrogen
Test gas	nitrogen
Test section size (h,w,l)	0.30 × 0.06 × 1.00 m
Mach range	up to 0.83
Contraction ratio	14:1
Stagnation pressure	up to 1.77 bars
Stagnation temperature	108 K - ambient
Running time	up to 150 minutes
Max. Reynolds number (1/m)	97 million
Drive motor	75 kW
Fan speed	up to 2250 rpm
LN_2 tank volume	4.9 m^3

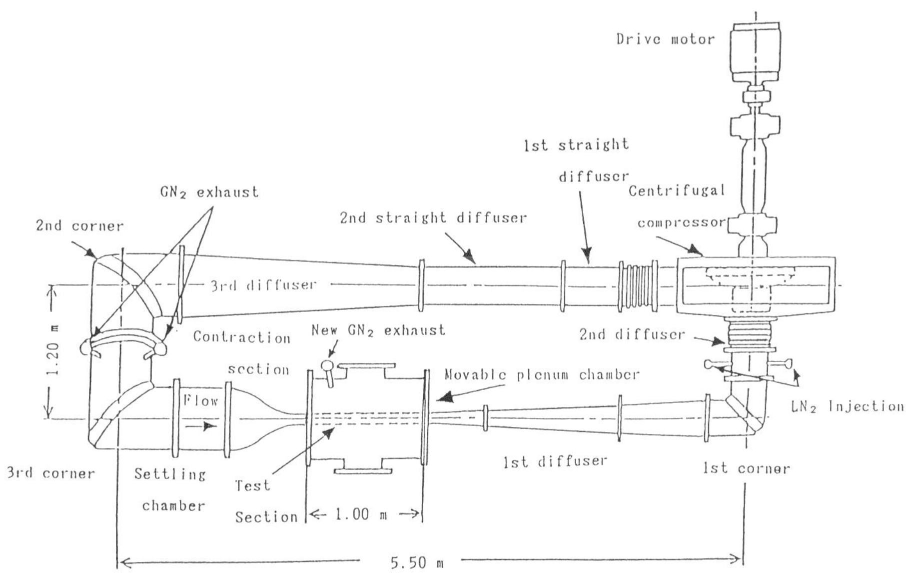

Figure 1. NDA cryogenic wind tunnel.

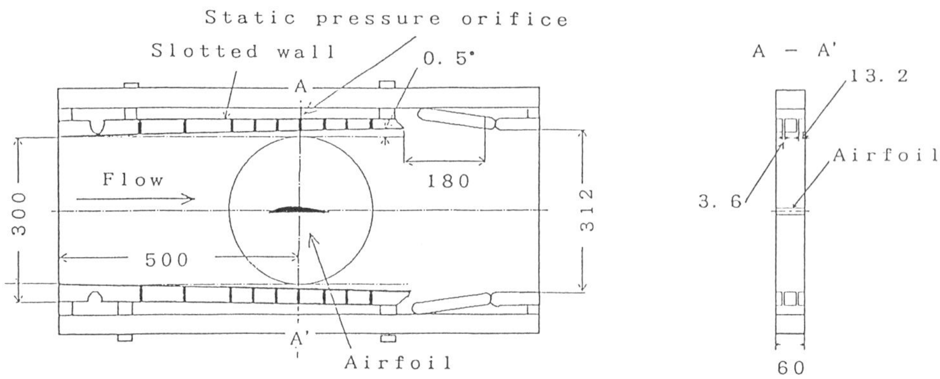

Figure 2. Test section.

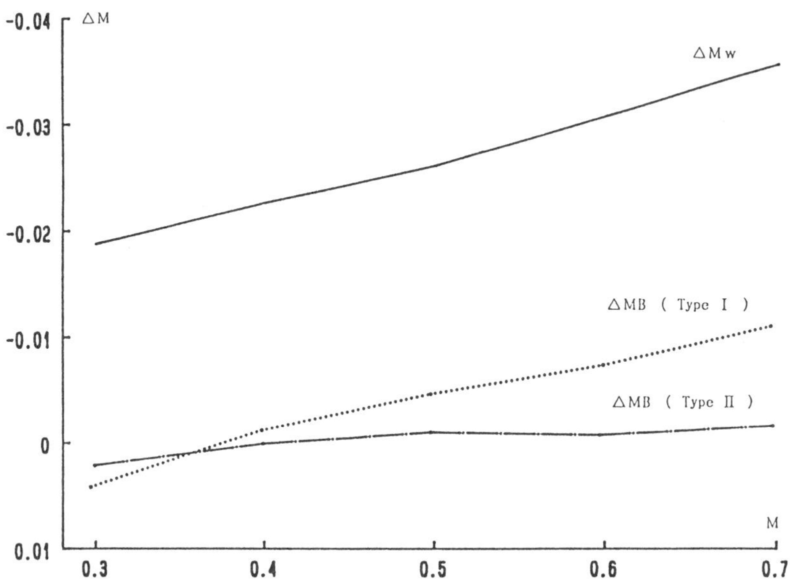

Figure 3. Freestream Mach number correction for the angle of attack of 0°.

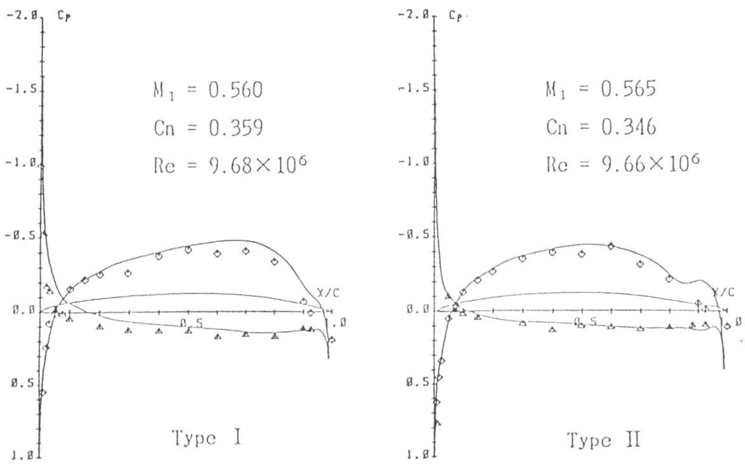

Figure 4. Comparison of experiment with computed results.

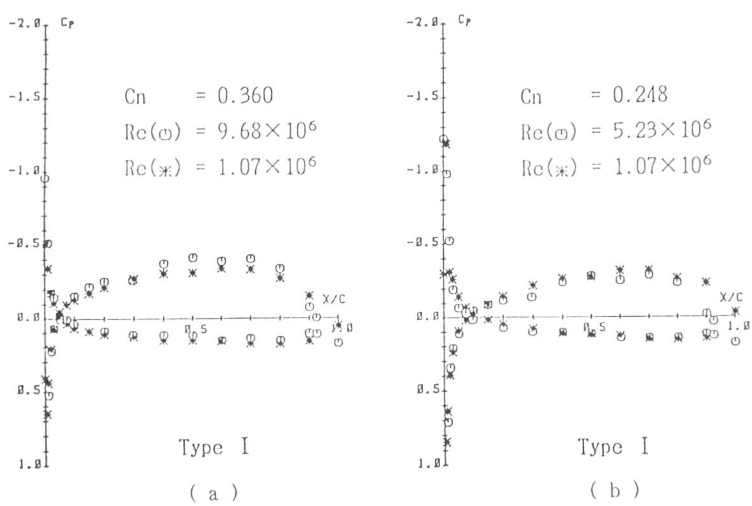

Figure 5. Comparison of experimental data of cryogenic tunnel with data of low speed ambient tunnel.

Experimental Study of a Low Reynolds Number Complex Airfoil Configuration

Masahide HATANO, Toshihiko OKAMOTO, and Osamu IKARI

Department of Aerospace Engineering, The National Defense Academy, Yokosuka, Kanagawa

An experimental study is performed on the specially designed complex split type two-element airfoil with closely coupled configurations at a low Reynolds number of 3.3×10^5. For the data presented here, the values of the stagger, gap and decalage angle are fixed at 1.0, 0.2 and -6 degree, respectively. Pressure distributions around the two airfoil components and the defect-velocity distributions in the far-wake are obtained, and the mean velocity and turbulence intensity data are measured in the flow region between the upper and lower airfoils. Flow visualization using kerosene vapor is performed under a perfectly identical condition for the quantitative measurement. The results show that the flow between two airfoils and in the cove region of upper airfoil characterize the total lift and drag perfomance of the complex airfoil system. The total lift increase results in a significant increase in the lift-to-drag ratio for the optimum configuration of the complex airfoil.

NOMENCLATURE

Cd = section drag coefficient, $d/(c\cdot q_\infty)$
Cl = section lift coefficient, $l/(c\cdot q_\infty)$
Cp = pressure coefficient, $1-(q/q_\infty)$
c = chord length
d = section drag force
De = decalage angle (in degree)
Ga = gap (in chord lengh)
l = section lift force
M = mid-point of chord line of airfoil
q = dynamic pressure, $(1/2)\times\rho U_\infty{}^2$
Re_c = Reynolds number basd on chord, $U_\infty\times c/\nu$
St = stagger (in chord length)
U = flow velocity
u = velocity fluctuation in free-stream direction
u' = rms value of u (turbulence intensity), $(\overline{u^2})^{1/2}$
x = distance in streamwise direction
y = distance normal to streamwise direction
α = angle of attack (of lower airfoil with respect to complex airfoil configuration)
ν = dynamic viscosity of air
ρ = density of air

Subscripts
C = complex airfoil configuration
l = lower airfoil
u = upper airfoil
∞ = free-stream condition

I. INTRODUCTION

In recent years, there has been an increasing interest in low Reynolds number aerodynamic in connection with the design of high altitude remotely piloted vehicles (RPVs), ultra-light aircraft, human powered aircraft and so on.[1,2] It is well-known that many significant aerodynamic problems occur below the chord Reynolds numbers of about 5×10^5, where the transition of boundary layer on airfoils is characterized by the formation of laminar separation bubble.

The aerodynamics of multi-element airfoil has also been investigated to obtain the performance of wing with high lift devices developed and to know the flow interaction mechanism around trailing edge flap and leading edge slat.[3,4] Most of the previous research involving these multi-element airfoils was in the Reynolds number range above 10^6, corresponding to the cruising condition of conventional aircraft.

Selberg et al.[5,6] looked over again the advantages of the dual-wing and biplane configurations. Mueller et al.[7,8] examined the canard tip-vortex / airfoil interactions and the performance of tandem airfoils. Their studies, however, were limited only to the double-wing systems consisited of two identical airfoils with fixed configurations.

Hatano et al.[9,10] designed a complex split-type airfoil system consisted of two non-identical airfoils whose relative position was variable, and measured its aerodynamic characteristics, in which the parameters used to describe the configuration of two airfoils were the stagger(St), gap(Ga) and decalage(De), as shown in Fig. 1. This specially designed airfoil system will be used as a biplane developed to upper and lower airfoils when it needs large lift in low speed take-off and landing regime, whereas it will be used as a closed monoplane for high speed cruise regime. They showed that the optimum configuration for the complex airfoil system was St = 1.0, Ga = 0.2 and De = 16 degree, where St and Ga were non-dimensionalized by chord length.

The present study is performed using several flow visualization techniques to obtain the flow mechanism in the regions around and between the upper and lower airfoils of the complex airfoil system, when it is in optimum configuration and at low Reynolds number regime. Then the results obtained are compared with those of the quantitative measurements of the lift and drag characteristics, which are obtained under a perfectly identical wind tunnel flow condition.

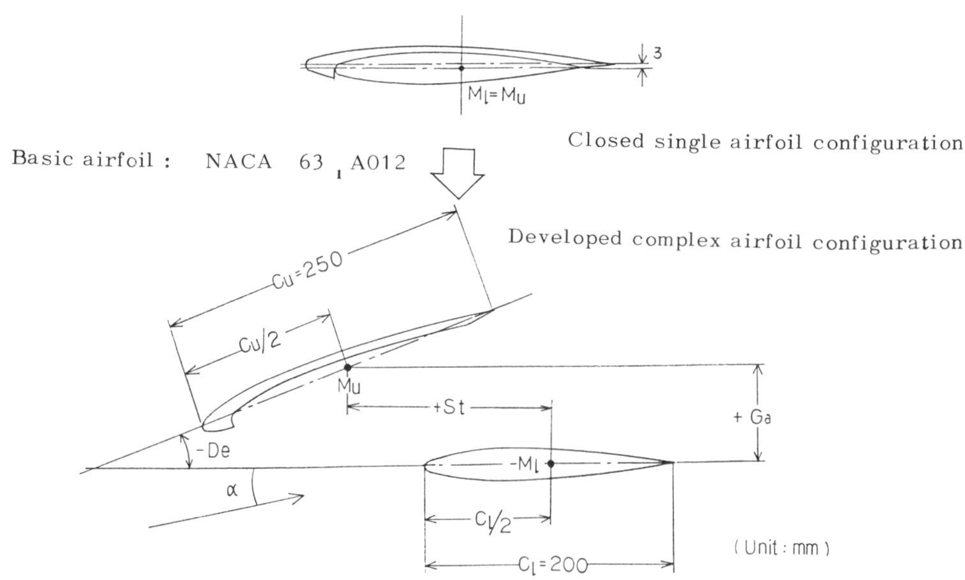

Fig. 1. Difinition of parameters for complex airfoil configuration.

II. EXPERIMENTAL APPARATUS AND PROCEDURE

All of the experiments were performed in a well-designed two-dimensional smoke tunnel at the National Defense Academy, which is shown schematically in Fig. 2. This tunnel is an open circuit, atmospheric exhaust type. The inlet nozzle of the tunnel has an area contraction ratio of 20 : 1, which permits a 0.15 m × 1.5 m rectangular test section. Clear smoke-lines can be observed up to the maximum free-stream velocity of U_∞ =26 m/s. The level of turbulence intensity in the test section is $u'/U_\infty < 0.4\%$. The wind tunnel free stream velocity for the present tests was 20 m/s, which corresponds to the chord Reynolds number of 3.33×10^5 .

The wind tunnel models of the specially designed complex airfoil system, which is designed through basic NACA63$_1$ A012 airfoil, have an overall chord length of 0.25 m when it is in the closed configuration. The upper airfoil, the chord length of which is 0.25 m, has a quite identical upper surface coordinate with the basic airfoil and a large cove region on its lower surface, then the section profile of this airfoil looks like a single surface cambered airfoil. The lower airfoil has a chord length of 0.20 m, whose surface coordinate is identical with the basic profile. Both upper and lower airfoils were equipped with surface pressure taps (0.5mmϕ) along the centerline of the models. The surface pressure distributions around the upper and lower surfaces of both airfoils were obtained using the U-type manometers, and the lift coefficients of the airfoils were caluculated from these data. Wake traverse of the U-type hot wire probe was performed to obtain the defect-velocity distributions in the far-wake of the complex airfoil configurations, and from which the drag coefficients were calculated. For the purpose of flow visualizations using oil flow, another couple of wind tunnel models without pressure taps were used. Several configurations were examined in the investigation, but the data presented here shows mainly for the configurations where the aerodynamic interaction was the most optimum, i.e. St = 1.0, Ga = 0.2 and De = -6 deg.

Measurements of the mean velocity and turbulence intensity distributions in the flow region between the upper and lower airfoils were taken using standard hot-wire anemometry. The measuring probe was attached to the end of a sting that forms part of two-dimensional, computer-controlled traversing mechanism inserted from the

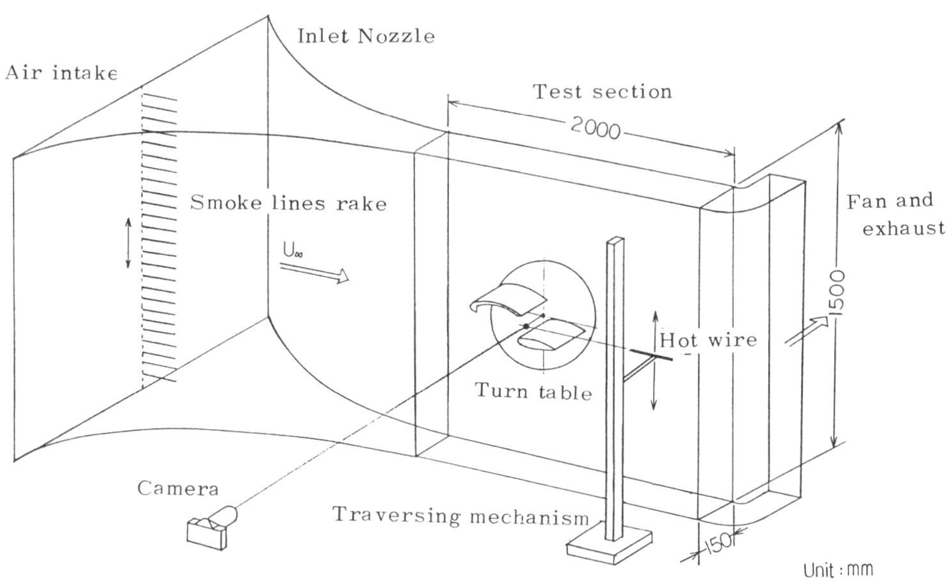

Fig. 2. Schematic view of experimental apparatus.

downstream wall of the test section. Measurements of the velocity and the turbulence intensity distributions in the cove regions were also performed using the same technique for complex airfoil at the angles of attack of 3.0°, 12.0° and 15.0°.

Flow visualizations were conducted to study flow configuration around these complex airfoils, using kerosene smoke lines and thin oil flow methods at angle of attack of $\alpha = -6^\circ \sim 15^\circ$. The oil used were a mixture of silicon-oil and fluorescent pigment. The visualization studies provided some valuable insight to the mechanism of flow field and added significant information to the quantitative data.[11]

III. RESULTS AND DISCUSSION

3.1 Flow Visualization Around Upper Airfoil

The geometrical profile of the upper airfoil is similar to a cambered single-surface airfoil. Therefore, it is required at first to confirm the performance of the upper airfoil by the methods of flow visualization. Fig. 3 shows a single streamline approaching the stagnation point and reattaching to the lower surface of the upper airfoil for the case of St = 1.0, Ga = 0.2 and De = -6 deg, and at an angle of attack of 12°. The recirculation region in the large cove region of the upper airfoil can be seen clearly. The separated shear layer from rearward-facing step reattaches at about 40% cu, and the point of reattachment is shifted toward upstream as the angle of attack is increased. The smaller closed recirculation region reduces the wake width of the upper airfoil and decreases the drag Cd_u, and therefore it decreases the total drag force of the complex airfoil.

The smoke-line patterns for the optimum complex configuration of St = 1.0, Ga = 0.2 and De = -6 deg, at angles of $\alpha = 3^\circ$ and 12° are shown in Figs. 4 (a) and (b), respectively. The basic airfoil, i.e. closed airfoil and single lower airfoil, stalls, and the flow on the upper surface fully separates at $\alpha = 12^\circ$. The flow on the upper surfaces of both airfoils, however, attaches for the optimum complex configuration, and which results in the increase of the total lift force, keeping the total drag force to be low.

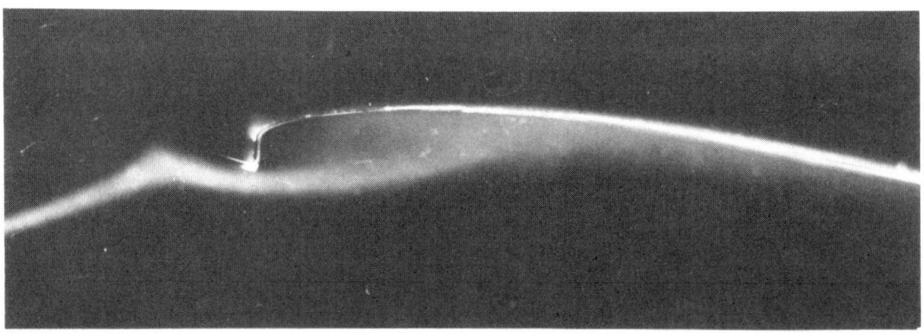

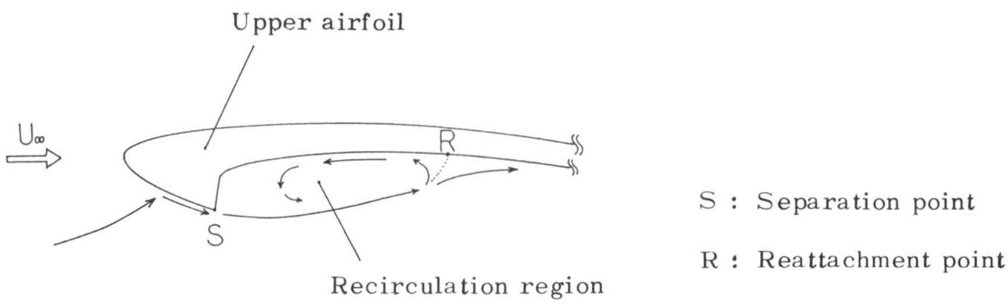

Fig. 3. Recirculation region in the cove on lower surface of upper airfoil; $\alpha = 12^\circ$.

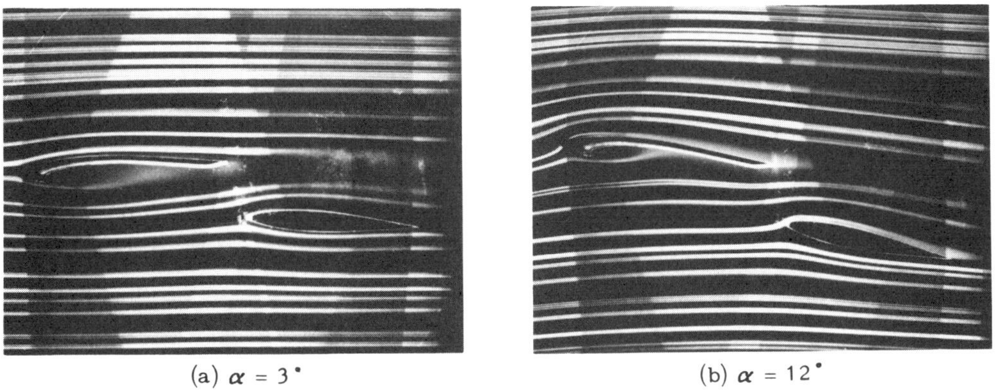

Fig. 4. Smoke-line patterns of complex airfoil configuration.

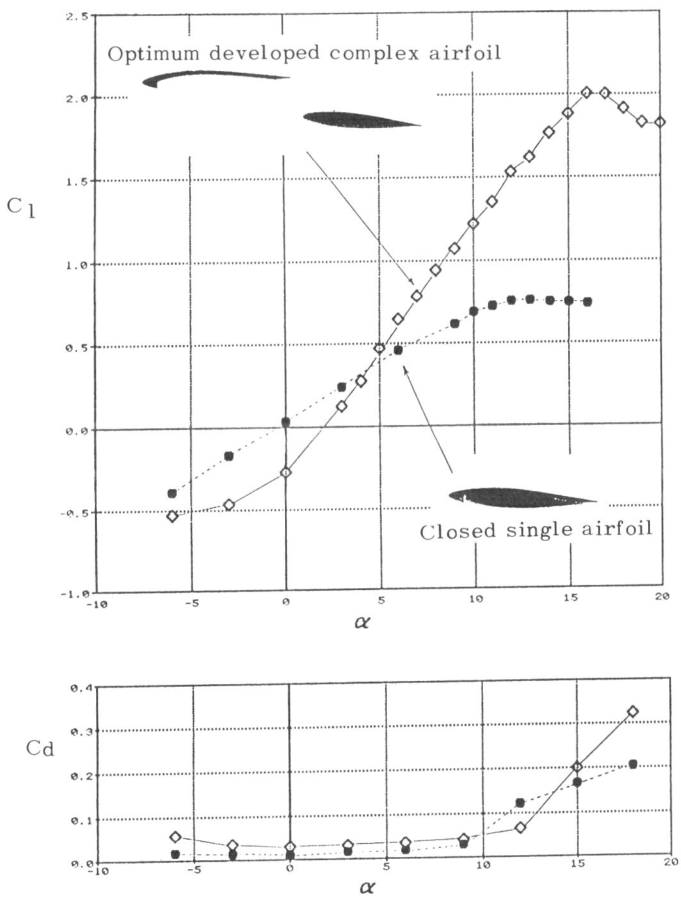

Fig. 5. Lift and drag curves for two airfoil configurations.

3.2 Lift and Drag Characteristics of Complex Airfoil

All data shown were uncorrected for the solid body blockage, the wake blockage and the wall interference effect. The overall section lift and the drag coefficient curves for two configurations are shown in Fig. 5. At higher angles of attack, the additional lift increments were generated on the upper surfaces of upper and lower airfoils. For the airfoil in closed configuration, the airfoil stalls at an angle of attack of approximately 12°, whereas the complex airfoil in optimum configuration does not stall untill the angle of attack of 16.5°. The section drag coefficient of the complex airfoil is characterized by the smaller value than the closed airfoil configuration at angles of attack $10° < \alpha < 14°$.

Cl/Cd to Cl curves for the several configurations are shown in Fig. 6, in comparison with the closed airfoil configuration. It illustrates the effect of stagger on lift-to-drag ratio, and optimum configurations is found to be St = 1.0, Ga = 0.2 and De = -6 deg.

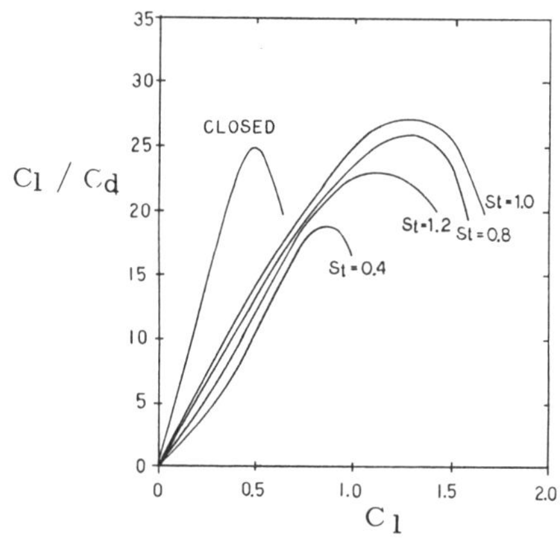

Fig. 6. Effect of stagger on lift-to-drag ratio (Ga=0.2, De=-6°).

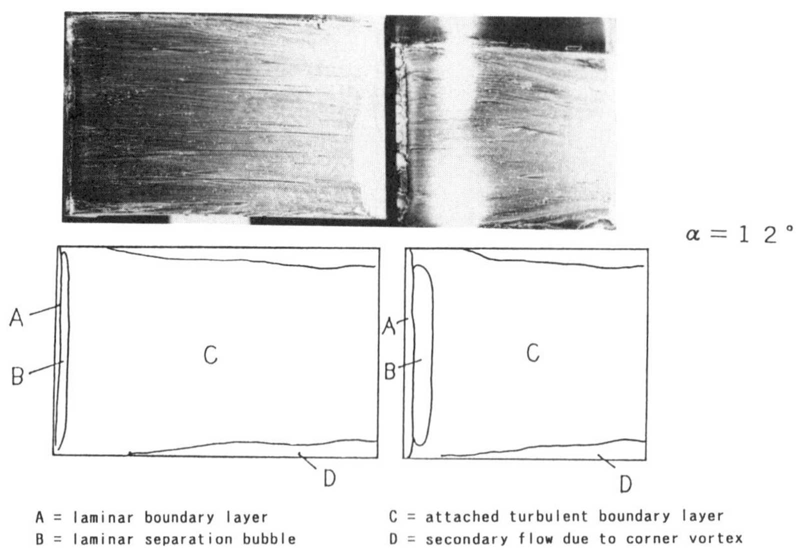

Fig. 7. Photograph and sketch of thin oil flow pattern on upper surface of optimum complex airfoil configuration (St=1.0, Ga=0.2, De=-6°).

3.3 Flow Visualization of Complex Airfoil

Thin oil flow technique was used to obtain the surface flow patterns of the complex airfoil configuration at an angle of attack of 12°. Fig. 7 shows the photograph and the sketch of the upper surface flow patterns for the complex airfoil with the optimum configuration. At $\alpha = 12°$, both upper and lower airfoils have an extensive attached turbulent boundary layer, whereas closed airfoil configuration has an extensive turbulent separation region at the same angle of attack. These phenomena are conidered to arise from the resulting effects of downwash flow upper airfoil and the wake / bounary layer interaction between two airfoils.

3.4 Mean Velocity and Turbulent Intensity Distributions in Gap Region

The mean velocity and the turbulence intensity distributions in the flow region between the upper and the lower airfoils of the optimum complex airfoil configuration at three angles of attack, 3.0°, 12.0°, 15.0°, are illustrated in Figs. 8 and 9, respectively. It should be emphasized that the hot-wire anemometry data presented

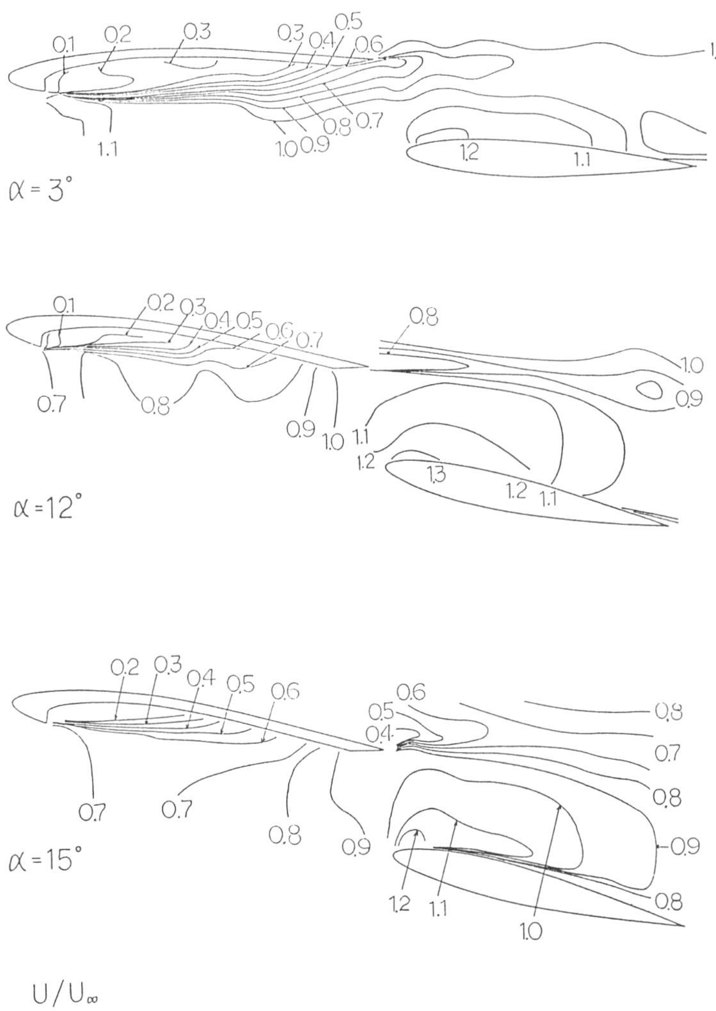

Fig. 8. Mean flow velocity contours between airfoils (St=1.0, Ga=0.2, De=-6°).

here must be regarded as qualitative due to highly turbulent separated and recirculating flow generated near the coves. [12, 13]

At the lowest angle of attack, $\alpha = 3.0^\circ$, for which relatively small amount of flow passes through the region between two airfoils, with the shear layer separating from the upstream corner of large cove region of the upper airfoil, partially reattaching near at the trailing edge, forming an extensive region of highly turbulent flow.

For increasing angle of attack, the serparated flow is directed toward the cove region, generating accelerated flow through the region between two airfoils producing velocities up to 20~30% greater than the free-stream. The only narrower area of high turbulence intensity is within the shear layer bounding the cove recirculation region on the underside of the upper airfoil.

These results obtained from the measurements of mean velocity and turbulence intensities agree with the previous results obtained, using several different visualization techniques.

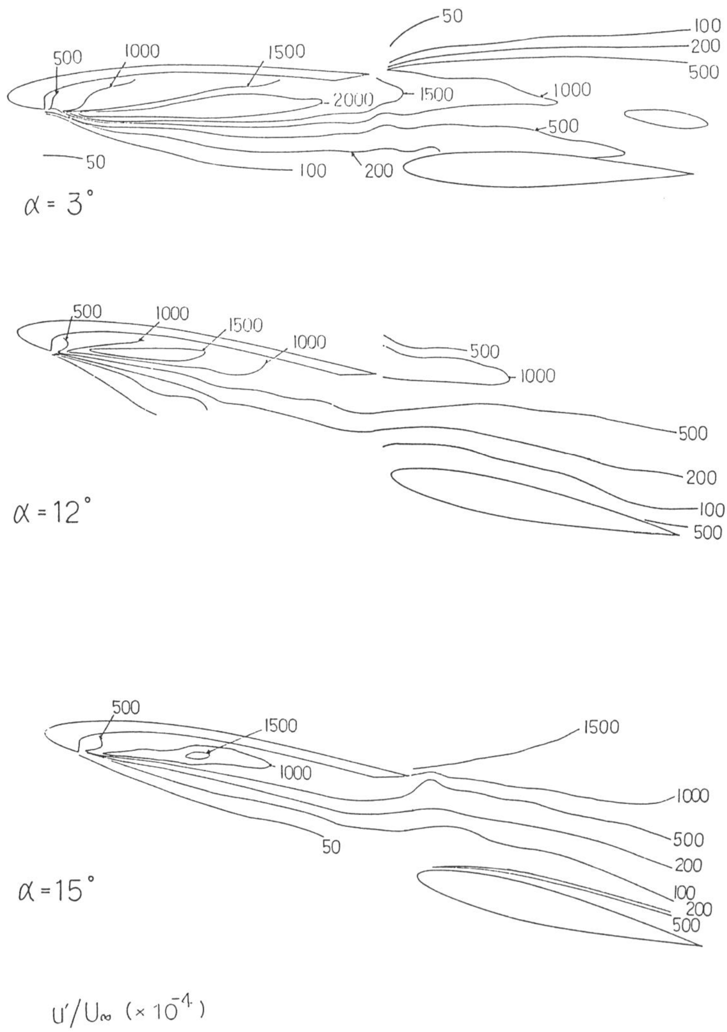

Fig. 9. Turbulence intensity contours between airfoils (St=1.0, Ga=0.2, De=-6°).

IV. CONCLUSION

From the present experimental study upon the specially designed complex airfoil configuration at a low Reynolds number in a smoke tunnel using quantitative and qualitative techniques under an identical flow condition, the following conclusions are obtained:

1) The complex airfoil, which has a relatively special section geometry compared with the conventional biplane, was investigated in a smoke tunnel, and it was found that the present airfoil shows well-defined and improved two-dimensional aerodynamic characteristics for the configuration, St = 1.0, Ga = 0.2 and De = -6° , due to the interaction between upper and lower airfoils.

2) The separated shear layer from the upstream corner of the cove of the upper airfoil is much associated with the overall performance of the complex airfoil, particularly with the aerodynamic interaction between two airfoils at $\alpha > 10°$.

3) Measurements using a hot-wire anemometery in the flow around the complex airfoil, together with the flow visualization using several techniques in an identical flow condition with the same wind tunnel, were usefull in understanding the aerodynamic characteristics. They provide some reliable information which can predict the complex flow field associated with low Reynolds number complex airfoil.

REFERENCES

1) Lissaman, P.B.S., Low-Reynolds-Number Airfoils, Annual Review of Fluid Mechanics, Vol. 15 (1983), pp. 223-239.
2) Brendel, M. and Mueller, T.J., Boundary-Layer Measurements on an Airfoil at Low Reynolds Numbers, Journal of Aircraft, Vol. 25, No. 7 (1988), pp. 612-617.
3) Savory, E., Toy, N., Tahouri, B. and Dalley, S., Flow Regimes in the Cove Regions Between a Slat and Wing and Between a Wing and Flap of a Multi-element Airfoil, Experimental Thermal and Fluid Science, Vol. 5, No. 3 (1992), pp. 307-316.
4) Workovitch, J., Subsonic VSTOL Aircraft Configuration with Tandem Wings, Journal of Aircraft, Vol. 16, No. 9 (1979), pp. 605-611.
5) Rhodes, M.D. and Selberg, B.P., Benefits of Dual Wings over Single Wings for High Performance Business Airplanes, Journal of Aircraft, Vol. 21, No. 2 (1984), pp. 116-127.
6) Olson, E.C. and Selberg, B.P., Experimental Determination of Improved Aerodynamic Characteristics Utilizing Biplane Wing Configurations, Journal of Aircraft, Vol. 13, No. 4 (1976), pp. 256-261.
7) Khan, F.A. and Mueller, T.J., Tip Vertex/Airfoil Interaction for a Low Reynolds Number Canard/ Wing Configuration, Journal of Aircraft, Vol. 28, No. 3 (1991), pp. 181-196.
8) Scharpf, D.F. and Mueller, T.J., Experimental Study of a Low Reynolds Number Tandem Airfoil Configuration, Journal of Aircraft, Vol. 29, No. 2 (1992), pp. 231-236.
9) Hatano, M. and Igarashi, M., Experimental Study of Split Wing, Proceedings of The 24th Aircraft Symposium (1986), pp. 298-301. (in Japanese)
10) Hatano, M., Okamoto, T. and Ikari, O., Flow Visualization around a Complex airfoil by the smoke-tube method, Journal of Visualization Society of Japan , Vol. 12, Suppl. No. 1 (1992), pp. 243-246. (in Japanese)
11) Yang, Handbook of Flow Visualization, Hemisphere Publishing Co., 1989.
12) Hatano, M., Okamoto, T. and Ikari, O., Measurement of a Flow Region between Two Airfoils in a Low Reynolds Number Complex Airfoil, Proceedings of The 29th JSASS Chubu / Kansai Branches Joint Annual Conference (1992), pp. 84-87. (in Japanese)
13) Bradshaw, P., An Introduction to Turbulence and Its Measurement, Pergamon Press (1971), p. 119.

VI
BIOMECHANICS AND BIOMATERIALS

Biomechanical Influences of the Cranio-Maxillo-Facial System on Its Form and Function

Sadami TSUTSUMI*, Yoshinobu MAEDA**, and Motofumi SOGO**

* *Research Center for Biomedical Engineering, Kyoto University, 53 Kawahara-cho, Shogoin Sakyo-ku, Kyoto*
** *Department of Removable Prosthodontics, Faculty of Dentistry, Osaka University, 1–8, Yamadaoka Suita, Osaka*

In this study the mechanisms controlling the biomechanical relationship between form and function in the Cranio-Maxillo-Facial (CMF) system were investigated using two-dimensional finite element analysis. Factors investigated included morphology and occlusion, the latter being an important attribute of this system (cervical spine, temporomandibular joint and mandible). Results indicated that pathogenesis and treatment of cervical spine deformity, change in temporomandibular joint morphology and trabecular formation and pattern in the mandible were closely related to the occlusion condition.

I. INTRODUCTION

Morphology and function are closely interrelated in many parts of the human body and the CMF system is no exception. In this study we attempted to elucidate the biomechanical mechanisms controlling the relationship between form and function in this region of biomechanics using two-dimensional finite element models [1,2].
The relationship between occlusion, one of the main attributes of this system, and morphological change in the cervical spine, temporomandibular joint and mandible were also examined.
In addition, the biomechanical influence of occlusion on the following aspects of the above mentioned regions were analyzed.
1) Pathogenesis and treatment of cervical spine deformity.
2) Temporomandibular joint morphology.
3) Trabecular formation and pattern in the mandible.

II. MATERIALS AND METHODS

2.1 Pathogenesis and Treatment of Cervical Spine Deformity

In this analysis a sagittal view of the cervical spine, maxilla and mandible was analyzed. These regions were divided into six slices with 1072 nodes and 2096 elements comprising the cortex and spongiosa of the vertebrae, endplates of the vertebral bodies, annulus fibrosus and nucleus pulposus of intervertebral discs, anterior and posterior longitudinal ligaments, upper and lower teeth, articular disc, and a portion of the cranium (Fig.1). Simulation of spinal deformities in the cervical region was performed with this model using a displacement incremental method in a pseudo three-dimensional finite element analysis based on the spanning element theory.
For simulation of the deformity process the displacement incremental

method was used to iteratively accumulate the displacements at each stage of the finite element analysis. Physiological loads were repeatedly applied to the displaced model to mimic the gradual development of the deformity. An unstable relationship between upper and lower teeth was used to simulate malocclusion.

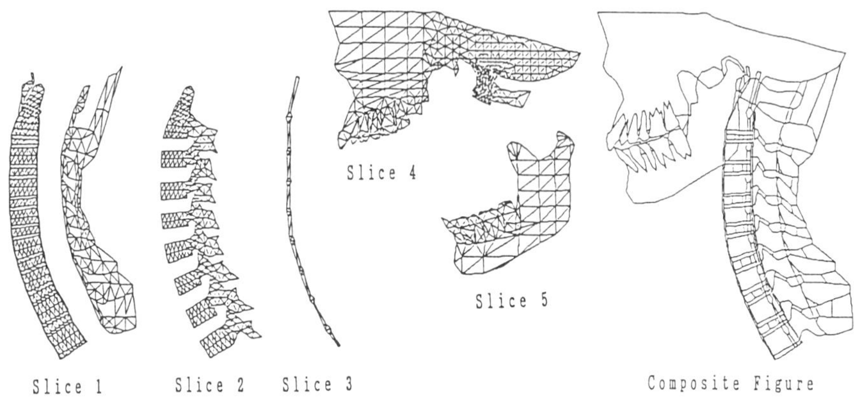

Fig.1. Six slices of CMF system model.

2.2 Temporomandibular Joint Morphology

To examine the influence of differences in the loading configuration due to occlusion on the condyle shape, models of four typical condyle head types (flat, convex, round and angled) were analyzed. These models included cortical and cancellous bone layers as shown in Fig.2. Each model was fixed on the upper surface of the condyle head and loaded on the lower surface in one of several different configurations. Constraints were placed on nodes on the upper surface to simulate the differences in disc location.

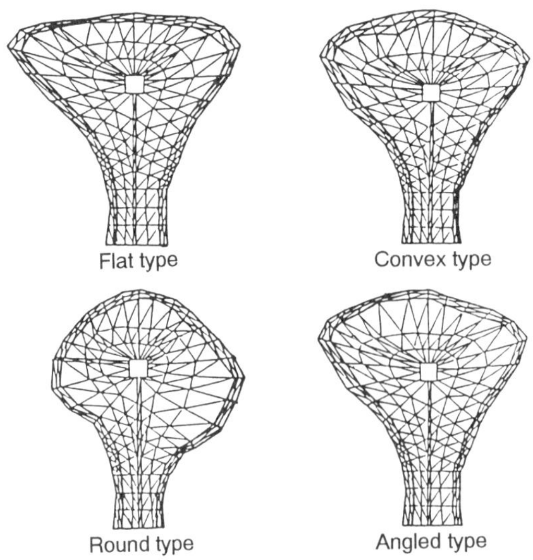

Fig.2. Four condyle models.

In these models a shape optimization theory[1] was applied. A node on the surface of the cortical bone layer was moved outward, if the tensile stress exceeded the tensile threshold level, an amount proportional to the difference between the calculated stress level and the threshold level. Likewise the node was moved inward, if the compressive stress exceeded the compressive threshold value, an amount proportional to the difference between the calculated stress level and the threshold level. The computations were then repeated and a new stress distribution was calculated.

2.3 Trabecular Formation and Pattern in the Mandible

This study was undertaken to test our hypothesis that the trabeculae around a tooth or dental implant are formed such that bone can distribute the applied loads most effectively.

Two dimensional finite element models were created of the frontal section of the molar region of the mandible. In conventional finite element models uniform material properties are assigned to the cancellous bone region. However, to simulate the physiological condition we divided this region into small elements to which were randomly assigned either bone or marrow material properties. The ratios of bone to marrow elements was set to obtain bone volume ratios (bone/(bone+marrow space)) of 50, 35 or 20% (Fig.3). This permitted evaluation of the influence of bone volume on trabeculation. A lower bone volume ratio (20%) was assumed to occur immediately after implantation or in the edentulous osteoporotic jaw.

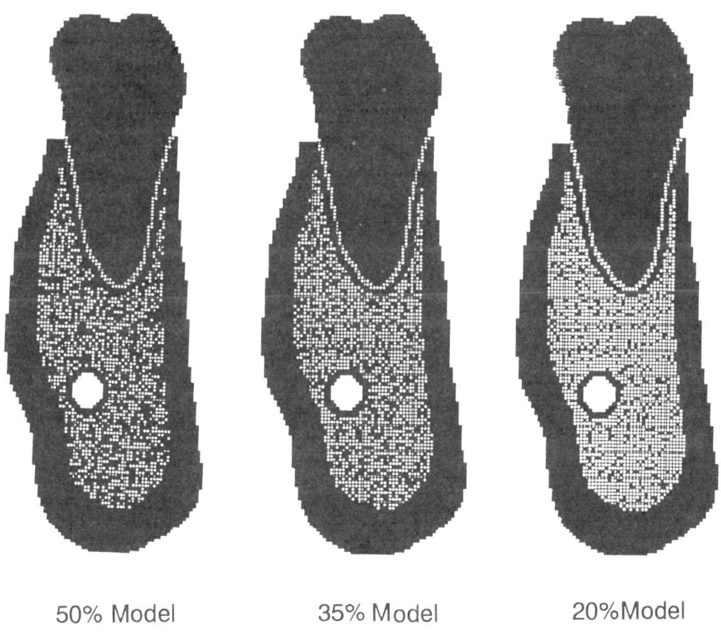

Fig.3. Models of the mandibular molar cross section with the three different bone volume ratio.

Initially, a vertical load was applied to the occlusal fossa and equivalent stress level in the elements determined. Based on the

stress level in the elements the material properties of the elements were altered. The mean stress in the bone elements was determined and those bone elements with stress levels lower than half of this figure were converted into marrow elements. To maintain the required bone volume ratio an equal number of marrow elements, those having the highest stress levels, were converted into bone elements (Fig. 4).
Then, the stress levels were recalculated and the alteration process repeated. This process was continued until the stress was evenly distributed in the bone elements.

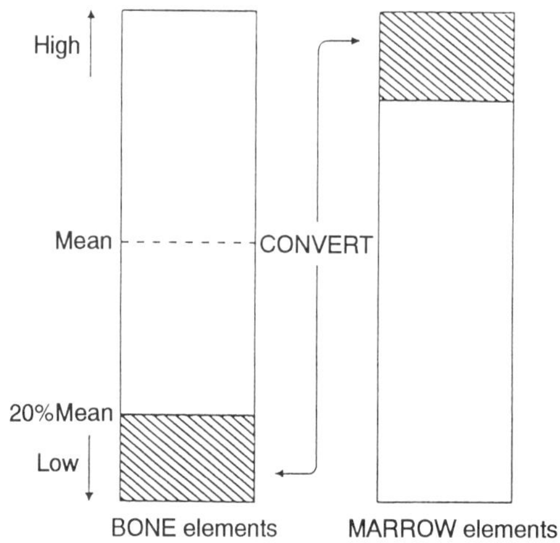

Fig.4. Rearrangement algorithm of bone and marrow elements based on their equivalent stress levels.

III. RESULTS AND DISCUSSION

3.1 Pathogenesis and Treatment of Cervical Spine Deformity

High principle stresses were found in the facet joint between C1 and C2 in the anterior and posterior regions, respectively. This resulted in shifting of vertebral bodies and facets (Fig.5).
In upper and middle cervical regions tensile stresses in the posterior portion of the facets and compressive stresses in the anterior region of the vertebral bodies caused a wedge-like deformation of the vertebral bodies. This wedge-like deformation may cause shoulder discomfort and facial, head and neck pains due to irritation of or injury to the transverse foramen and spinal cord (Fig.6). The wedge-like deformation was improved by the use of a bite plane or splint applied between upper and lower teeth suggesting changes in occlusion using an occlusal splint may be an effective treatment.

3.2 Temporomandibular Joint Morphology

When constraints were placed on the whole upper surface of the condyle vertical loading caused minimal change in the convex condyle, whereas a significant shape change was observed in the other types of condyles. Under non-vertical loading with constraints only on a portion of the upper surface changes to the condyle generally occurred only on the side of the head (Fig.7). These changes converted convex, angled, round type condyles into flat-shaped condyles. These results suggest that alteration of the loading configuration may lead to morphological changes in the condyle.

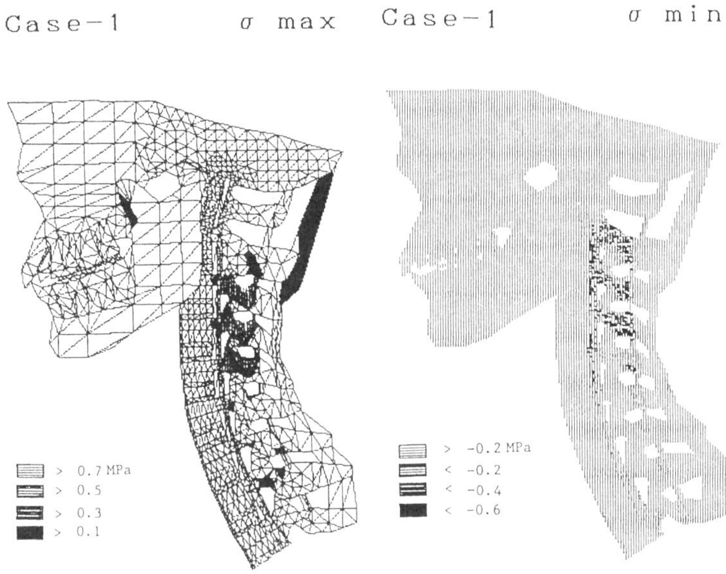

Fig.5. Example results for CMF system model for indicating maximum and minimum principle stress levels.

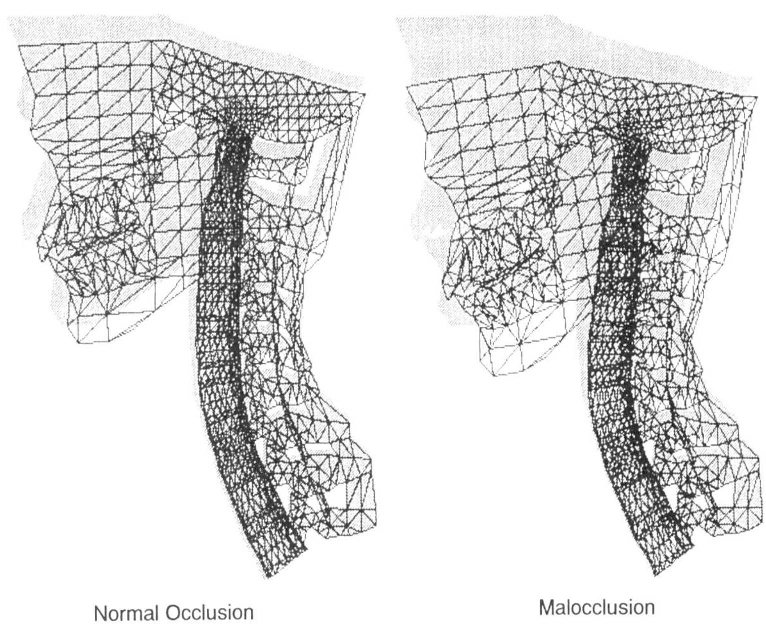

Fig.6. Differences in deflection between the normal- and malocclusion models.

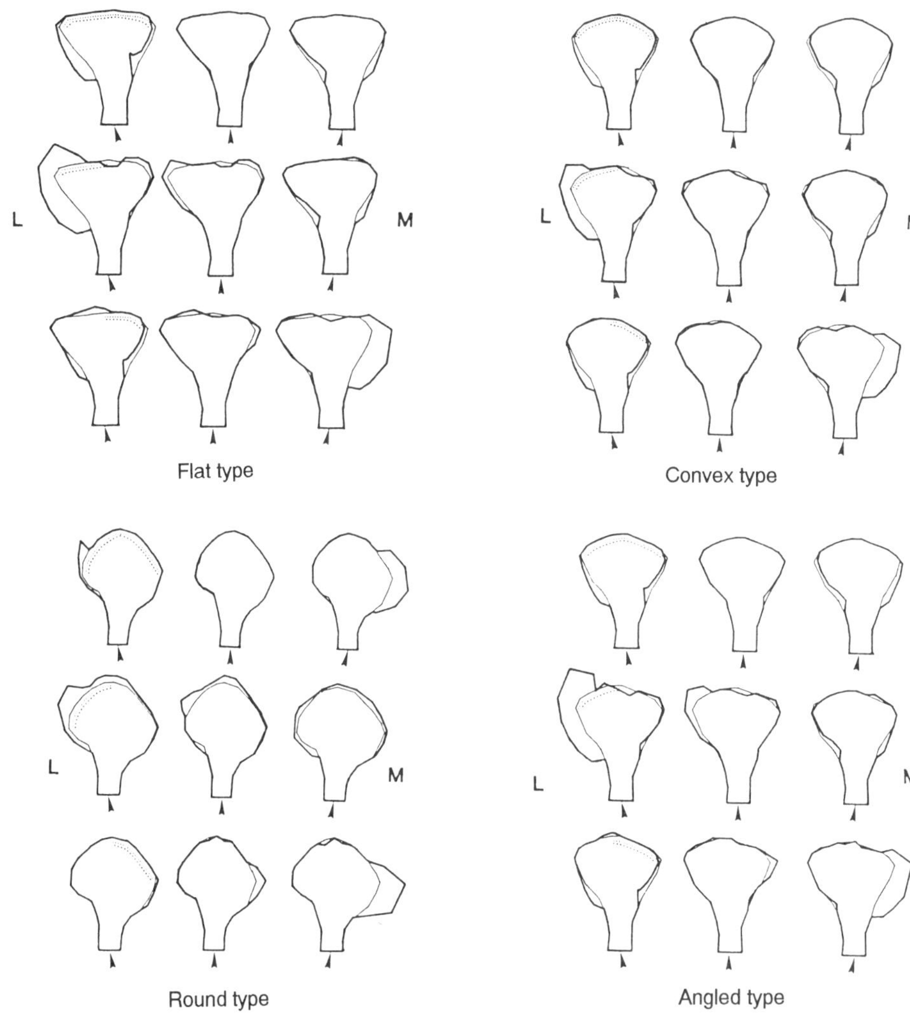

Fig.7. Shape change in four condyle models with variations in the loading angles and constraint configuration. (L: lateral, and M: mesial side)

3.3 Trabecular Formation and Pattern in the Mandible

Similarity between the pattern of the principle stresses in conventional two dimensional finite element models, indicated by vectors in Fig. 8a, and actual trabeculae in the human mandible, shown in cross section in Fig. 8b, shows that it is reasonable to suggest the trabecular pattern coincides stress distribution.
The adapted trabecular pattern (Fig.9) was greatly influenced by the bone volume ratio. In addition, the final "Total Strain Energy" (so-called 'stiffness') for each bone volume ratio (50%, 35%, and 20%) was approximately equal. From these results high incidents of bone resorption and bone mircrofractures can be anticipated if the bone volume ratio is low.

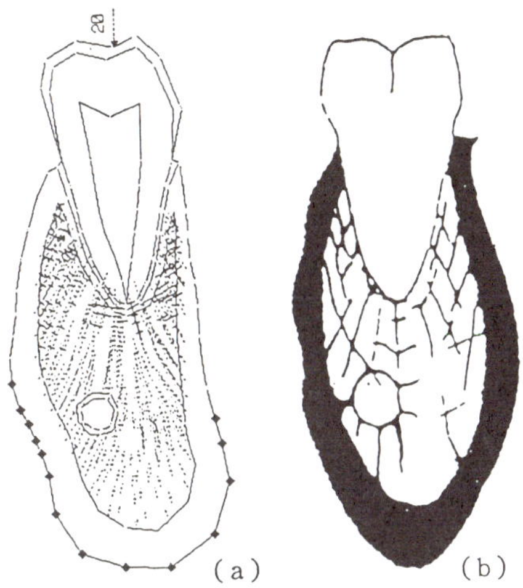

Fig.8. The stress distribution pattern from a conventional two-dimensional finite element model with 14 fixed nodes around the mandibular base (a) and an example of the natural physiological trabecular pattern (b).

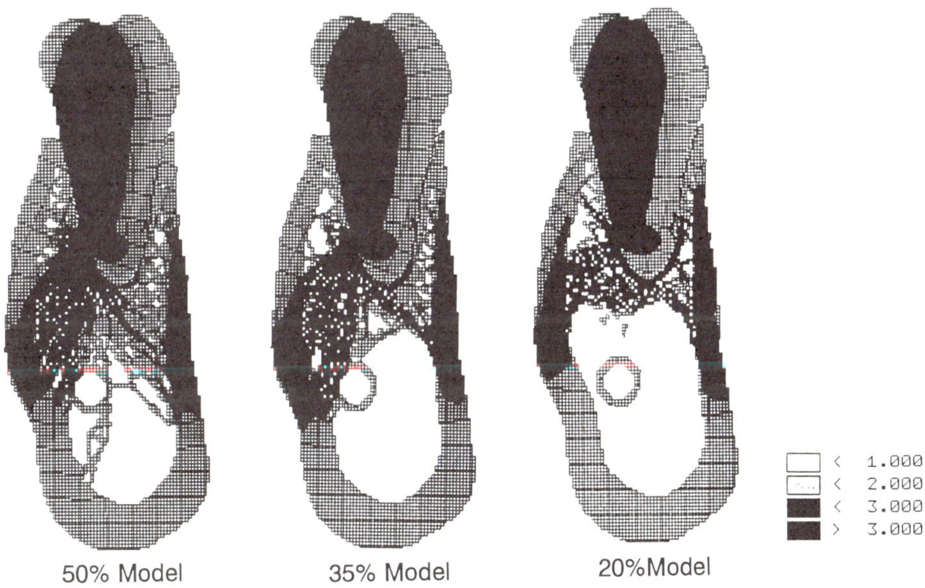

Fig.9. Results of trabeculae patter in the tooth model for different bone volume ratios.

REFERENCES

1) Tsutsumi,S., Biomechanical simulation of bone remodeling in dentistry. ASME Tissue Engineering, BED-14(1989):59-62.

2) Maeda,Y. and Wood,W.W., Finite element method simulation of bone resorption beneath a complete denture, J. Dent. Res, 68(9)(1989): 1370-1373.

The Effects of Preservation on AE Properties of Bone

Shigeo TANAKA*, Emiko WATANABE*, Toshiaki HARA**,
Kazuhiro HASEGAWA***, and Hideaki E. TAKAHASHI***

* *Graduate School of Science and Technology, Niigata University, Niigata*
** *Department of Mechanical Engineering, Faculty of Engineering, Niigata University, Niigata*
*** *Department of Orthopedic Surgery, School of Medicine, Niigata University, Niigata*

To reveal the effects of formalin preservation on the mechanical properties, we performed quasi-static three-point bending test on the fresh rooster tibia group and on the formalin preserved rooster tibia group. In addition, AE from these tibiae under the bending test was measured to discuss the effects of formalin preservation on the microscopic fracture mechanism of bone. Formalin preservation did not significantly affect the stiffness and maximum load, however, it produced a significant reduction in the absorption energy. And there were a decreasing tendency of cumulative AE event count and a significant increase of average AE energy with formalin preservation. From low vacuum SEM examination after testing, the rough fracture surface of fresh tibia and the flat one of formalin preserved tibia were observed. We concluded that formalin preservation invites the change in quality of collagen fiber, so that, bone becomes more brittle.

1. INTRODUCTION

In the case of mechanical properties investigation of bone, we can not always get a fresh bone and are often forced to use the bone preserved by various ways. It is important to understand sufficiently the effects of the preservation on the mechanical properties and fracture behavior of bone to obtain the essential properties of bone. However, the work about the effects of preservation on mechanical properties of bone is limited, and only a few researches on the effects of freezing have been reported[1)-4)]. On the other hand, formalin preservation is not fully investigated about its effects on mechanical properties of bone though it is conventional preservation method in medical field.

Acoustic emission (AE) may be defined as the elastic stress wave generated by microcracks , dislocation movements and so on under loading in a metal or a rock. Analyzing the information of AE detected by an AE sensor is a good way on understanding the microscopic fracture behavior in the material under loading. The AE from bone under loading has been first reported by Hanagud et al. in 1973[5)]. In the past, the AE technique was used to study contribution of mineral and collagen to permanent deformation of compact bone[6)]. And the influence of strain rate on AE characteristics in compact bone was discussed[7)]. Clinical application of AE technique has been tried to monitor the degree of osteoporosis[8)-9)] and fracture healing[10)].

In this study, in order to investigate the effects of formalin preservation on mechanical properties of bone, quasi-static three-point bending test was performed on the fresh bone group which consists of left tibiae of seven roosters and on the formalin preserved bone group which consists of right tibiae of same roosters. Simultaneously, AE from these tibiae during the three-point bending test was measured. From the change of AE properties with formalin preservation and the observation of fracture surfaces of both groups by low vacuum SEM (LV SEM) after testing, we also discussed the effects of formalin preservation on the microscopic fracture behavior of bone.

2. MATERIAL AND METHOD

Pairs of tibial bones used in this study were obtained from seven male Hi-line strain roosters aged about 48 weeks and weighing an average of 2166 grams (±166SD). The tibiae were harvested at sacrifice, and all soft tissue was removed by sharp dissection. Left tibiae were used as fresh group and were tested at once after the sacrifice and the dissection. On the other hand, right tibiae were used as formalin preservation group and were tested with storing in 10% formalin solution for 1 week after the sacrifice and the dissection. Before testing, the longitudinal lengths of the tibiae were recorded and marked the longitudinal center of the tibia on the anterior surface. All preparation processes and the mechanical tests were carried out at a constant temperature (20°C) and a constant humidity (60%).

Quasi-static three-point bending test was performed on an Instron type machine at a cross-head speed of 0.5mm/min. The loading site was corresponded to the marker on anterior surface which shows the longitudinal center of the tibia and 70mm was introduced as span. We determined the stiffness, maximum load and absorption energy from load-deformation curve obtained. Stiffness was defined by the slope of the initial linear region of load-deformation curve decided by least square method.

AE was monitored by a flat response piezoelectric transducer attached to the upper loading point. AE signals detected by a transducer were amplified and filtered by AE tester. Two signals were offered by AE tester. One was for AE event counts signal, another for real AE wave form signal. The former was recorded by means of a data recorder and utilized to determine cumulative AE event counts. To determine the average AE energy the latter was recorded by digital storage scope and transferred to personal computer and utilized. A schematic diagram of the test set-up is shown in Fig.1. The AE energy for 1 event was defined by the following equation.

$$E = \int_0^T f(t)^2 dt \quad \cdots \quad (1)$$

where, T is the duration of 1 event and f(t) is the instantaneous value of real AE wave form signal. Fig.2 shows the typical real AE wave form signal from a rooster tibia during loading. All AE waves from tibia observed in this study was burst type as shown in Fig.2.

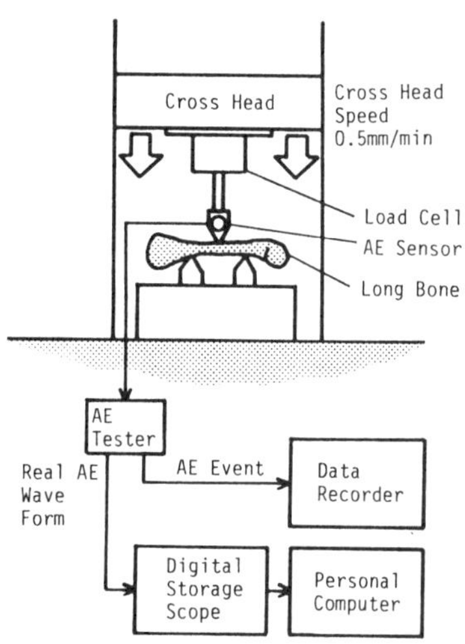

Fig.1. Schematic diagram of the test set-up.

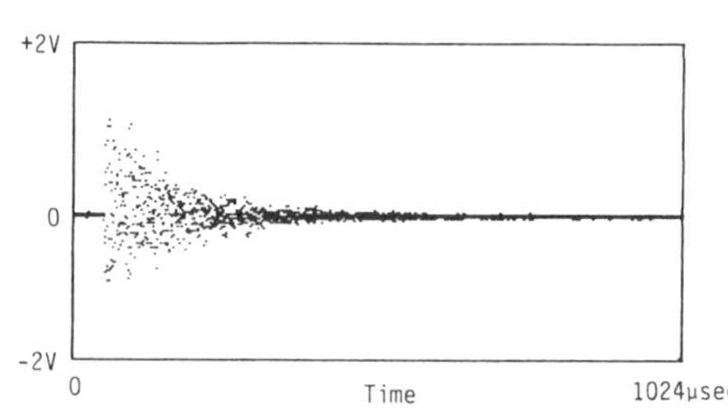

Fig.2. Typical real AE wave form signal.

After testing, we compared the fracture surfaces of both groups by LV SEM and discussed the effects of formalin preservation on microscopic fracture behavior of bone. LV SEM is a available mean to observe the living bodies as plants, insects and bone which don't have conductivity, because LV SEM don't have to do the metal coating and the specified chemical processes for specimens.

Table 1 shows the name of the manufacturers and the products for the measurement devices used.

Table 1. Measurement devices

Measurement Device	Manufacturer and Product
Mechanical Testing Machine	Shimadzu Corporation, Kyoto Japan, AUTOGRAPH AG-25TD
AE Sensor	NF Electronic Instruments, Yokohama Japan, AE-901S-WB
AE Tester	NF ELectronic Instruments, Yokohama Japan, 9501
Data Recorder	NEC San-ei Instruments, Ltd., Tokyo Japan, OMUNILIGHT 8M36
Digital Storage Scope	Kawasaki Electronica, Tokyo Japan, KDS-102
Personal Computer	NEC, Tokyo Japan, PC-9801VX
LV SEM	JEOL, Tokyo Japan, JSM-5000LV

3. RESULTS

Figs.3 and 4 show the typical relationships between load, cumulative AE event count, AE energy and deformation for a fresh tibia and a formalin preserved tibia respectively which were taken from the same rooster. From these figures, we can see that stiffness and maximum load for both tibia are not so different, but deformation at break and cumulative AE event count until break for the formalin preserved tibia are smaller than those for the fresh tibia, and there are many events with larger AE energy in the formalin preserved tibia. In both tibiae, AE started with the end of initial linear region of load-deformation curve.

Statistical comparisons between the fresh group and the formalin preserved group were accomplished using the paired t test. The results for mechanical parameters are depicted in Fig.5. There were no statistically significant differences in the stiffness and maximum load. The absorption energy significantly decreased in the formalin preserved group. The results for AE parameters are shown in Fig.6. There was a tendency of decreasing of cumulative AE event count in the formalin preserved group but this tendency was not statistically significant. And the significant increasing of average AE energy was seen in the formalin preserved group.

Figs.7(a) and (b) show the microphotographs by LV SEM of typical fracture surfaces for a fresh tibia and a formalin preserved tibia respectively. From these microphotograph, we can see that there are many Haversian canal in the both fracture surfaces i.e. the compact bone of the diaphysis of tibiae tested is Haversian bone. The fracture surface of the fresh tibia was rough, on the other hand, that of the formalin preserved tibia was flat.

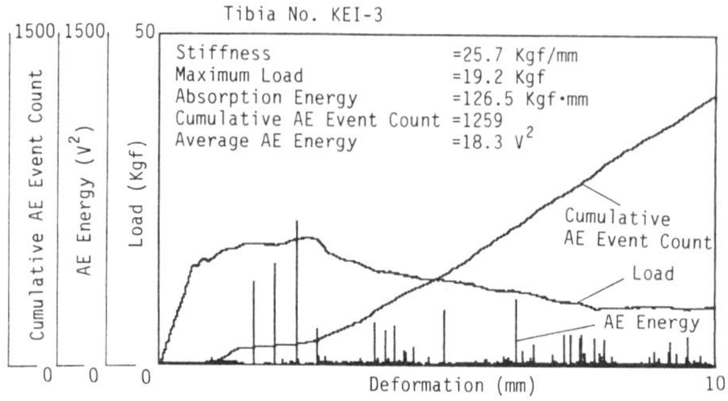

Fig.3. The relationship between load, cumulative AE event count, AE energy and deformation for the fresh tibia.

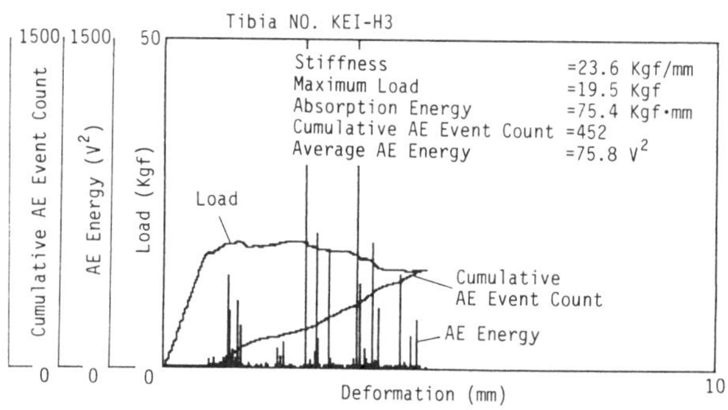

Fig.4. The relationship between load, cumulative AE event count, AE energy and deformation for the formalin preserved tibia.

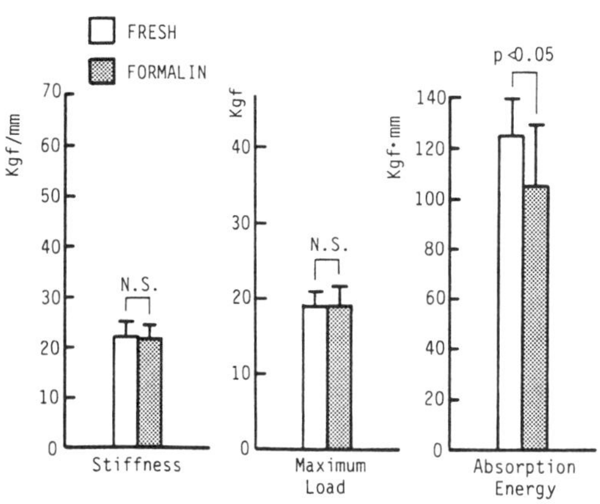

Fig.5. The mechanical properties for the fresh and the formalin preserved group#.

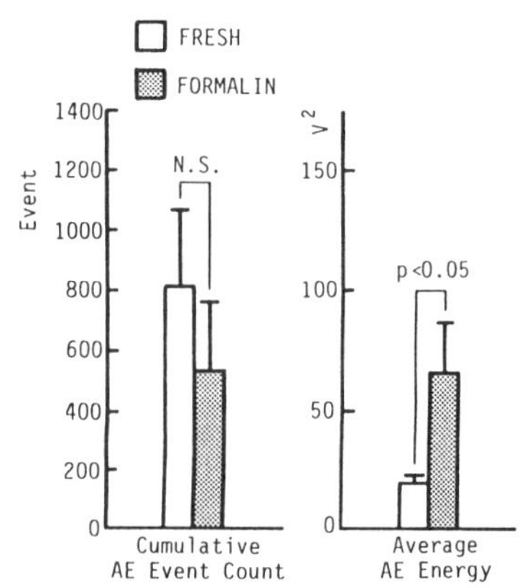

Fig.6. The AE properties for the fresh and the formalin preserved group#.

#Paired t test was used for statistically significant difference test between the the fresh and the formalin preserved group.

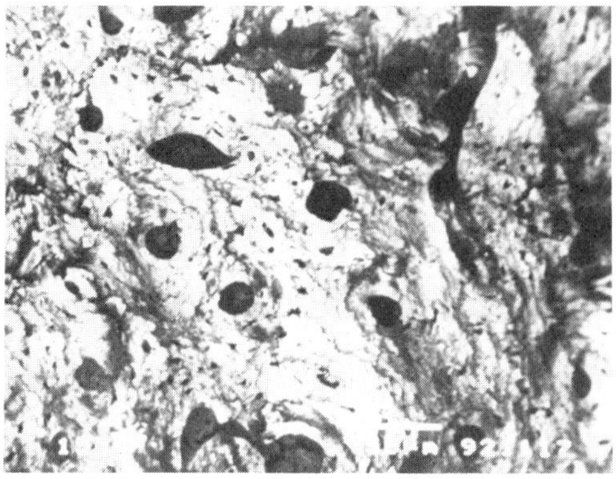

50μm

Fig.7(a). The microphotograph of the typical fracture surface of the fresh tibia by LV SEM.

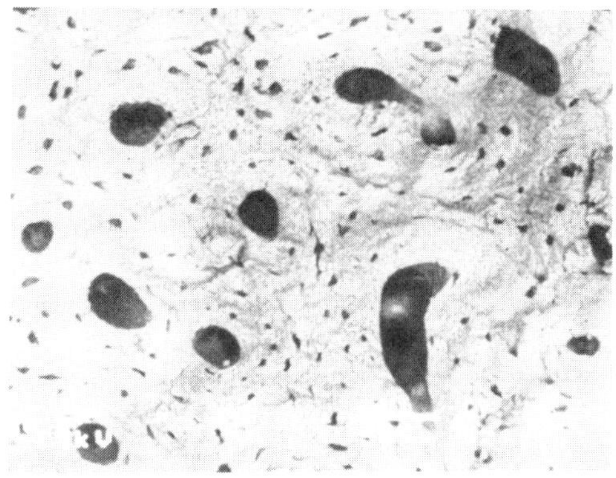

50μm

Fig.7(b). The microphotograph of the typical fracture surface of the formalin preserved tibia by LV SEM.

4. DISCUSSION

Bone consists of bone cell, bone matrix and moisture. Bone matrix bears the mechanical support and is constituted of the mineral phase (75%) and the organic phase (25%) as shown in Fig.8. The mineral phase is the hydroxyapatite($[Ca_{10}(PO)_4]_6(OH)_2$) and 90% of the organic phase is collagen fiber (Type I) i.e. bone is regarded as a composite material made of the hydroxyapatite and the collagen fiber.

We considered that the formalin preservation changed in the quality of the collagen fiber in the bone matrix and so that deprived the collagen fiber of the original tensile strength. In such a bone, the available bridging effect of collagen fiber when cracks open or progress can not be expected. Namely, bone may became more brittle by the formalin preservation. The flat fracture surface of the formalin preserved tibia shown in Fig.7(b) suggests the above consideration. And the decreasing of absorption energy for the formalin preserved group in our results also supports that. For the formalin preserved group, the collagen fiber can not arrest the crack propagation. In this case, the microcracks can propagate through the bone matrix and the crack opening releases large energy which transmits through the bone matrix with a high energy AE wave. In our results, the high average AE energy was observed on the formalin preservation group.

On the other hand, the hydroxyapatite may not be affected by the formalin preservation as much as the collagen fiber and bear the mechanical properties of bone as the stiffness and the maximum load. Therefore, the stiffness and maximum load didn't be changed by the formalin preservation in our results.

In this work, the period of preservation was limited to 1 week. In the future, more period of preservation should be tried.

5. CONCLUSION

To reveal the effects of the formalin preservation on the mechanical properties, we performed quasi-static three-point bending test on the fresh rooster tibia group and on the 1 week-formalin preserved rooster tibia group. Simultaneously, AE from these tibiae during the three-point bending test was measured and after testing, carried out the observation of the fracture surfaces for the both groups by LV SEM to discuss the effects of the formalin preservation on the microscopic fracture behavior of bone. The results obtained from above experiments may be summarized as follows.

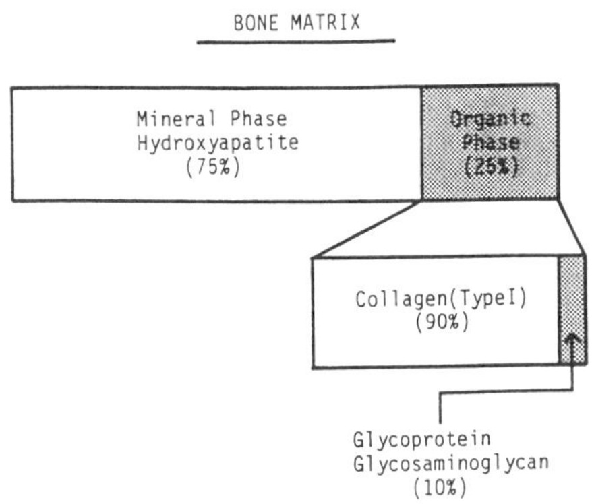

Fig.8.The components of bone matrix.

1.The stiffness and the maximum load of tibia did not be changed by the 1 week-formalin preservation.
2.The absorption energy of tibia significantly decreased by the 1 week-formalin preservation.
3.The cumulative AE event count of tibia had the decreasing tendency with the 1 week-formalin preservation.
4.The average AE energy of tibia was significantly increased by the 1 week-formalin preservation.

REFERENCES

1) Sedlin,E., A rheologic model for cortical bone. J. Bone Joint Surg.[Am], Vol.55(1973),pp.322-330.
2) Friedlaender,G.E. and Mankin,H.J., Bone banking: current methods and suggested guidelines. in:American Academy of Orthopaedic Surgeons Instructional Course Lectures, Vol.30, Murray,D.(ed), St.Louis, C.V.Mosby Co.(1981),pp.36-51.
3) Malinin T.I., Wu,N.M. and Flores,A., Freeze-drying of bone for allotransplanta tion. in:Osteochondral Allografts: Biology, Banking and Clinical Applications, Friedlaender,G.E., Mankin,H.J. and Sell,K.W.(eds), Boston, Little, Brown and Company(1983),pp.181-192.
4) Richard,R.P.,et al., Effects of Freezing and Freeze-Drying on the Biomechanical Properties of Rat Bone. J.Orthop.Res., Vol.1, No.4(1984),pp.405-411.
5) Hanagud,S., Clinton,R. and Lopez,J.P., Acoustic Emission in Bone Substance, Proceedings of Biomechanics Symposium of the American Society of Mechanical Engineering(1973),p.79.
6) Wright,T.M.,et al., Permanent Deformation of Compact Bone Monitored by Acoustic Emission, J.Biomechanics, Vol.14(1981),pp.450-409.
7) Fischer,R.A.,et al., Analysis of the Effect of Using Two Different Strain Rates on the Acoustic Emission in Bone, J.Biomechanics, Vol.19, No.2(1986),pp.119-127.
8) Hanagud,S.,et al., Acoustic Emission and Diagnosis of Osteoporosis, Pro. IEEE Symposium on Ultrasonics(1975),pp.77-80.
9) Thomas,R.A.,et al., Acoustic Emission from Fresh Bovine Femora, Ultrasonics Symposium Proceedings, IEEE Cat.No.77CH12G4-ISU(1977),pp.237-240.
10) Nicholls,P.J.,et al., Acoustic Emission Properties of Callus, Med. & Biol.Eng. & Comput., Vol.19(1981),pp.416-418.

Identification of the Dynamic Properties of Human Compact Bone

Koichi KOBAYASHI*, Yuji TANABE**, Yoshio KOGA***, and Toshiaki HARA**

* *Graduate School of Science and Technology, Niigata Universtiy, Niigata*
** *Department of Mechanical Engineering, Niigata University, Niigata*
*** *Niigata Kobari Hospital, Niigata*

A series of impact compression tests on human tibial compact bone has been performed using the split–Hopkinson pressure–bar (SHPB) technique. Methods for the identification of the viscoelastic properties of bone are presented based on the experimental information obtained from the SHPB test. Application of the methods to the bone has shown that its mechanical behaviour in the longitudinal and radial directions can be described by the three–element standard linear solid model and the model with a nonlinear spring and nonlinear dashpot in parallel, respectively. The orientational dependence of the viscoelastic characteristics such as the elastic modulus and damping loss is also determined.

I. INTRODUCTION

In order to develop bone–analogue materials which combine suitable mechanical compatibility with a favorable bioactive response and to predict injury or fracture of bone from the clinical point of view, the evaluation of mechanical properties of bone is necessary. Many investigations concerning the mechanical properties of compact bone have been reported in the field of biomechanics[1-3] but only a limited amount of work has been done on its dynamic response under high strain rates. These studies have utilized the split–Hopkinson pressure–bar (SHPB) technique to investigate the dependence of stiffness on orientation under impact compression[4], the dynamic constitutive relation[5] and variation of the viscoelastic response with post–mortem age[6]. The SHPB technique is an impact test on a reasonable consideration of wave propagation and can be used as a concise method to determine the unknown parameters in the dynamic constitutive relations of various kinds of viscoelastic materials[7]. However, application of such a method is very scarce in the field of biomechanics.

This paper deals with the dynamic constitutive equation of human compact bone. Based on impact compression tests using the SHPB technique, viscoelastic models applicable to the bone are discussed. The orientational dependence of its viscoelastic characteristics such as the stress–strain relation, stiffness and damping loss is also determined.

II. PROCEDURES FOR THE IDENTIFICATION

Two methods for the identification of the viscoelastic characteristics are used in this study. The solution procedures are briefly described below.

2.1 Method I

This method can be applied to viscoelastic materials with linear stress–strain relation (linear viscoelastic solid). The method is based on the measurement of transient response information obtained from an impact compression test using the SHPB technique and combines solution procedures of two problems, i.e. the prediction problem of wave propagation in the SHPB with a viscoelastic specimen and the identification problem. The prediction problem can be solved by means of the method of Laplace transformation. Predictions of the transmitted stress history at a point on the output bar of the SHPB depend on the viscoelastic parameters Q in the constitutive equation of the specimen. Hence the identification problem can be reduced to finding the best Q so that the least squares distance function F attains its minimum value and it can be explicitly stated as:

$$\min_{Q} F = \min_{Q} \{ \sum_{k=1}^{N} [\sigma_T(t_k;Q) - \sigma_m(t_k)]^2 \} \quad (1)$$

where

$$F = \sum_{K=1}^{N} [\sigma_T(t_k;Q) - \sigma_m(t_k)]^2$$

σ_T=theoretical prediction of transmitted stress history,
σ_m=measured transmitted stress history,
t_k=discrete time point (k=1,...,N) and
N=number of data.

Using Gauss-Newton iterative scheme, the minimization problem Eq.(1) can be successfully carried out. Convergence is taken to be reached if F changes by less than 1% of its value. A detailed description of the solution can be found in Ref.(8).

2.2 Method II

This method can be applied to nonlinear viscoelastic materials. The method is based on the stress-strain relations obtained by the SHPB and quasi-static tests and needs the specific constitutive equation to express nonlinear viscoelastic behaviour. Hence we introduce the following form to represent bone's mechanical behaviour.

$$\sigma = f(\varepsilon, \partial\varepsilon/\partial t) \quad (2)$$

where σ=longitudinal stress in the specimen,
ε=longitudinal strain in the specimen and
t=time.

Once we obtain the stress-strain relations on the same specimen both under impact and quasi-static loading, the parameter values in Eq.(2) can be identified using the similar procedure of "Method I".

III. CHARACTERIZATION OF THE VISCOELASTIC PROPERTIES OF HUMAN COMPACT BONE

3.1 Specimen

A right tibial bone of a 16-year-old male was used. Its diaphysis was divided into six samples (sample length=25mm). The samples were numbered from 1(Proximal side) to 6(Distal side) as shown in Fig.1(a). Two cylindrical specimens were taken from the cortex of the anterior part of each sample (diameter=6mm, length=6mm). A Cartesian coordinate system was assumed to refer the orientation of the specimens in the cortex (Fig.1(b)). The bone axis was parallel to the long axis of the tibia. The tangential axis was in the circumferential direction while the radial axis was in the endosteal-periosteal direction. These axes were specified by eye and denoted by BA, TA and RA, respectively. Figures 1(c) and 1(d) show the typical optical micrographs of the specimen's surfaces perpendicular to the BA and RA, respectively. The specimens have many secondary Haversian system(Haversian bone). The specimens were refrigerated at -20℃ during storage prior to testing.

3.2 Experimental Method

Impact compression tests were carried out at strain rate of about $200s^{-1}$ and at room temperature (20℃) using the SHPB technique. Figure 2 shows the schematic diagram of the SHPB apparatus used in the present study. The apparatus consists basically of an air gun, a striker bar, two Hopkinson (or one input and one output) pressure bars and recording equipment. The specimen was held in place between the input and output bars. During testing a lubricant was placed on the contact surfaces to reduce the frictional constraint and to ensure contact at the interfaces to permit stress wave transmission. The incident and transmitted stress waves were detected by foil strain gages bonded on the input and output bars, respectively. Signals of these waves were stored in a digital storage oscilloscope with a sampling frequency of 4MHz. Subsequent operation of the recorded data was performed with the help of a personal computer. Typical oscilloscope records for the SHPB test on the specimen oriented parallel to the BA-direction are shown in Fig.3. Strain rate $\dot{\varepsilon}$, strain ε and stress σ at any time τ in the specimen were evaluated by the following equations.

$$\dot{\varepsilon}=\frac{2}{\rho_0 c_0 l}(\sigma_I-\sigma_T) \quad (3)$$

$$\varepsilon=\frac{2}{\rho_0 c_0 l}\int_0^{\tau}(\sigma_I-\sigma_T)\,dt \quad (4)$$

$$\sigma=\frac{S_0}{S}\sigma_T \quad (5)$$

where subscripts I and T refer to the incident and transmitted stress waves, respectively and
l=initial length of the specimen,
ρ_0=density of the input and output bars,
c_0=longitudinal elastic wave velocity of the input and output bars,
A=cross-sectional area of the specimen and
A_0=cross-sectional area of the input and output bars.

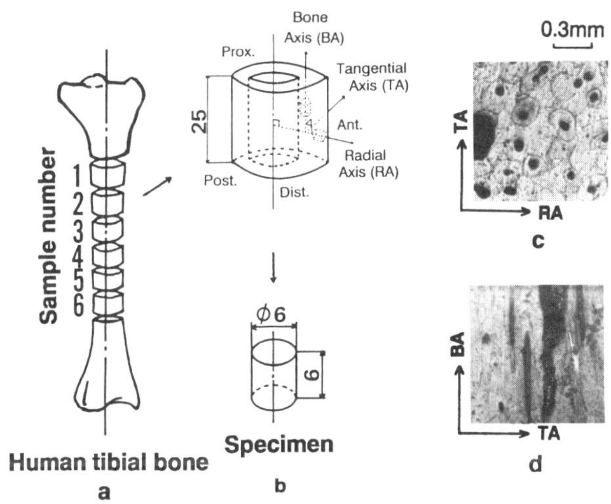

Fig.1. Specimen preparation procedure and optical observation of microstructure of specimens. a:Position of specimen; b:Geometry of specimen; c,d:Microstructure of bone samples oserved at c:BA-surface and RA-surface.

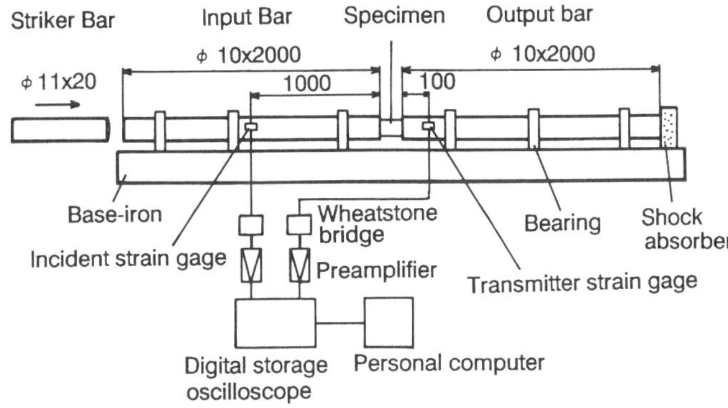

Fig.2. Schematic illustration of the SHPB apparatus.

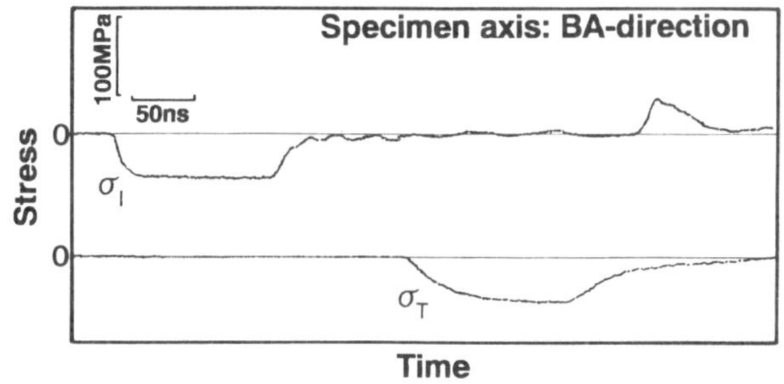

Fig.3. Typical oscilloscope records of stress wave profiles of human compact bone on the SHPB test. Top trace indicates incident stress wave profile and bottom trace indicates transmitted stress wave one.

3.3 Results and Discussion

3.3.1 Stress–strain relation

Figure 4 shows the stress–strain curves of all specimens. Stress–strain relations in the BA–direction are approximately linear while those in the RA–direction are nonlinear. The stiffness in the BA–direction are obviously larger than that in the RA–direction regardless of the site of the specimen. This is probably due to that small pores such as Haversian canals have more significant contributions to the deformation of the whole specimen in the RA–direction than in the BA–direction. The stress–strain curves of bovine femoral compact bone in both directions are also shown in the same figure (plotted by broken line). Bovine bone also has linear stress–strain relation in the BA–direction and nonlinear relation in the RA–direction.

3.3.2 Viscoelastic properties

Figure 4 emphasizes that the mechanical behaviour of the bone can be represented by the linear viscoelastic model in the BA–direction and by the nonlinear viscoelastic model in the RA–direction. Hence the three–element standard linear solid model (Fig.5(a)) and the model with a nonlinear spring and a nonlinear dashpot in parallel (Fig.5(b)) were adopted in this study. Their constitutive equations are given as follows:

$$\sigma+\frac{\eta}{E_1+E_2}\frac{\partial\sigma}{\partial t}=\frac{E_1E_2}{E_1+E_2}\varepsilon+\frac{E_1\eta}{E_1+E_2}\frac{\partial\varepsilon}{\partial t} \qquad (6)$$

$$\sigma=E\varepsilon-\alpha\varepsilon^2+\mu\left(\frac{\partial\varepsilon}{\partial t}/\frac{\partial\varepsilon_0}{\partial t}\right)^p \qquad (7)$$

where $\varepsilon_0=1$ and

E_1, E_2=stiffnesses of the springs as shown in Fig.5(a),

η=viscosity coefficient of the dashpot as shown in Fig.5(a) and

E, α, μ, p=viscoelastic constants as shown in Fig.5(b).

The parameter values, i.e. $Q_L=(E_1, E_2, \eta)$ in the BA–direction and $Q_{NL}=(E, \alpha, \mu, p)$ in the RA–direction were determined using "Method I" and "Method II" as described in Chapter II, respectively. The transmitted stress wave profile measured on the output bar, the stress–strain relation obtained from the SHPB test and the corresponding predictions based on the identified parameter values are shown in Figs.6(a) and 6(b). The predictions of transmitted stress pulse profile and stress–strain relation agree well with the experimental results. The similar results were also obtained for other specimens. We can, therefore, conclude that the mechanical behaviours in the BA– and RA–directions can be described by the three–element standard linear solid model and the proposed nonlinear model as shown in Fig.5(b), respectively. Figures 7(a) and 7(b) show the values of identified viscoelastic parameters for both directions plotted against sample number (or sample's site). The results of bovine bone are also shown in the same figures. The parameter values of human bone are independent of sample's location in the diaphyseal region, and no significant difference in parameter values between human and bovine bone can be observed.

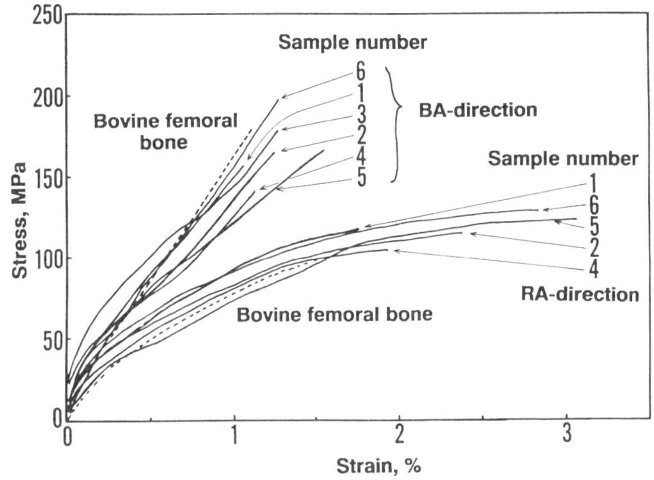

Fig.4. Stress–strain curves of human tibial compact bone. Broken lines represent stress–strain curves of bovine femoral bone.

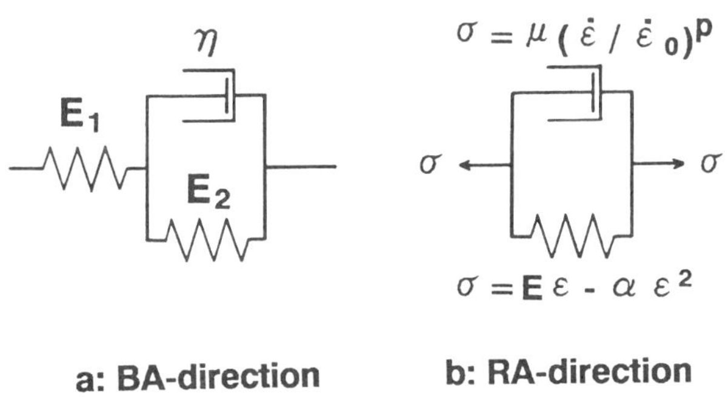

Fig.5. Proposed viscoelastic models for human compact bone. a:three–element standard linear solid model; b:the model with a nonlinear spring and a nonlinear dashpot in parallel.

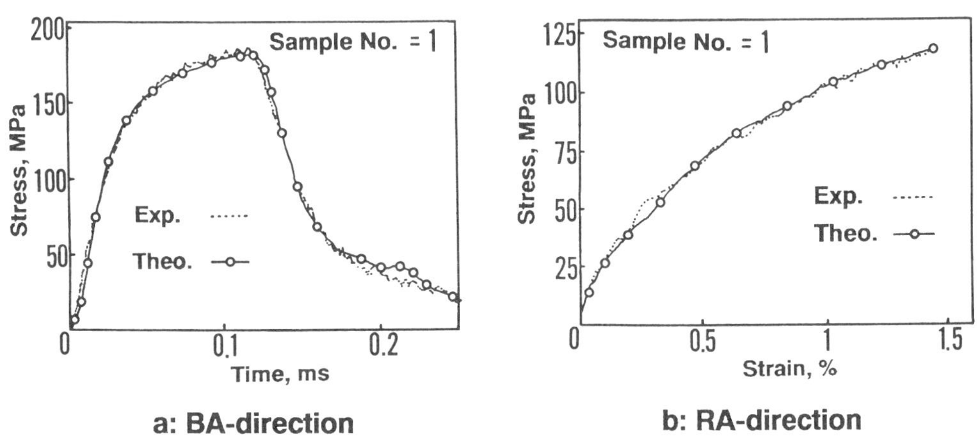

Fig.6. Results of identification. a:predicted and measured transmitted stress wave profiles; b:predicted and experimental stress–strain curves.

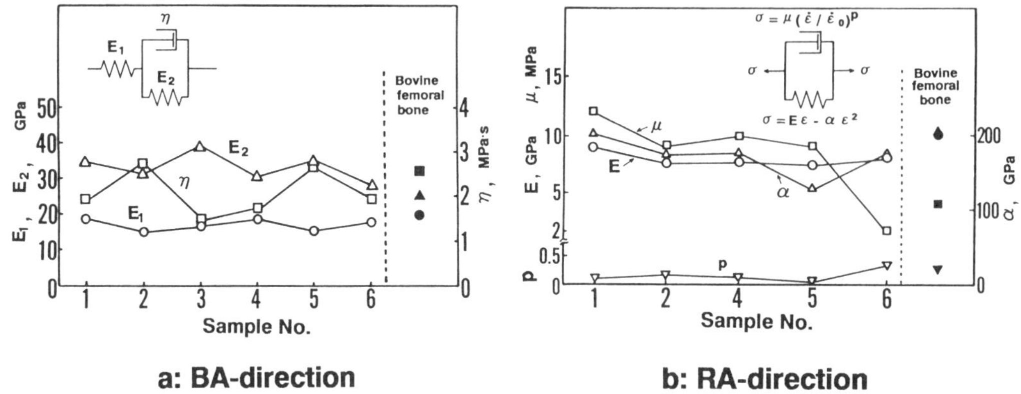

Fig.7. Viscoelastic parameters of both directions as a function of sample number.

IV. CONCLUSIONS

Impact compression tests on human tibial compact bone have been performed using the split-Hopkinson pressure-bar technique. The methods for the identification of the linear and nonlinear viscoelastic properties of the bone have been also presented. The results obtained are summarized as follows:
(1) The linear viscoelastic model can be applied to the bone in the BA-direction while the nonlinear viscoelastic model has to be required in the RA-direction.
(2) The stiffness in the BA-direction is greater than that in the RA-direction.
(3) No significant difference in the stress-strain relation and viscoelastic characteristics between human and bovine compact bone can be found.

Acknowledgment
One of the authors (Y. Tanabe) wishes to thank the support of a Grant-in-Aid for Scientific Research No. 04750058 by the Ministry of Education, Science and Culture.

REFERENCES

1) Reilly,D.T. and Burstein,A.H., The elastic and ultimate properties of compact bone tissue, J.Biomech., Vol.8(1975), pp393-405.
2) Bonfield,W. and Grynpas,M.D., Anisotropic of the Young's modulus of bone, Nature, Vol.270(Dec.1977), pp453-454.
3) Raftpoulos,D., Katsamanis,E., Saul, F., Liu,W. and Saddemi,S., An intermediate loading rate technique for the determination of mechanical properties of human cortical bone, J.Biomed.Eng.,Vol.15(Jan.1993), pp.60-66.
4) Tateishi,T., Shirasaki,Y., Kinoshita,Y. and Tateishi,K., Trans.Jpn.Soc.Mech.Eng., Ser.A, 46(1980), pp438-448(in Japanese).
5) Yang,G., Wu,W. and Fan,X., Dynamic properties of bone under high rate of strain, Progress and New Directions of Biomechanics,1989, pp343-347.
6) Tennyson,R.C., Ewert,R. and Niranjan,V., Dynamic viscoelastic response of bone, Experimental Mechanics, Vol.12(1972), pp502-507.
7) Lin,I.H. and Sackman,J.L., Identification of the dynamic properties of nonlinear viscoelastic materials and associated wave propagation problem, Int.J.Solid.Struct., Vol.11(1975), pp.1145-1149.
8) Tanabe,Y., Kobayashi,K., Sakamoto,M., Hara,T. and Takahashi,H., Identification of the dynamic properties of bone using split-Hopkinson pressure-bar technique, ASTM STP, to be submitted.

Nonuniform Deformation and Its Mechanical Model of Knee Ligament

Masao TANAKA*, Masahiro INOUE**, Yoshihiko KANNENN*, and Tomonori MIZUTANI*

* *Department of Mechanical Engineering, Faculty of Engineering Science, Osaka University, Toyonaka, Osaka*
** *Department of Orthopaedic Surgery, Osaka University Medical School, Suita, Osaka*

The nonuniform deformation of the anterior cruciate ligament is demonstrated by using healthy Japanese white rabbits by paying attention to the reference state independent on the testing conditions. Then, a mechanical model is proposed in order to describe the variable behaviors of the ACL under the different stretching conditions. The model is effective to predict the strength under the various conditions and the stress distribution over the cross section during the loading until the failure quantitatively.

I. INTRODUCTION

The femur–anterior cruciate ligament (ACL)–tibia complex (FATC) is often used for the bio-mechanical test of the ACL, in order to overcome difficulties of insufficient tissue size and of the adjustment of the fixity conditions. It has been reported that the mechanical behavior of the ACL as a whole deeply depends on the direction of the stretching, the knee flexion and so on[1]. In such cases, the mechanical state of the ACL is not uniform due to the complex ACL structure consisting of several bundles having different lengths arranged in the helical configuration coupled with the testing conditions. In fact, the local tissue deformation has been reported and its importance is pointed out[2), 3)], although the attention has been mainly paid to the mechanical properties as a whole. By paying attention to the reference state of the deformation, the nonuniform deformation of the ACL is emphasized first in the range of nondestructive uniaxial tension. Then a mathematical model is proposed to describe the variation of the mechanical response of the ACL as a whole and to estimate the distribution of the stress over the cross-section of the ACL under the various testing conditions.

II. NONUNIFORM DEFORMATION OF ACL

2.1 Material and Method

The femur–whole knee joint–tibia complexes are acquired from healthy Japanese white rabbits of 3.0±0.2 kg in weight, and stored at −30 degrees centigrade until the test. It is thawed to the room temperature and the FATC specimen is prepared. The medial condyle of the femur is removed to make the anteromedial bundle (AMB) of the ACL visible from the medial side, and a lattice of small dots is drawn on the surface of AMB. Then the specimen is held at several conditions of knee flexion and tensile axes. Tibial rotation and the medial–lateral bending are kept in the anatomical position. The tensile load up to 98 N (10 kgf) is applied from the preload of 1.96 N (0.2 kgf) after preconditioning. Other details of the procedure are decided by referring to the literature[1]. The behavior of the dot lattice on the AMB is recorded by a video cassette recorder through a TV camera. The surface deformation is analyzed based on the recorded image by using image processing technique. The precise procedure is found in the previous reports.[3), 4)] In this study, the reference state for the tissue deformation is the preloaded state at the unwinded position rotated with respect to the tibial axis at the shallow knee flexion, since the fibers of AMB surface become almost in parallel and uniformly stretched.

2.2 Nonuniform Surface Strain Distribution

The tissue deformations at the 1.96 N preload and 98 N tensile load are evaluated for three different knee flexion angles, 10, 45 and 90 degrees and for two tensile directions along the ACL and tibia. As is mentioned above, the strain is evaluated referring to the common state independent on the individual conditions. Figure 1 shows the strain distribution along a fiber segment of the posterior region of the AMB surface. Under the 1.96N preload along the ACL axis, the surface of AMB is stretched in the tibial side but remains relaxed in the femoral side. Under the preload along the tibial axis, the surface of AMB is kept to be relaxed, especially near both of the femur and tibia. At the 98 N tensile load along the tibial axis, the tensile strain appears at the tibial side but the femoral side is still relaxed. In the case of the ACL direction loading, the fiber segment shows the tensile strain over the fiber segment at shallow knee flexion, while the femoral side remains relaxed slightly at the deep knee flexion. These results show the nonuniformity of the tissue deformation is critical to understand the mechanical behavior of the ACL. It is also implied that the AMB surface is not so active as to carry the tensile load in these testing conditions, and the some other cross-sectional part of ACL is strongly responsible to resisting to the load.

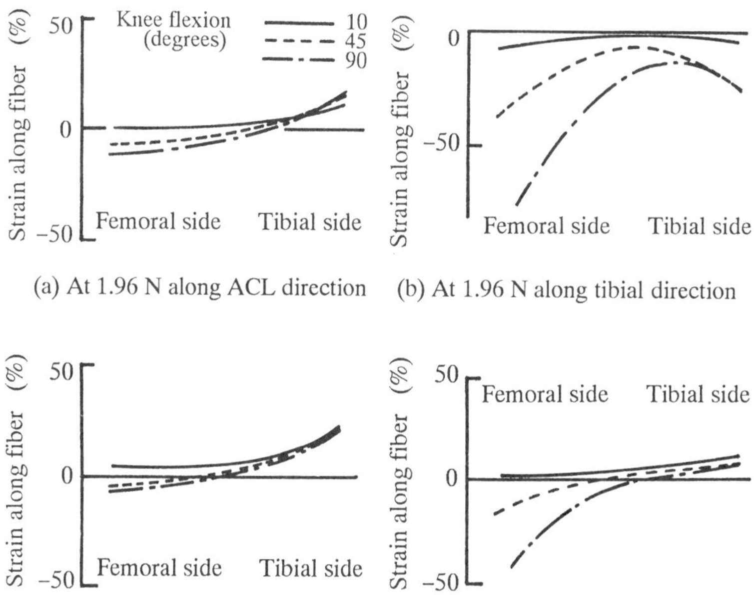

Figure 1. Tissue strain along a fiber segment on the AMB surface of ACL.

III. MODEL OF VARIABLE RESPONSE

It is possible to link the nonuniform response over the ACL cross-section and the variation of the load vs. tissue elongation due to the testing conditions. This section discusses this problem from a mathematical model-based point of view.

3.1 Effective Cross Sectional Area and Area Ratio

Since some of the cross-sectional part is not responsible to the applied load depending on the stretching conditions, the area ratio A_r is defined to represent the effective cross-sectional area A_{ef}. As far as any part of ACL does not failure, the effective cross-sectional area increases monotonously as to the progress of stretching. The total tissue elongation x is employed here to describe the progress of the stretching of the ACL. Then, the effective area is written as

$$A_{ef}(x)=A\,A_r(x) \tag{1}$$

where A stands for the physical cross-sectional area of the ACL.

Consider the uniform response of the ACL in which the area ratio is identically unity and the whole cross-section shares the applied load uniformly. Let the load vs. elongation curve be $P^*(x)$ for the uniform response, then its stress becomes

$$\sigma(x)=P^*(x)/A \tag{2}$$

giving the stress characteristics of the tissue. The increase of effective cross-sectional area A_{ef} is written as

$$dA_{ef}(u)=A_r'(u)\ A\ du \tag{3}$$

corresponding to the stretching from x=u to u+du, and its stress at the stretched state x is characterized as

$$\sigma(x;u)=\begin{cases} P^*(x-u)/A_0 & x > u \\ 0 & \text{others} \end{cases} \tag{4}$$

where ( )' stands for the derivative.

3.2 Load vs. Elongation and Area Ratio

Because of Eq.(4), the load vs. elongation curve is represented as the integral form of

$$P(x)=\int \sigma(x;u)\ dA_{ef}(u) = \int_0^x P^*(x-u)\ A_r'(u)\ du \tag{5}$$

in general. This relates the load P and the area ratio A_r. Since it is difficult to identify the zero load state in the experiment, the tissue elongation is measured from the small preloaded state, P(0)=p as shown in the experimental section. Therefore, Eq.(5) should be modified as

$$P(x)=P^*(x)A_r(0)+\int_0^x P^*(x-u)\ A_r'(u)\ du \tag{5'}$$

by taking the area ratio $A_r(0)$ at the preloaded state.

In the nondestructive range, the load vs. elongation curve has the nonlinear toe part and the linear part and could be modeled as

$$P(x)=\begin{cases} a[\exp(bx)-1] & x<x_0 \\ ab\exp(bx_0)\ x + a[(1-bx_0)\exp(bx_0)-1] & x>x_0 \end{cases} \tag{6}$$

where x_0 is the border between the toe and linear parts. Table I gives the average values of these parameters obtained by the least square curve fitting for the experimentally observed load vs. elongation curves up to 98 N (10 kgf).

The area ratio $A_r(x)$ is expected to increase monotonously and to be asymptotic to the unity. This study employs an exponential expression

$$A_r(x)= 1 - c\exp(dx) \tag{7}$$

for the area ratio, and the value $A_r(0)$ at the preloaded state becomes

$$A_r(0)= 1 - c. \tag{7'}$$

Thus, the parameters c and d characterize the load vs. elongation curve of Eq.(5'), and they are are identified so as to minimize the difference between the experimental load vs. elongation curve $P_{ex}(x)$ and P(x) calculated by Eqs.(5') and (7). That is, the minimization problem

$$\text{minimize} \int [P_{ex}(x)-P(x)]^2\,dx \quad \text{with respect to c and d} \tag{8}$$

gives the mathematical formulation of the problem to identify the variable response of the ACL.

3.3 Identified Responses

In the parameter identification, Eq.(6) with coefficients in the TABLE I is used as the experimental curve $P_{ex}(x)$ (Fig.2), and the problem (8) is solved by the bisection method. The curve $P^{*}(x)$ for the uniform response is approximated by the experimental curve under the ACL direction loading with the 10 degrees knee flexion, since it is the stiffest among the experimental conditions examined in this study and the uniform response has to give the most stiffest load vs. elongation curve. This is identical to that the area ratio is assumed to be A_r=1 for the ACL direction loading with 10 degrees knee flexion.

The identified parameters for the variable response are listed in Table 2, and Fig.3 shows the increase in area ratio. At the preloaded state, the area ratio under the ACL direction loading is larger than that under the tibial direction loading. In the case of the shallow knee flexion, the area ratio under the tibial

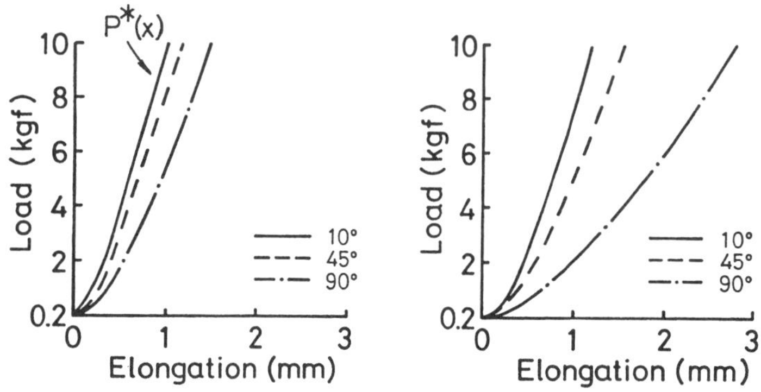

(a) ACL direction loading (b) Tibial direction loading
Figure 2. Load vs. tissue elongation curve obtained by biomechanical test.

Table 1. Parameters for load vs. elongation curves.

Loading Direction	Knee Flexion	a	b	x_0
ACL	10	1.20	3.01	0.39
	45	0.76	3.84	0.33
	90	0.50	3.10	0.54
Tibia	10	0.60	3.52	0.46
	45	0.71	2.40	0.68
	90	0.09	3.03	0.92

Table 2. Identified parameters of area ratio and ultimate load.

Loading Direction	Knee Flexion	c	d	$A_r(0)$	A_r at 98N(10kgf)	Ultimate load N(kgf)
ACL	10	0	–	1	1	343 (35.0)
	45	0.10	−0.01	0.89	0.90	304 (31.0)
	90	0.57	−0.93	0.43	0.85	272 (27.8)
Tibia	10	0.51	−3.73	0.49	0.99	322 (32.9)
	45	0.67	−1.06	0.34	0.87	270 (27.5)
	90	0.84	−0.19	0.16	0.51	119 (12.1)

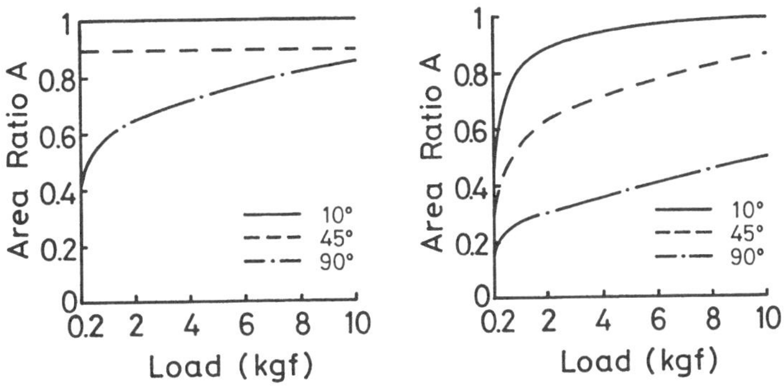

(a) ACL direction loading (b) Tibial direction loading
Figure 3. Identified area ratio.

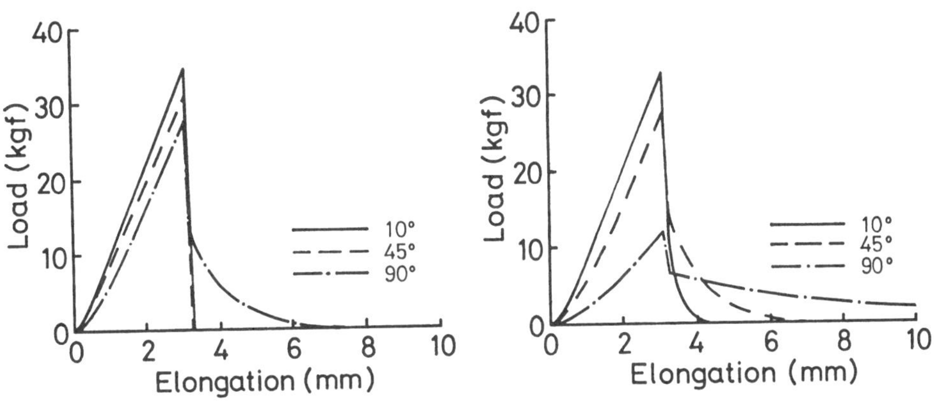

(a) ACL direction loading (b) Tibial direction loading
Figure 4. Calculated load vs. elongation to the failure.

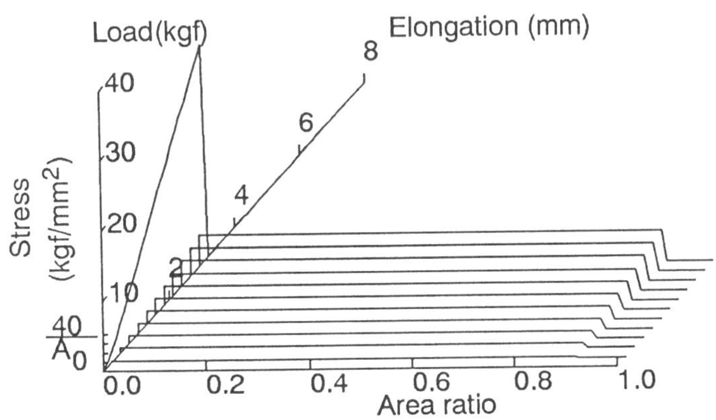

(a) ACL direction loading with 45 degrees knee flexion
Figure 5. Calculated stress distribution during the loading(continue).

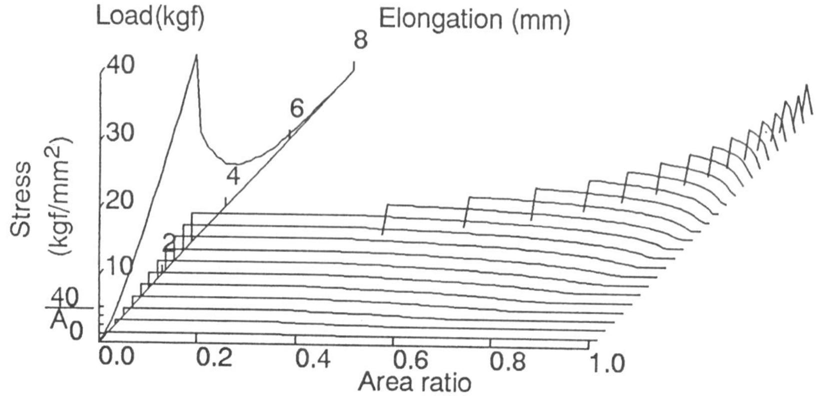

(b) Tibial direction loading with 45 degrees knee flexion
Figure 5. Calculated stress distribution during the loading.

direction loading increases rapidly in the early stage of the loading and reaches almost the same as that under the ACL direction loading at the load of 98 N. On the contrary, the area ratio under the tibial direction loading increases slowly in the case of the deep knee flexion. The area ratio at the load of 98 N is 0.51 under the tibial direction loading and remains only a half of that under the ACL direction loading. This feature of area ratio is useful to interpret the variation in the load vs. elongation due to the testing conditions.

The calculated load vs. elongation curves are shown in Fig.4 for the failure test by using the identified area ratio for the variable response, where the failure stress is calculated from the ultimate load 343 N (35 kgf)[1] for the uniform response. The calculated results are reasonable compared with the experimental results in the literature[1], showing the remarkable decrease of ultimate load due to the increase of knee flexion angle in the case of tibial direction loading. The stress distribution over the cross-section is also calculated based on the proposed model as shown in Fig.5. This enables us to interpret the dependency of the load history on the testing conditon.

IV. CONCLUSIONS

The nonuniform deformation of the ACL is demonstrated by using Japanese white rabbits and it is observed that the anteromedial surface of the ACL is not responsible to carry the load under some stretching conditions. This nonuniform response of the ACL is modeled as the variable response model by means of the area ratio. The proposed model is effective to understand and to predict the mechanical behavior of the ACL. The model can be extended to understand the behavior of the knee ligament in substance. This work was supported in part by Grant-in-Aid for Scientific Research from The Ministry of Education, Science and Culture of Japan.

REFERENCES

1) Woo, S.L-Y., Mollis, R.D., Roux, R.D., Gomez, M.A., Inoue M., Kleiner, J.B. and Akeson, W.H., Effect of knee flexion on the structural properties of the rabbit femur-anterior cruciate ligament-tibia complex(FATC), J. Biomechanics, **20**(1987), pp. 557-563.

2) Noyes, F.R., Butler, D.L., Grood E.S., Zernicke, R.F., and Hefzy, M.S., Biomechanical analysis of human ligament grafts used in knee-ligament repairs and reconstructions, J. Bone Joint Surgery, **66A**(1984), pp.334-352.

3) Tanaka, M., Seguchi, Y., Mizutani T. and Inoue, M., A note on the ACL deformation under tensile loading along ligamentous and/or tibia direction, Tissue Engineering, **BED-14**(1989), pp.115-118, ASME.

4) Tanaka, M., Seguchi, Y., Sonoda, Y. and Inoue, M., Feasibility study of measurement of deformation distribution of knee ligament by using image processing, Advances in Bioengineering, **BED-8**(1988), pp.61-64, ASME.

Biomechanics of Arterial Architecture

Takashi MATSUO and Riki OKEDA

Medical Research Institute, Tokyo Medical and Dental University, 1-5-45 Yushima, Bunkyo-ku, Tokyo

We have constructed arterial casts of the cat brain and kidney, and measured parent and branch vessel diameters and branching angles at arterial bifurcations. The results were compared with theoretical predictions based on the optimality principle in order to understand the physiological principle governing arterial bifurcation geometries. The relationships between branching angles and diameter ratios of branch vessels showed statistically significant correlations, which were in qualitative agreement with the theoretical predictions. However, the arteries were found to be well winding, and therefore the branching angles were defined only in the neighborhood of branch points, suggesting that the branching angle of arterial bifurcation is governed by local and mechanical factors.

I. INTRODUCTION

The arterial system which carries blood from the heart to all peripheral parts of the body is incredibly complex in architecture. It has long been a subject in the the field of biomechanics to understand physiological significance of organization and structure of the vascular system. Murray has applied minimum work concept to a theoretical analysis of blood flow through a cylindrical segment of blood vessel and derived a rule that the diameters of blood vessels D are proportional to the cube root of the flow F which they convey[1], i.e. $D \propto F^{1/3}$. With similar consideration, the angle of vascular branching has been predicted to be a function of relative sizes of the parent trunk and its branch vessels[2]. Since Murray's work, various theories based on the minimum work concept have been proposed for the study on the relationship between branching angles and diameters of parent and branch vessels[3-7]. However, quantitative data relating to the branching angles and vessel diameters in the cardiovascular system is too scarce to confirm these theoretical predictions.

In this study, we have measured the branch geometries of arteries distributed in the brain and kidney, and compared the measured data with the theoretical results based on the minimum work concept in order to provide information on the physiological principles governing the arterial architecture.

II. METHOD

The arteries with diameters >10 μm were visualized by using corrosion compound. Details of the casting procedure of vessels have been described elsewhere[8]. The diameters and branching angles were determined for randomly selected branch points with parent vessel diameters larger than 100 μm. As shown in Fig. 1, an arterial bifurcation is specified in terms of diameters D_0, D_1 and D_2 of the parent vessel, the larger branch and smaller branch, respectively; as well as the angles θ_1 and θ_2 which the larger and smaller branches make with the direction of the parent vessel, respectively. The elliptic cross section of the vessels is not taken into account here.

A problem encountered in the measurement is that most arteries are not straight but winding, and thereby the branching angles were defined only in the neighborhood of branching points. We, therefore, determined the region of measurement where the vessels were considered to be enough straight. As shown in Fig. 1, the diameters of vessels were measured at the most proximal point of the parent vessel and the most distal points of each branch vessel within the region of measurement. The distances from the origin of branching to the measuring points of vessel diameter, denoted l_i (i = 0, 1 and 2) in Fig.1, were also ascertained.

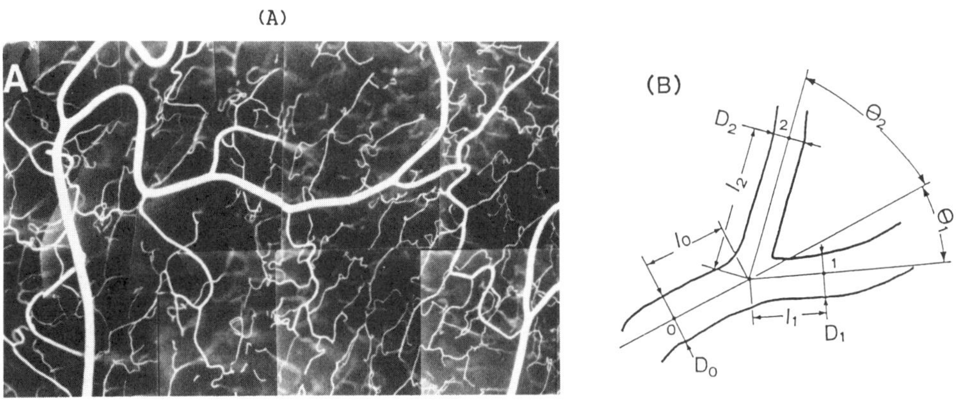

Fig. 1. (A) Photograph of arterial tree of the cat brain. (B) Schematic diagram of an arterial bifurcation. Subscripts 0, 1, and 2 are used to denote the parent vessel, the larger branch and the smaller branch, respectively.

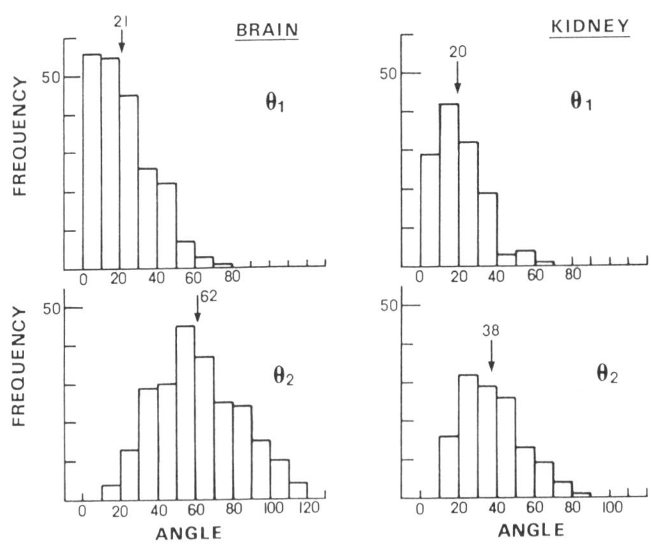

Fig. 2. Frequency distributions of branching angles; θ_1 = angle of the larger branch and θ_2 = angle of the smaller branch. The numbers with an arrow indicate mean values of the branching angle.

III. RESULTS AND DISCUSSION

3.1 Frequency Distributions of Branching Angles

Frequency distributions of the branching angles (θ_1 and θ_2) measured for the brain and kidney are shown in Fig. 2. The branching angles were distributed in a wide range of 0-120 degrees. It should be noted that the angle distributions are considerably different in these two organs; in the brain, the frequencies of the angles θ_1 of the larger branch decrease as the angle increases, meanwhile they shows a convex distribution in the kidney. Moreover, the mean values of the angles θ_2 of the smaller branch are markedly different between brain (62°) and kidney (38°). Such differences might suggest that the architecture of arteries is strongly dependent on the structure and function of the organ to which they are belonging.

3.2 Relationship between Angle Ratio and Diameter Ratio

In order to grasp the correlation between two branching angles and vessel diameters at a bifurcation, the relationship between the angle ratio ($\sin\theta_1/\sin\theta_2$) and diameter ratio ($D_2/D_1$) is shown graphically in Fig 3. Although the data points are largely scattered, it can be noted that they are dominantly distributed in the region of ($\sin\theta_1/\sin\theta_2$)<1, indicating that the larger branch generally makes smaller branching angle than does the smaller branch. The diverse of the data is not ascribed to the method of the measurements, because the photographs were clear and the geometries were measured rather accurately.

In order to see a general tendency of the relationship between angle ratios and diameter ratios, the regression analysis has been conducted with assuming linear relationship, $Y=aX+b$, and the results are also indicated by solid lines in Fig. 3. The regression analysis revealed that the angle ratio has a statistically significant ($P<0.01$) positive relationship to the diameter ratio in both brain and kidney. The regression line reaches roughly unity at an asymptote of diameter ratio=1, showing that if an artery bifurcates into two branches with equal diameters, these branches tend to make equal angles.

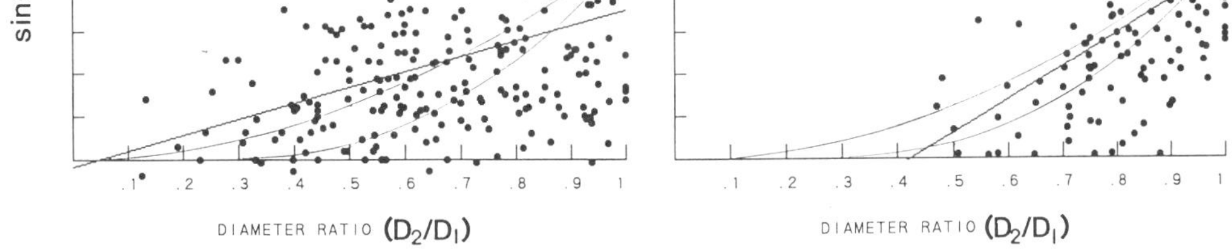

Fig. 3. Relationship between the angle ratio ($\sin\theta_1/\sin\theta_2$) and the diameter ratio (D_2/D_1) at arterial bifurcations in the brain (A) and kidney (B). The results of linear regression analysis, (aX+b, r: correlation coefficient) are indicated by solid lines. The solid curves represent theoretical results based on the principle of minimum work.

3.3 Comparison with Theoretical Predictions

In the theories based on the optimality principle, the vascular architectures are supposed to be optimally designed in a sense that they minimize their energy consumption (economical structure), and a cost functional is postulated as an analytical expression of the energy consumption. Various cost functionals have been proposed so far: the work required to circulate the blood and maintain the blood[1], the total volume of blood with a given pressure drop[3] and so on, although the predictions derived from such optimal theories are quite similar. In Fig. 3, the results based on the optimum theory[6] are also indicated by solid curves for a comparison.

The present data seem to be in qualitative agreement with the theoretical results. and confirm the measured data for the human retina by Zamir et al.[9] and the empirical laws by Wilhelm Roux, which were cited in the work of d'Arcy Thompson[5]. Then, one may consider that arterial branchings are indeed understandable in terms of optimization with respect to certain cost functionals.

However, we call attention to another important aspects of branching geometries found in the present study. First, there exists considerable number of branching angles larger than 90 degrees, although the theoretical upper limit of the branching angle is 90 degrees. About 10 percent of branching angles measured exceeded 90 degrees in the brain. Secondly, the branching angles were defined only in the neighborhood of branch points because most vessels are curved. Since this causes scatter of the data, we determined the distance l_i between the points of measurements and the origin of branching, where the vessels were enough straight, as described in Sec. 2. The mean values of the distances are about 3.0×diameter and 3.3×diameter for brain and kidney, respectively. These values of the vessel lengths leave a doubt about the cost functional assumed in the theories based on the principle of minimum work which always results in a requirement of straight vessels. Thus, the agreement between the measured data for branching angles and the theoretical predictions seem to be extrinsic. The present findings for the vessel lengths indicates that the geometries of vessel branching are considered to be determined by local and probably mechanical factors.

IV. SUMMARY

The geometry (diameters of vessels and branching angles) of arteries distributed in the cat brain and kidney has been measured. The relationships between branching angles and diameter ratios of branched vessels were found to be in qualitative agreement with the theoretical predictions based on the principle of minimum work. It was, however, found that the most arteries were curved and the branching angles were well defined only in the neighborhood of the branching point, showing that the angles are determined by local factors. Further theoretical studies and experimental data from other parts of the cardiovascular system will be required for understanding the physiological principles governing arterial branching geometries.

Acknowledgment

This work was supported in part by a Grant-in-Aid for Scientific Research for the Priority Areas (BIOMECHANICS) from the Ministry of Education, Science and Culture of Japan (No. 04237210, 1992 and No. 05221214, 1993).

REFERENCES

1) Murray, C. D., The physiological principle of minimum work I. Proc. Natl. Acad. Sci., 12 (1926), pp. 207-214.
2) Murray,C. D., The physiological principle of minimum work applied to the angle of branching of arteries. J. Gen. Physiol., 9 (1926), 835-841.
3) Kamiya, A. and Togawa, T., Optimal branching of the vascular tree. Bull. Math. Biophys., 34 (1972), pp. 431-438.
4) Rosen, R., Optimality principles in biology. (Butterworth, London, 1967), pp. 41-60.

5) Thompson, d'A. W., On growth and form. (Macmillan, New York, 1945) pp. 948-957.
6) Zamir, M., Nonsymmetrical bifurcations in arterial branching. J. Gen. Physiol. 72 (1978), pp. 837-845.
7) Oka, S. and Nakai, M., Optimality principles in vascular bifurcation. Biorheology 24 (1987), pp. 737-751.
8) Matsuo,T., Takahashi,M., Funata, M. and Okeda, R., Characterization of bifurcating structure of blood vessels using Fractal dimensions. FROMA 5 (1990), pp. 19-27.
9) Zamir,M., Medeiros J. A. and Cunningham, T. K., (1979) Arterial bifurcation in the human retina. J. Gen. Physiol.74 (1979), pp.537-548.

Computational Fluid Mechanical Study of the Blood Flow with Moving Walls in the Cardiovascular System

Takami YAMAGUCHI and Tad W.TAYLOR

Department of Bio-Medical Engineering, School of High-Technology for Human Welfare, Tokai University, Numazu, Shizuoka

We applied computational fluid mechanics (CFM) method with moving boundary condition to two different bio-fluid problems, namely, a dynamically changing coronary spasm model, and a dynamically changing left ventricular heart model. In the coronary spasm model study, detailed fluid mechanical structures and the macroscopic flow rate changes due to the dynamic alterations of the stenosis severity were shown and examined. A significant decrease in the average flow and a phase delay of the flow behind the stenotic shape changes were clearly observed.

In the study of the left ventricle of the heart, detailed flow and pressure gradient behavior was obtained. The framework of the present study can be applicable to more realistic model studies based on measured left ventricular coordinates. These studies will elucidate the realistic flow behavior in the heart and will make it possible to investigate the effects of the blood flow on various kinds of cardiovascular diseases, such as myocardial infarction.

I. INTRODUCTION

The major mortality in westernized society is due to the ischemic heart disease, a disease of the coronary arteries, which are the most important blood flow tract supporting the heart. It is well known that these kind of vascular or arterial diseases, a major part of which is frequently referred as athero-sclerosis, occurs, develops, and even regresses under the influence of the blood flow[1].

The occurrence of the atherosclerosis has been considered as being related with the shear stress exerted on the vascular wall by the local blood flow[2]. It is, therefore, important to know the detail of the blood flow from the physiological, pathological, and clinical viewpoints. In spite of recent great advances in medical instrumentation techniques, the details of the blood flow are not yet fully understood to estimate the localized forces or the shear stress. A much higher spatial resolution is necessary to evaluate the distribution of the shear stress and other mechanical factors.

This is partly why we have been conducting computational fluid mechanical (CFM) studies of the blood flow. By using CFM method with a sufficient computer power available, we can build as complex a model as we need and can carry out an accurate flow calculation using the model with a high spatial and temporal resolution.

There are several major difficulties in computational bio-fluid mechanics of the blood flow. Of most mechanical interests are; modeling of a complex 3D geometry, nature of blood vessels, such as visco-elasticity of the wall, and nature of blood as a non-Newtonian fluid.

The only driving force generating mechanism in the whole cardiovascular system is the heart, to be exact, its two ventricles.The left and the right ventricles consist of very strong muscle walls which contract approximately every second. The 3D configuration of the inner cavity of those ventricles is complex and not just a simple saccular space. The three dimensional configuration of the arterial system, particularly that of the branching is also very complex and no simple mathematical expression could so far successfully describe the actual picture. The only way to reconstruct the arterial configuration might be based on measurements and we reported some methods to perform this and preliminary results obtained[3].

The second difficulty we encounter when we apply the computational method to arterial flows is that of the visco-elastic properties of the blood vessel walls. This is very true in the case of veins, which should be treated as just thin-walled floppy tubes. In contrast to this, the artery is a rigid pipe. However, its diameter changes nearly 10% during a heart beat due to pulsatile blood pressure. This implies that we have to take the wall movement into account particularly when we try to determine near-wall phenomena.

The third difficulty is the nature of the blood as a fluid. This is because the blood is a suspension with many different constituents, such as the red blood cells, the white blood cells, and other smaller components in the electrolytic aqueous solution of large molecule proteins. Moreover, those constituents are not stable but actively change its physical and/or chemical properties. The non-Newtonian viscosity of the blood might have to be incorporated when we consider the low-shear flows and small-scale flows, such as the capillary flow.

It is therefore quite a difficult task to model the blood flow field properly for the computational fluid mechanical studies even if high-powered computers become available today. We have been trying to build models of the blood flow incorporating as many biological factors as possible, as mentioned above. Based on these results, an attempt to incorporate the wall motion to our system of CFM study, namely the moving boundary problem, is discussed in the present article.

II. CLASSIFICATION OF THE PROBLEM

Before we discuss the application of the moving boundary problems to bio-fluid mechanics, it is convenient to classify those problems in a rather general sense. We can consider at least two classes of moving boundary problems in terms of the cross reaction of the flow and the wall in the living system.

The first and simpler class is that the wall voluntarily moves with no respect to the fluid motion nearby. Some examples of this class in the biological field is a spastic movement of the artery which will be discussed later in this article. What we have to do in this class of problems is to solve the movement of the boundary and the fluid motion separately. The movement of the wall is always treated as the boundary condition in the fluid flow computation with no reverse influence from the fluid to the wall. In this simple class, we can distinguish two subcategories regarding the treatment of the velocity of the wall motion. We will discuss these subcategories in detail later with respect to the coronary spasm model and the left ventricle model.

The second class is that which, in contrast to the former one, the wall passively moves due to fluid forces. More complex problems can be derived by combining the voluntary wall motion at the same time. It is therefore necessary to solve the equations of motion of the wall at the same time in this class of problems. This is, as a matter of fact, the proper moving boundary problem or the fluid-structure interaction problem.

In the present study, it is our purpose to apply our computational scheme of fluid flow to the class of moving wall problems. We, therefore, started from a simpler class and tried to apply it to some bio-fluid mechanical problems relevant to physiological or pathophysiological phenomena. In the following, we discuss two types of applications of moving boundary problems in the cardiovascular system, namely, the coronary spasm and the ejection phase blood flow of the left ventricle of the heart. These flow fields basically belong to the first class explained above. In the coronary spasm case, the time scale of the wall movement was thought of as being much longer than that of the fluid motion nearby in the real situation, so that the wall motion was assumed to be quasi-steady. In this case the wall was assumed as acting just a fixed non-slip boundary at each time step for the calculation of the fluid flow. On the other hand, the left ventricular wall of the heart in the present study was assumed to move by its muscular contraction and gives a momentum to the fluid nearby. The basic assumption about the wall motion was fundamentally the same as that of the coronary spasm, that is, the wall moves much slower than the fluid motion inside the heart distant from the wall. The coronary spasm case and the left ventricle case consist of two slightly different subcategories in the first class. We will discuss this point in detail later.

III. THEORETICAL APPROACH AND COMPUTATIONAL SCHEME

3.1 Mathematical formulations

The fluid mechanical equations were the full 3D unsteady Navier-Stokes equations of fluid motion[4].

$$\rho \frac{D\mathbf{v}}{Dt} = -\nabla p + \mu \nabla^2 \mathbf{v} \quad (1)$$

$$(\nabla \cdot \mathbf{v}) = 0 \quad (2)$$

where, $\mathbf{v}$ is the velocity vector, ρ is the density, p is the pressure and μ is the viscosity. The fluid was assumed homogenous, incompressible and Newtonian.

The wall motions were prescribed and no passive movement of the wall was assumed. The wall movement was implemented as the reconstruction of the geometrical model with proper computational grids. Typically, we assumed that the time scale of the flow development was much shorter than that of the wall movement. In other words, we introduced two different time scales, namely, the time scale of flow development, t_f, and the time scale of the wall movement t_w. Since we assume $t_f << t_w$, the fluid mechanical computation time step was taken much smaller than that of the model reconstruction, and the flow development was calculated using steady computational grids for the time step of the wall reconstruction. When the model was reconstructed at each time step t_w, the velocity field in the older model was mapped to the new grids. Though an interpolation of the velocity at each grid point is necessary, in general, it can be avoided if the t_w is short enough. Details of how to deal with the velocity perpendicular to the wall are slightly different in the coronary spasm case and the left ventricle case, which will be discussed later.

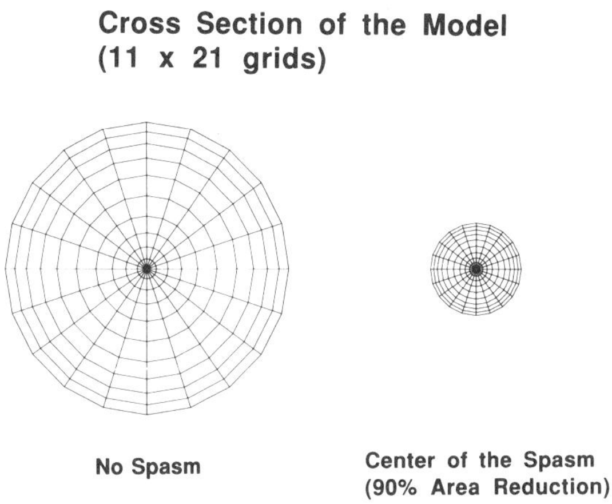

Figure 1.Cross-sectional view of the coronary spasm model showing the symmetrical grid spacing. Left: the initial structure, Right: the maximally reduced structure.

3.2 Computational Scheme and the Grid System

The numerical technique employed was a finite volume method for solving the unsteady three-dimensional viscous flows. To handle the complex geometry, we used a boundary-fitted coordinate (BFC) system[5]. The contravariant physical velocity components were selected as the dependant variables in the momentum equations and a staggered grid arrangement was employed. The convective terms of the momentum equations were discretized using the first order up-wind differencing and the time-dependant term and other terms were discretized by the central differencing. The unsteady term was treated by a fully implicit scheme. A pressure correction was performed by correcting the pressure itself initially, and the predicted velocities were then corrected to satisfy the mass conservation. The resultant Poisson equations were solved iteratively until a convergence obtained. The computational code used to solve the fluid motion was SCRYU system by Software Cradle (Osaka, Japan) which consists of a pre-processor or grid generating system, a solver for solving the equations of motions as well as energy equations, mass transport equations, etc., and simple graphics programs.

Wall motions and the reconstruction of the model were calculated using our own programs incorporated with the SCRYU pre-processor and the solver using a UNIX C-Shell script program facilities[6]. Details of the wall movement are described fully in the following sections.

IV. PARTICULAR PROBLEM

4.1 The Coronary Spasm

1) Introduction

Coronary spasms are clinically drawing an attention as one of the possible causes of unstable angina pectoris and acute myocardial infarction. The most significant difference between coronary spasms and an ordinary arterial stenosis is that its stenosis severity or the area reduction varies with time[7-8].

The speed of the spasm is known to be of the order of one radius/second, which implies that the time scale of the wall movement is comparable with the heart rate or the fundamental cycle of the flow inside the coronary artery. This consideration leads

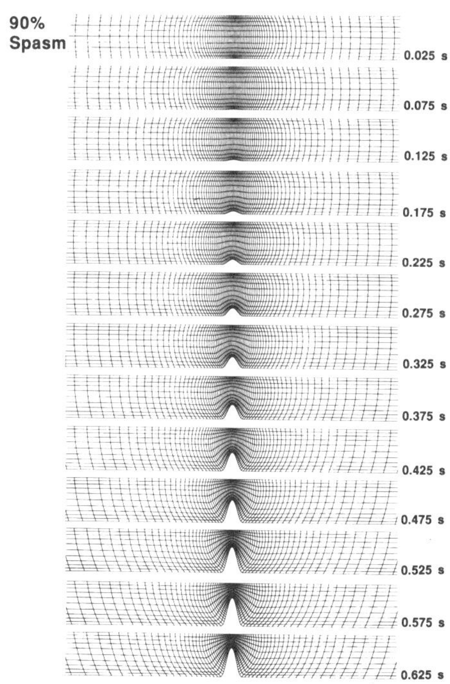

Figure 2.Transverse view of the coronary spasm grid system showing the time-varying pattern.

to the hypothesis that even the coronary spasm of very short duration may affect the average coronary flow rate, if it occurs in a specific phase of cardiac cycle.

To our knowledge, the effects of dynamically changing stenosis in arteries has not been investigated in detail. Considering the nature of the coronary flow, particularly its occurrence in diastole, and also that the substantial flow development cannot be expected in moving stenosis, flow behavior may significantly differ from that in the steady stenosis case.

2) Model Construction

One of the basic assumptions of the present study is that the wall movement due to the spasm can be assumed quasi-steady for the time scale within which the velocity development is calculated. Stenosis severity values of 75% and 90% of the coronary spasm were implemented. The severity of the stenosis was expressed by the area reduction in the vessel. The time course of the spastic changes were assumed to have a trigonometrically varying nature with time as given by:

$$ar(t) = ar_{max}(1 - \cos(2\pi t / P)) / 2 \qquad (3)$$

where ar(t) is the area reduction at time t, ar_{max} is the maximum area reduction, and P is the period of the spasm contraction.

The longitudinal shape of the artery spasm was assumed to be a cosine bell, which had the basement length of a diameter of the pipe. Typically, the stenosis development with time was divided into 25-50 steps according to the time course given by Equation (3). The computational grids were 11 in the radial direction, 21 in the circumferential direction and 101 in the longitudinal direction with 23,311 total node points, and the radius without stenosis was assumed to be 0.25 cm. The cross-sectional view of the coronary spasm model shows the symmetric grid structure (Figure 1). In Panel A, the initial structure is shown with a radius of 0.25 cm and area 0.196 cm^2, and the reduced structure in the 90% spasm case at 0.625 seconds is shown in

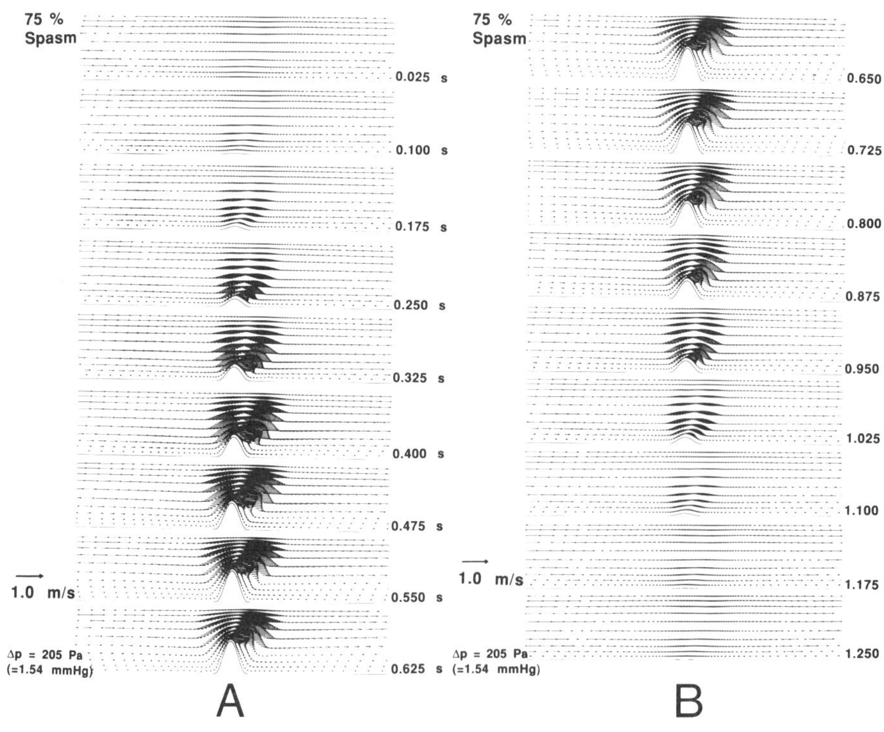

Figure 3.Time-varying flow patterns in the coronary spasm with 75% area reduction at the peak stenosi at 0.625 seconds. Panel A: the contraction stage at times 0.025-0.625 seconds, Panel B: the expanding stage at times 0.625-1.25 seconds.

Panel B with a radius of 0.0791 cm and area 0.0196 cm^2.

Figure 2 shows (halves of) several longitudinal sections of the model for the 90% spasm at various phases of the arterial spasm at intervals of 0.05 seconds. The spasm vessel was 5 cm in total length, and the spasm constriction occurred over 0.125 cm length. At 0.625 seconds, the maximum constriction is reached. There were 50 steps used in the entire spasm constriction cycle.

The boundary conditions applied were as follows: The walls were assumed rigid and non-slip, and the pressure at the inlet and the outlet cross-sections were prescribed in order to give a steady driving pressure. The velocity gradients at the outlet cross sections were zero. Initial velocities were given as zero, and the computation was carried out until a fully developed flow was observed using non-stenotic models. It took several thousands computational steps to obtain such a steady state before the spasm started.

3) Results

The viscosity value of 4.0 x 10^{-4} Pa·s (4.0 cP) was used, and the pressure differences used were 51.2 and 205 Pa, i.e., approximately 0.384 and 1.54 mm Hg, respectively. These driving pressures could produce approximately 3.8 ml/s and 12.5 ml/s flow rate when the flow was fully developed without stenosis. The time step of the iteration for the finite volume integration was 0.5 msec. This was within the limit that the Courant number be of the order of one.

Time-varying flow patterns in the coronary spasm with 75% area reduction at the peak stenosis at 0.075 second intervals showed that at 0.325 seconds flow separation starts and increases to the peak constriction at 0.625 seconds (Figure 3, Panel A). Here one half of the vessel is shown to enhance the clarity. In this case a total pressure drop of 205 Pa is input. As the spasm contracts after 0.625 seconds, the flow separation decreases until it ceases at 1.025 seconds (Figure 3, Panel B). Even at the maximum constriction, the flow separation was not as large as that predicted with steady case simulations. This may be caused by the fact that there was insufficient time for fully separated flow to develop.

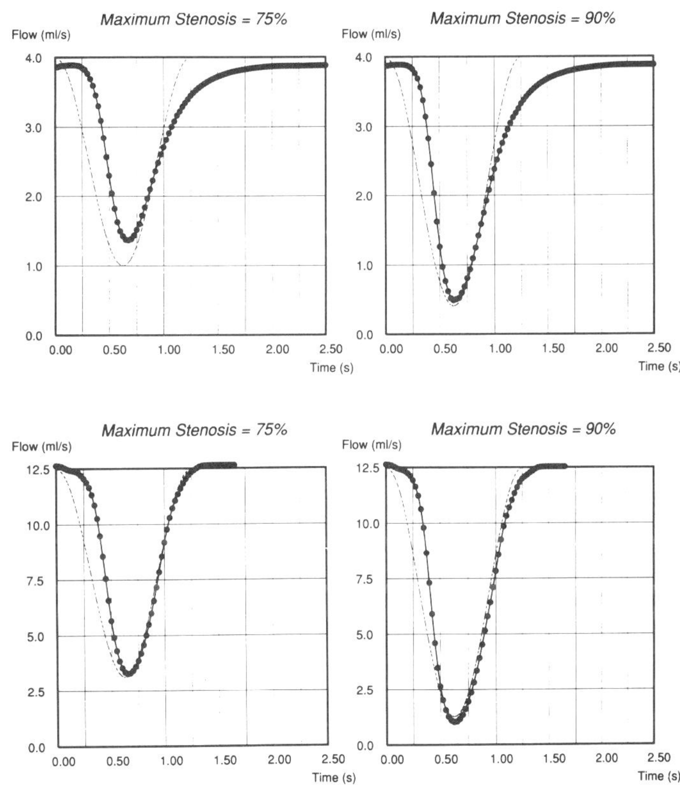

Figure 4. Average inlet flow and area reduction versus time in the coronary spasm. The dotted line represents the mean flow at the spasm inlet and the dashed line represents the area reduction.

Flow predictions at 51.2 and 205 Pa pressure drop showed that as the flow is increased, the flow rate pattern approaches that of the coronary spasm (Figure 4). The dotted line represents the mean flow at the spasm inlet, and the dashed line represents the area reduction. With flow at 75% and 90% stenosis and 51.2

Pa total pressure drop from the spasm vessel inlet to outlet, it took almost 2.0 seconds for the flow rate to return to normal (Panels A and C). The maximum flow was approximately 4 ml/sec with zero stenosis and 1.5 and 0.5 ml/s at peak stenosis in the 75 and 90% stenosis cases.

However, with flow at 75% and 90% stenosis and 205 Pa total pressure drop from the spasm vessel inlet to outlet, the flow pattern does not lag behind the wall motion as much in Panels A and C, and the maximum flow is approximately 12.5 ml/s with zero stenosis and 3.5 and 1.0 ml/s at peak stenosis in the 75 and 90% stenosis cases. This would be due to the increase of fluid mixing at the higher flow rates.

4.2 The Left Ventricle of the Heart

1) Introduction

Heart disease is a leading cause of mortality in developed countries. In this study, it is our goal to develop a computational scheme in which realistic three-dimensional flow simulations can be achieved[9]. As some recent studies[10] show, computational fluid mechanics techniques can be valuable tool for simulating heart flow conditions. They have used a method similar to the method that we are developing here; however, their computations were based on axisymmetric flow. In the present study, we developed a complete three-dimensional flow model of left ventricular contraction. In the computation, the blood flow was computed using the prescribed wall motion as the boundary condition.

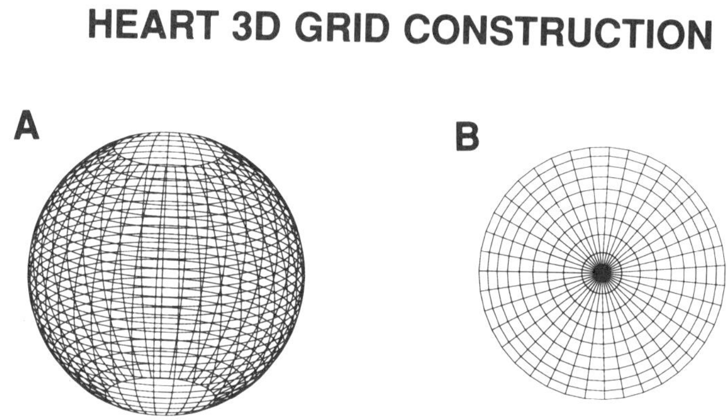

Figure 5.Three-dimensional spherical heart model grid representations. Panel A: Three-dimensional view, Panel B: Cross sectional view.

2) Model Construction

The computational grids were constructed by using our own user-developed system and the preprocessor system of the SCRYU system. Figure 5 illustrates the model. Initially, the total height was 5.5 cm, the center radius was 3.0 cm, and the entrance and exit radius were 1.12 cm. In Panel A, the three-dimensional representation has from top to bottom (apex of the heart) 30 lateral grids, and there are 40 longitudinal grids. In the cross-sectional view of the heart model (Panel B) there are 10 radial grids which gives 12,000 total nodes.

One of the basic assumptions of the present study, as discussed in the foregoing, is that the time scale of the movement of the contracting ventricular wall is much slower than the time scale of the temporal resolution of the computation. In other words, the left ventricular wall moves much slower than the fluid particle nearby moves, and could be assumed quasi-steady for the time scale of the velocity development used. By assuming this, we can introduce two different time scales, namely the “flow time scale” and the “cardiac wall time scale”. What we assumed is that the “cardiac wall time scale” is much longer than the “flow time scale”. If we assume this, we can compute the evolution of the velocity field using a stationary computational grid in each “cardiac wall time step”. Usually we used the “flow time scale” of an order of magnitude of 10^{-4} seconds and the “cardiac wall scale” of 10^{-2} seconds.

The left ventricular wall change was assumed to follow a trigonometrically varying function of time:

$$red(t) = red_{max} [1 - \cos(2\pi t / P)] / 2 \qquad (4)$$

where red(t) is the radial reduction at time t, red_{max} is the maximum radial reduction, and P is the period. For the ventricular contraction, the maximum time input was P/2, or half the period. Another important assumption in the present study is that the movement of the ventricular wall can be regarded equivalent for the flow across the ventricular wall within the short duration of time in which the wall does not move significantly. Then the wall movement can be substituted by the flow across the wall. A brief description of the method to compute the flow with a moving boundary condition using the flow simulation package follows. At the first step, the radius of the model at a given stage was calculated according to Equation 4, and the volume of the left ventricle was calculated using the radius at the stage. At the second step, the wall velocity was calculated from the volume reduction calculated from the former stage and the current stage of the contraction. To perform this, we assumed that the center of gravity of each element of the wall evenly moves towards the center of the outflow orifice. The outflow orifice was assumed to keep the same cross sectional area during the contraction. At the third step, the calculated wall velocity was put into the solver program of the SCRYU system as a set of flux boundary condition, that is, we assumed that there was a flow across the wall instead of the wall motion. Finally, after a short period of time, the model of the left ventricle was reconstructed incorporating the accumulated movement of the wall during the "cardiac wall time scale" step of the time.

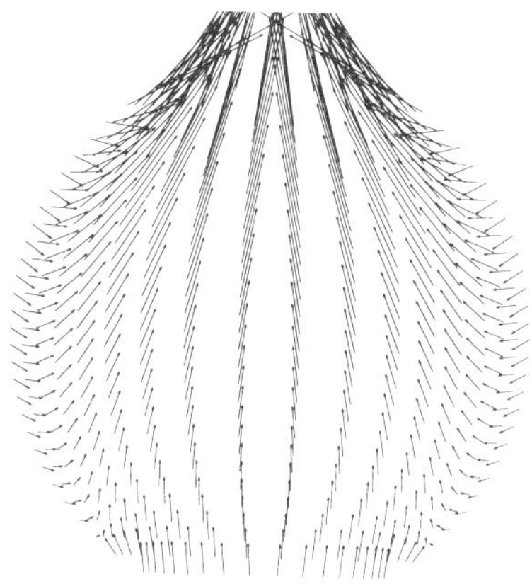

Figure 6. Centerline velocity vectors computed at t=0.10 seconds, which is the 10th velocity calculation. Here the radius has been reduced by 17%, and the wall velocity is 19.3 cm/s.

Typically, the ventricular wall contraction with time was divided into 25-50 steps according to Equation 4. The number of steps for the ventricular wall motion was selected so that at each step the fluid motion never traveled beyond the minimum mesh length. Using this precaution, we were able to assume that the movement of the grid points was within one grid distance throughout the entire model throughout the cycle, and it was not necessary to interpolate the change in the shape between each time step. This was within the limit that the Courant number be on the order of one. The feasibility of this method was checked by decreasing the time step for the new wall grids and observing no change in the computed results.

3) Velocity Flow Fields

Figure 6 shows centerline velocity vectors computed at t = 0.10 seconds, which corresponds to the 10th step. Here the radius has been reduced by 17%, and the wall velocity is approximately 20 cm/sec. The apex of the heart moves upward vertically, while the side walls are moving inward towards the base. There is symmetric flow with the maximum flow velocity achieved at the aortic outlet. The results in Figure 6 were obtained using 25 different grids for the computations, but results using 50 different grids were within a few percent of the results using 25 grids and the graphs were indistinguishable.

Cross-sectional velocity vectors were also computed at t = 0.10 seconds (Figure 7). Here 4 cross-sectional velocity profiles are given which illustrate velocities at the heart apex (K = 1) and at positions increasing in distance from the apex (K = 10, 20, 25), where K = 30 represents the aortic outlet. As the fluid flows toward the aortic outlet, the fluid velocity

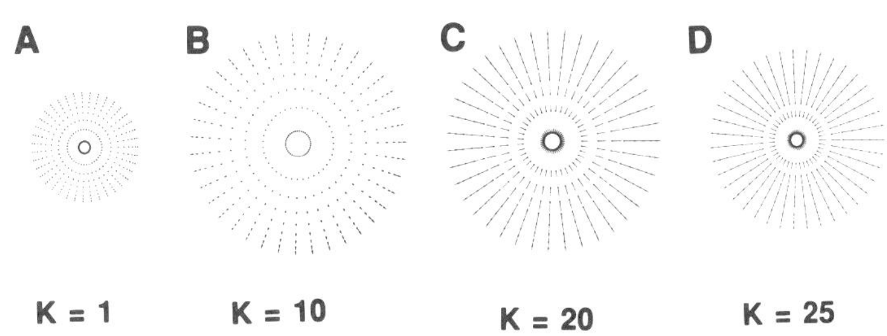

Figure 7. Cross-sectional velocity vectors computed at t=0.10 seconds. There are 4 cross-sectional velocity profiles from the apex of the heart (K=1) to the aortic outlet (K=30).

towards the center increases. In Panel A, at the apex of the ventricle, there is a small radial flow away from the ventricle center, and the fluid velocity changed directions as it flowed towards the aortic outlet.

V. CONCLUDING REMARKS

In the first part of the application of our method, a dynamically changing coronary spasm model for the computational fluid mechanical study was developed. Detailed structures as well as the macroscopic flow rate changes due to the dynamic alterations of the stenosis severity were examined. Other factors were also considered using steady-state boundary conditions. The application of the dynamically moving boundary model in order to study the blood flow under the time-varying constriction was demonstrated to be a useful tool in order to investigate the mechanism of an ischemic event provoked by a coronary spasm.

Although many studies have been carried out on the flow in stenotic arteries, the effects of dynamically changing stenosis in arteries have not been investigated in detail. In the computations, significant decreases in the average flow rate were found to affect the overall flow structures. It was observed that typical vortex formation predicted by non-moving boundary computations and experiments was considerably decreased.

In the second part, a dynamically changing left ventricular heart model of three-dimension was developed. Detailed flow and pressure gradient behavior can be obtained from this modeling structure and methodology. Although the developed technique at the moment can only deal with the flow inside the ventricle and cannot be applied to the mechanics of the ventricle wall itself, this framework will be applied to measured left ventricular coordinates to investigate realistic flow behavior.

In conclusion, the moving-wall three-dimensional computational technique should allow comparison of the coronary spasm and the left ventricular heart simulation results with clinical monitoring techniques. These could be a useful tool in order to investigate the effects that various kinds of cardiovascular diseases, particularly cardiac infarction, have on blood flow.

Acknowledgment

Part of this study was supported by the Grant-in-Aid for Scientific Research "Biomechanics of Structure and Function of Living Cells, Tissues, and Organs" (#04237101) and Grant #04454537 from the Ministry of Education, Science and Culture of Japan. We also wish to thank the Japan Science and Technology Agency for an STA Fellowship supporting TWT.

REFERENCES

1) Yoshida,Y., Yamaguchi,T., Caro, C.G., Glagov, S., and Nerem, R.M. (eds.), Role of Blood Flow in Atherogenesis (Springer-Verlag, 1988).
2) Caro, C.G., Fitz-Gerald, J.M., and Schroter, R.C., Atheroma and arterial wall shear observation, correlation, and proposal of a shear dependent mass transfer mechanism for atherogenesis. Proc. Roy. Soc. Lond. B., 177 (1971), pp.109-159.
3) Taylor, T.W. et al.,Three-dimensional graphics and computational model construction of vascular chambers from physiological casts. Bulletin School. High-Tech. Human Welfare Tokai Univ. 2 (1992), 231-240.
4) Bird,R.B., Stewart,W.E., and Lightfoot, E.N., Transport Phenomena, (Wiley, 1960).
5) Thompson, J.F., Warsi, Z.U.A., and Mastin, C.W., Boundary-fitted coordinate systems for numerical solutions of partial differential equations - A review. J. Comp. Physics, 47 (1982) , pp.1-108.
6) Anderson,G. and Anderson, P., The Unix C Shell Field Guide, (Prentice Hall, 1986).
7) Maseri, A., L'Abbate, A., Baroldi, G. et.al., Coronary vasospasm as a possible cause of myocardial infarction: a conclusion derived from the study of "preinfarction" angina. New Engl. J. Med., 299 (1978), pp.1271-1272.
8) Epstein, S.E., and Talbot,T.L., Dynamic coronary tone in precipitation, exacerbation and relief of angina pectoris. Am. J. Cardiol., 48 (1981), pp.797-803.
9) Wendt, J.F. (ed.), Computational Fluid Mechanics, (Springer-Verlag, 1992).
10) Georgiadis, J.G., Wang, M., and Pasipoularides, A., Computational fluid dynamics of left ventricular ejection. Annals Biomed. Eng., 20 (1992), pp.81-97.

INDEX OF AUTHORS